Elementary
Linear Algebra
with
Applications

Elementary Linear Algebra with Applications

FRANCIS G. FLOREY

Department of Mathematics
University of Wisconsin-Superior

Prentice-Hall, Inc.,
Englewood Cliffs, New Jersey 07632

Library of Congress Cataloging in Publication Data

FLOREY, FRANCIS G. (date)
  Elementary linear algebra.

  Bibliography: p.
  Includes index.
  1. Algebras, Linear. I. Title.
QA184.F56   512'.5   78-9412
ISBN 0-13-258251-1

Editorial/production supervision and interior
    design by Eleanor Henshaw Hiatt
Cover design by Jorge Hernandez
Manufacturing buyer: Phil Galea

Printed in the United States of America

10 9 8 7 6 5 4 3 2

Prentice-Hall International, Inc., *London*
Prentice-Hall of Australia Pty. Limited, *Sydney*
Prentice-Hall of Canada, Ltd., *Toronto*
Prentice-Hall of India Private Limited, *New Delhi*
Prentice-Hall of Japan, Inc., *Tokyo*
Prentice-Hall of Southeast Asia Pte. Ltd., *Singapore*
Whitehall Books Limited, *Wellington, New Zealand*

To my wife, Maxine; and to my children, Kevin, Todd, and Pam

Contents

Preface

This book is written for students in a first course in linear algebra at the sophomore level and is intended primarily for mathematics majors, engineers, science students, and business and economics majors.

The question may properly be asked: Why another textbook in linear algebra? I believe that many of the present books in this subject are too abstract, others are computationally oriented to the point that the mathematics is ignored, and many, if not most, have omitted the *applications* of linear algebra. I have written this book with the idea of achieving a balance among computational skills, applications, and the theory of linear algebra. At the same time I have tried to keep the reading level of the text at a sophomore or even a freshman level.

Chapters 1 and 2 are geometric in nature. The discussion begins with line vectors. The operations on line vectors are used to motivate the definitions of operations on 2- and 3-tuples and these definitions are extended to n-tuples. Proofs in the first two chapters are less rigorous and are geometrically oriented. Much of the material in the first two chapters may have already been covered in a calculus and analytic geometry course. I have started with this material because I believe it is the least abstract and provides the most concrete path to the ideas of linear algebra.

Chapter 3 discusses reduction methods for solving systems of linear equations and also the algebra of matrices. Numerous applications are integrated with the material. Chapter 4 is a short chapter on determinants. I believe that the method I have chosen for defining determinants, which avoids permutations, leads more easily to the theory and also to the techniques for calculating determinants.

Chapter 5 deals with the theory of abstract (real) vector spaces. The reader has already seen numerous examples of vector spaces in the first three chapters. Various function spaces are introduced so that the reader will realize that there are important vector spaces other than n-tuples. Proofs of most of the theorems are given but are often illustrated first with an example. This gives the instructor the option of discussing the example or the proof of the theorem, which largely imitates the example.

Chapter 6 discusses linear transformations and their relationship to matrices. Chapter 7 discusses eigenvalues and eigenvectors and the diagonalization process. Chapter 8 discusses the general inner product. Applications are included in Chapters 6, 7, and 8.

I have tried to motivate concepts through examples, applications, and other means. For instance, I show that the definition of matrix multiplication arises in a natural way as the result of a certain substitution process involving systems of linear equations. This is what led Arthur Cayley to the definition of matrix multiplication in 1858.

Over 165 examples are given in the text. These examples include applications and methods for solving problems. Answers to all the odd-numbered computational exercises are given at the end of the text. The answers to the even-numbered problems are available to instructors in a separate answer book.

It is assumed that the reader has a knowledge of precalculus mathematics including the idea of a function. (A brief review of functions is offered in Section 6.1.) With the exception of some examples and problems (that can be omitted) and the final section of the text, which requires a knowledge of integration techniques, a knowledge of elementary calculus *is not needed.*

This book is an outgrowth of lecture notes that were used during the years 1974–1977 at the University of Wisconsin–Superior. I have taught all the material at a leisurely pace in five semester hours and believe that the entire text can be covered in four semester hours. By a judicious selection of topics, most of the book can be covered in three semester hours. Excluding Sections 4.3, 6.8, 7.5, and 8.3 (which are optional) there are 39 sections. Most of these sections can be covered in a single class period. If a shorter course is desired, Sections 2.4 and 2.5 could be omitted as well as some of the applications. Various options are possible, some of which follow:

1. Start with Chapter 1 and continue through the text. This would work well if the text is used concurrently with or prior to a third semester of calculus.
2. If the text is used following a third semester of calculus, the instructor may wish to cover the first two chapters rapidly or omit Chapters 1 and 2 altogether (allowing students to review this material on their own) and start with Chapter 3.

3. Start with Chapter 3 and continue through Section 3.3; then pick up Chapter 1 and Sections 2.1–2.2, leaving the remainder of Chapter 2 until just before Chapter 8.

I gratefully acknowledge the helpful comments and suggestions made by the reviewers: Professor Jan Jaworowski of Indiana University, Professor Jack Goldberg of the University of Michigan, and Professor Robert Weber of Yale University. I am especially appreciative for the many suggestions and the encouragement provided by Professor David E. Kullman of Miami University, Oxford, Ohio, and for the constructive criticism and perceptive comments of Professor Robert E. Mosher, formerly of California State University at Long Beach. Both Professors Kullman and Mosher carefully read the entire manuscript and are responsible for many improvements in the text. I wish also to thank my friend and colleague, Professor Robert E. Dahlin at the University of Wisconsin-Superior, who provided helpful comments and with whom I have had many mathematical discussions.

For typing the first draft I wish to express appreciation to Grace Collins and Hondoko Tingsantoso. I also wish to thank John J. McCanna, regional editor for Prentice-Hall, who gave me encouragement when I first started this project. Finally, I express appreciation to my mathematics editor, Harry H. Gaines; my production editor, Eleanor Henshaw Hiatt; and the production staff at Prentice-Hall.

FRANCIS G. FLOREY

Elementary
Linear Algebra
with
Applications

1

Line Vectors
and
Coordinate Vectors

Linear algebra is, in a sense, a study of vectors. From a mathematical point of view, vectors come in many different forms, but all must share certain common properties. Probably the first idea of vector that most of us are exposed to is the notion that a vector is a quantity that has a length and a direction. It is with this informal idea that we begin. As we progress further through the text, we shall give a precise mathematical definition of vector. It will become clear that our initial introduction to vector as a quantity having length and direction is only one example of a much broader classification of objects that we shall call vectors.

1.1 VECTOR ADDITION
AND SCALAR MULTIPLICATION

In science, physical quantities such as force, velocity, displacement (movement of a particle from one point to another), and acceleration are described by a magnitude and a direction. The term *vector* is used to identify such a quantity.

Geometrically, a vector can be represented by a directed line segment or arrow. The length of the line segment denotes the magnitude of the vector; the direction of the arrow denotes the direction of the vector. For example, a force of 8 lb could be represented by an arrow 8 units long in the direction of the force (Figure 1.1).

One can draw many arrows 8 units long and in the same direction as the force. All such directed line segments represent the same vector. As

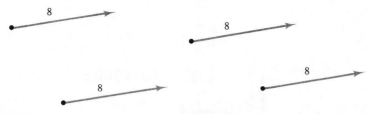

Figure 1.1   An 8-pound force

geometric entities they are different sets of points, but as vector representations they are equal.

Definition 1.1.   Two **directed line segments** (arrows) of nonzero length **represent the same vector** if and only if they have the same length and the same direction. The term **line vector** will also be used to refer to the vector represented by a directed line segment.

In the text we shall denote a line vector by a letter with a bar over it, for example, $\bar{v}$ denotes the "vector v" (see Figure 1.2). If the *tail* and *head* of $\bar{v}$ are points A and B, respectively, we shall also write $\bar{v} = \overrightarrow{AB}$.

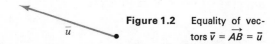

Figure 1.2   Equality of vectors $\bar{v} = \overrightarrow{AB} = \bar{u}$

Addition of Vectors

There are two equivalent procedures that can be used for adding vectors. As Figure 1.3 suggests, from the head of $\bar{u}$ draw $\bar{v}$. The vector $\bar{u} + \bar{v}$ is the vector from the tail of $\bar{u}$ to the head of $\bar{v}$. This method of vector addition is called the **triangle rule of addition**.

An alternative but equivalent procedure is the parallelogram rule for addition (Figure 1.4). Using the **parallelogram rule** to obtain $\bar{u} + \bar{v}$, we draw representations for $\bar{u}$ and $\bar{v}$ from the same point (the tails of $\bar{u}$ and $\bar{v}$ coincide), and then complete the parallelogram. The diagonal drawn from the common point represents $\bar{u} + \bar{v}$.

It is clear from Figure 1.4 and the triangle rule that vector addition is **commutative**, that is,

$$\bar{u} + \bar{v} = \bar{v} + \bar{u}$$

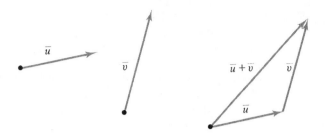

Figure 1.3   Addition of vectors (triangle rule)

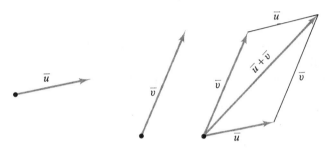

Figure 1.4   Addition of vectors (parallelogram rule)

The **length** or **magnitude** of $\bar{v}$ will be denoted by $\| \bar{v} \|$. The **direction** of $\bar{v}$ will be denoted by dir $\bar{v}$. If $\bar{u}$ and $\bar{v}$ are vectors such that the angle between them is 180° when they are drawn from the same point, then we shall write dir $\bar{u} = -$dir $\bar{v}$ and read "the direction of $\bar{u}$ is opposite the direction of $\bar{v}$" (Figure 1.5).

The **zero vector**, denoted by $\bar{0}$, is a vector of 0 length. We do not assign a direction to $\bar{0}$. Note that

$$\bar{v} + \bar{0} = \bar{0} + \bar{v} = \bar{v}$$

by the triangle rule for addition.

For any vector $\bar{v}$ we define $-\bar{v}$ (read "minus v" or "the opposite of v" or "the **additive inverse** of $\bar{v}$"; see Figure 1.6) to be the vector such that

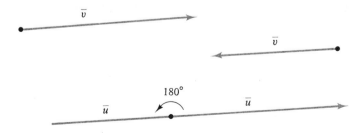

Figure 1.5   Dir $\bar{u} = -$ dir $\bar{v}$

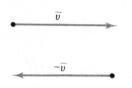

Figure 1.6   The additive inverse of $\bar{v}$

$$\| -\bar{v} \| = \| \bar{v} \|$$

and $\text{dir } -\bar{v} = -\text{dir } \bar{v}$

(If $\bar{v} = \bar{0}$, then $-\bar{0} = \bar{0}$.) From the triangle rule for addition it is clear that

$$\bar{v} + -\bar{v} = \bar{0}$$

If line vectors $\bar{u}$ and $\bar{v}$ have representations that are equal in length and parallel, then either $\bar{u} = \bar{v}$ or $\bar{u} = -\bar{v}$. *We shall assume that we can tell from the directions that the arrows point whether or not $\bar{u} = \bar{v}$ or $\bar{u} = -\bar{v}$.*

Example 1.1.   In the parallelogram below, decide (a) which arrows represent the same vector, and (b) which arrows represent opposite vectors.

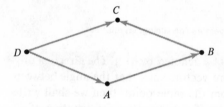

Solution: (a) *Line segments* $\overline{AB}$ and $\overline{DC}$ are opposite sides of the parallelogram. Therefore, $\| \overrightarrow{AB} \| = \| \overrightarrow{DC} \|$ and $\overline{AB}$ and $\overline{DC}$ are parallel. From the figure, dir $\overrightarrow{AB} =$ dir $\overrightarrow{DC}$. Therefore, $\overrightarrow{AB} = \overrightarrow{DC}$.

(b) Line segments $\overline{BC}$ and $\overline{DA}$ are opposite sides of the parallelogram. So $\overline{BC}$ and $\overline{DA}$ are parallel and $\| \overrightarrow{BC} \| = \| \overrightarrow{DA} \|$. From the figure, dir $\overrightarrow{BC} = -\text{dir } \overrightarrow{DA}$. Hence $\overrightarrow{BC} = -(\overrightarrow{DA})$.

Since every line vector has an additive inverse, we can define **subtraction of vectors** by

$$\bar{u} - \bar{v} = \bar{u} + (-\bar{v})$$

Example 1.2.   Given representations for $\bar{u}$ and $\bar{v}$, use the definition of subtraction and the triangle rule for addition to draw $\bar{u} - \bar{v}$.

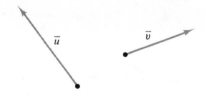

Solution

1. Draw $\bar{u} = \overrightarrow{AB}$ and $\bar{v} = \overrightarrow{AC}$ from the same point A.

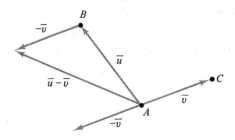

2. Draw $-\bar{v}$.
3. Using the triangle rule for addition, draw $\bar{u} - \bar{v} = \bar{u} + (-\bar{v})$.

Example 1.3.   Show that $\bar{u} - \bar{v}$ **is the vector which when added to** $\bar{v}$ **is** $\bar{u}$.

Solution

1. Draw representations for $\bar{u}$ and $\bar{v}$ from some common point A. Let $\bar{u} = \overrightarrow{AB}$
   and $\bar{v} = \overrightarrow{AC}$.

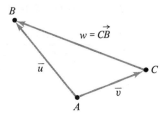

2. Complete the triangle by drawing $\bar{w} = \overrightarrow{CB}$.

By the triangle rule for addition, $\bar{v} + \bar{w} = \bar{u}$. Writing $\bar{v} + \bar{w}$ as $\bar{w} + \bar{v}$
and adding $-\bar{v}$ to both sides, we have $(\bar{w} + \bar{v}) + (-\bar{v}) = \bar{u} + (-\bar{v})$. Reassociating parentheses (Problem 2, Exercise 1.1) and using the definition of
subtraction, we have

$$\bar{w} + (\bar{v} + (-\bar{v})) = \bar{u} - \bar{v}$$

Since $\bar{v} + (-\bar{v}) = \bar{0}$ and $\bar{w} + \bar{0} = \bar{w}$, we get $\bar{w} = \bar{u} - \bar{v}$.

Note: The reader should compare the result of Example 1.2 with the
result of Example 1.3.

Exercise 1.1

In the problems that call for drawings the reader will find that a transparent ruler is convenient for drawing parallel segments.

1. Let $\vec{u} = \overrightarrow{AD}$, $\vec{v} = \overrightarrow{DC}$, $\vec{w} = \overrightarrow{CB}$, and $\vec{x} = \overrightarrow{BA}$, as the figure indicates. Assume that $ABCD$ is a parallelogram with $\overline{AD}$ parallel to $\overline{BC}$.

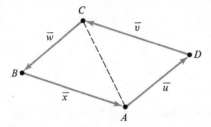

(a) Explain why $\overrightarrow{AD} = \overrightarrow{BC}$.

(b) Explain why $\vec{x} = -\vec{v}$.

(c) Write $\overrightarrow{AC}$ in terms of $\vec{u}$ and $\vec{v}$.

(d) Write $\overrightarrow{AC}$ in terms of $\vec{x}$ and $\vec{w}$.

(e) Find the sums $\vec{x} + \vec{u}, (\vec{x} + \vec{u}) + \vec{v}$, and $((\vec{x} + \vec{u}) + \vec{v}) + \vec{w}$, and express the answers in terms of the vertices of the parallelogram.

2. Show geometrically that vector addition is associative; that is,

$$(\vec{u} + \vec{v}) + \vec{w} = \vec{u} + (\vec{v} + \vec{w})$$

Hint: Use the triangle rule for addition.

3. Label the unlabeled arrows in the figure below in terms of $\vec{v} = \overrightarrow{PA}$ and $\vec{w} = \overrightarrow{PB}$. You may assume that segments which look parallel are parallel. All arrows have their tails at the point P in the center of the parallelogram.

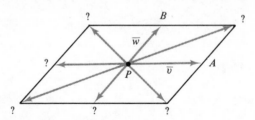

4. (a) In the figure below find $((\vec{u} + \vec{v}) + \vec{w}) + \vec{x}$ and $\vec{u} + (\vec{v} + (\vec{w} + \vec{x}))$.

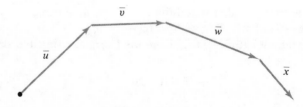

(b) Would it be appropriate to label the vector from the tail of $\vec{u}$ to the head of $\vec{x}$ as $\vec{u} + \vec{v} + \vec{w} + \vec{x}$? Why?

5. In each figure two of the vectors are labeled $\vec{u}$ and $\vec{v}$. Use these labels to label

each of the vectors called for. You may assume that segments that look parallel
are parallel.

(a) $\overrightarrow{BC} =$ _____

   $\overrightarrow{CB} =$ _____

   $\overrightarrow{CA} =$ _____

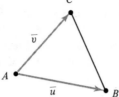

(b) $\overrightarrow{CD} =$ _____

   $\overrightarrow{DA} =$ _____

   $\overrightarrow{DB} =$ _____

   $\overrightarrow{AC} =$ _____

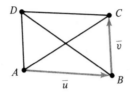

6. (a) If dir $\bar{v} = -$dir $\bar{u}$ and dir $(\bar{v} + \bar{u}) =$ dir $\bar{v}$, how does $\|\bar{v}\|$ compare with
    $\|\bar{u}\|$?
  (b) If $\|\bar{v}\| < \|\bar{u}\|$ and dir $\bar{v} = -$dir $\bar{u}$, what is dir $(\bar{u} + \bar{v})$? (Give your answer
    in terms of either dir $\bar{u}$ or dir $\bar{v}$.)

7. Given line vectors $\bar{u}$, $\bar{v}$, and $\bar{w}$ as in Figure 1.7. Draw (a) $\bar{u} + \bar{v}$; (b) $(\bar{u} + \bar{v}) + \bar{w}$;
   (c) $\bar{v} + \bar{w}$; (d) $\bar{u} + (\bar{v} + \bar{w})$.

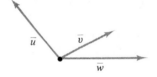

Figure 1.7

8. For line vectors $\bar{u}, \bar{v}$, and $\bar{w}$ as in Figure 1.7, draw (a) $\bar{u} - \bar{v}$; (b) $\bar{v} - \bar{w}$; (c)
   $\bar{u} - \bar{w}$; (d) how is the line vector in (c) related to the vectors in (a) and (b)?

9. Draw a triangle with sides $\bar{a}$, $\bar{b}$, and $\bar{a} + \bar{b}$.
   (a) Why is $\|\bar{a} + \bar{b}\| \leq \|\bar{a}\| + \|\bar{b}\|$?
   (b) For arbitrary line vectors $\bar{a}$ and $\bar{b}$, under what condition(s) is it true that
   $\|\bar{a} + \bar{b}\| = \|\bar{a}\| + \|\bar{b}\|$?

10. Draw a triangle with sides $\bar{a}$, $\bar{b}$, and $\bar{a} - \bar{b}$.
   (a) Why is $\|\|\bar{a}\| - \|\bar{b}\|\| \leq \|\bar{a} - \bar{b}\|$? (The outside vertical lines on the left
    side of the inequality denote absolute value.)
   (b) For arbitrary line vectors $\bar{a}$ and $\bar{b}$, under what condition(s) is it true that
   $\|\bar{a}\| - \|\bar{b}\| = \|\bar{a} - \bar{b}\|$?

11. Verify that $-(-\bar{v}) = \bar{v}$ by showing that the two vectors have the same length
   and the same direction.

Applications

Example 1.4. The water in a river is flowing from north to south at a speed of 4 km/h (1 kilometer $= 0.6214$ miles). A boat is being propelled across the river at a water speed of 12 km/h in a direction 30° south of east. Use a scale diagram to approximate the direction and land speed of the boat.

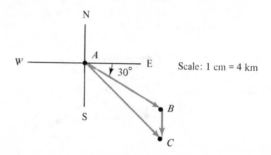

Solution: In the figure, $\overrightarrow{AB}$ and $\overrightarrow{BC}$ represent the velocities of the boat (with respect to water) and water, respectively. $\overrightarrow{AC} = \overrightarrow{AB} + \overrightarrow{BC}$ represents the direction and land speed of the boat. Actual measurement shows that $\|\overrightarrow{AC}\| \doteq 3.6$ cm and dir $\overrightarrow{AC} \doteq 44°$ south of east. Since the scale is 1 cm $= 4$ km, we have 14.4 km for the approximate land speed of the boat.

Example 1.5. A town T is located 600 km in a direction of 30° west of north from city C. Flying at a constant maximum speed, a plane takes 4 h to fly from C to T with the wind blowing 30 km/h from east to west. Assuming that the wind velocity remains constant, what is the shortest possible time for the return trip?

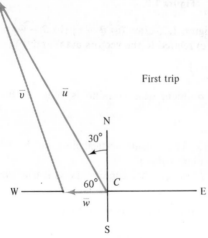

Solution: In planning a flight, there are two velocity vectors that the pilot of a small plane must consider, the air velocity vector $\vec{v}$ and the ground velocity vector $\vec{u}$. The magnitude of $\vec{v}$ is the plane's air speed, and the direction of $\vec{v}$ is the heading that the pilot must take to arrive at his destination. The ground velocity vector $\vec{u}$ has magnitude equal to the ground speed of the plane, and its direction is the path of flight over the ground.

The plan is to find the air speed and then use it to find the ground speed on the return trip. The ground speed on the trip out is $600/4 = 150$ km/h $= ||\bar{u}||$. Furthermore, dir $\bar{u} = 30°$ west of north. If $\bar{w}$ represents the wind velocity, then $||\bar{w}|| = 30$ and, dir $\bar{w} =$ west. The ground velocity vector $\bar{u}$ is the sum of the wind velocity vector $\bar{w}$ and the air velocity vector $\bar{v}$. So $\bar{v} = \bar{u} - \bar{w}$. Using the law of cosines, we get

$$||\bar{v}||^2 = ||\bar{u}||^2 + ||\bar{w}||^2 - 2||\bar{u}||\,||\bar{w}||\cos 60°$$

To simplify computation, we let 1 unit $= 30$ km. Then $||\bar{w}|| = 1$, $||\bar{u}|| = 5$, $||\bar{v}||^2 = 25 + 1 - 5 = 21$, and $||\bar{v}|| = \sqrt{21}$. Hence the air speed is $||\bar{v}|| = 30\sqrt{21} \doteq 137.5$.

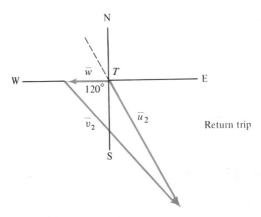

Return trip

On the return trip, the wind velocity vector remains the same; the new ground speed vector $\bar{u}_2$ has direction $= -$dir $\bar{u}$, and the new air velocity vector $\bar{v}_2$ has magnitude equal to $||\bar{v}|| = \sqrt{21}$. We have $||\bar{v}_2||^2 = ||\bar{u}_2||^2 + ||\bar{w}||^2 - 2||\bar{u}_2||\,||\bar{w}||\cos 120°$ or $21 = ||\bar{u}_2||^2 + 1 + ||\bar{u}_2||$. Hence $||\bar{u}_2||^2 + ||\bar{u}_2|| - 20 = 0$ and $||\bar{u}_2|| = 4$. Thus the ground speed on the return trip is $30(4) = 120$ km/h. The time for the return trip is $600/120 = 5$ h.

Exercise 1.1

12. Solve Example 1.4 by using the law of cosines and the law of sines.

13. (a) An airplane has a maximum airspeed of 500 km/h. If the plane is flying at its maximum speed with a heading 40° west of north, and the wind is blowing from north to south at 40 km/h, use a scale drawing to approximate the ground speed of the plane and its direction of travel over the ground.

   (b) Check your accuracy by calculating the ground speed and direction using the law of cosines and the law of sines.

14. (a) A force of 30 N (1 newton = 0.2248 lb) is acting on an object. A second force of 20 N is acting on the object at an angle of 90° from the first force. Use a scale drawing to approximate the magnitude and direction of the resultant force (relative to the first force).

(b) Calculate the direction and magnitude of the resultant force by using the law of cosines and the law of sines.

15. Solve Example 1.5 by making use of the diagram below. *Hint:* Show that $\|\bar{u}_2\| = \|\bar{u}\| - 1$ by using the law of cosines and the fact that $\|\bar{v}\| = \|\bar{v}_2\|$.

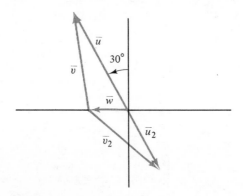

Scalar Multiplication

From the definition of vector addition (triangle rule), it is clear that $\bar{v} + \bar{v}$ is a vector with magnitude twice that of $\bar{v}$ and having the same direction. If we write $2\bar{v} = \bar{v} + \bar{v}$, then $2\bar{v}$ ought to be a vector such that $\|2\bar{v}\| = 2\|\bar{v}\|$ and dir $2\bar{v}$ = dir $\bar{v}$. Similarly, $\bar{v} + \bar{v} + \bar{v}$ is a vector with magnitude three times that of $\bar{v}$ and having the same direction as $\bar{v}$. Again, if we write $3\bar{v} = \bar{v} + \bar{v} + \bar{v}$, then $3\bar{v}$ should be a vector such that $\|3\bar{v}\| = 3\|\bar{v}\|$ and dir $3\bar{v}$ = dir $\bar{v}$. Applying the triangle rule for vector addition also tells us that $-\bar{v} + -\bar{v}$ is a vector twice the length of $\bar{v}$ and having the opposite direction of $\bar{v}$. Again, it would be appropriate to write $(-2)\bar{v} = -\bar{v} + -\bar{v}$, if by $(-2)\bar{v}$ we mean the vector such that $\|(-2)\bar{v}\| = 2\|-\bar{v}\| = 2\|\bar{v}\|$, and dir $(-2)\bar{v}$ = dir $-\bar{v}$ = $-$dir $\bar{v}$ (see Figure 1.8).

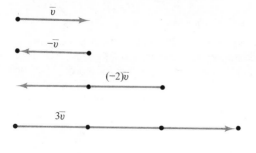

Figure 1.8   Scalar multiplication

With these examples in mind, the following definition of multiplication of a vector by a real number should seem reasonable.

Definition 1.2. For any real number r and for any vector $\bar{v}$,

i. If $r > 0$, $r\bar{v}$ is a vector such that $\|r\bar{v}\| = r\|\bar{v}\|$, and dir $r\bar{v} = $ dir $\bar{v}$.
ii. If $r = 0$, $r\bar{v} = \bar{0}$.
iii. If $r < 0$, $r\bar{v}$ is a vector such that $\|r\bar{v}\| = |r|\,\|\bar{v}\| = (-r)\|\bar{v}\|$, and dir $r\bar{v} = -$dir $\bar{v}$.

We will not assign a meaning to $\bar{v}r$. In other words, we will always write the real number on the left of the vector when we multiply a vector by a real number. It is a common convention to call the real number a **scalar** and to refer to the multiplication in Definition 1.2 as **scalar multiplication** (see Figure 1.9). The reader is probably convinced that

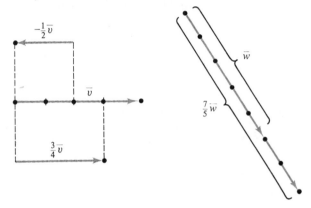

Figure 1.9 Scalar multiplication (again)

$$(2\cdot 3)\bar{v} = 2(3\bar{v}) = 3(2\bar{v})$$
$$(3 + 5)\bar{v} = 3\bar{v} + 5\bar{v}$$
$$\tfrac{2}{3}(\bar{u} + \bar{v}) = \tfrac{2}{3}\bar{u} + \tfrac{2}{3}\bar{v}$$

are true statements. In fact, you may have guessed at the truth of the following generalized statements:

$$(rs)\bar{v} = r(s\bar{v}) = s(r\bar{v})$$
$$(r + s)\bar{v} = r\bar{v} + s\bar{v}$$
$$r(\bar{u} + \bar{v}) = r\bar{u} + r\bar{v}$$

These statements are true for arbitrary scalars r and s and for arbitrary line vectors $\bar{u}$ and $\bar{v}$.

We pause to make a formal list of some of the properties of line vectors that have been mentioned. The importance of these properties is difficult to overemphasize. In fact, these properties will be used later to *define* a more abstract notion of vector.

Properties of Vector Addition

For arbitrary line vectors $\bar{u}$, $\bar{v}$, and $\bar{w}$ (in a fixed plane),

1A. The sum $\bar{u} + \bar{v}$ is a uniquely determined line vector. (This is what is meant when we say that $+$ is an operation.)

2A. $\bar{u} + \bar{v} = \bar{v} + \bar{u}$. Vector addition is commutative.

3A. $(\bar{u} + \bar{v}) + \bar{w} = \bar{u} + (\bar{v} + \bar{w})$. Vector addition is associative. (As a consequence of this property there is no ambiguity in omitting the parentheses and writing $\bar{u} + \bar{v} + \bar{w}$.)

4A. $\bar{u} + \bar{0} = \bar{u} = \bar{0} + \bar{u}$. $\bar{0}$ is an identity for the operation $+$.

5A. $\bar{u} + -\bar{u} = \bar{0} = -\bar{u} + \bar{u}$. $-\bar{u}$ is an inverse for $\bar{u}$ with respect to vector addition.

Properties of Scalar Multiplication

For arbitrary line vectors $\bar{u}$ and $\bar{v}$ and scalars r and s,

6M. The product $r\bar{v}$ of the scalar r and the vector $\bar{v}$ is a uniquely determined line vector. (This is what is meant when we say that the product is a scalar mulplication.)

7M. $(rs)\bar{v} = r(s\bar{v}) = s(r\bar{v})$.

8M. $(r + s)\bar{v} = r\bar{v} + s\bar{v}$.

9M. $r(\bar{u} + \bar{v}) = r\bar{u} + r\bar{v}$.

10M. $1\bar{u} = \bar{u}$.

Exercise 1.1

16. Interpret the various operations that appear in each of the properties 7M–9M. For example, in property 8M

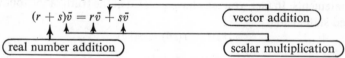

$(r + s)\bar{v} = r\bar{v} + s\bar{v}$

vector addition

real number addition

scalar multiplication

17. Use the definition of scalar multiplication to show that property 7M holds for the case $r > 0$ and $s < 0$ (you should be able to verify property 7M for the other cases as well). *Hint:* Show that dir $(rs)\bar{v} = $ dir $r(s\bar{v}) = $ dir $s(r\bar{v})$ and $\|(rs)\bar{v}\| = \|r(s\bar{v})\| = \|s(r\bar{v})\|$.

18. Use a scale drawing to illustrate that $(r + s)\bar{v} = r\bar{v} + s\bar{v}$ when $r = 2$ and $s = -3$.

19. Prove property 8M. *Hint:* If r and s have opposite signs, the direction is determined by the sign of that scalar which has the largest absolute value. For lengths, show that $|r + s| \|\bar{v}\| = |r| \|\bar{v}\| + |s| \|\bar{v}\|$ if r and s have the same sign, and $|r + s| \|\bar{v}\| = \|r| - |s\| \|\bar{v}\|$ if r and s have opposite signs.

20. Use a scale drawing to illustrate property 9M when $r = \frac{2}{3}$.

21. Given Figure 1.10, verify that $r(\bar{u} + \bar{v}) = r\bar{u} + r\bar{v}$ for $r > 0$. *Hint:* Show that $\triangle ABD \sim \triangle ACE$. Then show that $\|r\bar{u} + r\bar{v}\| = r\|\bar{u} + \bar{v}\|$.

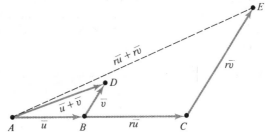

Figure 1.10

22. Given the vector $\bar{v} = \overrightarrow{OA}$:

(a) Sketch $\frac{3}{2}\bar{v}, -\frac{1}{2}\bar{v}, 2\bar{v}, \sqrt{2}\bar{v}, -\bar{v}$.
(b) What geometric figure is given by the collection of heads of all the scalar multiples of $\bar{v}$ if the tails are all drawn from 0?

In Problems 23–24, use any of the properties 1A–5A and 6M–10M to prove the following results:

23. $0\bar{v} = \bar{0}$. *Hint:* Write $0\bar{v}$ as $(0 + 0)\bar{v}$.
24. $(-1)\bar{v} = -\bar{v}$.
25. If $k\bar{v} = \bar{0}$ and $k \neq 0$, then $\bar{v} = \bar{0}$.

1.2 LINE VECTORS IN MECHANICS AND PLANE GEOMETRY

In this section, line vectors are used to derive two theorems, the *ratio theorem* and the *basis theorem*. The ratio theorem is used to solve a problem in an area of science called *mechanics*. We use both the ratio theorem and the basis theorem to derive some familiar results from plane geometry. Before stating the ratio theorem, we need to discuss parallel vectors and position vectors.

Consider two parallel line segments $\overline{CD}$ and $\overline{AB}$ (Figure 1.11). The corresponding vectors $\bar{v} = \overrightarrow{CD}$ and $\bar{w} = \overrightarrow{AB}$ have either the same direction or the opposite direction (in Figure 1.11, dir $\overrightarrow{CD} = -$dir $\overrightarrow{AB}$). Therefore, there is a scalar k such that $\bar{v} = k\bar{w}$ ($k = -\frac{2}{3}$ in Figure 1.11). Conversely, if

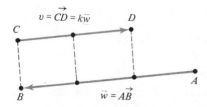

Figure 1.11   Parallel vectors

two nonzero vectors have the property that one is a scalar multiple of the other, then their geometric representations, when viewed as line segments, are parallel (we allow a segment to be parallel to itself). Definition 1.3 should now seem reasonable.

Definition 1.3.   Nonzero **vectors** $\bar{v}$ **and** $\bar{w}$ **are parallel** if and only if there is a scalar $k \neq 0$ such that $\bar{v} = k\bar{w}$. We write $\bar{v} \parallel \bar{w}$.

We now indicate how a one-to-one correspondence between points in a plane (or space) and line vectors can be obtained.

Choose a fixed point O. The **position vector** of a point A relative to the **origin** O is the vector $\overrightarrow{OA}$ with tail at O and head at A. Given A, the position vector $\overrightarrow{OA}$ is uniquely determined; conversely, any line vector with tail at O uniquely determines a point at the head of the vector. Thus *there is a one to one correspondence between points and position vectors. We will write the position vector of a point A as $\bar{a}$; that is, $\bar{a} = \overrightarrow{OA}$.*

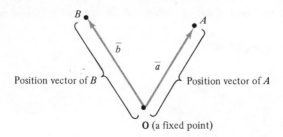

We are now ready to state and prove the ratio theorem.

Theorem 1.1 (The Ratio Theorem).   If A and B are points and P is a point on $\overrightarrow{AB}$ which divides the segment in the ratio $r : s$ so that $\overline{AP}/\overline{PB} = r/s$, then,

$$\bar{p} = \frac{s}{r + s}\bar{a} + \frac{r}{r + s}\bar{b}$$

where the origin O can be chosen as any fixed point not collinear with A and B.

Proof: Through P draw a line parallel to $\overline{OB}$ intersecting $\overline{OA}$ at Q (see Figure 1.12). By the triangle rule for addition we have $\bar{p} = \overrightarrow{OQ} + \overrightarrow{QP}$.

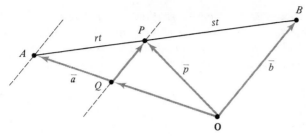

Figure 1.12

Since $\overrightarrow{OQ} \| \bar{a}$ and $\overrightarrow{QP} \| \bar{b}$, we can write $\overrightarrow{OQ} = x\bar{a}$ and $\overrightarrow{QP} = y\bar{b}$ for some scalars x and y. Furthermore, x and y are positive, since dir $\bar{a} = $ dir $\overrightarrow{OQ}$ and dir $\bar{b} = $ dir $\overrightarrow{QP}$. So

$$\bar{p} = x\bar{a} + y\bar{b} \quad \text{for some positive scalars } x \text{ and } y$$

From the similar triangles, $\triangle APQ$ and $\triangle ABO$ we have

$$\frac{\|\overrightarrow{QP}\|}{\|\bar{b}\|} = \frac{\|\overrightarrow{AP}\|}{\|\overrightarrow{AB}\|} = \frac{r}{r+s}$$

Therefore,

$$\|\overrightarrow{QP}\| = \frac{r}{r+s}\|\bar{b}\|$$

Also

$$\|\overrightarrow{QP}\| = \|y\bar{b}\| = y\|\bar{b}\|$$

Since $\|\bar{b}\| \neq 0$, we have

$$y = \frac{r}{r+s}$$

We also have

$$\frac{\|\overrightarrow{OQ}\|}{\|\bar{a}\|} = \frac{\|\overrightarrow{PB}\|}{\|\overrightarrow{AB}\|} = \frac{s}{r+s}$$

So

$$\|\overrightarrow{OQ}\| = \frac{s}{r+s}\|\bar{a}\|$$

But $\|\overrightarrow{OQ}\| = \|x\bar{a}\| = x\|\bar{a}\|$, and $\|\bar{a}\| \neq 0$. Thus

$$x = \frac{s}{r+s}$$

and

$$\bar{p} = x\bar{a} + y\bar{b} = \frac{s}{r+s}\bar{a} + \frac{r}{r+s}\bar{b} \qquad \text{Q.E.D.}$$

Note: Q.E.D. is an abbreviation for "quod erat demonstrandum," which is Latin for "which was to be shown" or "proved."

The restriction in Theorem 1.1 that the origin O be a point not collinear with A and B is not a necessary restriction. The reader is asked to prove this in Problem 13 of Exercise 1.2. We obtain the following corollary to Theorem 1.1 by setting both r and s equal to 1.

Corollary 1.1.   If A and B are points and P is the midpoint of $\overline{AB}$, then $\bar{p} = \frac{1}{2}\bar{a} + \frac{1}{2}\bar{b}$.

The ratio theorem will now be used to derive a result in *mechanics*.

Applications: Center of Mass

Example 1.6.   The **center of mass** of two particles of masses m_1 and m_2 located at points A and B, respectively, is the point G that divides the segment $\overline{AB}$ in the ratio $m_2 : m_1$. Show that (relative to some fixed origin O)

$$\bar{g} = \frac{1}{m_1 + m_2}(m_1\bar{a} + m_2\bar{b})$$

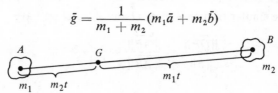

Solution: We may think of the center of mass G as the point where the system will balance if a knife edge is placed at G and the line segment (thin rod) is considered as weightless. Relative to a fixed origin O we see from the ratio theorem that

$$\overrightarrow{OG} = \bar{g} = \frac{m_1}{m_1 + m_2}\bar{a} + \frac{m_2}{m_1 + m_2}\bar{b} = \frac{1}{m_1 + m_2}(m_1\bar{a} + m_2\bar{b})$$

Example 1.7.   Suppose that particles of masses m_1, m_2, and m_3 are located at points A, B, and C, respectively. If G is the center of mass of the system of masses, find the position vector $\bar{g}$ (relative to some fixed origin O).

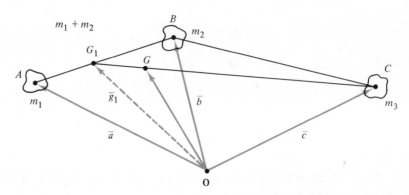

Solution: If G_1 is the center of mass of the system of masses m_1 and m_2 located at A and B, respectively, then this system of masses is equivalent to a single mass $m_1 + m_2$ located at G_1. It then follows from Example 1.6 that the center of mass G of the given system has the position vector

$$\bar{g} = \frac{m_1 + m_2}{m_1 + m_2 + m_3}\bar{g}_1 + \frac{m_3}{m_1 + m_2 + m_3}\bar{c}$$

But

$$\bar{g}_1 = \frac{m_1}{m_1 + m_2}\bar{a} + \frac{m_2}{m_1 + m_2}\bar{b}$$

Therefore,

$$\bar{g} = \left(\frac{m_1 + m_2}{m_1 + m_2 + m_3}\right)\left(\frac{m_1}{m_1 + m_2}\bar{a} + \frac{m_2}{m_1 + m_2}\bar{b}\right) + \frac{m_3}{m_1 + m_2 + m_3}\bar{c}$$

$$= \frac{1}{m_1 + m_2 + m_3}(m_1\bar{a} + m_2\bar{b} + m_3\bar{c})$$

Using mathematical induction (see Section 3.7), one can prove the following generalization: the center of mass of a system of particles of masses $m_1, m_2,$ $\ldots, m_k$ located at $A_1, A_2, \ldots, A_k$ is G, where the position vector of G relative to an arbitrary origin O is

$$\bar{g} = \frac{1}{m_1 + m_2 \cdots + m_k}(m_1\bar{a}_1 + m_2\bar{a}_2 + \cdots + m_k\bar{a}_k) \qquad (1.1)$$

We now introduce a rectangular coordinate system so that the origin O has coordinates $(0, 0)$ [In space the coordinates of O are $(0, 0, 0)$]. Let $\bar{i}$ be the position vector with head at $(1, 0)$, and let $\bar{j}$ be the position vector with head at $(0, 1)$. Then if P is a point with coordinates (x, y), we can write the position vector of P as

$$\bar{p} = x\bar{i} + y\bar{j}$$

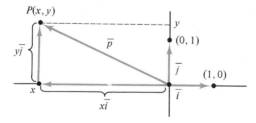

In our next example we consider the masses as lying in a plane that has been coordinatized so that the fixed point O is the origin $(0, 0)$.

Example 1.8. Find the coordinates of the center of mass of a system of four particles of masses 2, 1, 4, and 2 located at points $A_1 = (1, 0)$, $A_2 = (2, 1)$, $A_3 = (0, 1)$, and $A_4(-2, -1)$, respectively.

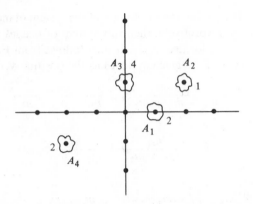

Solution: From Equation (1.1), G is the point with position vector $\bar{g} = \frac{1}{9}(2\bar{a}_1 + \bar{a}_2 + 4\bar{a}_3 + 2\bar{a}_4)$. We can find the coordinates of G as follows: if $\bar{i}$ is the position vector with head at $(1, 0)$ and $\bar{j}$ is the position vector with head at $(0, 1)$, then $\bar{a}_1 = \bar{i}$, $\bar{a}_2 = 2\bar{i} + 1\bar{j}$, $\bar{a}_3 = \bar{j}$, and $\bar{a}_4 = -2\bar{i} + -1\bar{j}$. So

$$\bar{g} = \tfrac{1}{9}(2\bar{i} + (2\bar{i} + \bar{j}) + 4\bar{j} + 2(-2\bar{i} - \bar{j})) = \tfrac{1}{9}(3\bar{j}) = \tfrac{1}{3}\bar{j}$$

Since the head of $\frac{1}{3}\bar{j}$ is at $(0, \frac{1}{3})$, we have $G = (0, \frac{1}{3})$.

Next we illustrate the use of line vectors in proving some familiar theorems from plane geometry.

Example 1.9.   The medians of a triangle intersect at a point, and this point is two thirds of the distance from a vertex to the opposite side along any median.

Proof: Given $\triangle ABC$ with M_1, M_2, and M_3 the midpoints of $\overline{AC}$, $\overline{BC}$, and $\overline{AB}$, respectively. Let T be a point on $\overline{AM_2}$ such that $\overrightarrow{AT} = \frac{2}{3}\overrightarrow{AM_2}$ (see Figure 1.13). Using the ratio theorem, we have, relative to some origin O, the following relationships among position vectors:

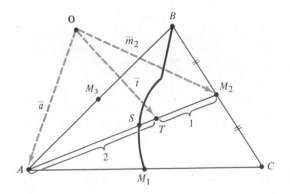

Figure 1.13

$$\bar{t} = \tfrac{1}{3}\bar{a} + \tfrac{2}{3}\bar{m}_2$$

$$\bar{m}_2 = \tfrac{1}{2}\bar{b} + \tfrac{1}{2}\bar{c}$$

Therefore,

$$\bar{t} = \tfrac{1}{3}\bar{a} + \tfrac{2}{3}(\tfrac{1}{2}\bar{b} + \tfrac{1}{2}\bar{c}) = \tfrac{1}{3}\bar{a} + \tfrac{1}{3}\bar{b} + \tfrac{1}{3}\bar{c}$$

Now let S be a point on $\overline{BM}_1$ such that $\overrightarrow{BS} = \tfrac{2}{3}\overrightarrow{BM}_1$, and let R be a point on $\overline{CM}_3$ such that $\overrightarrow{CR} = \tfrac{2}{3}\overrightarrow{CM}_3$. We obtain

$$\bar{s} = \tfrac{1}{3}\bar{b} + \tfrac{2}{3}\bar{m}_1$$

$$\bar{m}_1 = \tfrac{1}{2}\bar{a} + \tfrac{1}{2}\bar{c}$$

Therefore,

$$\bar{s} = \tfrac{1}{3}\bar{a} + \tfrac{1}{3}\bar{b} + \tfrac{1}{3}\bar{c}$$

Similarly,

$$\bar{r} = \tfrac{1}{3}\bar{a} + \tfrac{1}{3}\bar{b} + \tfrac{1}{3}\bar{c}$$

Since $\bar{t}$, $\bar{s}$, and $\bar{r}$ are the position vectors of the points T, S, and R, respectively, we conclude that $T = S = R$. Consequently, the medians are concurrent and meet at a point that is two thirds the distance from a vertex to the opposite side along a median.   Q.E.D.

Before we state and prove the basis theorem, we introduce the following term. A **linear combination** of the vectors $\bar{u}$ and $\bar{v}$ is any vector $\bar{w}$ of the form $\bar{w} = r\bar{u} + s\bar{v}$.

Theorem 1.2 (The Basis Theorem).   If $\bar{u}$ and $\bar{v}$ are nonzero, nonparallel vectors, then any vector $\bar{w}$ in the plane determined by $\boldsymbol{O}$, $\bar{u}$, and $\bar{v}$ is a linear combination of $\bar{u}$ and $\bar{v}$. The linear combination is unique in the sense that if $\bar{w} = r_1\bar{u} + s_1\bar{v} = r_2\bar{u} + s_2\bar{v}$ then $r_1 = r_2$ and $s_1 = s_2$.

Proof: Consider representations of $\bar{u}$, $\bar{v}$, and $\bar{w}$ that start from the same point $\boldsymbol{O}$. Let $\bar{u} = \overrightarrow{OU}$, $\bar{v} = \overrightarrow{OV}$, and $\bar{w} = \overrightarrow{OW}$. Through W, draw a line $\mathcal{L}$ parallel to $\bar{u}$ (see Figure 1.14). Extend $\bar{v}$ in both directions. $\mathcal{L}$ and the line

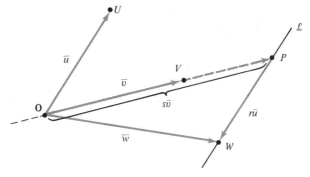

Figure 1.14

through O determined by $\bar{v}$ must intersect at some point P, because $\bar{u}$ is not parallel to $\bar{v}$. $\overrightarrow{PW} = r\bar{u}$ for some scalar r, and $\overrightarrow{OP} = s\bar{v}$ for some scalar s. Also, $\bar{w} = \overrightarrow{OP} + \overrightarrow{PW} = s\bar{v} + r\bar{u} = r\bar{u} + s\bar{v}$. So $\bar{w}$ is a linear combination of $\bar{u}$ and $\bar{v}$.

Now suppose that $\bar{w} = r_1\bar{u} + s_1\bar{v} = r_2\bar{u} + s_2\bar{v}$. We then have $r_1\bar{u} - r_2\bar{u} = s_2\bar{v} - s_1\bar{v}$ or $(r_1 - r_2)\bar{u} = (s_2 - s_1)\bar{v}$. If either $r_1 - r_2$ or $s_2 - s_1$ is not zero, we could then write $\bar{v}$ as a scalar multiple of $\bar{u}$, or vice versa. (Multiply each side of the equation by the multiplicative inverse of the nonzero scalar.) Since $\bar{v}$ is not parallel to $\bar{u}$, we are forced to conclude that $r_1 - r_2 = 0$ and $s_1 - s_2 = 0$. Hence $r_1 = r_2$ and $s_1 = s_2$. Q.E.D.

The final example of this section indicates how the basis theorem can be used to verify a familiar result from plane geometry.

Example 1.10. Given parallelogram $ABCD$ with M the midpoint of $\overline{CD}$ (see Figure 1.15). Prove that the diagonal $\overline{BD}$ and line segment $\overline{AM}$ intersect at a point of trisection for each segment.

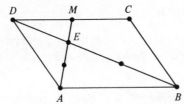

|Figure 1.15

Plan: Let E be the point of intersection of $\overline{BD}$ and $\overline{AM}$. Write $\overrightarrow{BE}$ as a linear combination of the vectors $\overrightarrow{BA}$ and $\overrightarrow{AM}$ in two different ways. Then use the basis theorem to equate coefficients.

Proof:
1. $\overrightarrow{BE} = \overrightarrow{BA} + \overrightarrow{AE} = \overrightarrow{BA} + r\overrightarrow{AM}$. ($\overrightarrow{AE} = r\overrightarrow{AM}$ for some scalar r.)
2. Also, $\overrightarrow{BE} = s\overrightarrow{BD} = s(\overrightarrow{BA} + \overrightarrow{AD}) = s\overrightarrow{BA} + s\overrightarrow{AD}$.
3. $\overrightarrow{AD} = \overrightarrow{AM} + \overrightarrow{MD} = \overrightarrow{AM} + \frac{1}{2}\overrightarrow{CD} = \overrightarrow{AM} + \frac{1}{2}\overrightarrow{BA}$. ($\overrightarrow{CD} = \overrightarrow{BA}$.)
4. Substituting from 3 into 2 we obtain, $\overrightarrow{BE} = s\overrightarrow{BA} + s(\overrightarrow{AM} + \frac{1}{2}\overrightarrow{BA}) = \frac{3}{2}s\overrightarrow{BA} + s\overrightarrow{AM}$.

Equating coefficients from 1 and 4 we obtain $\frac{3}{2}s = 1$ and $r = s$. Hence $s = \frac{2}{3} = r$, $\overrightarrow{AE} = \frac{2}{3}\overrightarrow{AM}$, and $\overrightarrow{BE} = \frac{2}{3}\overrightarrow{BD}$. Q.E.D.

Exercise 1.2

1. Find the position vector of the center of mass G of a system of particles of masses $m_1 = 5$, $m_2 = 3$, and $m_3 = 8$ located at points A, B, and C, respectively.

2. Find the coordinates of the center of mass G of a system of five particles of masses 3, 4, 7, 9, and 5 located at points $(3, 2)$, $(1, 4)$, $(2, -1)$, $(3, 2)$, and $(0, 2)$, respectively.

3. Find the position vector of the center of mass G of a system of particles of masses $m_1 = 3$, $m_2 = 2$, $m_3 = 5$, $m_4 = 4$, and $m_5 = 6$ located at points A, B, C, D, and E, respectively.

4. In Problem 3, if $A = (2, 0)$, $B = (3, 2)$, $C = (4, -2)$, $D = (-6, 3)$, and $E = (-4, 2)$, find the coordinates of G.

5. If a system of particles of masses $m_1, m_2, \ldots, m_k$ are located at points $A_1 = (x_1, y_1)$, $A_2 = (x_2, y_2), \ldots, A_k = (x_k, y_k)$, respectively, show that the center of mass G has coordinates (x^*, y^*), where

$$x^* = \frac{m_1 x_1 + m_2 x_2 + \cdots + m_k x_k}{m_1 + m_2 + \cdots + m_k}$$

and

$$y^* = \frac{m_1 y_1 + m_2 y_2 + \cdots + m_k y_k}{m_1 + m_2 + \cdots + m_k}$$

Hint: Write $\bar{a}_n = x_n \bar{i} + y_n \bar{j}$, $n = 1, 2, \ldots, k$, and use Equation (1.1).

Note: In studying center-of-mass problems in calculus, one usually accepts the result of Problem 5 as a starting point.

For further problems on vector methods in mechanics, the reader may consult R. C. Smith and P. Smith, *Mechanics* (New York: John Wiley & Sons, Inc., 1968) or Hans Liebeck, *Algebra for Scientists and Engineers* (New York: John Wiley & Sons, Inc., 1969).

6. If $\bar{a} = \overrightarrow{OA}$ and $\bar{b} = \overrightarrow{OB}$, write the vector $\overrightarrow{AB}$ in terms of the position vectors $\bar{a}$ and $\bar{b}$.

7. In quadrilateral $ABCD$, write each of the vectors $\overrightarrow{AB}$, $\overrightarrow{BC}$, $\overrightarrow{CD}$, and $\overrightarrow{DA}$ in terms of the position vectors $\bar{a}$, $\bar{b}$, $\bar{c}$, and $\bar{d}$ (relative to some origin O).

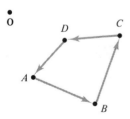

8. Verify the result in Example 1.10 by using the ratio theorem.

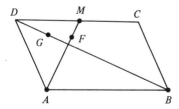

Hint: Let F and G be points on $\overrightarrow{AM}$ and $\overrightarrow{BD}$, respectively, so that $\overrightarrow{AF} = \frac{2}{3}\overrightarrow{AM}$ and $\overrightarrow{BG} = \frac{2}{3}\overrightarrow{BD}$. Show that the position vectors of F and G are equal, and consequently $F = G$.

9. Verify the result in Example 1.9 by using the basis theorem.

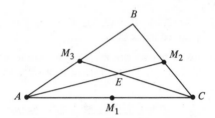

Hint: Let $\overrightarrow{AE} = r\overrightarrow{AM_2}$ and let $\overrightarrow{CE} = s\overrightarrow{CM_3}$. Write $\overrightarrow{AC}$ as a linear combination of $\overrightarrow{AM_2}$ and $\overrightarrow{CM_3}$ in two different ways.

In Problems 10–11, use either the ratio theorem or the basis theorem to verify the given statements.

10. The line segment that joins the midpoints of two sides of a triangle is parallel to and one half the length of the third side.

11. The midpoints of the sides of a quadrilateral are the vertices of a parallelogram.

12. Given a parallelogram $ABCD$ and an origin O. If $\bar{a}, \bar{b}$, and $\bar{c}$ are the position vectors of A, B, and C, respectively, find the position vector of D in terms of $\bar{a}, \bar{b}$, and $\bar{c}$.

13. Verify that the ratio theorem is true even if A, B, and O are collinear. *Hint:* Consider as separate cases the various possible locations of O relative to A, B, and P.

1.3 COORDINATE VECTORS

The reader is already aware of the fact that the real numbers can be viewed as points on a line, that ordered pairs of real numbers can be viewed as points in a plane, and that ordered triples of real numbers can be viewed as points in space (see Figure 1.16). Although we cannot give a *geometric interpretation* for ordered n-tuples $(x_1, x_2, \ldots, x_n)$ for $n > 3$, there are *useful interpretations* for them. For example, the 4-tuple $(2, 0, -1, 1)$ can be interpreted as a solution to the linear equation $3x + 4y + 5w + 12z = 13$. Just as we refer to ordered pairs as points in 2-space, we shall refer to ordered n-tuples as points in n-space. In this section we shall define an addition of n-tuples and a multiplication of an n-tuple by a real number (scalar multiplication). This addition and scalar multiplication will satisfy the properties 1A–5A and

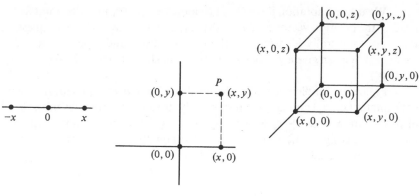

Figure 1.16

6M–10M listed in Section 1.2. It is because of this that n-tuples can and will be considered vectors as well as points in n-space.

Definition 1.4. An n-**tuple of real numbers** is denoted by $(x_1, x_2, \ldots, x_n)$, where each x_i is a real number. The n-tuples are **ordered** n-**tuples** in the sense that $X = (x_1, x_2, \ldots, x_n)$ and $Y = (y_1, y_2, \ldots, y_n)$ are equal if and only if $x_1 = y_1, x_2 = y_2, \ldots, x_n = y_n$.

The fact that the n-tuples are ordered means that $X = Y$ if and only if X and Y have the same entries in the same order. For example, $(1, 2, 3, 4)$ $\neq (3, 2, 1, 4)$. Before defining addition and scalar multiplication of n-tuples, we first indicate what the definition ought to be for ordered pairs and ordered triples of real numbers.

Suppose that a plane has been coordinatized with a rectangular coordinate system, that all points in the plane are viewed as ordered pairs of real numbers, and consider all line vectors in the plane as position vectors having their tails at the origin. Then every point $P = (x, y)$ in the plane is the head of a unique position vector $\bar{p} = x\bar{i} + y\bar{j}$; conversely, the head of each position vector $\bar{p} = x\bar{i} + y\bar{j}$ determines a unique point $p = (x, y)$.

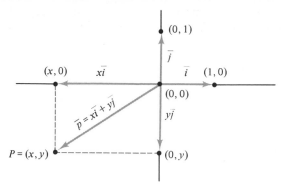

We use the notation $\bar{p} \leftrightarrow (x, y)$ to indicate that (x, y) is the point at the head of the position vector $\bar{p}$. If $\bar{p} \leftrightarrow (x_1, y_1)$ and $\bar{q} \leftrightarrow (x_2, y_2)$, then a question that arises naturally is: what are the coordinates x and y such that $\bar{p} + \bar{q} \leftrightarrow (x, y)$? We can give a geometric argument to show that $\bar{p} + \bar{q} \leftrightarrow (x_1 + x_2, y_1 + y_2)$.

Figure 1.17 indicates that if the heads of $\bar{p}$ and $\bar{q}$ are in quadrant I with coordinates (x_1, y_1) and (x_2, y_2) respectively, then the coordinates of the head of $\bar{p} + \bar{q}$ are $x = x_1 + x_2$ and $y = y_1 + y_2$ (The shaded triangles in each diagram are congruent.) Similarly, one can show that this result is correct for any position vectors $\bar{p}$ and $\bar{q}$.

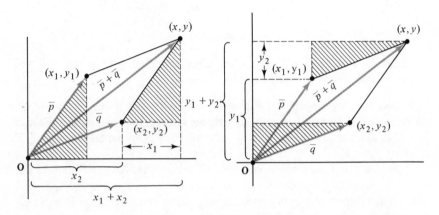

Figure 1.17 $\bar{p} + \bar{q} \leftrightarrow (x_1 + x_2, y_1 + y_2)$

A simpler way of finding the coordinates (x, y) would be to use the properties already discussed for line vectors. Using these properties, we have $\bar{p} + \bar{q} = (x_1\bar{i} + y_1\bar{j}) + (x_2\bar{i} + y_2\bar{j}) = (x_1 + x_2)\bar{i} + (y_1 + y_2)\bar{j}$. Thus the coordinates at the head of $\bar{p} + \bar{q}$ are $(x, y) = (x_1 + x_2, y_1 + y_2)$. Using this method it is clear that, in 3-space if the heads of $\bar{p}$ and $\bar{q}$ are at points with coordinates (x_1, y_1, z_1) and (x_2, y_2, z_2), respectively, then the head of $\bar{p} + \bar{q}$ has coordinates (x, y, z), where $x = x_1 + x_2$, $y = y_1 + y_2$, and $z = z_1 + z_2$. This then is the motivation for the following definition of addition of n-tuples.

Definition 1.5. If $X = (x_1, x_2, \ldots, x_n)$ and $Y = (y_1, y_2, \ldots, y_n)$ are arbitrary ordered n-tuples of real numbers, we define the **sum** $X + Y$ to be the n-tuple

$$X + Y = (x_1 + y_1, x_2 + y_2, \ldots, x_n + y_n)$$

We say that addition of n-tuples is defined *componentwise*.

Example 1.11. Let $P = (x_1, y_1)$, $Q = (x_2, y_2)$, and let $\bar{v}$ be the position vector such that $\bar{v} = \overrightarrow{PQ}$. Show that $\bar{v} \leftrightarrow (x_2 - x_1, y_2 - y_1)$.

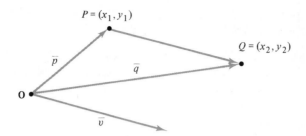

Solution: Since $\overrightarrow{PQ}$ is the line vector which when added to $\bar{p}$ is $\bar{q}$, we have

$$\overrightarrow{PQ} = \bar{q} - \bar{p} = (x_2\bar{i} + y_2\bar{j}) - (x_1\bar{i} + y_1\bar{j}) = (x_2 - x_1)\bar{i} + (y_2 - y_1)\bar{j}$$

Therefore, $\bar{v} = \bar{q} - \bar{p}$ and

$$\bar{v} \longleftrightarrow (x_2 - x_1, y_2 - y_1)$$

Next we consider multiplication by a scalar. If $\bar{p} = (x_1, y_1)$ and r is a scalar, we ask the question: what are the coordinates x and y such that $r\bar{p} \longleftrightarrow (x, y)$? Again, we can give a geometric argument to show that

$$r\bar{p} \longleftrightarrow (rx_1, ry_1)$$

In Figure 1.18 the head of $\bar{p}$ is in quadrant I and $r > 1$. From the similar

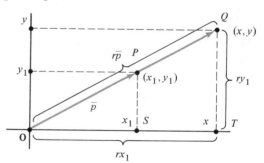

Figure 1.18 $r\bar{p} \longleftrightarrow (rx_1, ry_1)$

triangles $\triangle OPS$ and $\triangle OQT$, it follows that $\| \overrightarrow{OT} \| = r\| \overrightarrow{OS} \| = rx_1$ and $\| \overrightarrow{QT} \| = r\| \overrightarrow{PS} \| = ry_1$. So the point Q, which is the head of $r\bar{p}$, has coordinates (rx_1, ry_1). Similarly, one can show that this is true for a position vector $\bar{p}$ with head in any quadrant and for any scalar r.

Alternatively, we could have found the coordinates of $r\bar{p}$ as follows. Using properties of line vectors, we have

$$r\bar{p} = r(x_1\bar{i} + y_1\bar{j}) = r(x_1\bar{i}) + r(y_1\bar{j}) = (rx_1)\bar{i} + (ry_1)\bar{j}$$

Thus the coordinates at the head of $r\bar{p}$ are (rx_1, ry_1). Using this approach, it is clear that, in 3-space if the head of $\bar{p}$ has coordinates (x_1, y_1, z_1) then the head of $r\bar{p}$ has coordinates (rx_1, ry_1, rz_1). The following definition of a scalar times an n-tuple should now seem reasonable.

Definition 1.6. If $X = (x_1, x_2, \ldots, x_n)$ is an arbitrary n-tuple of real numbers and if r is any scalar, we define the **scalar multiplication** rX by $rX = (rx_1, rx_2, \ldots, rx_n)$. We say that scalar multiplication of ordered n-tuples is defined *componentwise*.

Example 1.12. If $X = (1, -1, 2, 4, 3)$ and $Y = (2, 4, -5, 1, 0)$, find $2X + (-3)Y$.

    Solution:

$$2X + (-3)Y = 2(1, -1, 2, 4, 3) + (-3)(2, 4, -5, 1, 0)$$
$$= (2, -2, 4, 8, 6) + (-6, -12, 15, -3, 0)$$
$$= (-4, -14, 19, 5, 6).$$

Since addition of ordered pairs (triples) and multiplication of an ordered pair (triple) by a scalar were defined to reflect the addition of line vectors and multiplication of a line vector by a scalar respectively, it should not be too surprising that properties 1A–5A and 6M–10M hold for ordered pairs (triples) as well as for line vectors. In fact, one can easily establish (although it may be tedious to do so) that properties 1A–5A and 6M–10M hold for the collection $\mathbf{R_n}$ of all ordered n-tuples. We state this result as our next theorem.

Theorem 1.3. The collection $\mathbf{R_n}$ of all ordered n-tuples of real numbers, together with addition and scalar multiplication defined componentwise as in Definitions 1.5 and 1.6, satisfies properties 1A–5A and 6M–10M.

We shall state and prove properties 2A, 4A, 5A, and 8M, and leave it to the reader to state and prove the remaining properties.

Let $X = (x_1, x_2, \ldots, x_n)$ and $Y = (y_1, y_2, \ldots, y_n)$ be arbitrary n-tuples in $\mathbf{R_n}$ and let r and s be any scalars. We wish to show the following:

2A. $X + Y = Y + X$.
4A. $0 = (0, 0, \ldots, 0)$ has the property $X + 0 = 0 + X = X$.
5A. $-X = (-x_1, -x_2, \ldots, -x_n)$ has the property $X + (-X) = (-X) + X = 0$.
8M. $(r + s)X = rX + sX$.

    Proof:

2A. $X + Y = (x_1 + y_1, x_2 + y_2, \ldots, x_n + y_n)$ (Definition 1.5)
$$= (y_1 + x_1, y_2 + x_2, \ldots, y_n + x_n) \quad \text{(scalar addition is commutative)}$$
$$= Y + X \quad \text{(Definition 1.5)}$$

4A.

$$X + 0 = 0 + X \quad \text{(property 2A)}$$
$$= (0 + x_1, 0 + x_2, \ldots, 0 + x_n) \quad \text{(Definition 1.5)}$$
$$= (x_1, x_2, \ldots, x_n) \quad \text{(0 is the identity for addition of scalars)}$$
$$= X$$

5A.

$$X + (-X) = (-X) + X \quad \text{(property 2A)}$$
$$= (-x_1 + x_1, -x_2 + x_2, \ldots, -x_n + x_n)$$
$$= (0, 0, \ldots, 0) \quad (-x_i + x_i \text{ is the scalar } 0)$$
$$= 0$$

8M.

$$(r + s)X = ((r + s)x_1, (r + s)x_2, \ldots, (r + s)x_n \quad \text{(Definition 1.6)}$$
$$= (rx_1 + sx_1, rx_2 + sx_2, \ldots, rx_n + sx_n) \quad \text{(for real numbers, multiplication distributes over addition)}$$
$$= (rx_1, rx_2, \ldots, rx_n) + (sx_1, sx_2, \ldots, sx_n) \quad \text{(Definition 1.5)}$$
$$= rX + rY \quad \text{(Definition 1.6)} \qquad \text{Q.E.D.}$$

It is because $\mathbf{R}_n$ satisfies the properties 1A–5A and 6M–10M that we shall refer to the members of $\mathbf{R}_n$ not only as n-tuples but also as *vectors*. We will use the term *coordinate vector* for n-tuples when we wish to distinguish them from line vectors.

Definition 1.7. For any coordinate vectors X and Y in $\mathbf{R}_n$, we define subtraction by $X - Y = X + (-Y)$ ($-Y$ is defined in the proof of property 5A of Theorem 1.3.)

Throughout the text we will encounter mathematical systems that satisfy properties 1A–5A and 6M–10M. In fact, this is the common thread that holds the subject of linear algebra together. We close this section with the following definition.

Definition 1.8. A **vector space** is a set $\mathbf{V}$ together with an operation called addition that satisfies 1A–5A and a scalar multiplication that satisfies 6M–10M. The objects in $\mathbf{V}$ are called **vectors**.

It is convenient to refer to "the vector space $\mathbf{V}$," although technically $\mathbf{V}$ is just the set of vectors and the vector space is the mathematical system consisting of the ordered triple $(\mathbf{V}, +, \text{sm})$. (Here sm denotes scalar multiplication.)

Exercise 1.3

1. (a) If $P = (1, 3)$ and $Q = (2, 2)$, sketch the position vectors $\vec{p}, \vec{q}, -\vec{q}$, and $\vec{p} - \vec{q}$. Draw $\overrightarrow{QP}$.
   (b) How is $\vec{p} - \vec{q}$ related to $P - Q$?
   (c) How is $\vec{p} - \vec{q}$ related to $\overrightarrow{QP}$?

2. (a) If $P = (1, 3, 2)$ and $Q = (2, 2, 1)$ sketch the position vectors $\vec{p}, \vec{q}, -\vec{q}$ and $\vec{p} - \vec{q}$. Draw $\overrightarrow{QP}$.
   (b) How is $\vec{p} - \vec{q}$ related to $P - Q$?
   (c) How is $\vec{p} - \vec{q}$ related to $\overrightarrow{QP}$?

3. Let $P = (x_1, y_1, z_1)$, $Q = (x_2, y_2, z_2)$, and let $\vec{v}$ be the position vector equal to $\overrightarrow{PQ}$. Follow Example 1.11 and show that $\overrightarrow{PQ} = \vec{v} \longleftrightarrow (x_2 - x_1, y_2 - y_1, z_2 - z_1)$.

4. (a) If $\vec{p} \longleftrightarrow (x, y)$, give a geometric argument to show that $-\vec{p} \longleftrightarrow (-x, -y)$.
   (b) Use properties of line vectors to show that $-\vec{p} \longleftrightarrow (-x, -y)$.

5. If $A = (1, 2)$, $B = (3, -1)$, and $C = (2, 6)$, use Example 1.11 to find the coordinates of $D = (x, y)$ if $\overline{AC}$ and $\overline{BD}$ are opposite sides of a parallelogram such that $\overrightarrow{AC} = \overrightarrow{BD}$.

6. If $A = (2, 3)$, $B = (4, 0)$, and $C = (-2, 2)$, use Example 1.11 to find the coordinates of $D = (x, y)$ if $\overline{AB}$ and $\overline{CD}$ are opposite sides of a parallelogram such that $\overrightarrow{AB} = \overrightarrow{CD}$.

7. Find the coordinates of the head Q of a line vector with tail at $P = (3, 2, 1)$ if $\overrightarrow{PQ}$ equals the position vector with head at $(4, 6, 3)$.

8. Find the coordinates of the head Q of a line vector with tail at $P = (2, 6, 1)$ if $\overrightarrow{PQ}$ equals the position vector with head at $(3, 0, 2)$.

9. If $P = (3, 2, 1)$, $Q = (-1, 2, 3)$, and $S = (2, -3, 4)$, find
   (a) $P + Q$.   (f) $2P + 3P$.
   (b) $Q - S$.   (g) $(2 + 3)P = 5P$.
   (c) $3P + 2S$.   (h) $3(P + Q)$.
   (d) $(P + Q) + S$.   (i) $3P + 3Q$.
   (e) $P + (Q + S)$.

10. If $P = (1, 2, 4, 3)$, $Q = (0, -1, 2, 5)$, and $S = (3, -2, 1, -4)$, find
   (a) $P + S$.   (f) $(2 + -6)Q = (-4)Q$.
   (b) $S - Q$.   (g) $5Q + 5S$.
   (c) $P - (Q - S)$.   (h) $5(Q + S)$.
   (d) $(P - Q) - S$.   (i) $2(5S)$.
   (e) $2Q + (-6)Q$.   (j) $5(2S)$.

11–15. In Theorem 1.3 verify properties 1A, 3A, 6M, 7M, 9M, and 10M.

16. The set of real numbers may be viewed as a collection of vectors if we make the following interpretations:

  (i) Addition of vectors is the ordinary addition of real numbers.
  (ii) Scalar multiplication is the ordinary multiplication of real numbers. If x and y are real numbers, we interpret xy as the scalar x times the vector y.
  (a) Check that properties 1A–5A and 6M–10M hold for the above system.
  (b) Interpret the real numbers (points on a line) as line vectors and draw a sketch for 2, 3, 2 + 3, 2 − 3, and 2(3).
  (c) Explain the different interpretations for 2(3) and 3(2).

Coordinate Geometry by Vector Methods

As is done in elementary science, we described a line vector in Chapter 1 by its magnitude and its direction. In contrast, coordinate vectors in n-space were called *vectors* because of the properties they possessed, that is, properties 1A–5A and 6M–10M. Although no mention of length or direction of coordinate vectors was made, the reader might suspect that there is a way to define these concepts for coordinate vectors, at least for 1-, 2-, and 3-space. This in fact can be done, and the notions are extendable to n-space.

In this chapter we define two products for coordinate vectors, a *dot product* and a *cross product*. The dot product is used to define both the length and the direction of a coordinate vector in n-space. Applications of the dot product and the cross product (defined only in 3-space) are also discussed.

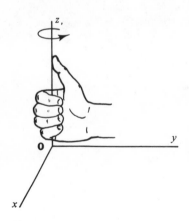

In choosing a coordinate system for 3-space we have and *will continue* to choose the axes so that the positive z-axis points up, the positive y-axis points to the right, and the positive x-axis points from the plane of the paper toward the reader. This coordinate system is called a **right-handed system** because, if one curls the fingers on his right hand from the positive x-axis to the positive y-axis, the thumb points in the direction of the positive z-axis.

If the axes are chosen so that, as one curls the fingers of his right hand from the positive x-axis to the positive y-axis, the thumb points in the direction of the negative z-axis, then the system is called a **left-handed system.**

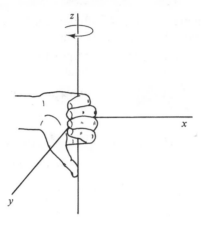

2.1 LENGTH AND DIRECTION OF A COORDINATE VECTOR

We can use the Pythagorean theorem to determine the length of a line vector in $\mathbf{R_3}$ (see Figure 2.1).

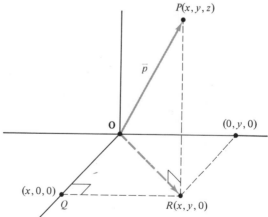

Figure 2.1

If $P = (x, y, z)$, then by the Pythagorean theorem we have $\|\overrightarrow{OP}\|^2$ $= \|\overrightarrow{OR}\|^2 + \|\overrightarrow{RP}\|^2 = \|\overrightarrow{OR}\|^2 + z^2$. A second application of the Pythagorean theorem gives $\| \overrightarrow{OR} \|^2 = \| \overrightarrow{OQ} \|^2 + \| \overrightarrow{QR} \|^2 = x^2 + y^2$. Hence $\| \overrightarrow{OP} \|^2$ $= x^2 + y^2 + z^2$. Since the length of a vector is not negative, we conclude that

$$\| \vec{p} \| = \| \overrightarrow{OP} \| = \sqrt{x^2 + y^2 + z^2}$$

Although Figure 2.1 shows P with the coordinates x, y, and z all positive (P in the first octant), the argument is valid for P in any of the eight octants. Since P is the coordinate vector corresponding to the line vector $\overrightarrow{OP}$, the following definition should seem reasonable.

Definition 2.1. The **length** (also **magnitude** or **norm**) of $P = (x, y, z)$, denoted by $\| P \|$, is $\sqrt{x^2 + y^2 + z^2}$. More generally, if $A = (a_1, a_2, \ldots, a_n)$ is a point in $\mathbf{R_n}$, we define

$$\| A \| = \sqrt{a_1^2 + a_2^2 + \cdots + a_n^2}$$

For $n = 1$ we have $\| A \| = \sqrt{a_1^2} = |a_1|$. In this case it is convenient to write $A = a_1$ instead of $A = (a_1)$.

For $n = 2$ we have $\| A \| = \sqrt{a_1^2 + a_2^2}$. Figure 2.2 gives the geometrical interpretation of $\| A \|$ for A in $\mathbf{R_1}$, $\mathbf{R_2}$, or $\mathbf{R_3}$.

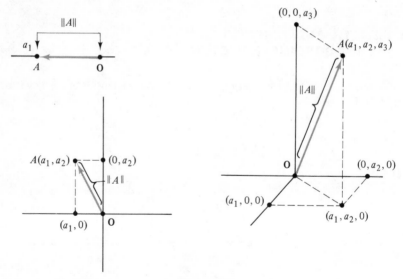

Figure 2.2

Example 2.1.
(a) If $A = -2$, $\| A \| = \sqrt{(-2)^2} = 2$.
(b) If $A = (-2, 3)$, $\| A \| = \sqrt{4 + 9} = \sqrt{13}$.
(c) If $A = (1, 3, -1)$, $\| A \| = \sqrt{1 + 9 + 1} = \sqrt{11}$.
(d) If $A = (1, -1, 2, 3)$, $\| A \| = \sqrt{1 + 1 + 4 + 9} = \sqrt{15}$.

Note that (a)–(c) have geometric interpretations, but that (d) does not. The direction of a line vector in 3-space is determined by the angles

α, β, and γ that the vector makes with the positive X, Y, and Z axes, respectively (Figure 2.3).

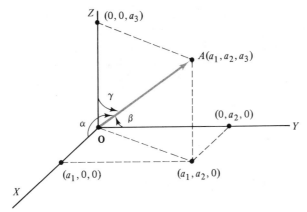

Figure 2.3

Definition 2.2. The **direction angles** of the line vector $\overrightarrow{OA}$ and also of the coordinate vector $A = (a_1, a_2, a_3) \neq (0, 0, 0)$ are the angles α, β, and γ, where α is the angle measured from the positive X-axis to $\overrightarrow{OA}$, β is the angle measured from the positive Y-axis to $\overrightarrow{OA}$, and γ is the angle measured from the positive Z-axis to $\overrightarrow{OA}$, and the measure of each of the angles α, β, and γ is between 0 and 180° inclusive.

Definition 2.3. The **direction cosines** of the line vector $\overrightarrow{OA}$ and also of the coordinate vector $A = (a_1, a_2, a_3)$ are $\cos \alpha$, $\cos \beta$, and $\cos \gamma$, where α, β, and γ are the direction angles of the vector $\overrightarrow{OA}$.

It is easy to find **formulas for the direction cosines** of $\overrightarrow{OA}$, which we now do.

In Figure 2.4, $\angle OQA$ is a right angle because $\overrightarrow{QA}$ lies in a plane that

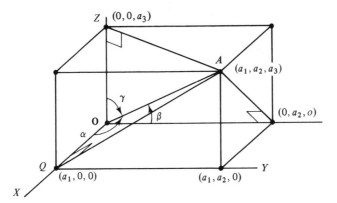

Figure 2.4

is perpendicular to $\overrightarrow{OQ}$. Hence

$$\cos \alpha = \frac{a_1}{\|\overrightarrow{OA}\|} = \frac{a_1}{\sqrt{a_1^2 + a_2^2 + a_3^2}}$$

Similarly,

$$\cos \beta = \frac{a_2}{\sqrt{a_1^2 + a_2^2 + a_3^2}}$$

and

$$\cos \gamma = \frac{a_3}{\sqrt{a_1^2 + a_2^2 + a_3^2}}$$

In Problem 8, Exercise 2.1, the reader is asked to verify that $\cos^2 \alpha + \cos^2 \beta + \cos^2 \gamma = 1$.

The **direction of** A is defined to be the direction of $\overrightarrow{OA}$, which in turn is completely determined by the direction cosines. The **direction angles** α, β, and γ may be thought of as being defined by the three formulas for the direction cosines.

For a point $A = (a_1, a_2, \ldots, a_n)$ in $\mathbf{R}_n$ one can define n **direction angles** $\alpha_1, \alpha_2, \ldots, \alpha_n$ by the formulas

$$\cos \alpha_i = \frac{a_i}{\sqrt{a_1^2 + a_2^2 + \cdots + a_n^2}}$$

for $i = 1, 2, \ldots, n$ and $0 \leq \alpha_i \leq 180°$. Also $\cos \alpha_i$, $i = 1, 2, \ldots, n$ are called the **direction cosines** of A.

Example 2.2. Find the direction cosines and the direction angles of $A = (3, 4, 5)$.

Solution:

$$\cos \alpha = \frac{3}{\sqrt{50}} = \frac{3\sqrt{2}}{10} \doteq 0.4243$$

$$\cos \beta = \frac{4}{\sqrt{50}} = \frac{2\sqrt{2}}{5} \doteq 0.5657$$

$$\cos \gamma = \frac{5}{\sqrt{50}} = \frac{\sqrt{2}}{2} \doteq 0.7071$$

From a table of trigonometric functions or possibly from a hand calculator, we obtain the direction angles of A:

$$\alpha \doteq 64°54', \qquad \beta \doteq 55°33', \qquad \gamma = 45°$$

We next define the angle between two coordinate vectors in $\mathbf{R}_3$ or in $\mathbf{R}_2$.

Definition 2.4. Let A and B be nonzero coordinate vectors in $\mathbf{R}_3$ (or $\mathbf{R}_2$).

The **angle θ between the coordinate vectors** A and B is the angle between the line vectors $\overrightarrow{OA}$ and $\overrightarrow{OB}$ where $0° \leq \theta \leq 180°$ ($0 \leq \theta \leq \pi$ if θ is given in radian measure).

We shall now derive a **formula for the cosine of the angle between two nonzero vectors in R_3**. From the law of cosines (see Figure 2.5), we have

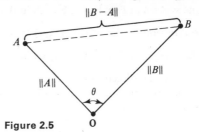

Figure 2.5

$$\| B - A \|^2 = \| A \|^2 + \| B \|^2 - 2 \| A \| \| B \| \cos \theta$$

Therefore,

$$\cos \theta = \frac{\| A \|^2 + \| B \|^2 - \| B - A \|^2}{2 \| A \| \| B \|}$$

If $A = (a_1, a_2, a_3)$ and $B = (b_1, b_2, b_3)$, then

$$\cos \theta =$$

$$\frac{(a_1^2 + a_2^2 + a_3^2) + (b_1^2 + b_2^2 + b_3^2) - (b_1 - a_1)^2 - (b_2 - a_2)^2 - (b_3 - a_3)^2}{2 \| A \| \| B \|}$$

which simplifies to

$$\cos \theta = \frac{a_1 b_1 + a_2 b_2 + a_3 b_3}{\| A \| \| B \|}$$

If $A = (a_1, a_2)$ and $B = (b_1, b_2)$ are coordinate vectors in R_2, the formula reads

$$\cos \theta = \frac{a_1 b_1 + a_2 b_2}{\| A \| \| B \|}$$

Example 2.3. Find the angle between $A = (2, -1, 3)$ and $B = (3, 1, -2)$.

Solution:

$$\cos \theta = \frac{6 - 1 - 6}{\sqrt{14}\sqrt{14}} = \frac{-1}{14}$$

$$\theta = \text{arc cos } \frac{-1}{14} \doteq 94°6'$$

Exercise 2.1

1. Find the length of the vector (a) $(1, 0, 1)$; (b) $(3, -1, 3)$; (c) $(-2, 1, 4)$; (d) $(3, 5, -2, -1, 4)$.

2. Find the norm of the vector (a) $(2, 1)$; (b) $(2, 0, 1)$; (c) $(3, 1, 2, -1)$; (d) $(1, 0, 3, -1, 2)$.

3. Find the direction cosines of (a) $(2, 1, 3)$; (b) $(2, 3, \sqrt{3})$; (c) $(-3, 4)$.

4. Find the direction angles of (a) $(\sqrt{3}, 1)$; (b) $(-1, \sqrt{3}, 0)$; (c) $(1, 0, 1)$.

5. Sketch the vectors in Problem 4. Sketch in the direction angles of each vector.

6. Find the direction angles of (a) $(1, \sqrt{3})$; (b) $(\sqrt{3}, 0, -1)$; (c) $(0, 1, 1)$.

7. Sketch the vectors in Problem 6. Sketch in the direction angles of each vector.

8. (a) Verify that the direction cosines of the coordinate vector $A = (a_1, a_2, a_3)$ have the property that the sum of their squares is 1; that is, verify that $\cos^2 \alpha + \cos^2 \beta + \cos^2 \gamma = 1$.

   (b) If $A = (a_1, a_2, \ldots, a_n)$ and $\alpha_1, \alpha_2, \ldots, \alpha_n$ are the direction angles of A, verify that $\cos^2 \alpha_1 + \cos^2 \alpha_2 + \cdots + \cos^2 \alpha_n = 1$.

9. (i) If θ is the angle between A and B, find $\cos \theta$ for
   (a) $A = (1, 1)$ and $B = (-1, 1)$.
   (b) $A = (1, 1, 1)$ and $B = (1, \sqrt{2}, 0)$.
   (c) $A = (1, 2, 3)$ and $B = (-2, 1, 0)$.

   (ii) Find θ in parts (a) and (c). If you have an appropriate calculator or a table of trigonometric functions find, θ in part (b). In each case sketch A, B, and θ.

10. If θ is the angle between A and B, find $\cos \theta$ for
    (a) $A = (-\sqrt{2}, 1)$ and $B = (\sqrt{2}, 1)$.
    (b) $A = (-1, 1, 1)$ and $B = (-\sqrt{2}, 0, 1)$.
    (c) $A = (-2, 1, 3)$ and $B = (2, 1, 1)$.
    (d) $A = \left[\cos \left(\alpha + \dfrac{\pi}{6} \right), \sin \left(\alpha + \dfrac{\pi}{6} \right) \right]$, $B = (\cos \alpha, \sin \alpha)$. Find θ.

11. If $A = (a_1, a_2, a_3)$ and $B = (b_1, b_2, b_3)$ are nonzero vectors and θ is the angle between them, determine in terms of the coordinates of A and B a necessary and sufficient condition that $\theta = 90°$.

12. If A and B are arbitrary nonzero vectors in $\mathbf{R}_3$ and θ is the angle between them, show that
    (a) If $\cos \theta = 1$, then A and B have the same direction.
    (b) If $\cos \theta = -1$, then A and B have opposite directions.

13. If $A = (a_1, a_2, \ldots, a_n)$, verify that $\| rA \| = | r | \| A \|$.

14. If $A = (a_1, a_2, \ldots, a_n) \neq \mathbf{0}$, and $r = 1/\| A \|$, find $\| rA \|$.

15. In deriving a formula for the cosine of the angle between two nonzero vectors in $\mathbf{R}_3$, we said that the expression for $\cos \theta$ simplifies to
$$\cos \theta = \frac{a_1 b_1 + a_2 b_2 + a_3 b_3}{\| A \| \| B \|}$$
Verify the simplification.

16. The vertices of a triangle are at $A = (1, 2, -1)$, $B = (3, 2, 0)$, and $C = (4, 6, -2)$. Find the cosine of each of the interior angles of the triangle. Is the triangle a right triangle?

2.2 THE DOT PRODUCT AND APPLICATIONS

If $A = (a_1, a_2, a_3)$ and $B = (b_1, b_2, b_3)$ are nonzero vectors and θ is the angle between them, we saw in Section 2.1 that

$$\cos \theta = \frac{a_1 b_1 + a_2 b_2 + a_3 b_3}{\|A\| \|B\|}$$

The number in the numerator is of considerable importance and is called the dot product of A and B. The notion of dot product can be generalized to coordinate vectors in $\mathbf{R_n}$.

Definition 2.5. If $A = (a_1, a_2, \ldots, a_n)$ and $B = (b_1, b_2, \ldots, b_n)$, the **dot product** $A \cdot B$ is defined as

$$A \cdot B = a_1 b_1 + a_2 b_2 + \cdots + a_n b_n = \sum_{i=1}^{n} a_i b_i$$

In 2-or 3-space the cosine of the angle between two nonzero vectors A and B can now be written as

$$\cos \theta = \frac{A \cdot B}{\|A\| \|B\|}$$

This formula for $\cos \theta$ is also used to define the angle between two nonzero vectors A and B in $\mathbf{R_n}$. One can show (Problem 16, Exercise 2.2) that

$$-1 \leq \frac{A \cdot B}{\|A\| \|B\|} \leq 1$$

and hence θ is always defined. Furthermore, θ is unique if it is in the range from 0 to 180°. We note that, for nonzero vectors A and B, $\cos \theta = 0$ if and only if $A \cdot B = 0$. Of course, if $\cos \theta = 0$, then $\theta = 90°$, and we say that the vectors are perpendicular or orthogonal.

Definition 2.6. Nonzero vectors A and B in $\mathbf{R_n}$ are **orthogonal** or **perpendicular** if and only if $A \cdot B = 0$.

Example 2.4. Find a vector orthogonal to $A = (1, 2, 1)$.

    Solution: One such vector is $B = (3, 0, -3)$, since $A \cdot B = 3 + 0 - 3 = 0$.

Example 2.5. If $A = (2, -5)$, $B = (1, 3)$, and $C = (-2, -3)$, use vector methods to show that $\triangle ABC$ is a right triangle.

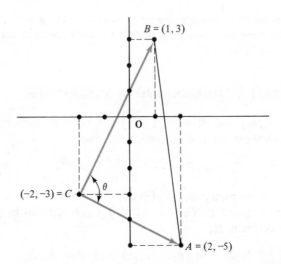

Solution: The sketch indicates that the angle θ between $\overrightarrow{CA}$ and $\overrightarrow{CB}$ may be a right angle. The angle θ between $\overrightarrow{CB} = \bar{b} - \bar{c}$ and $\overrightarrow{CA} = \bar{a} - \bar{c}$ is also the angle between $B - C$ and $A - C$ (Definition 2.4). So

$$\cos \theta = \frac{(A - C)\cdot(B - C)}{\|A - C\|\|B - C\|} = \frac{(4, -2)\cdot(3, 6)}{\|A - C\|\|B - C\|} = 0$$

Therefore, $\theta = 90°$ and $\triangle ABC$ is a right triangle.

The dot product defined in Definition 2.5 is easily seen to have the following properties:

1I. The dot product of two vectors A and B in $\mathbf{R_n}$ is a uniquely determined scalar.
2I. $A\cdot B = B\cdot A$.
3I. $rA\cdot B = A\cdot rB = r(A\cdot B)$ [consequently, $rA\cdot sB = rs\,(A\cdot B)$].
4I. $(A + B)\cdot C = A\cdot C + B\cdot C$.
5I. $A\cdot A > 0$ if A is not the zero vector. $A\cdot A = 0$ if $A = \mathbf{0}$.

The above properties often make calculation of the dot product easier. Properties **1I–5I** also serve as the defining properties of a more general *inner product* in Chapter 8.

Example 2.6. Find $C\cdot D$ if $C = (11, -33, 44)$ and $D = (4, 8, 16)$.

Solution:

$$C\cdot D = 11(1, -3, 4)\cdot 4(1, 2, 4) = 44(1, -3, 4)\cdot(1, 2, 4)$$
$$= (44)(11) = 484.$$

Example 2.7. Show by example that $A \cdot C = B \cdot C$ and $C \neq 0$ does not imply $A = B$.

Solution: In the following solution the symbol $\Longleftrightarrow$ may be read "is equivalent to."

$$A \cdot C = B \cdot C \Longleftrightarrow A \cdot C - B \cdot C = 0 \Longleftrightarrow (A - B) \cdot C = 0$$

Let $A - B = (2, -1, 0)$ and $C = (1, 2, 1)$. We have $(A - B) \cdot C = 0$. If $B = (1, 0, 0)$, then $A = (A - B) + B = (3, -1, 0)$. We see that $A \cdot C = (3, -1, 0) \cdot (1, 2, 1) = B \cdot C = (1, 0, 0) \cdot (1, 2, 1) = 1$, $C \neq 0$ and $A \neq B$.

Example 2.8. As an illustration of how properties 1I–5I can be verified, we prove that $rA \cdot B = r(A \cdot B)$.

Proof: Let $A = (a_1, a_2, \ldots, a_n)$ and $B = (b_1, b_2, \ldots, b_n)$.

$rA \cdot B = (ra_1, ra_2, \ldots, ra_n) \cdot (b_1, b_2, \ldots, b_n)$ (definition of rA)

$= (ra_1)b_1 + (ra_2)b_2 + \cdots + (ra_n)b_n$ (Definition 2.5)

$= r(a_1b_1 + a_2b_2 + \cdots + a_nb_n)$ (properties of real numbers)

$= r(A \cdot B)$ (Definition 2.5)

Therefore, $rA \cdot B = r(A \cdot B)$.   Q.E.D.

The length or norm of a vector A in $\mathbf{R_n}$ as defined in Definition 2.1 can be written as

$$\|A\| = \sqrt{A \cdot A}$$

The norm of a vector in $\mathbf{R_n}$ has the following properties:

1N. $\|A\| \geq 0$. $\|A\| = 0$ if and only if $A = 0$.
2N. $\|rA\| = |r| \|A\|$.
3N. $\|A + B\| \leq \|A\| + \|B\|$.

Properties 1N and 2N follow easily from the definition of norm. A method of proof for property 3N is outlined in Exercise 2.2 (see Problems 24–26).

Example 2.9. Use property 2N to calculate the length of $B = (-12, -16, 0)$.

Solution: $\|B\| = \| -4(3, 4, 0)\| = 4 \|(3, 4, 0)\| = 4(\sqrt{25}) = 20$.

Example 2.10. Find a vector of length 1 that is a scalar multiple of $A = (4, -2, 1)$.

Solution: $\|A\| = \sqrt{21}$. By property 2N, the length of $(1/\sqrt{21})A$ is $(1/\sqrt{21}) \|A\| = 1$.

The norm $\|A\| = \sqrt{A \cdot A}$ is used to define a **distance function** d on $\mathbf{R_n}$ as follows.

Definition 2.7.   For any two points A and B in $\mathbf{R_n}$,

$$d(A, B) = \| A - B \|$$

Read $d(A, B)$ as "the distance from A to B."

One can show that the following properties hold:

1D. $d(A, B) \geq 0$. Also, $d(A, B) = 0$ if and only if $A = B$.

2D. $d(A, B) = d(B, A)$ (symmetry).

3D. $d(A, B) \leq d(A, C) + d(C, B)$ (the triangle inequality).

The first two properties of the distance function are easily proved consequences of the definition of dot product. Property 1D is equivalent to 1N, and property 3D is equivalent to 3N (see Problems 26–28, Exercise 2.2).

Geometrically, if we interpret the coordinate vectors as points, all three properties are clear. The inequality in property 3D simply states that the sum of the lengths of two sides of a triangle is greater than the length of the third side. Equality in property 3D can occur if the three points are collinear (see Figure 2.6).

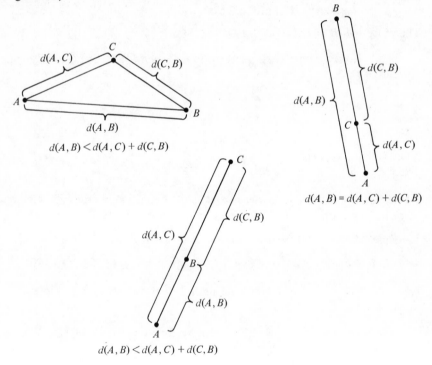

Figure 2.6   The triangle inequality

In Section 1.2 we saw that two nonzero line vectors were parallel if and only if one was a scalar multiple of the other. If the scalar was positive, then the vectors had the same direction. With this in mind we make the following definition.

Definition 2.8. Nonzero coordinate vectors A and B in $\mathbf{R}_n$ have the **same direction** if and only if $A = rB$ for some positive scalar r. A and B have **opposite directions** if and only if $A = rB$ for some negative scalar r.

Definition 2.9. A vector U in $\mathbf{R}_n$ such that $\|U\| = 1$ is called a **unit vector.** If $A \neq 0$, $(1/\|A\|)A$ is a unit vector that is said to be a **unit vector** in **the direction of** A. The process of finding a unit vector in the direction of a given vector is called **normalizing** the given vector.

Exercise 2.2

1. Find the dot product $A \cdot B$ if
   (a) $A = (1, 2)$, $B = (3, 2)$.
   (b) $A = (-6, 12, 6)$, $B = (15, 10, -10)$.
   (c) $A = (8, 16, -24, 16, 8)$, $B = (15, 10, -10, 0, 20)$.

2. Find the dot product $A \cdot B$ if
   (a) $A = (-1, 3)$, $B = (-1, 5)$.
   (b) $A = (14, -7, 21)$, $B = (0, 3, 0)$.
   (c) $A = (-18, 27, 0, 9)$, $B = (14, -7, 7, 0)$.

3. Find a vector orthogonal to
   (a) $A = (1, 2)$.
   (b) $A = (-1, 3, 2)$.

4. Find a vector orthogonal to $A = (-1, 3, 2)$ with second coordinate 4.

5. If $A = (-8, 6)$, $B = (-3, 4)$, and $C = (-1, 9)$, use vector methods to show that $\triangle ABC$ is a right triangle.

6. If $A = (4, 2, 1)$, $B = (7, 5, 3)$, and $C = (2, 2, 4)$, use vector methods to show that $\triangle ABC$ is a right triangle.

7. Find coordinate vectors in $\mathbf{R}_3$ such that $A \neq 0$, $B \neq 0$, and $A \cdot B = 0$.

8. Use property 2N to calculate the length of $(-28, -56, 21)$.

9. For $A = (3, 2, 1, 2)$ and $B = (-2, 0, 1, -3)$, verify that $\|A + B\| \leq \|A\| + \|B\|$.

10. Sketch the points $A = (1, 3, 2)$, $B = -2A$, and $C = 2A$. Compare dir A, dir B, and dir C.

11. Does it make any sense to ask the question, "Is $(A \cdot B) \cdot C = A \cdot (B \cdot C)$?" Explain your answer.

12. Find a unit vector in the direction of $A = (2, 4, 3)$.

13. Normalize each of the vectors (a) $A = (3, 4)$; (b) $A = (3, 4, -2)$.

14. Find a vector in the direction opposite to $A = (0, 4, 3)$ and having length 7.

15. Compute the distance between the following points:
    (a) $A = (1, 3)$, $B = (3, 4)$.
    (b) $A = (3, 3, 1)$, $B = (-4, 1, 6)$.
    (c) $A = (2, 3, -1, 1)$, $B = (4, 0, 2, 3)$.

16. Find $\cos \theta$ if θ is the angle between the vectors A and B in Problem 15, parts (a), (b), and (c).

17. Use property 3I to calculate $C \cdot D$ if $C = (325, 50, 150)$ and $D = (63, 77, 91)$.

18. With C and D as in Problem 17, use property 2N to calculate the lengths of both C and D. Find a unit vector in the direction of C.

19. Verify that the dot product in $\mathbf{R_n}$ satisfies (a) 1I; (b) 2I; (c) $rA \cdot B = A \cdot rB$; (d) 4I; (e) 5I.

20. Verify that $rA \cdot sB = (rs)(A \cdot B)$ for any scalars r and s and coordinate vectors A and B in $\mathbf{R_n}$.

21. Verify that the norm of a vector in $\mathbf{R_n}$ has the properties (a) 1N; (b) 2N (see Problem 11, Exercise 2.1).

22. Verify that the $d(A, B)$ defined in Definition 2.7 has the properties (a) 1D; (b) 2D.

23. Verify the following properties for the dot product of vectors in $\mathbf{R_n}$: (a) $A \cdot (B + C) = A \cdot B + A \cdot C$; (b) $(A + B) \cdot (A + B) = A \cdot A + 2(A \cdot B) + B \cdot B$.

Problems 24–29 are sequential and lead to proofs of the triangle inequality in $\mathbf{R_n}$ and of the fact that if the angle between two nonzero vectors A and B in $\mathbf{R_n}$ is defined to be θ, where

$$\cos \theta = \frac{A \cdot B}{\|A\| \|B\|} \quad \text{and} \quad 0° \le \theta \le 180°$$

then θ is well defined. A shorter and more elegant proof will be given later in Chapter 8.

24. For arbitrary vectors $A = (a_1, a_2, \ldots, a_n)$ and $B = (b_1, b_2, \ldots, b_n)$, show that

$$(a_1^2 + a_2^2 + \cdots + a_n^2)(b_1^2 + b_2^2 + \cdots + b_n^2) - (a_1 b_1 + a_2 b_2 + \cdots + a_n b_n)^2$$
$$= [(a_1 b_2 - a_2 b_1)^2 + (a_1 b_3 - a_3 b_1)^2 + \cdots + (a_1 b_n - a_n b_1)^2]$$
$$+ [(a_2 b_3 - a_3 b_2)^2 + (a_2 b_4 - a_4 b_2)^2 + \cdots + (a_2 b_n - a_n b_2)^2]$$
$$+ \cdots + (a_{n-1} b_n - a_n b_{n-1})^2$$

Using summation symbols, this may be written

$$\left(\sum_{i=1}^{n} a_i^2 \sum_{i=1}^{n} b_i^2 \right) - \left(\sum_{i=1}^{n} a_i b_i \right)^2 = \sum_{1 \le i < j \le n} (a_i b_j - a_j b_i)^2$$

Hint: Expand each side and cancel out like terms on the left side.

25. Use Problem 24 to establish the **Schwarz inequality**

$$\left(\sum_{i=1}^{n} a_i b_i \right)^2 \le \sum_{i=1}^{n} a_i^2 \sum_{i=1}^{n} b_i^2$$

which may be written as $(A \cdot B)^2 \le \|A\|^2 \|B\|^2$.

Hint: Note that the right side of the equality in Problem 24 is greater than or equal to zero.

26. Use Problem 25 to establish the **Minkowski inequality**

$$\sqrt{\sum_{i=1}^{n} (a_i + b_i)^2} \le \sqrt{\sum_{i=1}^{n} a_i^2} + \sqrt{\sum_{i=1}^{n} b_i^2}$$

which may be written $\|A + B\| \le \|A\| + \|B\|$, as in property 3N.

27. Use Problem 26 to establish property 3D:

$$d(A, B) \le d(A, C) + d(C, B)$$

Hint: Property 3D is equivalent to $\|A - B\| \le \|A - C\| + \|C - B\|$.

28. Use property 3D to establish property 3N. Thus properties 3N and 3D are equivalent.

29. Use the Schwarz inequality (Problem 25) to show that

$$-1 \le \frac{A \cdot B}{\|A\| \|B\|} \le 1$$

for nonzero vectors A and B. From this result it follows that, for nonzero vectors A and B, the angle θ between them is always defined by $\cos \theta = (A \cdot B)/\|A\| \|B\|$. Since there is exactly one angle θ, $0° \le \theta \le 180°$, such that $\cos \theta$ is a specific number between -1 and 1, it follows that θ is uniquely determined by A and B.

Applications of Coordinate Vectors

Example 2.11. Price Vectors. A toy manufacturer makes a toy that requires glass, steel, rubber, and plastic. We assume that to make a toy requires 1 unit of glass, $2\frac{1}{2}$ units of steel, 2 units of rubber, and 3 units of plastic. The *materials vector* $A = (1, 2\frac{1}{2}, 2, 3)$ is a convenient record-keeping device for this information. The *price vector* $P = (1.50, 2.00, 1.60, 0.80)$ records the unit price of glass, steel, rubber, and plastic, respectively. The dot product $A \cdot P = 1(1.50) + 2\frac{1}{2}(2.00) + 2(1.60) + 3(0.80) = \$12.10$ is the cost to the manufacturer for the raw materials needed to produce one toy.

Example 2.12. Price Vectors and Cost Vectors. Suppose that the toy manufacturer in Example 2.11 makes three different kinds of toys, which we label as toy 1, toy 2, and toy 3. Let A_i be the coordinate vector giving the number of units of glass, steel, rubber, and plastic, respectively, which are required in the manufacture of toy i. Suppose that $A_1 = (1, 2\frac{1}{2}, 2, 3)$, $A_2 = (\frac{1}{2}, 3, 1, 2)$, $A_3 = (1\frac{1}{2}, 2, 1\frac{1}{2}, 3)$, and let $P = (1.50, 2.00, 1.60, 0.80)$ be the price vector.

The *cost vector* $C = (A_1 \cdot P, A_2 \cdot P, A_3 \cdot P) = (12.10, 9.95, 11.05)$ gives the cost of raw materials needed to manufacture toy 1, toy 2, and toy 3, respectively. The information given by A_1, A_2, and A_3 can be arranged in an array

$$
\begin{array}{c}
\\
\text{toy 1} \\
\text{toy 2} \\
\text{toy 3}
\end{array}
\begin{array}{cccc}
\text{glass} & \text{steel} & \text{rubber} & \text{plastic} \\
\left[\begin{array}{cccc}
1 & 2\frac{1}{2} & 2 & 3 \\
\frac{1}{2} & 3 & 1 & 2 \\
1\frac{1}{2} & 2 & 1\frac{1}{2} & 3
\end{array}\right]
\end{array}
$$

The array is called a *matrix*, which is a main topic of Chapter 3, where we define the product of a matrix and a column vector so that

$$
\begin{bmatrix}
1 & 2\frac{1}{2} & 2 & 3 \\
\frac{1}{2} & 3 & 1 & 2 \\
1\frac{1}{2} & 2 & 1\frac{1}{2} & 3
\end{bmatrix}
\begin{bmatrix}
1.50 \\
2.00 \\
1.60 \\
.80
\end{bmatrix}
=
\begin{bmatrix}
A_1 \cdot P \\
A_2 \cdot P \\
A_3 \cdot P
\end{bmatrix}
=
\begin{bmatrix}
12.10 \\
9.95 \\
11.05
\end{bmatrix}
$$

which is the cost vector C (written as a column vector).

Example 2.13. Recipe Vectors. A hot dish called *hamburger pie* requires the ingredients onion, ground beef, green beans, condensed tomato soup, potatoes, milk, and egg. The vector $A = (\frac{1}{36}, \frac{1}{6}, \frac{1}{12}, \frac{11}{96}, \frac{1}{6}, \frac{1}{2}, 1)$, which we call the *recipe vector*, gives the number of units of each ingredient needed to make one serving of hamburger pie. The price in cents of one unit of each of the respective ingredients is given by $P = (30, 89, 78, 32, 10, 20, 7)$. The cost per serving of hamburger pie is

$$A \cdot P = (\tfrac{1}{36})30 + (\tfrac{1}{6})89 + (\tfrac{1}{12})78 + (\tfrac{11}{96})32 + (\tfrac{1}{6})10 + (\tfrac{1}{2})20$$
$$+ (1)7 = 44.5 \text{ cents}$$

Exercise 2.2

30. Suppose that the raw materials used in the manufacture of a certain standard-model snowmobile are aluminum, copper, glass, plastic, rubber, and steel. We assume that the production of one snowmobile requires $1\frac{1}{2}$ units of aluminum, $\frac{1}{4}$ unit of copper, $\frac{1}{2}$ unit of glass, 3 units of plastic, 2 units of rubber, and 3 units of steel. (The units of measure are not necessarily the same.) Assume further that the respective unit costs are \$20, \$30, \$10, \$10, \$15, and \$80. Find a materials vector A and a price vector P, and use the dot product to compute the cost of raw materials needed to produce one snowmobile.

31. Suppose that in addition to the standard-model snowmobile in Problem 30 the manufacturer produces an economy model and a superdeluxe model. The matrix

$$
\begin{array}{r}
\text{economy model} \\
\text{standard model} \\
\text{deluxe model}
\end{array}
\begin{array}{ccccccc}
\text{aluminum} & \text{copper} & \text{glass} & \text{plastic} & \text{rubber} & \text{steel} \\
\left[\begin{array}{c} 1 \\ \frac{3}{2} \\ 2 \end{array}\right. & \begin{array}{c} \frac{1}{4} \\ \frac{1}{4} \\ \frac{1}{4} \end{array} & \begin{array}{c} \frac{3}{8} \\ \frac{1}{2} \\ \frac{3}{4} \end{array} & \begin{array}{c} 2 \\ 3 \\ 4 \end{array} & \begin{array}{c} \frac{3}{2} \\ 2 \\ 2 \end{array} & \left.\begin{array}{c} 2 \\ 3 \\ 4 \end{array}\right]
\end{array}
$$

gives the number of units of the various raw materials needed to produce each of the snowmobiles. With the cost of raw materials given as in Problem 30, follow Example 2.12 to write a cost vector that gives the cost of raw materials needed to produce one snowmobile of each model.

32. According to one recipe for the casserole barbecued spareribs, the ingredients required are 1 T. chili powder, 1 T. salt, 1 T. celery seed, $\frac{1}{4}$ c. brown sugar (packed), 1 tsp. paprika, 2 lb. spareribs, $\frac{1}{2}$ c. vinegar, and 1 can tomato soup. The price in cents of one unit of each of the respective ingredients is 1, 1, 2, 25, 1, 150, 10, and 20. Find a recipe vector A and a price vector P, and use $A \cdot P$ to calculate the cost of the barbecued spareribs casserole.

33. A manufacturer makes a product requiring n different raw materials. Let m_1, $m_2, \ldots, m_n$ represent the number of units, respectively, of the raw materials that are needed to produce one unit of the product. Let $p_1, p_2, \ldots, p_n$ be the respective prices of one unit of each of the given raw materials. Write the materials vector M and the price vector P. Find $M \cdot P$. What does $M \cdot P$ represent?

34. (a) A recipe for ham loaf requires 1 lb of ground smoked ham, 1 lb of ground lean fresh pork, 3 cups of Wheaties, 2 eggs, and 1 cup of milk. The recipe makes six servings. Write a recipe vector A that gives the number of units of each ingredient needed to make one serving of ham loaf.

    (b) Write a price vector P for this recipe if ham costs \$1.60/lb, pork costs \$1.10/lb, Wheaties cost 10 cents a cup, eggs are 84 cents per dozen, and milk is 11 cents a cup.

    (c) Find $A \cdot P =$ cost of one serving of ham loaf.

2.3 VECTOR PROJECTIONS AND APPLICATIONS

The reader may already be familiar with the notion of projecting a point onto a line. The idea is this: if P is a point and $\mathcal{L}$ is a line, then the projection of P onto $\mathcal{L}$ is the point P' at the foot of the perpendicular from P to $\mathcal{L}$.

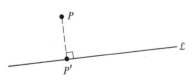

What we wish to do in this section is to extend the notion of projection of a point onto a line to the idea of projection of a vector onto (or along) a vector. We shall see that the idea of vector projection has both mathematical and physical applications

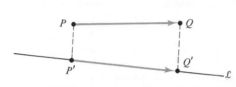

To project a vector $\overrightarrow{PQ}$ onto a line $\mathcal{L}$, we project the point Q at the head of $\overrightarrow{PQ}$ to the point Q' on $\mathcal{L}$, and the point P at the tail of $\overrightarrow{PQ}$ to the point P' on $\mathcal{L}$. $\overrightarrow{P'Q'}$ is the projection of $\overrightarrow{PQ}$ along $\mathcal{L}$.

Now let O be a fixed origin, and let $\bar{a}$ and $\bar{b}$ be the nonzero position vectors of points A and B, respectively. The **(vector) projection of $\bar{a}$ along $\bar{b}$** is the vector $\bar{v} = k\bar{b}$, where V is the projection of the point A onto the line determined by O and B (see Figure 2.7). The **(vector) projection of $\bar{a}$ orthogonal to $\bar{b}$**

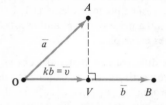
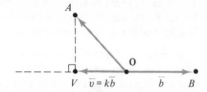

Figure 2.7   The projection of $\bar{a}$ along $\bar{b}$

is the vector $\bar{w}$, where W is the projection of the point A onto the line through O and perpendicular to the line determined by O and B (Figure 2.8).

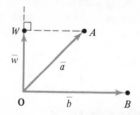
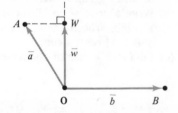

Figure 2.8   The projection of $\bar{a}$ orthogonal to $\bar{b}$

For any two nonzero position vectors $\bar{a}$ and $\bar{b}$, we can write $\bar{a} = \bar{v} + \bar{w}$, where $\bar{v}$ is the projection of $\bar{a}$ along $\bar{b}$, and $\bar{w}$ is the projection of $\bar{a}$ orthogonal to $\bar{b}$ (Figure 2.9). The following facts completely determine the projections $\bar{v}$ and $\bar{w}$:

1P. $\bar{v} = k\bar{b}$ for some scalar k ($\bar{v}$ is parallel to $\bar{b}$ or $\bar{v} = \mathbf{0}$).
2P. $\bar{a} = \bar{w} + \bar{v}$ (Parallelogram rule for addition).
3P. $\bar{w} \cdot \bar{b} = 0$†.

†We may define the dot product of two line vectors $\bar{a}$ and $\bar{b}$ to be the dot product $A \cdot B$ of their corresponding coordinate vectors. Alternatively, we could use the "coordinate-free" definition, $\bar{a} \cdot \bar{b} = \|\bar{a}\| \|\bar{b}\| \cos \theta$, where θ is the angle between $\bar{a}$ and $\bar{b}$. In regard to this alternative definition, see Problems 9–13, Exercise 2.3.

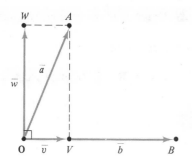 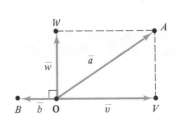

Figure 2.9   Vector projections

We have

$$\bar{a}\cdot\bar{b} = (\bar{w} + \bar{v})\cdot\bar{b} \qquad \text{(from 2P)}$$
$$= \bar{w}\cdot\bar{b} + \bar{v}\cdot\bar{b} \qquad \text{(property 4I)}$$
$$= \bar{v}\cdot\bar{b} \qquad \text{(from 3P)}$$
$$= (k\bar{b})\cdot\bar{b} \qquad \text{(from 1P)}$$
$$= k(\bar{b}\cdot\bar{b}) \qquad \text{(from 3I)}$$

Therefore, $\bar{a}\cdot\bar{b} = k(\bar{b}\cdot\bar{b})$ for some k. Solving for k, we have

$$k = \frac{\bar{a}\cdot\bar{b}}{\bar{b}\cdot\bar{b}}$$

The vector projection of $\bar{a}$ along $\bar{b}$ can now be written as

$$k\bar{b} = \frac{\bar{a}\cdot\bar{b}}{\bar{b}\cdot\bar{b}}\bar{b}$$

and the projection of $\bar{a}$ orthogonal to $\bar{b}$ as

$$\bar{w} = \bar{a} - k\bar{b} = \bar{a} - \frac{\bar{a}\cdot\bar{b}}{\bar{b}\cdot\bar{b}}\bar{b}$$

The length of $k\bar{b}$ is

$$\left|\frac{\bar{a}\cdot\bar{b}}{\bar{b}\cdot\bar{b}}\right| \|\bar{b}\| = \pm\frac{\bar{a}\cdot\bar{b}}{\|\bar{b}\|}$$

where the sign is plus if the direction of the projection $k\bar{b}$ is the same as the direction of $\bar{b}$. The number $(\bar{a}\cdot\bar{b})/\|\bar{b}\|$ is called the **scalar projection** of $\bar{a}$ along $\bar{b}$. Since $\bar{a}\cdot\bar{b} = \|\bar{a}\|\,\|\bar{b}\|\cos\theta$, the *scalar projection of $\bar{a}$ along $\bar{b}$ can be written* *as* $\| \bar{a} \| \cos \theta$, where θ is the angle (actually the measure of the angle) between $\bar{a}$ and $\bar{b}$.

We generalize the notion of vector projection to $\mathbf{R_n}$.

Definition 2.10. For any vectors A and $B \neq 0$ in $\mathbf{R_n}$, the **(vector) projection of A along B** is the vector

$$\frac{A \cdot B}{B \cdot B} B$$

The vector

$$A - \frac{A \cdot B}{B \cdot B} B$$

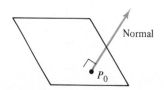

that is orthogonal to B is called the **projection of A orthogonal to B.**

A *plane* in 3-space is determined by (1) a point in that plane and (2) a nonzero vector perpendicular to the plane. Such a vector is called a *normal* to the plane.

Example 2.14. Show that an equation for a plane through the point $P_0 = (x_0, y_0, z_0)$ and perpendicular to the nonzero vector $\bar{v} = a\bar{i} + b\bar{j} + c\bar{k}$, where $\bar{i}, \bar{j}$, and $\bar{k}$ are the position vectors of $(1, 0, 0)$, $(0, 1, 0)$, and $(0, 0, 1)$, respectively, is $a(x - x_0) + b(y - y_0) + c(z - z_0) = 0$.

Solution: A point $P = (x, y, z)$ is in the plane determined by P_0 and the normal $\bar{v}$ if and only if $\bar{w} = \overrightarrow{P_0P}$ is perpendicular to $\bar{v}$. This means that $\overrightarrow{P_0P} \cdot \bar{v} = 0$. Since $\overrightarrow{P_0P} = (x - x_0)\bar{i} + (y - y_0)\bar{j} + (z - z_0)\bar{k}$, we have P in the plane if and only if $a(x - x_0) + b(y - y_0) + c(z - z_0) = 0$.

As an illustration, the plane through $(2, 0, 1)$ and perpendicular to $2\bar{i} + 4\bar{j} - \bar{k}$ is $2(x - 2) + 4(y - 0) + -1(z - 1) = 0$. This equation simplifies to $2x + 4y - z = 3$.

Example 2.15. Show that a plane with equation $ax + by + cz = d$ has $a\bar{i} + b\bar{j} + c\bar{k}$ as a normal.

Solution: Let $\mathcal{L}$ be any line in the given plane. Let $P_0 = (x_0, y_0, z_0)$ and $P_1 = (x_1, y_1, z_1)$ be any two points on $\mathcal{L}$. Then $ax_0 + by_0 + cx_0 = d = ax_1 + by_1 + cz_1$. Therefore, $a(x_1 - x_0) + b(y_1 - y_0) + c(z_1 - z_0) = 0$. This means that $\bar{v} = a\bar{i} + b\bar{j} + c\bar{k}$, and $\overrightarrow{P_0P_1} = (x_1 - x_0)\bar{i} + (y_1 - y_0)\bar{j} + (z_1 - z_0)\bar{k}$ are perpendicular. Hence $a\bar{i} + b\bar{j} + c\bar{k}$ is perpendicular to any line in the given plane and is consequently a normal to the given plane.

Example 2.16. Find the distance from the point $P = (3, 1, 4)$ to the plane $6x + 3y - 2z = 3$.

Solution: Find any point Q in the plane, for example, $Q = (0, 1, 0)$. (The coordinates of Q must satisfy the equation of the plane.)

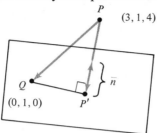

According to Example 2.15, $\bar{n} = 6\bar{\imath} + 3\bar{\jmath} - 2\bar{k}$ is a normal to the plane. If we project $\overrightarrow{PQ}$ along the vector $\bar{n}$, the resulting projection $\overrightarrow{PP'}$ is a vector from P perpendicular to the plane. Therefore, the desired distance is $\| \overrightarrow{PP'} \|$, where

$$\overrightarrow{PP'} = \frac{\overrightarrow{PQ} \cdot \bar{n}}{\bar{n} \cdot \bar{n}} \bar{n}$$

and

$$\overrightarrow{PQ} = -3\bar{\imath} + -4\bar{k}$$

Therefore,

$$\overrightarrow{PP'} = \frac{-18 + 8}{36 + 9 + 4} \bar{n}$$

$$= \frac{-10}{49} \bar{n}$$

and

$$\| \overrightarrow{PP'} \| = \frac{10}{49} \| \bar{n} \| = \frac{10}{7}$$

In physics the *component of a constant force in a given direction* is defined to be the scalar projection of the vector representing the force along the vector representing the given direction (see Figure 2.10).

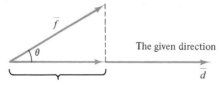

$$\frac{\bar{f} \cdot \bar{d}}{\| \bar{d} \|} = \| \bar{f} \| \cos \theta = \quad \text{The component of the force } \bar{f} \text{ in the given direction}$$

Figure 2.10

Work done by a force is defined as the distance the particle at the point of application of the force moves, multiplied by the component of the force in the direction of motion. In Figure 2.11 $\bar{f}$ represents the force, and the displacement vector $\bar{d}$ represents the distance and direction that an object is moved. The work done is

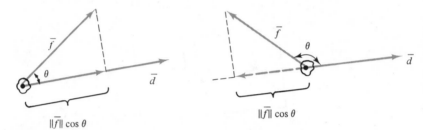

Positive work: the component of force $\bar{f}$ along $\underline{d}$ is in the direction of displacement $\bar{d}$.
$W = \|\bar{d}\| \|\bar{f}\| \cos \theta = \bar{f} \cdot \bar{d} > 0$

Negative work: the component of force $\bar{f}$ along $\bar{d}$ is opposite to the direction of displacement d.
$W = \|\bar{d}\| \|\bar{f}\| \cos \theta = \bar{f} \cdot \bar{d} > 0$

Figure 2.11

$W =$ (the distance the particle at the point of application of the force moves)

$\times$ (the component of the force along the direction of motion)

$$= \|\bar{d}\|\left(\frac{\bar{f} \cdot \bar{d}}{\|\bar{d}\|}\right) = \bar{f} \cdot \bar{d} = \|\bar{d}\| \|\bar{f}\| \cos \theta$$

where θ is the angle between $\bar{f}$ and $\bar{d}$.
In words, the work done is the dot product of the force vector and the displacement vector.

Work is a scalar quantity that can be positive or negative (see Figure 2.11). Negative work may be done if there are other forces acting on the object or when the object has some velocity to begin with. If the unit of force is the *newton* (N) and the unit of distance is a *meter* (m), then the product, newton-meter, is the corresponding unit of work called a *joule* (J) of work.

Example 2.17. A force $\bar{f}$ has a magnitude of 18 N (1 N = 0.2248 lbs) and $\pi/6$ is the radian measure of the angle giving its direction. Find the work done in moving an object along a straight line from the origin to the point (6, 1) where the distance is measured in meters (1 m = 3.281 ft).

Solution

$$W = \bar{f} \cdot \bar{d}, \quad \text{where } \bar{f} = (18 \cos (\pi/6))\bar{i} + (18 \sin (\pi/6))\bar{j}$$
$$= 9\sqrt{3}\,\bar{i} + 9\bar{j} \quad \text{and} \quad \bar{d} = 6\bar{i} + 1\bar{j}$$

So $W = 54\sqrt{3} + 9 \doteq 102.5$ J.

If in Example 2.17 we change the direction of the force $\bar{f}$ to $2\pi/3$ and keep everything else the same, we obtain a negative value for the work as follows (see Figure 2.12):

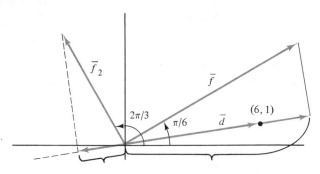

The component of $\bar{f_2}$ is in the direction opposite to the direction of motion. Therefore, the work is negative.

The component of $\bar{f}$ in the direction of motion is positive. Therefore, the work is positive.

Figure 2.12

$$W = \bar{f_2} \cdot \bar{d} = (18 \cos (2\pi/3)\bar{i} + 18 \sin (2\pi/3)\bar{j}) \cdot (6\bar{i} + \bar{j})$$

$$= -54 + 9\sqrt{3} \doteq -38.4$$

The reason for the negative answer is that the component of force is opposite to the direction of motion.

Exercise 2.3

1. Find an equation for the plane (a) passing through $P = (1, 2, 3)$ and perpendicular to $3\bar{i} + 4\bar{j} + 2\bar{k}$, and (b) passing through $P = (0, 1, 2)$ and perpendicular to $2\bar{i} - 2\bar{j} + \bar{k}$.

2. Find an equation for the plane (a) passing through $P = (0, 2, 1)$ with normal $\bar{i} - \bar{j}$, and (b) passing through $P = (3, -4, 2)$ with normal $\bar{i} + 3\bar{j} + 2\bar{k}$.

3. Find a normal for the plane (a) $3x + 4y - 6z = 10$, and (b) $3x + 2y = 6z - 2$.

4. Find a normal for the plane (a) $6x - 4z + 3y - 4 = 0$, and (b) $2x - 3z = 4y + 2$.

5. Find the distance from the point $P = (1, -3, 4)$ to the plane $2x - 6y + 3z = 5$.

6. Find the distance from the point $P = (2, 1, 2)$ to the plane $x + 3y - 4z = 5$.

7. Find the distance between the parallel planes $3x + 2y - 4z = -2$ and $3x + 2y - 4z = 8$.

8. Find the distance between the parallel planes $6x + 2y - 3z = 1$ and $6x + 2y - 3z = 5$.

In Problems 9–12, use the following *definition* of dot product for line vectors: the dot product of any line vectors $\bar{a}$ and $\bar{b}$ is $\bar{a} \cdot \bar{b} = \|\bar{a}\| \|\bar{b}\| \cos \theta$, where θ is the angle between $\bar{a}$ and $\bar{b}$ and $0° \leq \theta \leq 180°$.

9. Verify each of the following:
  (a) 2I: $\bar{a} \cdot \bar{b} = \bar{b} \cdot \bar{a}$.
  (b) 3I: $r\bar{a} \cdot \bar{b} = \bar{a} \cdot r\bar{b} = r(\bar{a} \cdot \bar{b})$.
  (c) 5I: $\bar{a} \cdot \bar{a} > 0$ if $\bar{a} \neq \bar{0}$. $\bar{a} \cdot \bar{a} = 0$ if $\bar{a} = \bar{0}$.

10. Use a geometrical argument and the figure to verify 4I: $\bar{a} \cdot (\bar{b} + \bar{c}) = \bar{a} \cdot \bar{b} + \bar{a} \cdot \bar{c}$.

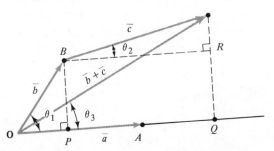

11. Verify (a) $\bar{i} \cdot \bar{i} = 1 = \bar{j} \cdot \bar{j} = \bar{k} \cdot \bar{k}$; (b) $\bar{i} \cdot \bar{j} = \bar{j} \cdot \bar{k} = \bar{i} \cdot \bar{k} = 0$.

12. If $\bar{a} = a_1\bar{i} + a_2\bar{j} + a_3\bar{k}$ and $\bar{b} = b_1\bar{i} + b_2\bar{j} + b_3\bar{k}$, verify that $\bar{a} \cdot \bar{b} = a_1b_1 + a_2b_2 + a_3b_3$ (use the results of Problems 9–11).

13. *Prove:* If $C = [(A \cdot B)/(B \cdot B)]B$ is the projection of A along B defined in Definition 2.10, and $D = A - C$ is the projection of A orthogonal to B, then properties 1P, 2P, and 3P hold for the vectors C and D.

Applications

14. Find the work done by a force $\bar{f} = 3\bar{i} + 2\bar{j}$ in moving a particle from the point $(0, 0)$ to $(1, 2)$.

15. Find the work done by a force $\bar{f} = 3\bar{i} - 2\bar{j} + 4\bar{k}$ in moving a particle from $(1, 2, 0)$ to $(3, -5, 6)$.

16. A force $\bar{f}$ has a magnitude of 20 N and 60° is the measure of the angle giving its direction. Find the work done in moving an object along a straight line from the origin to the point $(4, 3)$, where the distance is measured in meters.

17. A 15-lb cube of metal is on a frictionless plane inclined to the horizontal at 30°. The inclined plane will produce a force, the *reaction* of the surface, equal to the negative of the component of weight along a vector $\bar{n}$ perpendicular to the inclined plane. What force $\bar{f}$ must be applied in order to keep the cube from sliding? How much work is done (foot-pounds) in moving the cube 40 ft up the incline? In the figure, $\bar{w}$ is a vector representing the weight of the cube.

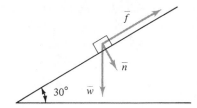

18. Let $ax + by + cz + d = 0$ be the equation of a plane. (Then at least one of the constants a, b, and c is not zero.) Let $P_1 = (x_1, y_1, z_1)$ be a point in 3-space not in the given plane. *Prove:* The (perpendicular) distance from P_1 to the given plane is

$$s = \frac{|ax_1 + by_1 + cz_1 + d|}{\sqrt{a^2 + b^2 + c^2}}$$

Hint: If $a \neq 0$, then the point $Q = (-d/a, 0, 0)$ is in the plane. If $a = 0$, then one of the points $(0, -d/b, 0)$ or $(0, 0, -d/c)$ is in the plane.

2.4 THE CROSS PRODUCT IN R₃

Consider the problem of finding a vector $X = (x, y, z)$ that is perpendicular to each of the nonzero, nonparallel vectors $A = (a_1, a_2, a_3)$ and $B = (b_1, b_2, b_3)$. Since $A \cdot X = B \cdot X = 0$, the problem is equivalent to solving the system of equations

$$\begin{aligned} a_1 x + a_2 y + a_3 z &= 0 \\ b_1 x + b_2 y + b_3 z &= 0 \end{aligned} \tag{E}$$

We can eliminate z by multiplying the first equation by b_3 and the second equation by $-a_3$ and then adding to obtain

$$(a_1 b_3 - a_3 b_1)x + (a_2 b_3 - a_3 b_2)y = 0 \tag{I}$$

Similarly, y is eliminated by multiplying the first equation by b_2 and the second equation by $-a_2$ and then adding to obtain

$$(a_1 b_2 - a_2 b_1)x + (a_3 b_2 - a_2 b_3)z = 0 \tag{II}$$

It is clear upon inspection that, for any constant k, $x = k(a_2 b_3 - a_3 b_2)$, $y = k(a_3 b_1 - a_1 b_3)$, and $z = k(a_1 b_2 - a_2 b_1)$ is a solution to the system consisting of (I) and (II). The reader may check (Exercise 2.4, Problem 1) that this is also a solution to the given system (E) for any constant k. On the other hand, there are no other solutions to the given system, because any two solution vectors have either the same or opposite directions (a solution is orthogonal to both A and B) and hence differ by only a constant multiple.

The solution when $k = 1$ is defined to be the *cross product* $A \times B$ (Figure 2.13). As the above discussion indicates, $A \times B$ is a vector that is perpendicular to each of the vectors A and B.

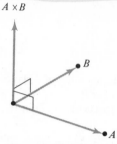

Figure 2.13 $A \times B$

Definition 2.11.   For any two coordinate vectors A and B in $\mathbf{R}_3$, the **cross product** $A \times B$ is defined as

$$A \times B = (a_2 b_3 - a_3 b_2, -(a_1 b_3 - a_3 b_1), a_1 b_2 - a_2 b_1)$$

We also define the cross product $\bar{a} \times \bar{b}$ of the position vectors to be the position vector of $A \times B$.

It is clear that, if either A or B is $\mathbf{0}$, then $A \times B = \mathbf{0}$. If A and B are nonzero and A is parallel to B, then $B = rA$ for some scalar r. Thus

$$A \times B = (a_2(ra_3) - a_3(ra_2), -a_1(ra_3) + a_3(ra_1), a_1(ra_2) - a_2(ra_1)) = \mathbf{0}$$

Conversely, if A and B are nonzero vectors and $A \times B = \mathbf{0}$, then it is true that A is parallel to B, although we shall not prove it here.

Determinants will be discussed in Chapter 4, but it is convenient at this time to introduce the 2 by 2 determinant. For arbitrary real numbers a, b, c, and d, the symbol $\begin{vmatrix} a & b \\ c & d \end{vmatrix}$ denotes the number $ad - bc$ and is called a 2 by 2 determinant. We shall use the determinant notation to aid us in remembering the definition of cross product. If $A = (a_1, a_2, a_3)$ and $B = (b_1, b_2, b_3)$, then

$$A \times B = \left(\begin{vmatrix} a_2 & a_3 \\ b_2 & b_3 \end{vmatrix}, -\begin{vmatrix} a_1 & a_3 \\ b_1 & b_3 \end{vmatrix}, \begin{vmatrix} a_1 & a_2 \\ b_1 & b_2 \end{vmatrix} \right)$$

Example 2.18.   A vector perpendicular to $A = (1, -1, 2)$ and $B = (-2, 0, 3)$ is

$$\left(\begin{vmatrix} -1 & 2 \\ 0 & 3 \end{vmatrix}, -\begin{vmatrix} 1 & 2 \\ -2 & 3 \end{vmatrix}, \begin{vmatrix} 1 & -1 \\ -2 & 0 \end{vmatrix} \right) = (-3, -7, -2)$$

Example 2.19.   Find an equation for the plane that passes through $P = (1, 2, 3)$, $Q = (0, -1, 1)$, and $R = (4, 3, -2)$.

Solution: We wish to find a *normal* to the plane. Let $\vec{v} = \overrightarrow{PQ} = -\vec{i} - 3\vec{j} - 2\vec{k}$ and $\vec{w} = \overrightarrow{QR} = 4\vec{i} + 4\vec{j} - 3\vec{k}$. A vector is a normal to the plane if and only if it is perpendicular to each of the vectors $\vec{v}$ and $\vec{w}$. Thus,

$$\vec{n} = \vec{v} \times \vec{w} = \begin{vmatrix} -3 & -2 \\ 4 & -3 \end{vmatrix}\vec{i} - \begin{vmatrix} -1 & -2 \\ 4 & -3 \end{vmatrix}\vec{j} + \begin{vmatrix} -1 & -3 \\ 4 & 4 \end{vmatrix}\vec{k}$$

$$= 17\vec{i} - 11\vec{j} + 8\vec{k}$$

is a normal to the plane. Using the normal $\vec{n}$ and the point Q, we obtain $17(x - 0) - 11(y + 1) + 8(z - 1) = 0$ or $17x - 11y + 8z = 19$ for the equation of the plane. As a check, we can use the fact that the coordinates of P and R must satisfy this equation.

Check:

$$17(1) - 11(2) + 8(3) = 19$$
$$17(4) - 11(3) + 8(-2) = 19$$

We now derive the following formula for $\| A \times B \|$.

$$\| A \times B \| = \| A \| \| B \| \sin \theta$$

where θ is the angle between A and B.

Proof
(1) $\| A \|^2 \| B \|^2 \sin^2 \theta = \| A \|^2 \| B \|^2 (1 - \cos^2 \theta)$
(2) $\qquad = \| A \|^2 \| B \|^2 - \| A \|^2 \| B \|^2 \cos^2 \theta$
(3) $\qquad = \| A \|^2 \| B \|^2 - (A \cdot B)^2 \qquad (A \cdot B = \| A \| \| B \| \cos \theta)$
(4) Also, $\| A \times B \|^2 = (a_2 b_3 - a_3 b_2)^2 + (a_1 b_3 - a_3 b_1)^2 + (a_1 b_2 - a_2 b_2)^2$
(5) $\qquad = (a_1^2 + a_2^2 + a_3^2)(b_1^2 + b_2^2 + b_3^2)$
$\qquad\qquad - (a_1 b_1 + a_2 b_2 + a_3 b_3)^2$
(6) $\qquad = \| A \|^2 \| B \|^2 - (A \cdot B)^2$
(7) Therefore, $\| A \times B \|^2 = \| A \|^2 \| B \|^2 \sin^2 \theta$ (comparing steps 3 and 6)
(8) $\| A \times B \| = \pm \| A \| \| B \| \sin \theta$ (taking the square root in 7)
(9) $\| A \times B \| = \| A \| \| B \| \sin \theta$ ($\sin \theta \geq 0$ for $0° \leq \theta \leq 180°$)
The calculation leading from step 4 to step 5 is not obvious but can be verified by the reader. Q.E.D.

The above formula for $\| A \times B \|$ has a geometric interpretation, for consider the parallelogram determined by A and B (see Figure 2.14). The

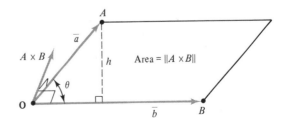

Figure 2.14

area of the parallelogram is given by base times height, where base $= \|B\|$ and height $= \|A\| \sin\theta$. So the area $= \|A\| \|B\| \sin\theta = \|A \times B\|$. Therefore, **the area of the parallelogram determined by** A **and** B (O and $A + B$ are the other vertices) **is** $\|\mathbf{A} \times \mathbf{B}\|$.

Exercise 2.4

1. Verify by direct substitution that any constant multiple k of the cross product $A \times B$ is a solution to the given system (E) at the beginning of Section 2.4.

2. If $A = (0, 1, 1)$ and $B = (1, 2, -1)$, find (a) a vector orthogonal to A and B, and (b) the area of the parallelogram determined by A and B.

3. Repeat Problem 2 for $A = (1, 0, -1)$ and $B = (3, -1, 2)$.

4. Find an equation for the plane passing through the points $P = (1, 1, 1)$, $Q = (2, 0, 3)$, and $R = (2, -1, 0)$.

5. Find the area of the triangle with vertices P, Q, and R in Problem 4.

6. Find an equation for the plane passing through the points $P = (-2, 4, 5)$, $Q = (3, 4, 6)$, and $R = (0, 1, 2)$.

7. Find the area of the triangle with vertices P, Q, and R in Problem 6.

8. Verify that $r(A \times B) = (rA) \times B = A \times (rB)$.

9. Verify that the cross product is *anticommutative*; that is, $A \times B = -(B \times A)$.

10. Verify that $A \times (B \times C) = (A \cdot C)B - (A \cdot B)C$.

11. Use Problems 9 and 10 to verify the *Jacobi identity:* $(A \times B) \times C + (B \times C) \times A + (C \times A) \times B = O$.

12. Verify that the cross product distributes over addition; that is, $A \times (B + C) = A \times B + A \times C$ and $(B + C) \times A = B \times A + C \times A$.

    A vector space $(\mathbf{V}, +, \mathrm{sm})$ that has a product defined on it satisfying the identities in Problems 8, 9, 11, and 12 is called a *Lie algebra* (see Exercise 3.6).

13. Use Problems 9 and 11 to verify that $(A \times B) \times C - A \times (B \times C) = B \times (C \times A)$. What can you conclude about the associativity of the cross product?

We have seen that, for nonzero, nonparallel vectors A and B, $A \times B$ is a vector orthogonal to both A and B. However, we have not determined which way $A \times B$ points. A *mechanical* procedure for deciding this question is as follows. Curl the figures of your right hand through the smaller angle from A to B. Your extended right thumb will point along $A \times B$. It is the direction a right-threaded screw would move if it were turned in the direction of the curled fingers (see Figure 2.15).

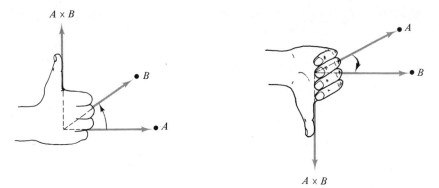

Figure 2.15   Direction of $A \times B$

14. Apply the above rule of thumb to decide the direction that $A \times B$ points. Verify by computing $A \times B$.
(a) $A = (1, 0, 0)$, $B = (0, 0, 1)$.
(b) $A = (0, 1, 0)$, $B = (1, 0, 0)$.
(c) $A = (0, 0, 1)$, $B = (-1, 0, 0)$.
(d) $A = (2, 2, 0)$, $B = (0, 1, 0)$.

Angular Velocity

Consider a rigid body (a propeller for example) rotating about a fixed axis L with a constant spin of ω radians/sec. Figure 2.16 represents a point P in the rigid body having position vector $\bar{p}$ relative to a point O on L. Point Q is the foot of the perpendicular from P to line L. The position vector $\bar{\Omega}$ along the axis L, having magnitude ω and pointing in the direction a right-threaded screw would advance under the given rotation, is called the *angular velocity* of the body (see Figure 2.16).

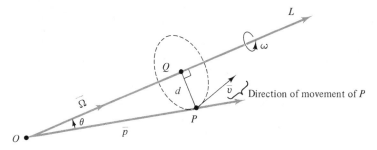

Figure 2.16

P moves in the direction of $\bar{\Omega} \times \bar{p}$, which is perpendicular to the plane determined by $\bar{\Omega}$ and $\bar{p}$. Let $\bar{v}$ be the vector representing the linear velocity of the point P. In 1 sec an angle of ω rad is swept out, and the point P travels a distance of s units along the circle from P to P'. We have

$$\frac{\text{arc length from } P \text{ to } P'}{\text{arc length of the circle}} = \frac{\omega \text{ rad}}{2\pi \text{ rad}}$$

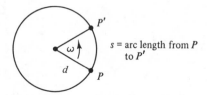

Therefore, $s/2\pi d = \omega/2\pi$ and $s = \omega d$. Hence the speed of the moving point P is ωd units/sec; that is, $\|\bar{v}\| = \omega d$, where d is the distance from P to Q. From Figure 2.16, $d = \|\bar{p}\| \sin \theta$. Therefore,

$$\|\bar{v}\| = \omega \|\bar{p}\| \sin \theta = \|\bar{\Omega}\| \|\bar{p}\| \sin \theta = \|\bar{\Omega} \times \bar{p}\|$$

So $\bar{v} = \bar{\Omega} \times \bar{p}$ is a vector representing the linear velocity of point P.

Exercise 2.4

15. A body spins about a line through the origin parallel to the vector $\bar{u} = 6\bar{i} - 3\bar{j} + 2\bar{k}$ at 20 rad/sec. The rotation appears clockwise if one looks in the direction of $\bar{u}$. Find the angular velocity $\bar{\Omega}$ for the body, and find the linear velocity of a point in the body with position vector $\bar{i} + 2\bar{j} + 3\bar{k}$. Find the linear speed (magnitude of the linear velocity vector) of the point.

16. A sphere with a radius of 3 cm and center at the origin is rotating at an angular speed $\omega = 6$ rad/sec about a fixed axis through the origin parallel to $\bar{u} = \bar{i} + 2\bar{j} + 2\bar{k}$. The rotation appears clockwise if one looks in the direction of $\bar{u}$. Find the linear velocity of each of the given points on the surface. Also find the linear speed: (a) $A = (3, 0, 0)$; (b) $B = (0, 3, 0)$; (c) $C = (1, 2, 2)$.

2.5 VECTOR AND PARAMETRIC EQUATIONS OF LINES AND PLANES

In this section we shall derive vector and parametric equations for lines in 2-space and for lines and planes in 3-space. In the discussion we shall move freely from line vectors to coordinate vectors, and vice versa (recall the correspondence between coordinate vectors and position vectors discussed in Section 1.3).

A line is determined by two points. It is also clear that a line is determined by a point on the line and a direction and consequently, by a point on the line and a vector parallel to the line.

Consider a line $\mathcal{L}$ in 3-space. Let A be a point on $\mathcal{L}$ and let $\bar{v}$ be a vector parallel to $\mathcal{L}$ (Figure 2.17). A point P different from A will be on $\mathcal{L}$ if and only if $\overrightarrow{AP} \parallel \bar{v}$, that is, if and only if

$$\overrightarrow{AP} = t\bar{v}$$

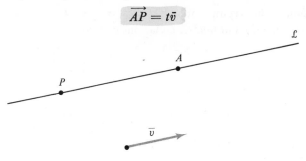

Figure 2.17

for some scalar $t \neq 0$. Note that $t = 0$ will give the point $P = A$. We coordinatize the space so that the origin coincides with the tail of $\bar{v}$. Using coordinate vectors, the equation $\overrightarrow{AP} = t\bar{v}$ may be written $P - A = tV$ or

$$P = A + tV \tag{1}$$

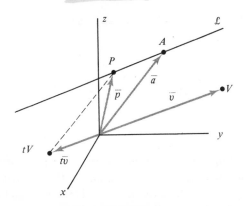

Equation (1) is called a **vector equation for line** $\mathcal{L}$. If $P = (x, y, z)$, $A = (a_1, a_2, a_3)$, and $V = (v_1, v_2, v_3)$, we have $(x, y, z) = (a_1, a_2, a_3) + t(v_1, v_2, v_3)$. Solving for x, y, and z, we obtain

$$
\begin{aligned}
x &= a_1 + tv_1 \\
y &= a_2 + tv_2 \\
z &= a_3 + tv_3
\end{aligned}
\tag{2}
$$

The equations in (2) are called **parametric equations of** $\mathcal{L}$ **with parameter** t. Note that one obtains a specific point $P = (x, y, z)$ on line $\mathcal{L}$ by assigning a specific value to t. As t ranges over the real numbers, P ranges over the line $\mathcal{L}$.

Example 2.20. Find vector and parametric equations for the line passing through the point $(1, 2, 3)$ and parallel to the line vector corresponding to $V = (2, 3, -2)$. Find two other points on $\mathcal{L}$.

Solution: The vector equation of $\mathcal{L}$ is $P = A + tV$ or $(x, y, z) = (1, 2, 3) + t(2, 3, -2)$. Parametric equations of $\mathcal{L}$ are

$$x = 1 + 2t$$
$$y = 2 + 3t$$
$$z = 3 + -2t$$

To find points on $\mathcal{L}$, we assign values to the parameter t. For example, $t = 1$ gives the point $B = (3, 5, 1)$ on $\mathcal{L}$, and $t = -1$ yields the point $C = (-1, -1, 5)$ on $\mathcal{L}$ (see Figure 2.18).

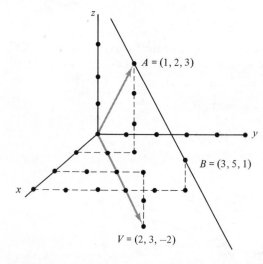

Figure 2.18

Example 2.21. Find vector and parametric equations for the line $\mathcal{L}$ passing through the point $A = (2, -1)$ and parallel to the line vector corresponding to $V = (1, 2)$. Eliminate the parameter to obtain a single linear equation for $\mathcal{L}$. Find two other points on $\mathcal{L}$.

Solution: The vector equation of $\mathcal{L}$ is $P = A + tV$ or $(x, y) = (2, -1) + t(1, 2)$. The parametric equations of $\mathcal{L}$ are

$$x = 2 + t$$
$$y = -1 + 2t$$

Letting $t = 1$ and $t = -1$, we get the points $(3, 1)$ and $(1, -3)$ on $\mathcal{L}$. To obtain a single linear equation for $\mathcal{L}$, we solve for t, obtaining $t = x - 2$ and $t = (y + 1)/2$. Eliminating the parameter t, we get $x - 2 = (y + 1)/2$ or $2x - y = 5$.

One advantage of a vector equation or the corresponding parametric equations for a line is that we can obtain *equations for specified line segments* along $\mathcal{L}$ by restricting the values of t. For example, the vector equation

$$(x, y) = (2, -1) + t(1, 2), \qquad 1 \le t \le 2$$

describes the line segment on $\mathcal{L}$ from (3, 1) to (4, 3).

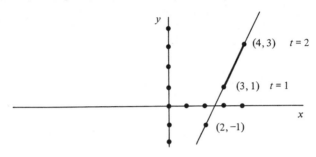

Eliminating the parameter in (2), we obtain

$$\frac{x - a_1}{v_1} = \frac{y - a_2}{v_2} = \frac{z - a_3}{v_3} \tag{3}$$

provided that the *direction numbers* v_1, v_2, and v_3 are not zero. These equations are called the **symmetric equations of a line.** They are not unique of course, since (a_1, a_2, a_3) can be any point on the line, and the direction numbers v_1, v_2, and v_3 can be replaced by any nonzero scalar multiples kv_1, kv_2, and kv_3.

Example 2.22. Find the vector equation and symmetric equations for the line $\mathcal{L}$ passing through the points $A = (1, 2, 1)$ and $B = (2, -2, 3)$.

Solution: The vector $\overrightarrow{AB}$ is parallel to $\mathcal{L}$. The corresponding coordinate vector is $B - A = (1, -4, 2)$. Thus a vector equation is $P = A + t(B - A)$ or $(x, y, z) = (1, 2, 1) + t(1, -4, 2)$. Since 1, -4, and 2 are direction numbers for $\mathcal{L}$ and $A = (1, 2, 1)$ is a point on $\mathcal{L}$, we have that

$$\frac{x - 1}{1} = \frac{y - 2}{-4} = \frac{z - 1}{2}$$

are symmetric equations for $\mathcal{L}$.

Equations (3) are not applicable if any of the direction numbers are zero. If one of the numbers is zero, say v_3, then the symmetric equations are

$$\frac{x - a_1}{v_1} = \frac{y - a_2}{v_2} \quad \text{and} \quad z = a_3$$

If two of the direction numbers are zero, say v_2 and v_3, then the symmetric equations are $y = a_2$ and $z = a_3$.

Next we consider the problem of finding a vector equation of a plane that is parallel to two given nonparallel vectors V and W and that passes through a fixed point A (see Figure 2.19). Let $P = (x, y, z)$ be any point in

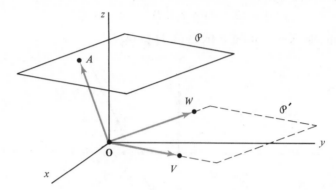

Figure 2.19

the plane through A and parallel to each of the nonparallel vectors V and W (V is not a scalar multiple of W). The plane $\mathcal{P}'$ determined by O, V, and W is the set of all points P' that are linear combinations of V and W, that is, all points of the form $P' = sV + tW$. The plane $\mathcal{P}$ parallel to $\mathcal{P}'$ and containing the point $A = (a_1, a_2, a_3)$ may be thought of as the plane $\mathcal{P}'$ translated to A. Thus a point P in the desired plane satisfies the equation

$$P = A + sV + tW \tag{4}$$

Using coordinates, this equation may be written as

$$(x, y, z) = (a_1, a_2, a_3) + s(v_1, v_2, v_3) + t(w_1, w_2, w_3)$$

or

$$x = a_1 + sv_1 + tw_1$$
$$y = a_2 + sv_2 + tw_2 \tag{5}$$
$$z = a_3 + sv_3 + tw_3$$

Equation (4) is called a **vector equation for the plane** $\mathcal{P}$, and the equations in (5) are called **parametric equations of** $\mathcal{P}$. The scalars s and t are called **parameters**. As s and t range over the real numbers, the point P ranges over the plane $\mathcal{P}$.

Example 2.23.   Find a vector equation and parametric equations for the plane $\mathcal{P}$ passing through $A = (1, 2, 4)$ and parallel to the plane determined by O, $V = (1, -2, 2)$, and $W = (2, 3, -1)$. Find two other points in the plane $\mathcal{P}$.

Solution: If $P = (x, y, z)$ is a point in the plane, then P satisfies the equation $(x, y, z) = (1, 2, 4) + s(1, -2, 2) + t(2, 3, -1)$. Parametric equations for this plane are

$$x = 1 + s + 2t$$
$$y = 2 - 2s + 3t$$
$$z = 4 + 2s - t$$

If $s = 1$ and $t = 0$, we get the point $(2, 0, 6) \in \mathcal{P}$. Setting $s = 0$ and $t = 1$, we obtain the point $(3, 5, 3) \in \mathcal{P}$. Note that $s = t = 0$ gives the point $A = (1, 2, 4)$.

Example 2.24. Find a vector equation and parametric equations for the plane determined by the three noncollinear points $A = (1, 0, 1)$, $B = (2, 2, 1)$, and $C = (2, 1, 0)$.

Solution: Let $V = B - A = (1, 2, 0)$ and $W = C - A = (1, 1, -1)$. Since the points A, B, and C are noncollinear, the vectors V and W are nonparallel. The plane through A and parallel to V and W has vector equation $P = A + sV + tW = A + s(B - A) + t(C - A)$ or $(x, y, z) = (1, 0, 1) + s(1, 2, 0) + t(1, 1, -1)$. The corresponding parametric equations are

$$x = 1 + s + t$$
$$y = 2s + t$$
$$z = 1 - t$$

Note that $t = s = 0$ gives the point A, $s = 1$ and $t = 0$ gives the point $B = A + 1(B - A)$, and $s = 0$, $t = 1$ gives the point $C = A + 1(C - A)$.

Exercise 2.5

1. Find vector and parametric equations for the line parallel to $V = (1, 3, 1)$ and passing through the point $A = (3, 4, 2)$. Find two other points on the line.

2. Find vector and parametric equations for the line parallel to $V = (1, 4)$ and passing through $A = (3, 2)$. Find a second point on the line.

3. Find symmetric equations for the line passing through the points $A = (1, 2, 3)$ and $B = (2, -1, 4)$.

4. Find vector and parametric equations for the line passing through the points $A = (-1, 0, 2)$ and $B = (2, 4, 3)$.

5. Find vector and parametric equations for the line passing through the points $A = (2, 4)$ and $B = (6, 1)$.

6. Find a vector equation and parametric equations for the plane passing through $A = (1, 2, 1)$ and parallel to the plane determined by O, $V = (0, 1, 3)$, and $W = (2, 0, 3)$. Find two points in this plane different from A.

7. Find a vector equation and parametric equations for the plane passing through (a) $A = (-1, 2, 2)$, $B = (0, 3, 1)$, and $C = (0, 0, 2)$; (b) $A = (3, 1, 0)$, $B = (2, 1, 1)$, and $C = (0, 2, -4)$.

8. For the plane in Problem 6, find an equation of the form $ax + by + cz = d$.

9. Determine whether or not the following planes are parallel $[P = (x, y, z)]$:
   (a) $P = (1, -2, 0) + s(1, 2, 1) + t(2, 3, 4)$,
   $\quad P = (2, 4, 1) + u(3, 5, 5) + v(1, 1, 3)$.
   (b) $P = (3, 2, 1) + s(1, 0, 0) + t(0, 1, 0)$,
   $\quad P = (1, 1, 1) + u(1, 0, 0) + v(0, 0, 1)$.

10. (a) Show that $(5, 10)$ and $(0, -10)$ are points on the line in Problem 2.
   (b) Use parametric equations and suitable restrictions on the parameter to describe the line segment from $(5, 10)$ to $(0, -10)$.

11. (a) Show that $(-1, 8, 1)$ and $(4, -7, 6)$ are points on the line in Problem 3.
   (b) Use parametric equations and suitable restrictions on the parameter to describe the line segment from $(-1, 8, 1)$ to $(4, -7, 6)$.

12. Find symmetric equations of a line $\mathcal{L}$ if
   (a) $A = (2, 3, 4) \in \mathcal{L}$ and $V = (3, 2, -1)$ is parallel to $\mathcal{L}$.
   (b) $A = (2, 3, 4) \in \mathcal{L}$ and $V = (0, 2, 1)$ is parallel to $\mathcal{L}$.
   (c) $A = (2, 3, 4) \in \mathcal{L}$ and $V = (0, 2, 0)$ is parallel to $\mathcal{L}$.

Systems
of Linear Equations
and Matrices

In this chapter we shall study systems of linear equations. The number of equations and the number of unknowns in a system will be completely arbitrary. A method is developed for determining the complete solution in the sense that (1) we can determine if a solution exists, and (2) if a solution does exist, we can determine all the solutions. The method used is attributed to the great German mathematician, Karl Friedrich Gauss (1777–1855). Our work is related to Chapter 2 in that the collection of all solutions to a homogeneous system of linear equations (see Definition 3.4) is a vector space.

3.1 THE GAUSS AND THE GAUSS–JORDAN METHODS OF REDUCTION

Given a system of linear equations, our goal will be to find an equivalent system from which the complete solution can be determined by inspection. By an equivalent system of linear equations we mean a system that has exactly the same solutions as the given system. The operations that we shall use to produce an equivalent system are the following three **elementary operations**:

Type I. The interchange of any two equations in the system.

Type II. Replacement of an equation in the system by a nonzero real multiple of itself (multiply both sides of an equation by a nonzero real number).

Type III. Replacement of an equation in the system by the sum of itself and a real multiple of another equation in the system.

It is quite easy to show that, if a given system of equations is transformed by any one of these elementary operations, the resulting system is equivalent to the given one. We shall illustrate the method first and then later outline a proof for the equivalence of the systems obtained through elementary operations.

The following notation will be used to indicate which operations have been performed.

OPERATION	SYMBOL	MEANING OF SYMBOL
Type I	E_{ij}	Interchange the i and jth equations in the system.
Type II	$kE_i, k \neq 0$	Replace the ith equation by the equation obtained by multiplying both sides of the ith equation by k.
Type III	$kE_i + E_j$	Replace the jth equation by k times the ith equation added onto the jth equation.

Example 3.1.   Find the complete solution to the system of equations

$$x + 2y + 7z = 1$$
$$-x + y - z = 2$$
$$3x - 2y + 5z = -5$$

The method of solution illustrates the Gauss–Jordan method and uses the three operations described above.

Solution

$$
\begin{array}{l}
x + 2y + 7z = 1 \\
-x + y - z = 2 \\
3x - 2y + 5z = -5
\end{array}
\quad\xrightarrow{E_1 + E_2}\quad
\begin{array}{l}
x + 2y + 7z = 1 \\
0x + 3y + 6z = 3 \\
3x - 2y + 5z = -5
\end{array}
$$

$$
\xrightarrow{-3E_1 + E_3}\quad
\begin{array}{l}
x + 2y + 7z = 1 \\
0x + 3y + 6z = 3 \\
0x - 8y - 16z = -8
\end{array}
$$

$$
\xrightarrow{(1/3)E_2}\quad
\begin{array}{l}
x + 2y + 7z = 1 \\
0x + y + 2z = 1 \\
0x - 8y - 16z = -8
\end{array}
$$

$$
\xrightarrow{-2E_2 + E_1}\quad
\begin{array}{l}
x + 0y + 3z = -1 \\
0x + y + 2z = 1 \\
0x - 8y - 16z = -8
\end{array}
$$

$$x + 0y + 3z = -1$$
$$\xrightarrow{8E_2 + E_3} \quad 0x + y + 2z = 1$$
$$0x + 0y + 0z = 0$$

The final system is equivalent to

$$x = -3z - 1$$
$$y = -2z + 1$$

Noting that x and y are completely determined by z and that there are no restrictions on z, we see that all solutions are of the form $x = -3s - 1$, $y = -2s + 1$, and $z = s$, where s is any real number. The number s is called a **parameter**. A *particular solution* can be found by assigning a value to s; for example, if $s = 0$, we obtain $x = -1$, $y = 1$, and $z = 0$. One can of course easily check the solution by substituting the values of x, y, and z back into the original system. If we consider a solution to be a 3-tuple (which we do), we see that every solution is of the form $(x, y, z) = (-3s - 1, -2s + 1, s)$.

In the process of going from one equivalent system to another, some of the labor can be saved by recording only the coefficients of the unknowns x, y, and z and the constants. For example, we can associate the array

$$\begin{bmatrix} 1 & 2 & 7 & 1 \\ -1 & 1 & -1 & 2 \\ 3 & -2 & 5 & -5 \end{bmatrix}$$

with the given system

$$x + 2y + 7z = 1$$
$$-x + y - z = 2$$
$$3x - 2y + 5z = -5$$

The array is called a *matrix*, and each number in the matrix is called an *entry*. The matrix

$$\begin{bmatrix} 1 & 2 & 7 & 1 \\ -1 & 1 & -1 & 2 \\ 3 & -2 & 5 & -5 \end{bmatrix}$$

is called the *augmented* matrix of the system; the matrix

$$\begin{bmatrix} 1 & 2 & 7 \\ -1 & 1 & -1 \\ 3 & -2 & 5 \end{bmatrix}$$

is called the *coefficient* matrix of the system. Instead of performing the operations on the equations in the system, we can accomplish the same result by performing the operations on the rows of the augmented matrix. Before illustrating with an example, we introduce another notation. Whereas E_{ij}, kE_i $(k \neq 0)$, and $kE_i + E_j$ denoted certain operations on the *equations* of the system, R_{ij}, kR_i $(k \neq 0)$, and $kR_i + R_j$ will denote the corresponding operations on the *rows* of a matrix. For instance, if we perform the operation denoted by $-2R_1 + R_2$ on

$$\begin{bmatrix} 2 & 3 & 2 \\ 4 & 1 & 6 \\ 0 & 1 & 3 \end{bmatrix}$$

we obtain

$$\begin{bmatrix} 2 & 3 & 2 \\ -2(2)+4 & -2(3)+1 & -2(2)+6 \\ 0 & 1 & 3 \end{bmatrix} = \begin{bmatrix} 2 & 3 & 2 \\ 0 & -5 & 2 \\ 0 & 1 & 3 \end{bmatrix}$$

In Example 3.2 we shall apply the reduction process to both the system and the corresponding augmented matrix of the system. The reader should compare and note the advantages of less labor and also the chance of making fewer mistakes with the matrix notation.

Example 3.2. Find all solutions to the system of equations

$$3y - 3z = 6$$
$$x - y + 4z = -3$$
$$x + 6z = 4$$

Solution

$$\begin{array}{ccc} 3y - 3z = 6 & & x - y + 4z = -3 \\ x - y + 4z = -3 & \xrightarrow{E_{12}} & 3y - 3z = 6 & \xrightarrow{-1E_1+E_3} \\ x + 6z = 4 & & x + 6z = 4 \end{array}$$

$$\begin{bmatrix} 0 & 3 & -3 & \vdots & 6 \\ 1 & -1 & 4 & \vdots & -3 \\ 1 & 0 & 6 & \vdots & 4 \end{bmatrix} \xrightarrow{R_{12}} \begin{bmatrix} 1 & -1 & 4 & \vdots & -3 \\ 0 & 3 & -3 & \vdots & 6 \\ 1 & 0 & 6 & \vdots & 4 \end{bmatrix} \xrightarrow{-R_1+R_3}$$

$$\begin{array}{ccc} x - y + 4z = -3 & & x - y + 4z = -3 \\ 3y - 3z = 6 & \xrightarrow{(1/3)E_2} & y - z = 2 & \xrightarrow{-E_2+E_3} \\ y + 2z = 7 & & y + 2z = 7 \end{array}$$

$$\begin{bmatrix} 1 & -1 & 4 & \vdots & -3 \\ 0 & 3 & -3 & \vdots & 6 \\ 0 & 1 & 2 & \vdots & 7 \end{bmatrix} \xrightarrow{(1/3)R_2} \begin{bmatrix} 1 & -1 & 4 & \vdots & -3 \\ 0 & 1 & -1 & \vdots & 2 \\ 0 & 1 & 2 & \vdots & 7 \end{bmatrix} \xrightarrow{-R_2+R_3}$$

$$x - y + 4z = -3 \qquad\qquad x - y + 4z = -3*$$
$$y - z = 2 \xrightarrow{(1/3)E_3} \qquad y - z = 2 \xrightarrow{1E_2+E_1}$$
$$3z = 5 \qquad\qquad z = \tfrac{5}{3}$$

$$\begin{bmatrix} 1 & -1 & 4 & \vdots & -3 \\ 0 & 1 & -1 & \vdots & 2 \\ 0 & 0 & 3 & \vdots & 5 \end{bmatrix} \xrightarrow{(1/3)R_3} \begin{bmatrix} 1 & -1 & 4 & \vdots & -3 \\ 0 & 1 & -1 & \vdots & 2 \\ 0 & 0 & 1 & \vdots & \tfrac{5}{3} \end{bmatrix} \xrightarrow{R_2+R_1}$$

$$x + 3z = -1 \qquad\qquad x + 3z = -1$$
$$y - z = 2 \xrightarrow{1E_3+E_2} \qquad y = \tfrac{11}{3} \xrightarrow{-3E_3+E_1}$$
$$z = \tfrac{5}{3} \qquad\qquad z = \tfrac{5}{3}$$

$$\begin{bmatrix} 1 & 0 & 3 & \vdots & -1 \\ 0 & 1 & -1 & \vdots & 2 \\ 0 & 0 & 1 & \vdots & \tfrac{5}{3} \end{bmatrix} \xrightarrow{R_3+R_2} \begin{bmatrix} 1 & 0 & 3 & \vdots & -1 \\ 0 & 1 & 0 & \vdots & \tfrac{11}{3} \\ 0 & 0 & 1 & \vdots & \tfrac{5}{3} \end{bmatrix} \xrightarrow{-3R_3+R_1}$$

$$x = -6$$
$$y = \tfrac{11}{3}$$
$$z = \tfrac{5}{3}$$

$$\begin{bmatrix} 1 & 0 & 0 & \vdots & -6 \\ 0 & 1 & 0 & \vdots & \tfrac{11}{3} \\ 0 & 0 & 1 & \vdots & \tfrac{5}{3} \end{bmatrix}$$

From the last system of equations or from the corresponding augmented matrix of this system, we see that there is a unique solution $(x, y, z) = (-6, \tfrac{11}{3}, \tfrac{5}{3})$.

This method of starting with the given system of equations and reducing it to the equivalent system $x = -6$, $y = \tfrac{11}{3}$, $z = \tfrac{5}{3}$, is called the **Gauss–Jordan reduction process** or **Gauss–Jordan elimination**. The corresponding matrix

$$\begin{bmatrix} 1 & 0 & 0 & \vdots & -6 \\ 0 & 1 & 0 & \vdots & \tfrac{11}{3} \\ 0 & 0 & 1 & \vdots & \tfrac{5}{3} \end{bmatrix}$$

is said to be in *reduced row echelon* form. The Gauss–Jordan process is a refinement of the **Gauss reduction** or **Gauss elimination** process, which we now illustrate with the same example. In Gauss elimination we proceed as above but stop with the system $x - y + 4z = -3$, $y - z = 2$, $z = \tfrac{5}{3}$, which we marked with an asterisk. The matrix

$$\begin{bmatrix} 1 & -1 & 4 & -3 \\ 0 & 1 & -1 & 2 \\ 0 & 0 & 1 & \frac{5}{3} \end{bmatrix}$$

corresponding to this system is said to be in *row echelon* form.

We solve the above system by *back substitution*. From the last of the three equations we have $z = \frac{5}{3}$. Substituting back into the second equation and solving, we obtain $y = \frac{5}{3} + 2 = \frac{11}{3}$. Knowing values for y and z, we now substitute back into the first equation and get $x = y - 4z - 3 = \frac{11}{3} - \frac{20}{3} - 3 = -6$.

It can be shown that the amount of arithmetic involved is less with the Gauss method then with the Gauss–Jordan method. In fact, to solve a system of n equations and n unknowns, the Gauss elimination method requires approximately $\frac{1}{3}n^3$ additions and multiplications, whereas the Gauss–Jordan method requires about $\frac{1}{2}n^3$ additions and multiplications [see Ben Noble, *Applied Linear Algebra* (Englewood Cliffs, N.J.: Prentice-Hall, Inc., 1969), pp. 211–215]. Even on a high—speed computer, this 50% additional work can be significant for large n (see Problem 12, Exercise 3.3). The Gauss–Jordan method is emphasized here because of its importance in the theory.

Example 3.3. Find all the solutions to the system of equations

$$x_1 + 2x_2 + 4x_3 = 2$$
$$2x_1 + 3x_2 + 7x_3 = 3$$
$$3x_1 - x_2 + 5x_3 = 1$$

Solution: This time we shall apply the elimination process to the augmented matrix of the system and not to the system of equations. Several operations are performed at each step as indicated. However, it is important to realize that *the operations are performed in sequence and not simultaneously.*

$$\begin{bmatrix} 1 & 2 & 4 & 2 \\ 2 & 3 & 7 & 3 \\ 3 & -1 & 5 & 1 \end{bmatrix} \xrightarrow[-3R_1+R_3]{-2R_1+R_2} \begin{bmatrix} 1 & 2 & 4 & 2 \\ 0 & -1 & -1 & -1 \\ 0 & -7 & -7 & -5 \end{bmatrix} \xrightarrow[\substack{-2R_2+R_1 \\ 7R_2+R_3}]{-1R_2} \begin{bmatrix} 1 & 0 & 2 & 0 \\ 0 & 1 & 1 & 1 \\ 0 & 0 & 0 & 2 \end{bmatrix}$$

The system corresponding to the last matrix is

$$x_1 + 2x_3 = 0$$
$$x_2 + x_3 = 1$$
$$0 = 2$$

Since $0 = 2$ is false no matter what x_1, x_2, and x_3 are, this system has no solutions and consequently the original system, which is equivalent to it, has no solutions.

Examples 3.1–3.3 illustrate the different possibilities for the solution to a nonhomogeneous system. There can be more than one solution, exactly one solution, or no solutions.

Example 3.4. Solve the system

$$-x_1 + 3x_2 + 3x_3 + 2x_4 = 0$$
$$2x_1 + 0x_2 + 6x_3 + x_4 = 0$$
$$-2x_1 + 4x_2 + 2x_3 + 4x_4 = 0$$

This system is called a *homogeneous system* because the constant terms are all zero. If we use matrix notation to solve the system, there is no need to carry the last column of zeros, because, under the three elementary row operations, the entries in the last column will remain all zeros. For this reason we use the coefficient matrix of the system rather than the augmented matrix. Note that a homogeneous system always has at least one solution, that is, $0 = x_1 = x_2 = \ldots$.

Solution

$$\begin{bmatrix} -1 & 3 & 3 & 2 \\ 2 & 0 & 6 & 1 \\ -2 & 4 & 2 & 4 \end{bmatrix} \xrightarrow[-2R_1+R_3]{2R_1+R_2} \begin{bmatrix} -1 & 3 & 3 & 2 \\ 0 & 6 & 12 & 5 \\ 0 & -2 & -4 & 0 \end{bmatrix} \xrightarrow[(1/6)R_2]{-R_1} \begin{bmatrix} 1 & -3 & -3 & -2 \\ 0 & 1 & 2 & \frac{5}{6} \\ 0 & -2 & -4 & 0 \end{bmatrix}$$

$$\xrightarrow[2R_2+R_3]{3R_2+R_1} \begin{bmatrix} 1 & 0 & 3 & \frac{1}{2} \\ 0 & 1 & 2 & \frac{5}{6} \\ 0 & 0 & 0 & \frac{5}{3} \end{bmatrix} \xrightarrow[\substack{(-5/6)R_3+R_2 \\ (-1/2)R_3+R_1}]{(3/5)R_3} \begin{bmatrix} 1 & 0 & 3 & 0 \\ 0 & 1 & 2 & 0 \\ 0 & 0 & 0 & 1 \end{bmatrix}$$

Remembering that the last column of zero constants was not included in the matrix, we see that the system corresponding to the last matrix is

$$\begin{aligned} x_1 + 3x_3 &= 0 & x_1 &= -3x_3 \\ x_2 + 2x_3 &= 0 \quad \text{or} \quad & x_2 &= -2x_3 \\ x_4 &= 0 & x_4 &= 0 \end{aligned}$$

We can parametrize the solutions with $x_3 = t$ as parameter. The solution set is

$$S = \{(-3t, -2t, t, 0) \,|\, t \in \mathbf{R}\}$$

The next example illustrates that more than one parameter may be needed to express the solutions to a given system of linear equations.

Example 3.5. Find the complete solution to the system of equations

$$2x_1 + 4x_2 + 8x_3 - 2x_4 = 4$$
$$x_1 + x_2 + 3x_3 = 5$$
$$4x_1 + 6x_2 + 14x_3 - 2x_4 = 14$$
$$x_1 - x_2 + x_3 + 2x_4 = 11$$

Solution

$$\begin{bmatrix} 2 & 4 & 8 & -2 & \vdots & 4 \\ 1 & 1 & 3 & 0 & \vdots & 5 \\ 4 & 6 & 14 & -2 & \vdots & 14 \\ 1 & -1 & 1 & 2 & \vdots & 11 \end{bmatrix} \xrightarrow[\substack{-R_1+R_2 \\ -4R_1+R_3 \\ -R_1+R_4}]{(1/2)R_1} \begin{bmatrix} 1 & 2 & 4 & -1 & \vdots & 2 \\ 0 & -1 & -1 & 1 & \vdots & 3 \\ 0 & -2 & -2 & 2 & \vdots & 6 \\ 0 & -3 & -3 & 3 & \vdots & 9 \end{bmatrix}$$

$$\xrightarrow[\substack{-1R_2 \\ 2R_2+R_3 \\ 3R_2+R_3}]{2R_2+R_1} \begin{bmatrix} 1 & 0 & 2 & 1 & \vdots & 8 \\ 0 & 1 & 1 & -1 & \vdots & -3 \\ 0 & 0 & 0 & 0 & \vdots & 0 \\ 0 & 0 & 0 & 0 & \vdots & 0 \end{bmatrix}$$

The equivalent system corresponding to this last matrix is

$$x_1 + 2x_3 + x_4 = 8 \quad \text{or} \quad x_1 = -2x_3 - x_4 + 8$$
$$x_2 + x_3 - x_4 = -3 \qquad x_2 = -x_3 + x_4 - 3$$

We see that x_1 and x_2 are completely determined by x_3 and x_4, and furthermore that there are no restrictions on x_3 and x_4. Thus x_3 and x_4 can be assigned arbitrary values, and we obtain the parametric solutions

$$x_1 = -2s - t + 8$$
$$x_2 = -s + t - 3$$
$$x_3 = s$$
$$x_4 = t$$

A particular solution is obtained by assigning values to s and t. For example, $s = 0$ and $t = 0$ gives the solution $(8, -3, 0, 0)$. The solution $(6, -4, 1, 0)$ is obtained by taking $s = 1$ and $t = 0$.

Exercise 3.1

1. Find the augmented matrix and the coefficient matrix of the following systems of equations.

(a) $2x + 3y + 4z = 1$
$\quad\quad x - 3z = 4$
$\quad 4x + y - z = 6$

(b) $\quad\quad x_1 + 3x_2 = 1$
$\quad -x_1 + 2x_2 - 7x_3 = 1$
$\quad 2x_1 + 3x_2 - x_3 = -2$

2. Find the augmented matrix and the coefficient matrix of the following systems of equations.

(a) $7x - 12y + 2z = 0$
$\quad 3x + 8y - 11z = 0$
$\quad \frac{1}{2}x + \frac{1}{4}y + 6z = 0$

(b) $\quad 2x_1 + 6x_3 = 9$
$\quad 5x_1 + 8x_2 + 4x_3 = 3$
$\quad -3x_1 + 7x_2 = 5$

3. Find a system of linear equations that has the augmented matrix

(a) $\begin{bmatrix} 1 & -1 & 3 & \vdots & 2 \\ 2 & 0 & -2 & \vdots & 1 \\ 3 & 1 & 4 & \vdots & 0 \\ 1 & -1 & 4 & \vdots & 5 \end{bmatrix}$

(b) $\begin{bmatrix} 1 & -2 & \vdots & 1 \\ 3 & 4 & \vdots & 0 \end{bmatrix}$

4. Find a linear system of equations that has

(a) The coefficient matrix $\begin{bmatrix} 3 & -4 & 1 & 7 \\ 2 & 0 & 3 & -5 \\ 1 & -1 & 6 & 0 \end{bmatrix}$.

(b) The augmented matrix $\begin{bmatrix} 3 & -4 & 1 & \vdots & 7 \\ 2 & 0 & 3 & \vdots & -5 \\ 1 & -1 & 6 & \vdots & 0 \end{bmatrix}$.

In Exercises 5–17, apply the Gauss–Jordan method of elimination to the given system and solve the system. If there is more than one solution, give a particular solution as well as the general or complete solution.

5. $x + 5y + 11z = -5$
    $2x + 3y + 8z = 4$
    $-x + 2y + 3z = -9$

6. $x + 5y + 11z = -5$
    $2x + 3y + 8z = 4$
    $-x + 2y + 3z = 3$

7. $2x_1 + 3x_2 + x_3 = 3$
    $x_1 + 2x_2 + x_3 = 1$
    $-x_1 + 4x_2 = -2$

8. $x_1 + 3x_2 = -6$
    $2x_1 - 2x_2 = 4$

9. $2x + 4y = 6$
    $3x + 6y = 5$

10. $2x + 4y = 6$
    $3x + 6y = 9$

11. $x_1 + 3x_2 + 5x_3 + 10x_4 = 2$
    $-x_1 - 2x_3 - 4x_4 = 4$
    $2x_1 + 4x_2 + 8x_3 + 16x_4 = 0$
    $x_2 + x_3 + 2x_4 = 2$

12. $x_1 + 3x_2 + 5x_3 + 10x_4 = 0$
    $-x_1 - 2x_3 - 4x_4 = 0$
    $2x_1 + 4x_2 + 8x_3 + 16x_4 = 0$
    $x_2 + x_3 + 2x_4 = 0$

13. $2x - y - z = 4$
    $3x - 2y + 4z = 11$
    $6x + 8y - 4z = 22$

14. $x_1 + x_2 - 3x_4 - x_5 = -3$
    $x_1 - x_2 + 2x_3 - x_4 = -1$
    $4x_1 - 2x_2 + 6x_3 + 3x_4 - 4x_5 = 3$
    $2x_1 + 4x_2 - 2x_3 + 4x_4 - 7x_5 = 4$

15. $2x_1 + x_2 - x_3 - x_4 + x_5 = 0$
    $1x_1 - x_2 + x_3 + x_4 - 2x_5 = 0$
    $3x_1 + 3x_2 - 3x_3 - 3x_4 + 4x_5 = 0$
    $4x_1 + 5x_2 - 5x_3 - 5x_4 + 7x_5 = 0$

16. $2x_1 + 2x_2 + 4x_3 + 4x_4 = 3$
    $8x_1 + 4x_2 + 3x_4 = 7$
    $x_1 + x_3 + x_4 = 1$
    $5x_1 + 4x_2 + 5x_3 + 6x_4 = 6$
    $x_1 + 2x_2 + 3x_3 + 3x_4 = 2$

17. $9x_1 + 12x_2 + 17x_4 = 4$
$6x_1 + 3x_2 + 3x_3 + 8x_4 = 4$
$3x_1 + 6x_2 + 6x_3 + 13x_4 = 2$
$x_1 + 6x_2 + 2x_3 + 9x_4 = 0$
$6x_1 + 5x_2 - x_3 + 7x_4 = 3$

3.2 THE REDUCTION METHOD APPLIED TO A GENERAL SYSTEM OF LINEAR EQUATIONS

We now define and generalize some of the concepts to which we have alluded in the previous section.

Definition 3.1.   An m by n **matrix** is a rectangular array of numbers

$$\begin{bmatrix} a_{11} & a_{12} & \cdots & a_{1n} \\ a_{21} & a_{22} & \cdots & a_{2n} \\ \cdot & \cdot & & \cdot \\ \cdot & \cdot & & \cdot \\ \cdot & \cdot & & \cdot \\ a_{m1} & a_{m2} & \cdots & a_{mn} \end{bmatrix}$$

The horizontal "lines" are called **rows** and the vertical "lines" are called **columns**. An $m \times n$ (m by n) matrix has m rows and n columns. The ith row of the given matrix is

$$[a_{i1} \quad a_{i2} \quad \cdots \quad a_{in}]$$

and the jth column of the given matrix is

$$\begin{bmatrix} a_{1j} \\ a_{2j} \\ \cdot \\ \cdot \\ \cdot \\ a_{mj} \end{bmatrix}$$

The numbers a_{ij} are called the **entries** of the matrix.

Definition 3.2.   Two matrices are **equal** if and only if they have the same number of rows and columns and the corresponding entries are equal.

Definition 3.3.   A **linear equation in n unknowns** $x_1, x_2, \ldots, x_n$ is an equation that can be written in the form $a_1x_1 + a_2x_2 + \cdots + a_nx_n = b$. The a's are called the **coefficients** of the x's, and the number b is called the **constant term**. It is assumed that the a's and b are known.

Definiton 3.4.   A system of m **linear equations in** n **unknowns** is a set of linear equations that can be written in the following form:

$$a_{11}x_1 + a_{12}x_2 + \cdots + a_{1n}x_n = b_1$$
$$a_{21}x_1 + a_{22}x_2 + \cdots + a_{2n}x_n = b_2$$
$$\cdots$$
$$a_{m1}x_1 + a_{m2}x_2 + \cdots + a_{mn}x_n = b_m$$

(3.1)

The **coefficients** a_{ij} for $1 \leq i \leq m$ and $1 \leq j \leq n$ are fixed numbers, as are the **constants** b_i, $i = 1, 2, \ldots, m$. The system is said to be **homogeneous** if $b_i = 0$ for $i = 1, 2, \ldots, m$.

Definition 3.5.   The matrix

$$\begin{bmatrix} a_{11} & a_{12} & \cdots & a_{1n} \\ a_{21} & a_{22} & \cdots & a_{2n} \\ \cdot & \cdot & & \cdot \\ \cdot & \cdot & & \cdot \\ \cdot & \cdot & & \cdot \\ a_{m1} & a_{m2} & \cdots & a_{mn} \end{bmatrix}$$

is called the **coefficient matrix** of the system (3.1).

Definition 3.6.   The matrix

$$\begin{bmatrix} a_{11} & a_{12} & \cdots & a_{1n} & b_1 \\ a_{21} & a_{22} & \cdots & a_{2n} & b_2 \\ \cdot & \cdot & & \cdot & \cdot \\ \cdot & \cdot & & \cdot & \cdot \\ \cdot & \cdot & & \cdot & \cdot \\ a_{m1} & a_{m2} & \cdots & a_{mn} & b_m \end{bmatrix}$$

is called the **augmented matrix** of the system (3.1).

Definition 3.7.   An ordered n-tuple $(c_1, c_2, \ldots, c_n)$ is a **solution** to the system (3.1) if and only if

$$a_{i1}c_1 + a_{i2}c_2 + \cdots + a_{in}c_n = b_i$$

for $i = 1, 2, \ldots, m$

Definition 3.8.   Two systems of linear equations are **equivalent** if and only if each solution of the first system is a solution to the second, and each solution of the second system is a solution to the first.

We shall now prove that the operations of Types I, II, and III produce equivalent systems.

Theorem 3.1. If a system of linear equations is transformed into another linear system by means of one of the elementary operations, the two systems are equivalent.

Proof: Suppose that (3.1) is our given system of equations. The effect of a Type I operation is to rearrange the equations in the system. Clearly, this does not alter the solutions.

Now suppose that a Type II operation kE_i has been performed on the given system. Here $k \neq 0$ is a fixed real number and i is a fixed positive integer less than or equal to m, the number of equations. The ith equation of the old system is

$$a_{i1}x_1 + a_{i2}x_2 + \cdots + a_{in}x_n = b_i$$

The ith equation of the new system is

$$ka_{i1}x_1 + ka_{i2}x_2 + \cdots + ka_{in}x_n = kb_i$$

Since, for $k \neq 0$,

$$a_{i1}c_1 + a_{i2}c_2 + \cdots + a_{in}c_n = b_i$$

if and only if

$$ka_{i1}c_1 + ka_{i2}c_2 + \cdots + ka_{in}c_n = kb_i$$

these two *equations* have the same solutions. It follows that the two *systems* must have the same solutions, because the remaining equations in each of the systems are identical. Note that if $k = 0$ then the proof breaks down, because $0a_{i1}c_1 + 0a_{i2}c_2 + \cdots + 0a_{in}c_n = 0b_i$ does not imply that $a_{i1}c_1 + a_{i2}c_2 + \cdots + a_{in}c_n = b_i$.

Next suppose that a Type III operation $kE_i + E_j$ has been performed on the given system (3.1). Here k, i, and j are fixed. All equations in the new and old systems are identical except that the jth equation of the old system is

$$a_{j1}x_1 + a_{j2}x_2 + \cdots + a_{jn}x_n = b_j$$

and the jth equation of the new system is

$$(ka_{i1} + a_{j1})x_1 + (ka_{i2} + a_{j2})x_2 + \cdots + (ka_{in} + a_{jn})x_n = kb_i + b_j$$

Suppose that $(c_1, c_2, \ldots, c_n)$ is a solution to the old system. Then *all* equations in the system are satisfied, and in particular $a_{i1}c_1 + a_{i2}c_2 + \cdots + a_{in}c_n = b_i$ and $a_{j1}c_1 + a_{j2}c_2 + \cdots + a_{jn}c_n = b_j$. Therefore,

$$k(a_{i1}c_1 + a_{i2}c_2 + \cdots + a_{in}c_n) + (a_{j1}c_1 + a_{j2}c_2 + \cdots + a_{jn}c_n) = kb_i + b_j$$

Consequently,

$$(ka_{i1} + a_{j1})c_1 + (ka_{i2} + a_{j2})c_2 + \cdots + (ka_{in} + a_{jn})c_n = kb_i + b_j$$

Thus $(c_1, c_2, \ldots, c_n)$ satisfies each equation of the new system *including* the jth equation. Consequently, $(c_1, c_2, \ldots, c_n)$ is a solution to the new system.

Conversely, if $(c_1, c_2, \ldots, c_n)$ is a solution to the new system, then

$$(ka_{i1} + a_{j1})c_1 + (ka_{i2} + a_{j2})c_2 + \cdots + (ka_{in} + a_{jn})c_n = kb_i + b_j$$

Also

$$a_{i1}c_1 + a_{i2}c_2 + \cdots + a_{in}c_n = b_i$$

because the solution must satisfy *each equation* of the new system. Multiplying the second equation by k and subtracting from the first equation, we see that $a_{j1}c_1 + a_{j2}c_2 + \cdots + a_{jn}c_n = b_j$ and $(c_1, c_2, \ldots, c_n)$ is a solution to the jth equation of the old system. It now follows that $(c_1, c_2, \ldots, c_n)$ is a solution to the old system and that the two systems are equivalent.   Q.E.D.

In Example 3.2 the augmented matrix of the system was reduced to

$$\begin{bmatrix} 1 & 0 & 0 & -6 \\ 0 & 1 & 0 & \frac{11}{3} \\ 0 & 0 & 1 & \frac{5}{3} \end{bmatrix}$$

In Example 3.4 the final matrix was

$$\begin{bmatrix} 1 & 0 & 3 & 0 \\ 0 & 1 & 2 & 0 \\ 0 & 0 & 0 & 1 \end{bmatrix}$$

Each of these matrices is said to be in reduced row echelon form. Another example is the final matrix

$$\begin{bmatrix} 1 & 0 & 2 & 1 & 8 \\ 0 & 1 & 1 & -1 & -3 \\ 0 & 0 & 0 & 0 & 0 \\ 0 & 0 & 0 & 0 & 0 \end{bmatrix}$$

in Example 3.5. We make the following definition.

Definition 3.9.   A matrix is in **reduced row echelon form** if and only if it satisfies the following conditions.

a. Any row with all zero entries must be below any row with a nonzero entry.
b. The first nonzero entry in each row that has nonzero entries must be a 1. These entries are called **leading** 1's.
c. The leading 1 in any nonzero row appears in a column to the right of any leading 1 in a preceding row.
d. The leading 1 in each nonzero row is the only nonzero entry in its column.

A matrix in reduced row echelon form is called a **reduced row echelon matrix**.
    In terms of this definition, we see that *to solve a system of linear equations by the Gauss–Jordan method we simply take the augmented matrix of the system*

(coefficient matrix in case the system is homogeneous) and perform appropriate elementary row operations on it until we have a matrix in reduced row echelon form.

The reader may have noted that in Example 3.3 the final matrix was

$$\begin{bmatrix} 1 & 0 & 2 & 0 \\ 0 & 1 & 1 & 1 \\ 0 & 0 & 0 & 2 \end{bmatrix}$$

while the reduced row echelon matrix is

$$\begin{bmatrix} 1 & 0 & 2 & 0 \\ 0 & 1 & 1 & 0 \\ 0 & 0 & 0 & 1 \end{bmatrix}$$

In this example it was apparent that there were no solutions to the system before we reached the reduced row echelon form.

We now make use of the reduced row echelon form of a matrix to discuss the general solution of an arbitrary system of linear equations.

Consider the system of linear equations (3.1) having m equations and n unknowns. To solve the system we take the augmented matrix

$$C = \begin{bmatrix} a_{11} & a_{12} & \cdots & a_{1n} & b_1 \\ a_{21} & a_{22} & \cdots & a_{2n} & b_2 \\ \cdot & \cdot & & \cdot & \cdot \\ \cdot & \cdot & & \cdot & \cdot \\ \cdot & \cdot & & \cdot & \cdot \\ a_{m1} & a_{m2} & \cdots & a_{mn} & b_m \end{bmatrix}$$

and apply appropriate elementary row operations on it until we have a reduced row echelon matrix E. That this can always be done is proved in Section 3.8 (Although we shall not prove it, the reduced row echelon matrix E is unique.) If we write $C = [A \mid B]$, where A is the $m \times n$ coefficient matrix of the system (3.1) and B is the column of constants on the right side of the system (3.1), then the reduced row echelon matrix E can be written as $E = [A' \mid B']$, where A' is the reduced row echelon form of A. The $m \times n$ matrix A' will have some number r of nonzero rows, with each nonzero row having a leading 1 as its first nonzero entry. The remaining $m - r$ rows will have all zero entries.

If we assume that at least one $a_{ij} \neq 0$, then $r \geq 1$, and since the number of nonzero rows of A' is less than or equal to the number of rows of E, we have $\mathbf{r} \leq \mathbf{m}$. Corresponding to each leading 1 of A' will be a leading variable called a **basic variable**. The number r of basic variables does not exceed the number of variables, and thus $\mathbf{r} \leq \mathbf{n}$. In the system of linear equations having augmented matrix E and coefficient matrix A', each basic variable will appear in one and only one equation because of Definition 3.9, part (d).

For example, if

$$E = \begin{bmatrix} 1 & 2 & 0 & -1 & 2 & 1 & 0 & 2 \\ 0 & 0 & 1 & 3 & -4 & 5 & 0 & 1 \\ 0 & 0 & 0 & 0 & 0 & 0 & 1 & 4 \\ 0 & 0 & 0 & 0 & 0 & 0 & 0 & b'_4 \\ 0 & 0 & 0 & 0 & 0 & 0 & 0 & b'_5 \end{bmatrix} = [A' \mid B']$$

the reduced system is

$$x_1 + 2x_2 - x_4 + 2x_5 + x_6 = 2$$
$$x_3 + 3x_4 - 4x_5 + 5x_6 = 1$$
$$x_7 = 4$$
$$0 = b'_4$$
$$0 = b'_5$$

The basic variables are x_1, x_3, and x_7, and each appears in exactly one of the equations. Note that $r = 3$, $m = 5$, and $n = 7$ in this example. We shall refer to the system with augmented matrix E as a *reduced system* of linear equations. To determine the solutions to this reduced system and consequently to the equivalent system (3.1), we consider two cases:

Case 1: A' has one or more zero rows and the corresponding b primes are not all zero.

Under this condition the reduced system has no solutions because it has at least one equation of the form $0 = b'_i$, where $b'_i \neq 0$. A system which has no solutions is said to be *inconsistent*.

Case 2: Either A' has no zero rows, or if there are zero rows in A', then the corresponding b primes are all zero.

If the number of basic variables r is less than the number of variables n, then each of the r basic variables can be solved in terms of the remaining $n - r$ nonbasic variables (remember that each basic variable appears in one and only one equation). Since the $n - r$ nonbasic variables can be assigned arbitrary values (the nonbasic variables are called *parameters*), we see that there are infinitely many solutions.

In our example, if $b'_4 = b'_5 = 0$, then solving for the basic variables we have

$$x_1 = -2x_2 + x_4 - 2x_5 - x_6 + 2$$
$$x_3 = -3x_4 + 4x_5 - 5x_6 + 1$$
$$x_7 = 4$$

For each assignment of values to the nonbasic variables x_2, x_4, x_5, and x_6, we obtain a solution. For instance, if $x_2 = 0$, $x_4 = 1$, $x_5 = -1$, and $x_6 = 0$, we obtain the particular solution

$$(x_1, x_2, x_3, x_4, x_5, x_6, x_7) = (5, 0, -6, 1, -1, 0, 4)$$

If the number of basic variables r is equal to the number of variables n, then it is clear that there will be exactly one solution. The solution is

$$(x_1, x_2, \ldots, x_n) = (b'_1, b'_2, \ldots, b'_n), \quad \text{where } B' = \begin{bmatrix} b'_1 \\ b'_2 \\ \cdot \\ \cdot \\ \cdot \\ b'_m \end{bmatrix}$$

We see from the above discussion that system (3.1) is inconsistent if and only if the reduced row echelon form of its augmented matrix has a row in which every entry is zero except the last entry. If the system (3.1) is consistent, then the character of the solutions depends on the number r of nonzero rows of A' which equals the number of nonzero rows of E. If $r < n$; there are infinitely many solutions, and if $r = n$ there is a unique solution.

Definition 3.10.   The **row rank of a reduced row echelon matrix** E is the number r of nonzero rows of E.

We summarize the above results in the following theorems.

Theorem 3.2.   The system (3.1) of m linear equations in n unknowns is inconsistent if and only if the reduced row echelon form of the augmented matrix of the system has a row in which each entry is zero except for the last entry in the row.

Theorem 3.3.   If the system (3.1) of m linear equations in n unknowns is consistent, and if r is the row rank of the reduced row echelon form of the augmented matrix of the system then
(i) If $r < n$, the system has infinitely many solutions. The solutions are parametrized by $n - r$ parameters.
(ii) If $r = n$, the system has a unique solution.
        If the system (3.1) is consistent and if the number of equations m is less than the number of unknowns n, then $r \le m < n$, and by Theorem 3.3(i) the system has infinitely many solutions. We state this result as a corollary.

Corollary 3.1.   If for a consistent system of linear equations the number of equations m is less than the number of unknowns n, then the system has infinitely many solutions.

We remark again that a homogeneous system of linear equations is consistent because we can obtain a solution by assigning each variable the value 0. Thus, if $m < n$ and the system is homogeneous, the solution set is infinite. We have proved the following corollary.

Corollary 3.2. If for a homogeneous system of linear equations the number of equations m is less than the number of unknowns n, then the system has infinitely many solutions.

Finally, we determine a condition for which a unique solution will exist if the number of equations is equal to the number of unknowns.

Suppose that the system (3.1) has a unique solution and $m = n$. Then the system is consistent and the row rank of $E = [A' \mid B']$ must be n, because $r < n$ would give a system with infinitely many solutions [Theorem 3.3(i)]. No leading 1 can be in the last column of E for then the system would be inconsistent. Therefore, A' has n leading 1's. Since A' is an $n \times n$ matrix, it is clear from Definition 3.9 that

$$A' = \begin{bmatrix} 1 & 0 & \cdots & 0 \\ 0 & 1 & \cdots & 0 \\ \cdot & \cdot & & \cdot \\ \cdot & \cdot & & \cdot \\ \cdot & \cdot & & \cdot \\ 0 & 0 & \cdots & 1 \end{bmatrix}$$

For reasons that will become apparent later, this matrix is called the $n \times n$ *identity matrix*, and it is denoted by I_n or I.

Conversely, suppose that $A' = I_n$ and $m = n$. Then the row rank of E is n. The system is not inconsistent because A' has no zero rows. By Theorem 3.3(ii), the system has a unique solution. We have proved the following theorem.

Theorem 3.4. A system of n linear equations in n unknowns has a unique solution if and only if the reduced row echelon form of the coefficient matrix is the identity matrix I_n.

Exercise 3.2

1. If $A = \begin{bmatrix} 3 & 2 & -1 & 7 & 5 & -4 \\ 4 & 6 & 0 & -2 & -3 & 8 \\ 1 & -5 & -6 & -7 & 9 & -8 \end{bmatrix}$

and a_{ij} denotes the entry in the ith row and jth column,
(a) Find $a_{13}, a_{26}, a_{11}, a_{24}$, and a_{35}.
(b) If we call the given matrix an $m \times n$ matrix, what are m and n?
(c) Write the fourth column of A.
(d) Write the second row of A.

2. If $A = \begin{bmatrix} 1 & 2 & 3 & 4 \\ 5 & 6 & 7 & 8 \end{bmatrix}$, $B = \begin{bmatrix} 1 & 5 \\ 2 & 6 \\ 3 & 7 \\ 4 & 8 \end{bmatrix}$

$C = \begin{bmatrix} 5 & 1 \\ 6 & 2 \\ 7 & 3 \\ 8 & 4 \end{bmatrix}$, $D = \begin{bmatrix} 1 & 2 & 3 & 4.1 \\ 5 & 6 & 7 & 8 \end{bmatrix}$

is (a) $A = B$? (b) $A = D$? (c) $B = C$? Explain your answers.

3. Given the equations
　(i) $2x + 3y - 4z = 2y$　　　　(iii) $2x_1^2 + 3x_1 + 4x_2 - 5x_3 = x_2 + 2x_1^2$
　(ii) $x_1 - 2x_2 + 3 = 6$　　　　(iv) $x_1 + 3x_2 + 4x_3 - 2x_4 = 7$
　(a) Which of the equations are linear equations?
　(b) What is the constant term in (ii)?
　(c) What is the coefficient of x_2 in (iii)?

4. Determine which of the following matrices are reduced row echelon matrices.

(a) $\begin{bmatrix} 1 & 2 & 3 & 0 \\ 0 & 1 & 2 & 0 \end{bmatrix}$.

(b) $\begin{bmatrix} 1 & \frac{1}{2} & 0 & 0 \\ 0 & 0 & 1 & 3 \\ 0 & 0 & 0 & 0 \end{bmatrix}$.

(c) $\begin{bmatrix} 1 & \frac{1}{2} & \frac{1}{4} & 0 \\ 0 & 0 & 0 & 1 \\ 0 & 0 & 0 & 0 \end{bmatrix}$.

(d) $\begin{bmatrix} 0 & 0 & 0 & 0 \\ 0 & 0 & 0 & 0 \\ 0 & 0 & 0 & 0 \end{bmatrix}$.

(e) $\begin{bmatrix} 0 & 1 & 0 & 0 \\ 0 & 0 & 1 & 0 \\ 0 & 0 & 0 & 0 \\ 0 & 0 & 0 & 0 \end{bmatrix}$.

(f) $\begin{bmatrix} 1 & 6 & 0 & 4 & 2 \\ 0 & 0 & 0 & 0 & 0 \\ 0 & 0 & 1 & 3 & 8 \\ 0 & 0 & 0 & 0 & 0 \end{bmatrix}$.

(g) $\begin{bmatrix} 1 & 46 & 0 & 17 & 0 & 32 & 0 & 19 & 0 \\ 0 & 0 & 1 & 23 & 0 & 19 & 6 & 25 & 0 \\ 0 & 0 & 0 & 0 & 1 & 29 & 3 & 2 & 0 \\ 0 & 0 & 0 & 0 & 0 & 0 & 0 & 0 & 1 \\ 0 & 0 & 0 & 0 & 0 & 0 & 0 & 0 & 0 \end{bmatrix}$.

(h) $\begin{bmatrix} 1 & 2 & 3 & 4 \\ 0 & 1 & 3 & 4 \\ 0 & 0 & 1 & 0 \\ 0 & 0 & 0 & 1 \end{bmatrix}$.

(i) $\begin{bmatrix} 1 & 0 & 6 & 8 & 4 & 0 & 4 & -6 & 0 & -4 & 1 & 0 \\ 0 & 1 & 9 & 3 & 1 & 0 & 9 & -3 & 0 & 0 & 0 & 0 \\ 0 & 0 & 0 & 0 & 0 & 1 & 3 & 1 & 0 & 7 & 3 & 0 \\ 0 & 0 & 0 & 0 & 0 & 0 & 0 & 0 & 1 & 8 & 6 & 0 \\ 0 & 0 & 0 & 0 & 0 & 0 & 0 & 0 & 0 & 0 & 0 & 1 \end{bmatrix}$.

5. Follow the proof of Theorem 3.1 through in detail for a Type II operation on the system (3.1) with $m = 4$, $n = 3$, and $i = 2$.

6. Follow the proof of Theorem 3.1 through in detail for a Type III operation on the system (3.1) with $m = 4$, $n = 3$, $i = 1$, and $j = 3$.

7. Assume that the given matrix E is the augmented matrix of a reduced system of linear equations.

$$E = \begin{bmatrix} 1 & 0 & 2 & 0 & -1 & -4 & -6 & 0 & \vdots & 3 \\ 0 & 0 & 0 & 1 & -2 & 3 & 1 & 0 & \vdots & 2 \\ 0 & 0 & 0 & 0 & 0 & 0 & 0 & 1 & \vdots & -1 \\ 0 & 0 & 0 & 0 & 0 & 0 & 0 & 0 & \vdots & 0 \\ 0 & 0 & 0 & 0 & 0 & 0 & 0 & 0 & \vdots & 0 \end{bmatrix}$$

(a) Write the reduced system of linear equations corresponding to the given matrix.

(b) What are m and n for the nonreduced system (the number of equations and the number of unknowns, respectively)?

(c) What is the row rank of E?

(d) Use Theorems 3.2 and 3.3 to determine if the system is inconsistent, has a unique solution, or has infinitely many solutions.

(e) Find the solutions to the system.

8. Same as Problem 7 with

$$E = \begin{bmatrix} 1 & -1 & 2 & 3 & -2 & 0 & \vdots & 6 \\ 0 & 0 & 0 & 1 & -4 & -1 & \vdots & 4 \\ 0 & 0 & 0 & 0 & 1 & 2 & \vdots & 3 \\ 0 & 0 & 0 & 0 & 0 & 0 & \vdots & 1 \\ 0 & 0 & 0 & 0 & 0 & 0 & \vdots & 0 \end{bmatrix}$$

9. (a–e) Same as Problem 7 with

$$E = \begin{bmatrix} 1 & 0 & 0 & \vdots & 1 \\ 0 & 1 & 0 & \vdots & 3 \\ 0 & 0 & 1 & \vdots & 2 \end{bmatrix}$$

(f) Does Theorem 3.4 apply to this system? Why or why not?

10. (a–e) Same as Problem 7 with

$$E = \begin{bmatrix} 1 & 0 & 0 & 0 & \vdots & 2 \\ 0 & 1 & 0 & 0 & \vdots & -1 \\ 0 & 0 & 1 & 0 & \vdots & 3 \\ 0 & 0 & 0 & 1 & \vdots & 2 \\ 0 & 0 & 0 & 0 & \vdots & 0 \\ 0 & 0 & 0 & 0 & \vdots & 0 \end{bmatrix}$$

(f) Does Theorem 3.4 apply to this system? Why or why not?

Definition 3.11. An $m \times n$ matrix B is **row equivalent** to an $m \times n$ matrix A if and only if B can be obtained from A by applying a finite sequence of row operations on A.

11. Restate Theorem 3.4 using the term "row equivalent."

12. (a) Verify that any matrix A is row equivalent to itself.
    (b) Verify that if B is row equivalent to A then A is row equivalent to B.
    (c) Verify that if A is row equivalent to B and B is row equivalent to C then A is row equivalent to C.

A solution to a system of linear equations in n unknowns is an n-tuple, that is, an element of $\mathbf{R_n}$.

13. Verify that the set S of all solutions to a homogeneous system of m linear equations in n unknowns together with addition and scalar multiplication defined as in Definitions 1.5 and 1.6 satisfies properties 1A–5A and 6M–10M. *Hint:* Several of the properties 1A–5A and 6M–10M are true for all n-tuples in S because $S \subseteq \mathbf{R_n}$. One must verify however, that the sum of two solutions is a solution and that a scalar multiple of a solution is a solution.

3.3 APPLICATIONS OF LINEAR SYSTEMS OF EQUATIONS

The mathematical model for each of the following problems is a system of linear equations. The problem is stated and the mathematical model of the problem is given along with the solution. The details of the solution are left for the reader.

Business and Economics

An industrialist has three machines that are used in the manufacture of four different products. To fully utilize the machines, they are to be kept operating 8 hours a day. The number of hours each machine is used in the production of one unit of each of the four products is given by

	prd. 1	prd. 2	prd. 3	prd. 4
machine 1	1	2	1	2
machine 2	2	0	1	1
machine 3	1	2	3	0

For example, in the production of one unit of product 1, machine 1 is used 1 h, machine 2 is used 2 h, and machine 3 is used 1 h.

Problem. Find the number of units of each of the four products that would be produced in one 8-h day under the assumption that each machine is used the full 8 h.

Solution: Let x_i be the number of units of product i produced during an 8-h day for $i = 1, 2, 3$, and 4. A solution to the system of equations

$$1x_1 + 2x_2 + 1x_3 + 2x_4 = 8$$
$$2x_1 + 0x_2 + 1x_3 + 1x_4 = 8$$
$$1x_1 + 2x_2 + 3x_3 + 0x_4 = 8$$

represents the production necessary to keep the machines fully utilized. The solutions to the first equation represent the production due to machine 1, the solutions to the second equation represent the production due to machine 2, and the solutions to the third equation represent the production due to machine 3. We solve this system using the Gauss–Jordan method of reduction. The augmented matrix of the system reduces to

$$\begin{bmatrix} 1 & 0 & 0 & 1 & \vdots & 4 \\ 0 & 1 & 0 & 1 & \vdots & 2 \\ 0 & 0 & 1 & -1 & \vdots & 0 \end{bmatrix}$$

Therefore, the solutions are

$$x_1 = -x_4 + 4$$
$$x_2 = -x_4 + 2$$
$$x_3 = x_4$$
$$x_4 = x_4$$

Each of the x_i must be nonnegative, because x_i is the number of units of production of product i during one 8-h period and $x_i < 0$ would have no meaning. Thus $x_2 = -x_4 + 2 \geq 0$ and $x_4 \leq 2$. If we insist that each machine produce a whole number of units of each product, then the possible solutions are $(x_1, x_2, x_3, x_4) = (4, 2, 0, 0)$, $(3, 1, 1, 1)$, or $(2, 0, 2, 2)$. If, in addition, it is necessary to produce at least one unit of each product, then the unique solution is $(3, 1, 1, 1)$. For this solution, we see from the first equation that machine 1 works 3 h producing product 1, 2 h producing product 2, 1 h producing product 3, and 2 h producing product 4. From the second equation, we see that machine 2 would work 6 h producing product 1, 1 h producing product 3, and 1 h producing product 4. The third equation indicates that machine 3 would work 3 h producing product 1, 2 h producing product 2, and 3 h producing product 3.

Electrical Networks

Analysis of an electrical network can lead to the need for solving a system of equations involving many equations and unknowns. The following two laws, which are needed for the analysis, are called *Kirchhoff's laws* (the laws are named after the German, Gustav Kirchhoff, who stated them in 1845).

1. Whenever there is a junction of conductors in a circuit so that the charges have a choice of more than one path away from the junction, the sum of all the currents at the junction must be zero. In short, all current flowing into a junction must flow out of it.

2. Around any closed path in a circuit, the sum of the voltage drops is equal to the sum of the electromotive forces (emf's).

Consider the network in Figure 3.1.

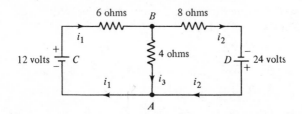

Figure 3.1

The symbol ——⊣|⊢—— represents a battery. We shall assume that current flows from the negative − to the positive +, as indicated in Figure 3.1. The symbol ——∧∧∧—— denotes a resistor. By Kirchhoff's first law, the sum of the currents at junction B is $i_1 - i_2 - i_3 = 0$. Also, at junction A we obtain the equivalent equation $-i_1 + i_2 + i_3 = 0$. According to Kirchhoff's second law, along the closed path $ACBA$ we have $6i_1 + 4i_3 = 12$, and along the closed path $BDAB$ we have $8i_2 - 4i_3 = 24$ (here we have also used Ohm's law that $V = IR$, where V is in volts, I is the current in amperes, and R is the resistance in ohms). The path $DACB$ gives the (redundant) equation $6i_1 + 8i_2 = 36$. To find the currents $i_1, i_2,$ and i_3, we solve the system

$$i_1 - i_2 - i_3 = 0$$
$$6i_1 + 4i_3 = 12$$
$$8i_2 - 4i_3 = 24$$
$$6i_1 + 8i_2 = 36$$

Solving the system, we see that the augmented matrix

$$\begin{bmatrix} 1 & -1 & -1 & 0 \\ 6 & 0 & 4 & 12 \\ 0 & 8 & -4 & 24 \\ 6 & 8 & 0 & 36 \end{bmatrix}$$

reduces to

$$\begin{bmatrix} 1 & 0 & 0 & \frac{30}{13} \\ 0 & 1 & 0 & \frac{36}{13} \\ 0 & 0 & 1 & -\frac{6}{13} \\ 0 & 0 & 0 & 0 \end{bmatrix}$$

The solutions are $i_1 = \frac{30}{13} = 2\frac{4}{13}$, $i_2 = 2\frac{10}{13}$, and $i_3 = \frac{-6}{13}$, where the unit of measure of current is the ampere. The negative value for i_3 indicates that the flow of current is opposite to the direction chosen.

Analysis of Traffic Flow

A network of streets in a large city consists of all one-way streets. We wish to analyze the traffic flow. The direction of traffic flow on each of the streets is given in Figure 3.2. Also, traffic counters have been set up at various

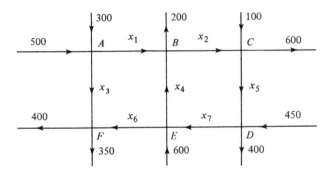

Figure 3.2

points and the average number of cars passing these points in a 1-h time period is given in Figure 3.2. The variables $x_1, x_2, \ldots, x_6$ and x_7 represent the number of cars per hour passing from intersection A to intersection B, from intersection B to intersection C, etc.

We first determine the possible values for each x_i. Assuming that there is no stoppage of traffic flow, the number of cars into an intersection must equal the number of cars exiting from an intersection. From this assumption we obtain the linear system

$$x_1 + x_3 = 800 \quad \text{(traffic flow at intersection } A\text{)}$$
$$x_1 - x_2 + x_4 = 200 \quad \text{(traffic flow at intersection } B\text{)}$$
$$x_2 - x_5 = 500 \quad \text{(traffic flow at intersection } C\text{)}$$
$$x_3 + x_6 = 750 \quad \text{(traffic flow at intersection } F\text{)}$$
$$x_4 + x_6 - x_7 = 600 \quad \text{(traffic flow at intersection } E\text{)}$$
$$-x_5 + x_7 = 50 \quad \text{(traffic flow at intersection } D\text{)}$$

Using the Gauss–Jordan method of reduction, the augmented matrix of this system reduces to

$$\begin{bmatrix} 1 & 0 & 0 & 0 & 0 & -1 & 0 & \vdots & 50 \\ 0 & 1 & 0 & 0 & 0 & 0 & -1 & \vdots & 450 \\ 0 & 0 & 1 & 0 & 0 & 1 & 0 & \vdots & 750 \\ 0 & 0 & 0 & 1 & 0 & 1 & -1 & \vdots & 600 \\ 0 & 0 & 0 & 0 & 1 & 0 & -1 & \vdots & -50 \\ 0 & 0 & 0 & 0 & 0 & 0 & 0 & \vdots & 0 \end{bmatrix}$$

The corresponding system is

$$x_1 = x_6 + 50$$
$$x_2 = x_7 + 450$$
$$x_3 = -x_6 + 750$$
$$x_4 = -x_6 + x_7 + 600$$
$$x_5 = x_7 - 50$$
$$x_6 = x_6$$
$$x_7 = x_7$$

Negative values for the x_i are not allowed, because all the streets are one way, and a negative x_i would be interpreted as the number of cars going in the wrong direction. With this restriction, we have $x_3 = 750 - x_6 \geq 0$. This means that $x_6 \leq 750$. Also, $x_5 = x_7 - 50 \geq 0$, and therefore $x_7 \geq 50$.

Now suppose that the road from D to E is going to be under construction and because of this we want the traffic along this stretch to be a minimum. This means that $x_7 = 50$. Consequently, $x_2 = 500$ and $x_5 = 0$. Conversely, if $x_5 = 0$, we have $x_7 = 50$. Therefore, closing off the stretch of road from C to D will result in the minimum flow of traffic on the road from D to E. The traffic flows x_1, x_3, x_4, and x_6 are not uniquely determined. If the entire stretch of road from D to F were under construction, we would want x_6 to be the minimum value of zero. In this case, $x_1 = 50$, $x_3 = 750$, and $x_4 = 650$.

A Leontief Input–Output Model

Suppose that a simple economy has three industries that are dependent on each other, but are not dependent on industries from the outside (closed model). The three industries are agriculture, building, and clothing. The fraction of each commodity consumed by each of the industries is given by

		Production		
		agriculture	building	clothing
	agriculture	$\frac{7}{16}$	$\frac{3}{6}$	$\frac{3}{16}$
Consumption	building	$\frac{5}{16}$	$\frac{1}{6}$	$\frac{5}{16}$
	clothing	$\frac{4}{16}$	$\frac{2}{6}$	$\frac{8}{16}$

The entry a_{ij} in the ith row and jth column denotes the fraction of goods produced by the people working in industry j (input) that is consumed by the people working in industry i (output). For example, the clothing industry uses $\frac{4}{16}$ of the total agriculture production, and the agriculture industry uses $\frac{3}{6}$ of the total production of the building industry.

Suppose that the incomes to the agriculture, building, and clothing industries are P_1, P_2, and P_3, respectively.

Problem. Under the equilibrium condition that output (expense due to consumption) is equal to input (income due to sales of the product), determine the incomes of each sector of the economy.

Solution: We obtain the linear system

$$\tfrac{7}{16}P_1 + \tfrac{3}{6}P_2 + \tfrac{3}{16}P_3 = P_1 \quad \text{(expense to ag. industry} = \text{income}$$
$$\text{to ag. industry)}$$

$$\tfrac{5}{16}P_1 + \tfrac{1}{6}P_2 + \tfrac{5}{16}P_3 = P_2 \quad \text{(expense to bldg. industry} = \text{in-}$$
$$\text{come to bldg. industry)}$$

$$\tfrac{4}{16}P_1 + \tfrac{1}{3}P_2 + \tfrac{1}{2}P_3 = P_3 \quad \text{(expense to clothing industry}$$
$$= \text{income to clothing industry)}$$

The coefficient matrix of the equivalent *homogeneous* system is

$$B = \begin{bmatrix} -\frac{9}{16} & \frac{1}{2} & \frac{3}{16} \\ \frac{5}{16} & -\frac{5}{6} & \frac{5}{16} \\ \frac{1}{4} & \frac{1}{3} & -\frac{1}{2} \end{bmatrix}$$

Using a variation of the Gauss–Jordan method, we reduce this matrix as follows:

$$B \xrightarrow[\substack{48R_2 \\ 12R_3}]{16R_1} \begin{bmatrix} -9 & 8 & 3 \\ 15 & -40 & 15 \\ 3 & 4 & -6 \end{bmatrix} \xrightarrow[-5R_3+R_2]{3R_3+R_1} \begin{bmatrix} 0 & 20 & -15 \\ 0 & -60 & 45 \\ 3 & 4 & -6 \end{bmatrix}$$

$$\xrightarrow[\substack{1/5R_1 \\ -R_1+R_3}]{3R_1+R_2} \begin{bmatrix} 0 & 4 & -3 \\ 0 & 0 & 0 \\ 3 & 0 & -3 \end{bmatrix} \xrightarrow[\substack{R_{12} \\ 1/3R_1 \\ 1/4R_2}]{R_{23}} \begin{bmatrix} 1 & 0 & -1 \\ 0 & 1 & -\frac{3}{4} \\ 0 & 0 & 0 \end{bmatrix}$$

The corresponding system is

$$P_1 = P_3$$
$$P_2 = \tfrac{3}{4}P_3$$
$$P_3 = P_3$$

Setting $P_3 = 4k$, we see that any solution (P_1, P_2, P_3) is of the form $k(4, 3, 4)$. Thus we see that there are infinitely many solutions. However, the incomes of the agriculture, building, and clothing industries are always in the ratio of $4:3:4$, respectively. [The preceding example is adapted from Ben Noble,

"Input–Output (Leontief Closed) Models," in *Proceedings, Summer Conference for College Teachers on Applied Mathematics*, University of Missouri, Rolla, Missouri, 1971 (supported by the National Science Foundation), pp. 111–113.]

The Leontief input–output (closed) model can be extended to n industries. The basic properties are that:

1. The matrix has entries a_{ij}, where $0 \leq a_{ij} \leq 1$.
2. The sum of the entries in any column is 1.
3. The equilibrium condition is satisfied.

Exercise 3.3

1. Calculate the currents i_1, i_2, and i_3 for the electrical network shown.

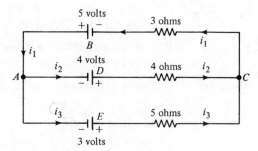

2. Calculate the currents in the given network.

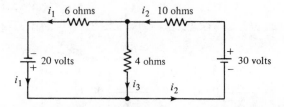

3. In the control of a certain plant disease, three chemicals are used in the following proportions: 10 units of chemical A, 12 units of chemical B, and 8 units of chemical C. Brand X, brand Y, and brand Z are commercial sprays sold on the market. One gallon of brand X contains each of the chemicals A, B, and C in the amounts 1, 2, and 1 units, respectively. One gallon of brand Y contains the chemicals in the amounts 2, 1, and 3 units, respectively, and 1 gallon of brand Z contains the chemicals in the amounts 3, 2, and 1 units, respectively. How much of each type of spray should be used to spread the chemicals in the exact amounts needed to control the disease?

4. The accompanying diagram indicates a network of one-way streets with traffic flow in the directions indicated. The traffic count is given as the average number of cars per hour.

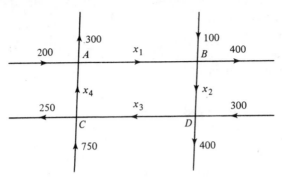

Assuming that the traffic flow into an intersection is equal to the traffic flow out of the intersection, construct a mathematical model of the traffic flow. If the road from C to A were placed under construction, what would be the minimum allowable traffic on this road if traffic is to continue flowing? How could this minimum be attained?

5. A trucking firm owns three different types of trucks, types A, B, and C. The trucks are equipped to haul different types of heavy machinery, type 1 and type 2. The number of machines of each type that the trucks can haul is

$$\text{Machines} \begin{cases} \text{type 1} \\ \text{type 2} \end{cases} \begin{array}{ccc} \overbrace{\text{type } A \quad \text{type } B \quad \text{type } C}^{\text{Trucks}} \\ \begin{bmatrix} 2 & 1 & 1 \\ 0 & 1 & 2 \end{bmatrix} \end{array}$$

The firm obtains an order for 32 machines of type 1 and 10 machines of type 2. Find the number of trucks of each type that can be used to fill the order, assuming that each truck must be fully loaded and that the exact number of machines is delivered. If it costs the firm the same amount of money to operate each of the three types of trucks, what would be the most economical solution for the firm?

6. Suppose that a simple economy has the four industries agriculture, building, clothing, and transportation, and that the conditions for the Leontief closed model are satisfied. The inputs and outputs are given by the matrix

$$\begin{array}{c} \\ \text{agriculture} \\ \text{building} \\ \text{clothing} \\ \text{transportation} \end{array} \begin{array}{cccc} \text{agricuture} \quad \text{building} \quad \text{clothing} \quad \text{transportation} \\ \begin{bmatrix} \frac{1}{3} & \frac{4}{9} & \frac{1}{3} & \frac{1}{3} \\ \frac{1}{3} & \frac{1}{3} & \frac{1}{6} & \frac{1}{6} \\ \frac{1}{4} & \frac{1}{9} & \frac{1}{4} & \frac{1}{4} \\ \frac{1}{12} & \frac{1}{9} & \frac{1}{4} & \frac{1}{4} \end{bmatrix} \end{array}$$

Suppose that the incomes to the agriculture, building, clothing, and transportation industries are P_1, P_2, P_3, and P_4, respectively. Assuming that the equilibrium condition is satisfied, determine the incomes of each sector of the economy.

7. The same as Problem 6 with the inputs and outputs given by

	agriculture	building	clothing	transportation
agriculture	$\frac{1}{2}$	$\frac{3}{8}$	$\frac{7}{16}$	$\frac{4}{16}$
building	$\frac{1}{4}$	$\frac{1}{4}$	$\frac{4}{16}$	$\frac{4}{16}$
clothing	$\frac{1}{8}$	$\frac{1}{4}$	$\frac{2}{16}$	$\frac{4}{16}$
transportation	$\frac{1}{8}$	$\frac{2}{16}$	$\frac{3}{16}$	$\frac{4}{16}$

Computer-Related Topics

As was mentioned in Section 3.1, for solving systems of equations a method called *Gaussian elimination* is more efficient than the Gauss–Jordan method because fewer arithmetic steps need to be performed. It is this method that is used by high-speed computers to solve large systems. In Example 3.2, we illustrated the Gaussian elimination method by solving the system

$$3y - 3z = 6$$
$$x - y + 4z = -3$$
$$x + 6z = 4$$

In that example we reduced the augmented matrix to

$$\begin{bmatrix} 1 & -1 & 4 & \vdots & -3 \\ 0 & 1 & -1 & \vdots & 2 \\ 0 & 0 & 1 & \vdots & \frac{5}{3} \end{bmatrix}$$

This last matrix is said to be in *row echelon form*. It satisfies all the requirements of Definition 3.9 except part (d). The corresponding system is

$$x - y + 4z = -3$$
$$y - z = 2$$
$$z = \frac{5}{3}$$

As mentioned earlier, this system can be solved by *back substitution*. We see that if $\frac{5}{3}$ is substituted for z in the second equation of this system, we obtain $y = \frac{11}{3}$. If these values for z and y are substituted into the first equation, we get $x = -6$. The back substitution is equivalent to applying the following operations on the row echelon matrix.

$$\begin{bmatrix} 1 & -1 & 4 & \vdots & -3 \\ 0 & 1 & -1 & \vdots & 2 \\ 0 & 0 & 1 & \vdots & \frac{5}{3} \end{bmatrix} \xrightarrow[-4R_3+R_1]{R_3+R_2} \begin{bmatrix} 1 & -1 & 0 & \vdots & -\frac{29}{3} \\ 0 & 1 & 0 & \vdots & \frac{11}{3} \\ 0 & 0 & 1 & \vdots & \frac{5}{3} \end{bmatrix} \xrightarrow{R_2+R_1} \begin{bmatrix} 1 & 0 & 0 & \vdots & -6 \\ 0 & 1 & 0 & \vdots & \frac{11}{3} \\ 0 & 0 & 1 & \vdots & \frac{5}{3} \end{bmatrix}$$

$$\begin{array}{ccc} x - y + 4z = -3 & x - y = -\frac{29}{3} & x = -6 \\ y - z = 2 & y = \frac{11}{3} & y = \frac{11}{3} \\ z = \frac{5}{3} & z = \frac{5}{3} & z = \frac{5}{3} \end{array}$$

We see that sweeping out column 3 is equivalent to substituting $\frac{5}{3}$ for z in the first and second equations, and sweeping out column 2 is equivalent to substituting $\frac{11}{3}$ for y in the "new" first equation.

In applying the last row operation, arithmetic computation was necessary to change columns 2 and 4. However, *no computation was necessary to change column* 3 because of the zero obtained in the preceding step. It is precisely for this reason that Gaussian elimination results in fewer arithmetic steps than the Gauss–Jordan method. For large n this can result in a significant difference in the time required for a computer to solve the problem.

Another factor to consider in machine calculations is the error due to rounding off. As illustrations of how serious this can be, we give the following examples.

Example 3.6.　Consider the system

$$x_1 + x_2 = 1$$
$$x_1 + 1.025x_2 = 6.$$

The unique solutions are $x_1 = -199$ and $x_2 = 200$. If the computer rounds up to three significant figures the system becomes

$$x_1 + x_2 = 1$$
$$x_1 + 1.03x_2 = 6$$

The actual solutions to this system are $x_1 = -165\frac{2}{3}$ and $x_2 = 166\frac{2}{3}$, which are quite different from the original solutions.

Example 3.7.　Consider the systems

$$x_1 + 1.03x_2 = 4$$
$$x_1 + 1.025x_2 = 4 \tag{1}$$

and

$$x_1 + 1.03x_2 = 4$$
$$x_1 + 1.025x_2 = 3 \tag{2}$$

System (1) has the unique solution $(x_1, x_2) = (4, 0)$; system (2) has the unique solution $(x_1, x_2) = (-202, 200)$. If, as in Example 3.6, the computer rounds up to three significant figures, system (1) becomes

$$x_1 + 1.03x_2 = 4$$
$$x_1 + 1.03x_2 = 4 \tag{1'}$$

which has infinitely many solutions of the form $(4 - 1.03k, k)$. System (2) becomes

$$x_1 + 1.03x_2 = 4$$
$$x_1 + 1.03x_2 = 3 \tag{2'}$$

which is inconsistent and has no solutions. Thus it is possible for round-off error to change the entire character of the solutions, for we passed from a system (1) having a unique solution to a system (1') having infinitely many solutions, and from the system (2) having a unique solution to the system (2'), which is inconsistent.

Systems such as those in Examples 3.6 and 3.7 that are very sensitive to small changes in the coefficients are called *ill-conditioned systems*. For a discussion of how this problem is handled, see Ben Noble, *Applied Linear Algebra* (Englewood Cliffs, N.J.: Prentice-Hall, Inc., 1969), pp. 231–241.

Exercise 3.3

8. Solve Problem 7 of Exercise 3.1 by using the Gaussian elimination method.

9. Solve Problem 11 of Exercise 3.1 by using the Gaussian elimination method.

10. Given the system $x_1 + \frac{1}{2}x_2 = 5$, $\frac{1}{2}x_1 + \frac{1}{3}x_2 = 4$.
   (a) Find the exact solution(s) to this system.
   (b) Represent in decimal form to three places the coefficients in the given system. Find the exact solution(s) of this new system.
   (c) Compare the solutions. Is the given system an ill-conditioned system?

11. Given the system $x_1 + x_2 = 3$, $x_1 + 1.0001x_2 = 2.9990$.
   (a) Find the exact solution(s) to this system.
   (b) Round off the coefficients in the given system to four places and find the exact solutions to the new system.
   (c) Compare the solutions. Is the given system an ill-conditioned system?

12. If a computer takes 25×10^{-6} s to perform a single arithmetic calculation, determine approximately how long in hours it would take the computer to solve a system of n linear equations and n unknowns by
   (a) The Gaussian elimination method if there are $n^3/3$ arithmetic calculations and $n = 1000$.
   (b) The Gauss–Jordan elimination method if there are $n^3/2$ arithmetic calculations and $n = 1000$.

13. Given the system $x_1 + x_2 = -2$, $x_1 + 1.015x_2 = 1$.
   (a) Solve the given system.
   (b) Round 1.015 off of 1.02 and solve the new system.
   (c) Round 1.015 off to 1.01 and solve the new system.
   (d) Is the given system ill-conditioned?

3.4 ADDITION AND SCALAR MULTIPLICATION OF MATRICES

We have already introduced matrices as a convenient notation to use in obtaining solutions to a system of linear equations. As we shall see later, there are many other uses for matrices. In this section we shall define addition for matrices of the same size and also multiplication of a matrix by a scalar.

These operations will be defined in such a way that the resulting mathematical system is a vector space.

When referring to a specific matrix, it is necessary to list all the entries and display the entire array. However, if the entries are not specified such as in

$$A = \begin{bmatrix} a_{11} & a_{12} & \cdots & a_{1n} \\ a_{21} & a_{22} & \cdots & a_{2n} \\ \cdot & \cdot & & \cdot \\ \cdot & \cdot & & \cdot \\ \cdot & \cdot & & \cdot \\ a_{m1} & a_{m2} & \cdots & a_{mn} \end{bmatrix}$$

it is convenient to write the abbreviation

$$A = [a_{ij}]_{(m,n)}$$

Here a_{ij} denotes the entry in the ith row and jth column and (m, n) indicates that the matrix has m rows and n columns. If the number of rows m and the number of columns n is either understood or not important we simply write $A = [a_{ij}]$. If $m = n$, the matrix is a **square matrix**, and we say that the matrix has **order** n. A matrix is called a **real matrix** if the entries are real numbers. *We shall assume throughout that all matrices under consideration are real matrices unless it is explicitly stated otherwise.*

An $m \times 1$ matrix is called a **column matrix** or a **column vector** because it consists of a single column. Likewise, a $1 \times n$ matrix is called a **row matrix** or a **row vector**.

Example 3.8. $B = \begin{bmatrix} 1 & 2 & 3 & -7 \end{bmatrix}$ is a 1×4 row vector, and $C = \begin{bmatrix} 2 \\ -3 \\ 0 \end{bmatrix}$

is a 3×1 column vector.

Definition 3.12. If $A = [a_{ij}]$ and $B = [b_{ij}]$ are $m \times n$ matrices, then $A + B$ is defined to be the $m \times n$ matrix

$$\mathbf{A} + \mathbf{B} = [\mathbf{c_{ij}}]$$

where $\mathbf{c_{ij}} = \mathbf{a_{ij}} + \mathbf{b_{ij}}$.

Important: Addition of matrices is defined only for matrices that have the same number of rows and columns.

Example 3.9

$$\begin{bmatrix} 1 & 2 & -3 \\ 4 & 6 & 8 \end{bmatrix} + \begin{bmatrix} -1 & 4 & 1 \\ 1 & 3 & -5 \end{bmatrix} = \begin{bmatrix} 0 & 6 & -2 \\ 5 & 9 & 3 \end{bmatrix}$$

$$\begin{bmatrix} \frac{1}{2} & 3 \\ 6 & -2 \end{bmatrix} + \begin{bmatrix} \frac{1}{2} & 1 \\ 2 & 7 \end{bmatrix} = \begin{bmatrix} 1 & 4 \\ 8 & 5 \end{bmatrix}$$

and

$$\begin{bmatrix} 1 & 2 & 3 \\ 4 & 5 & 6 \end{bmatrix} + \begin{bmatrix} 7 & 8 \\ 9 & 0 \end{bmatrix}$$

is undefined.

It is an easy exercise to prove that under this definition of addition the collection $\mathbf{R}_{m \times n}$ of all real m by n matrices satisfies the addition properties 1A–5A mentioned in Chapter 1. We illustrate by verifying properties 2A, 4A, and 5A in the following examples. The proofs of 1A and 3A are left as Problems 10 and 11, Exercise 3.4.

Example 3.10.   Prove that for any $m \times n$ matrices $A = [a_{ij}]$, and $B = [b_{ij}]$, $A + B = B + A$.

Proof: Let $A + B = C = [c_{ij}]$ and $B + A = D = [d_{ij}]$. By Definition 3.12, C and D are both $m \times n$ matrices. Also, $c_{ij} = a_{ij} + b_{ij}$ and $d_{ij} = b_{ij} + a_{ij}$. Since addition of real numbers is commutative, we have $c_{ij} = d_{ij}$. Hence $C = D$. Q.E.D.

Example 3.11.   Prove that $\mathbf{R}_{m \times n}$ has an additive identity.

Proof: Let $A = [a_{ij}]$ be any $m \times n$ matrix. Define the matrix $Z = [z_{ij}]$ by $z_{ij} = 0$. From the definition of addition, $A + Z = [c_{ij}]$, where $c_{ij} = a_{ij} + z_{ij} = a_{ij} + 0 = a_{ij}$. Therefore, $A + Z = A$. $A + Z = Z + A$ from Example 3.10. Hence Z is an additive identity. Q.E.D.

Example 3.11 shows that the $m \times n$ matrix in which every entry is 0 is an additive identity for $\mathbf{R}_{m \times n}$. It can be shown that no other $m \times n$ matrix can have this property (Problem 18, Exercise 3.4). This matrix is called the **zero** matrix and is denoted by **0**. If it is important to emphasize the size we denote it by $\mathbf{0}_{(m, n)}$.

Example 3.12.   Prove that every $m \times n$ matrix $A = [a_{ij}]$ has an additive inverse.

Proof: Given the $m \times n$ matrix $A = [a_{ij}]$, define $B = [b_{ij}]$ by $b_{ij} = -a_{ij}$. Then $a_{ij} + b_{ij} = 0$. Hence $A + B = \mathbf{0} = B + A$. Q.E.D.

The additive inverse of a matrix is unique (Problem 19, Exercise 3.4) and is denoted by $-A$. We can now define subtraction of $m \times n$ matrices.

Definition 3.13.   For any matrices A and B in $\mathbf{R}_{m \times n}$, **subtraction** is defined by $A - B = A + (-B)$. In words, A minus B is A plus the additive inverse of B. Our next step is to define scalar multiplication of matrices.

Definition 3.14.   For any $m \times n$ matrix $A = [a_{ij}]$ and any real number r, rA is defined to be the matrix $[ra_{ij}]$ obtained by multiplying each entry of A by r.

Example 3.13.   If

$$
A = \begin{bmatrix} 2 & 8 & 6 & 1 \\ 4 & -6 & -1 & 8 \\ 0 & 2 & -8 & -4 \end{bmatrix} \quad \text{and} \quad r = \tfrac{1}{2}
$$

then

$$
rA = \begin{bmatrix} 1 & 4 & 3 & \tfrac{1}{2} \\ 2 & -3 & -\tfrac{1}{2} & 4 \\ 0 & 1 & -4 & -2 \end{bmatrix}
$$

The multiplication in Definition 3.14 is called **multiplication of a matrix by a scalar** or simply **scalar multiplication.** This scalar multiplication satisfies 6M–10M. Thus, according to Definition 1.8, $\mathbf{R}_{m \times n}$ together with addition and scalar multiplication as defined in Definitions 3.12 and 3.14 is a vector space. We shall verify property 7M in the next example and leave the proofs of 6M and 8M–10M as Problems 12–15, Exercise 3.4.

Example 3.14.   Prove that, for any scalars r and s and for any $m \times n$ matrix $A = [a_{ij}]$, $(rs)A = r(sA) = s(rA)$.

Proof: Let $B = (rs)A$, $C = r(sA)$, and $D = s(rA)$. We first note that B, C, and D are all $m \times n$ matrices. Also $b_{ij} = (rs)a_{ij}$, $c_{ij} = r(sa_{ij})$ (applying Definition 3.14 twice), and $d_{ij} = s(ra_{ij})$. Since real number multiplication is both associative and commutative, it follows that $b_{ij} = c_{ij} = d_{ij}$. Hence $B = C = D$.   Q.E.D.

We state the results of Examples 3.10–3.12, 3.14, and Problems 10–15, Exercise 3.4, as the next theorem.

Theorem 3.5.   The collection $\mathbf{R}_{m \times n}$ of all $m \times n$ real matrices with addition and scalar multiplication as defined in Definitions 3.12 and 3.14 is a vector space.

We close this section with some definitions. The **main** or **principal diagonal** of a matrix $A = [a_{ij}]_{(m, n)}$ is the ordered set of entries $\{a_{11}, a_{22}, \ldots, a_{kk}\}$, where $k = \text{minimum } \{m, n\}$. If A is a square matrix, the principal diagonal is basically the line of entries from the upper left corner of the matrix to the lower right corner.

Definition 3.15.   The **transpose** of a matrix $A = [a_{ij}]_{(m, n)}$ is the matrix A^T, where $A^T = [b_{ij}]_{(n, m)}$ and $b_{ij} = a_{ji}$. In case $m = n$, it can happen that $A^T = A$. A square matrix A such that $A^T = A$ is called a **symmetric** matrix.

Example 3.15.   The main diagonal of

$$A = \begin{bmatrix} 1 & 2 & 3 & 4 \\ 5 & 6 & 7 & 8 \\ 9 & 0 & 8 & 6 \end{bmatrix}$$

is the ordered set $\{1, 6, 8\}$. The transpose of A is

$$A^T = \begin{bmatrix} 1 & 5 & 9 \\ 2 & 6 & 0 \\ 3 & 7 & 8 \\ 4 & 8 & 6 \end{bmatrix}$$

The main diagonal of the square matrix

$$B = \begin{bmatrix} 1 & 2 & 3 \\ 4 & 5 & -2 \\ 6 & 3 & 7 \end{bmatrix}$$

is the ordered set $\{1, 5, 7\}$.

$$B^T = \begin{bmatrix} 1 & 4 & 6 \\ 2 & 5 & 3 \\ 3 & -2 & 7 \end{bmatrix}$$

The matrix

$$C = \begin{bmatrix} 1 & 3 & 2 & 4 \\ 3 & -2 & -1 & 6 \\ 2 & -1 & 5 & 0 \\ 4 & 6 & 0 & 1 \end{bmatrix}$$

is symmetric because $C^T = C$. The term symmetric is used because geometrically the matrix is symmetric about the main diagonal. The main diagonal of C is the ordered set $\{1, -2, 5, 1\}$. Note that, in the ordered set, 1 is listed twice. If the set were not ordered, there would be no distinction between this set and $\{1, -2, 5\}$. [For a precise definition of ordered set, see Paul R. Halmos, *Naive Set Theory*, (New York: Van Nostrand Reinhold Company, 1960).]

Exercise 3.4

1. If $A = \begin{bmatrix} 1 & 3 & 8 & 7 \\ 2 & 4 & 5 & 6 \end{bmatrix}$,

    (a) What is the main diagonal of A?
    (b) Write A^T.

2. If $A = \begin{bmatrix} 2 & 0 & -1 & 3 \\ 3 & 1 & 4 & 6 \\ -5 & 7 & 2 & 1 \end{bmatrix}$,

(a) What is the main diagonal of A?

(b) Write A^T.

3. Find $A + B$, where

(a) $A = \begin{bmatrix} 1 & 2 & -1 & 1 \\ 2 & 3 & 0 & 2 \\ 2 & 2 & -1 & -4 \end{bmatrix}$, $\quad B = \begin{bmatrix} -2 & 0 & 1 & 2 \\ 3 & 2 & 0 & -1 \\ -1 & -1 & 1 & 0 \end{bmatrix}$.

(b) $A = \begin{bmatrix} 1 & 2 \\ -3 & 4 \end{bmatrix}$, $\quad B = \begin{bmatrix} -3 & 1 \\ 0 & 1 \end{bmatrix}$.

4. With A and B as in Problem 3(a), find (a) $3A$; (b) $2B$; (c) $3A + 2B$.

5. With A and B as in Problem 3(b), find (a) $0A$; (b) $3B$; (c) $4A$; (d) $3B + 4A$.

6. If $A = \begin{bmatrix} 1 & 3 & 0 \\ -2 & 1 & 4 \end{bmatrix}$, $B = \begin{bmatrix} 2 & -1 & 1 \\ 3 & 7 & 2 \end{bmatrix}$, and $C = \begin{bmatrix} -3 & 0 & 5 \\ -2 & 4 & 6 \end{bmatrix}$, find (a) $A - B$; (b) $A + 2C - 3B$; (c) $2A - 3B + 4C$; (d) $3A - 2C$.

7. If A is symmetric and $A = \begin{bmatrix} -1 & \_ & \_ \\ 3 & 2 & 5 \\ -4 & \_ & 2 \end{bmatrix}$ fill in the missing entries.

8. If A is symmetric and $A = \begin{bmatrix} 21 & 3 & 7 & 4 \\ \_ & 5 & 6 & \_ \\ \_ & \_ & 8 & 2 \\ \_ & 1 & \_ & 9 \end{bmatrix}$ fill in the missing entries.

9. (a) Write out the matrices $[a_{ij}]_{(n,n)}$, $[a_{rs}]_{(n,n)}$, and $[a_{ji}]_{(n,n)}$, where $n = 2$. What do you conclude about the matrices $[a_{ij}]_{(n,n)}$ and $[a_{ji}]_{(n,n)}$?

(b) Given a matrix $A = [a_{ij}]_{(n,n)}$, is there a difference between the matrix B defined by $B = [b_{ij}]_{(n,n)}$, where $b_{ij} = a_{ji}$ and the matrix C defined by $C = [a_{ji}]_{(n,n)}$? Find B and C if $A = \begin{bmatrix} 1 & 2 & 3 \\ 4 & 5 & 6 \\ 7 & 8 & 9 \end{bmatrix}$.

10. Verify property 1A for addition of matrices in $\mathbf{R}_{m \times n}$; that is, show that if A and B are in $\mathbf{R}_{m \times n}$, the sum $A + B$ is a uniquely determined matrix in $\mathbf{R}_{m \times n}$.

11. Verify property 3A for addition of matrices in $\mathbf{R}_{m \times n}$.

In Problems 12–15, the matrices are in $\mathbf{R}_{m \times n}$ and scalar multiplication is defined as in Definition 3.14.

12. Verify property 6M; that is, show that rA is a uniquely determined matrix in $\mathbf{R}_{m \times n}$ for any scalar r and matrix A.

13. Verify property 8M; that is, $(r + s)A = rA + sA$.

14. Verify property 9M; that is, $r(A + B) = rA + rB$.

15. Verify property 10M; that is, $1A = A$.

16. Verify that $(A^T)^T = A$.

17. Prove that the additive identity in $\mathbf{R}_{m \times n}$ is unique; that is, show that if $\mathbf{0}$ and $\mathbf{0}'$ are additive identities for $\mathbf{R}_{m \times n}$ then $\mathbf{0} = \mathbf{0}'$.

18. Prove that the additive inverse of a matrix A in $\mathbf{R}_{m \times n}$ is unique; that is, if B and C are additive inverses for A, then $B = C$.

   Note: The properties in Problems 17 and 18 are true of any mathematical system satisfying 1A–5A, and the proofs depend only on these properties.

19. *Prove:* If A and B are $m \times n$ matrices, then $(A + B)^T = A^T + B^T$.

20. *Prove:* If A is an $m \times n$ matrix and r is a scalar, then $(rA)^T = rA^T$.

3.5 MULTIPLICATION OF MATRICES AND APPLICATIONS

To motivate the definition of product of matrices, we begin by considering a system of m equations and n unknowns:

$$a_{11}x_1 + a_{12}x_2 + \cdots + a_{1n}x_n = y_1$$
$$a_{21}x_1 + a_{22}x_2 + \cdots + a_{2n}x_n = y_2$$
$$\cdots$$
$$a_{m1}x_1 + a_{m2}x_2 + \cdots + a_{mn}x_n = y_m$$

The constants on the right side of the system can be represented by the column matrix

$$Y = \begin{bmatrix} y_1 \\ y_2 \\ \cdot \\ \cdot \\ \cdot \\ y_m \end{bmatrix}$$

Let

$$X = \begin{bmatrix} x_1 \\ x_2 \\ \cdot \\ \cdot \\ \cdot \\ x_n \end{bmatrix}$$

be the column matrix of unknowns, and let $A = [a_{ij}]_{(m,n)}$ be the coefficient matrix of the system. If we define a matrix product AX in such a way that the matrix equation $AX = Y$ is equivalent to the given system, then the product AX should be defined by

$$AX = \begin{bmatrix} a_{11} & a_{12} & \cdots & a_{1n} \\ a_{21} & a_{22} & \cdots & a_{2n} \\ \cdot & \cdot & & \cdot \\ \cdot & \cdot & & \cdot \\ \cdot & \cdot & & \cdot \\ a_{m1} & a_{m2} & \cdots & a_{mn} \end{bmatrix} \begin{bmatrix} x_1 \\ x_2 \\ \cdot \\ \cdot \\ \cdot \\ x_n \end{bmatrix} = \begin{bmatrix} a_{11}x_1 + a_{12}x_2 + \cdots + a_{1n}x_n \\ a_{21}x_1 + a_{22}x_2 + \cdots + a_{2n}x_n \\ \cdot \\ \cdot \\ \cdot \\ a_{m1}x_1 + a_{m2}x_2 + \cdots + a_{mn}x_n \end{bmatrix}$$

The $m \times 1$ matrix on the right is equal to Y if and only if X is a column vector such that its entries satisfy the given system of linear equations. Thus the matrix equation $AX = Y$ is equivalent to the given system. We now have a reasonable way to define the product AX of an $m \times n$ matrix A and an $n \times 1$ matrix X.

Example 3.16. If

$$A = \begin{bmatrix} 1 & 2 & -1 & 0 \\ 2 & 0 & -2 & 3 \\ 5 & -2 & 3 & -4 \end{bmatrix}, \quad C = \begin{bmatrix} 2 & 1 & 0 & -1 \\ 3 & 4 & 2 & 0 \end{bmatrix}, \quad D = \begin{bmatrix} 1 \\ 2 \\ -1 \\ 3 \end{bmatrix}$$

and multiplication is defined as suggested above, find AD and CD.

Solution

$$AD = \begin{bmatrix} 1(1) + 2(2) + (-1)(-1) + 0(3) \\ 2(1) + 0(2) + (-2)(-1) + 3(3) \\ 5(1) + (-2)(2) + 3(-1) + (-4)(3) \end{bmatrix} = \begin{bmatrix} 6 \\ 13 \\ -14 \end{bmatrix}$$

$$CD = \begin{bmatrix} 2(1) + 1(2) + 0(-1) + (-1)(3) \\ 3(1) + 4(2) + 2(-1) + 0(3) \end{bmatrix} = \begin{bmatrix} 1 \\ 9 \end{bmatrix}$$

Next we consider a motivation for defining a more general product of matrices AB. Suppose that we have the systems of linear equations

$$\begin{aligned} x_1 &= a_{11}y_1 + a_{12}y_2 \\ x_2 &= a_{21}y_1 + a_{22}y_2 \\ x_3 &= a_{31}y_1 + a_{32}y_2 \end{aligned} \tag{1}$$

$$\begin{aligned} y_1 &= b_{11}z_1 + b_{12}z_2 + b_{13}z_3 \\ y_2 &= b_{21}z_1 + b_{22}z_2 + b_{23}z_3 \end{aligned} \tag{2}$$

and we wish to express x_1, x_2, and x_3 as functions of z_1, z_2, and z_3. By direct substitution we obtain

$$\begin{aligned} x_1 &= a_{11}(b_{11}z_1 + b_{12}z_2 + b_{13}z_3) + a_{12}(b_{21}z_1 + b_{22}z_2 + b_{23}z_3) \\ x_2 &= a_{21}(b_{11}z_1 + b_{12}z_2 + b_{13}z_3) + a_{22}(b_{21}z_1 + b_{22}z_2 + b_{23}z_3) \\ x_3 &= a_{31}(b_{11}z_1 + b_{12}z_2 + b_{13}z_3) + a_{32}(b_{21}z_1 + b_{22}z_2 + b_{23}z_3) \end{aligned}$$

Expanding and collecting like terms, we get

$$x_1 = (a_{11}b_{11} + a_{12}b_{21})z_1 + (a_{11}b_{12} + a_{12}b_{22})z_2 + (a_{11}b_{13} + a_{12}b_{23})z_3$$
$$x_2 = (a_{21}b_{11} + a_{22}b_{21})z_1 + (a_{21}b_{12} + a_{22}b_{22})z_2 + (a_{21}b_{13} + a_{22}b_{23})z_3 \quad (3)$$
$$x_3 = (a_{31}b_{11} + a_{32}b_{21})z_1 + (a_{31}b_{12} + a_{32}b_{22})z_2 + (a_{31}b_{13} + a_{32}b_{23})z_3$$

Now let's try to reach the same result but by writing the systems as equivalent matrix equations. System (1) can be written as

$$\begin{bmatrix} x_1 \\ x_2 \\ x_3 \end{bmatrix} = \begin{bmatrix} a_{11} & a_{12} \\ a_{21} & a_{22} \\ a_{31} & a_{32} \end{bmatrix} \begin{bmatrix} y_1 \\ y_2 \end{bmatrix} \quad \text{or} \quad X = AY$$

System (2) can be written as

$$\begin{bmatrix} y_1 \\ y_2 \end{bmatrix} = \begin{bmatrix} b_{11} & b_{12} & b_{13} \\ b_{21} & b_{22} & b_{23} \end{bmatrix} \begin{bmatrix} z_1 \\ z_2 \\ z_3 \end{bmatrix} \quad \text{or} \quad Y = BZ$$

Since we wish to write the x's in terms of the z's, we substitute BZ for Y and obtain $X = A(BZ)$. If this matrix equation is going to be equivalent to the system of linear equations (3), and if $A(BZ)$ is to equal $(AB)Z$, then the matrix equation $X = A(BZ)$ should be of the form $X = CZ$, and $C = AB$ must be defined by

$$AB = \begin{bmatrix} a_{11} & a_{12} \\ a_{21} & a_{22} \\ a_{31} & a_{32} \end{bmatrix} \begin{bmatrix} b_{11} & b_{12} & b_{13} \\ b_{21} & b_{22} & b_{23} \end{bmatrix}$$

$$= \begin{bmatrix} a_{11}b_{11} + a_{12}b_{21} & a_{11}b_{12} + a_{12}b_{22} & a_{11}b_{13} + a_{12}b_{23} \\ a_{21}b_{11} + a_{22}b_{21} & a_{21}b_{12} + a_{22}b_{22} & a_{21}b_{13} + a_{22}b_{23} \\ a_{31}b_{11} + a_{32}b_{21} & a_{31}b_{12} + a_{32}b_{22} & a_{31}b_{13} + a_{32}b_{23} \end{bmatrix}$$

In our illustration, system (1) had the 3×2 coefficient matrix A, and system (2) had the 2×3 coefficient matrix B. After the substitution process, we came up with system (3), which had a 3×3 coefficient matrix that we called AB. More generally, if we start with systems of the form (1) and (2) having coefficient matrices $A_{(m,n)}$ and $B_{(n,r)}$, respectively, then the system obtained after the substitution process will have an $m \times r$ coefficient matrix, which we shall call AB.

Historically, this substitution process led to the definition of matrix multiplication. In 1858, the great English mathematician Arthur Cayley defined matrix multiplication so that this substitution process would work. The reader might be interested in consulting *The Collected Mathematical Papers of Arthur Cayley*, Volume II (Cambridge: At the University Press, 1889, pp. 475–479) and reading directly from Cayley's work.

Definition 3.16. Let $A = [a_{ij}]$ be an $m \times n$ matrix and let $B = [b_{ij}]$ be an $n \times r$ matrix. The **product** AB is defined to be the $m \times r$ matrix $AB = [c_{ij}]$, where the element c_{ij} in the ith row and jth column of AB is $c_{ij} = a_{i1}b_{1j} + a_{i2}b_{2j} + \cdots + a_{in}b_{nj}$ for $i = 1, 2, \ldots, m$ and $j = 1, 2, \ldots, r$. *This product is defined if and only if the number of columns in the matrix A is equal to the number of rows in the matrix B.*

We shall say that A and B are **conformable** for multiplication if the product AB is defined (i.e., if the number of columns of A is equal to the number of rows of B).

$$
\begin{bmatrix}
a_{11} & a_{12} & \cdots & a_{1n} \\
a_{21} & a_{22} & \cdots & a_{2n} \\
\cdot & \cdot & & \cdot \\
\cdot & \cdot & & \cdot \\
\cdot & \cdot & & \cdot \\
\boxed{a_{i1} \quad a_{i2} \quad \cdots \quad a_{in}} \\
\cdot & \cdot & & \cdot \\
\cdot & \cdot & & \cdot \\
\cdot & \cdot & & \cdot \\
a_{m1} & a_{m2} & \cdots & a_{mn}
\end{bmatrix}
\begin{bmatrix}
b_{11} & b_{12} & \cdots & \boxed{b_{1j}} & \cdots & b_{1r} \\
b_{21} & b_{22} & \cdots & b_{2j} & \cdots & b_{2r} \\
\cdot & \cdot & & \cdot & & \cdot \\
\cdot & \cdot & & \cdot & & \cdot \\
\cdot & \cdot & & \cdot & & \cdot \\
b_{n1} & b_{n2} & \cdots & b_{nj} & \cdots & b_{nr}
\end{bmatrix}
$$

$$
=
\begin{bmatrix}
c_{11} & c_{12} & \cdots & c_{1j} & \cdots & c_{1r} \\
c_{21} & c_{22} & \cdots & c_{2j} & \cdots & c_{2r} \\
\cdot & \cdot & & \cdot & & \cdot \\
\cdot & \cdot & & \cdot & & \cdot \\
\cdot & \cdot & & \cdot & & \cdot \\
c_{i1} & c_{i2} & \cdots & \boxed{c_{ij}} & \cdots & c_{ir} \\
\cdot & \cdot & & \cdot & & \cdot \\
\cdot & \cdot & & \cdot & & \cdot \\
\cdot & \cdot & & \cdot & & \cdot \\
c_{m1} & c_{m2} & \cdots & c_{mj} & \cdots & c_{mr}
\end{bmatrix}
$$

A mechanical procedure for computing c_{ij} is to use a finger on the left hand to run along the elements of the ith row of A and to use a finger on the right hand to run along the corresponding elements of the jth column of B, and then calculate the sum of the products as one proceeds across the ith row of A and down the jth column of B.

Example 3.17. Examples of matrix multiplication.

(a) $\begin{bmatrix} 3 & -1 & 4 & 0 \end{bmatrix} \begin{bmatrix} 1 \\ -2 \\ 3 \\ 1 \end{bmatrix} = [17]$ (b) $\begin{bmatrix} 1 & -1 & 0 & 2 \\ 0 & 1 & 3 & 4 \\ 2 & 0 & -1 & 3 \end{bmatrix} \begin{bmatrix} 2 \\ -1 \\ 0 \\ 3 \end{bmatrix} = \begin{bmatrix} 9 \\ 11 \\ 13 \end{bmatrix}$

(c) $\begin{bmatrix} 1 \\ 2 \\ -1 \\ 0 \\ 3 \end{bmatrix} \begin{bmatrix} 0 & -1 & 2 & 4 & -2 & 3 \end{bmatrix} = \begin{bmatrix} 0 & -1 & 2 & 4 & -2 & 3 \\ 0 & -2 & 4 & 8 & -4 & 6 \\ 0 & 1 & -2 & -4 & 2 & -3 \\ 0 & 0 & 0 & 0 & 0 & 0 \\ 0 & -3 & 6 & 12 & -6 & 9 \end{bmatrix}$

(d) $\begin{bmatrix} 1 & 0 & -1 & 2 \\ 3 & 2 & 1 & -2 \\ 4 & 2 & 0 & -3 \end{bmatrix} \begin{bmatrix} 1 & 3 \\ 0 & -1 \\ 2 & 2 \\ 1 & 0 \end{bmatrix} = \begin{bmatrix} 1 & 1 \\ 3 & 9 \\ 1 & 10 \end{bmatrix}$

Example 3.18. Given the systems of linear equations

$$x_1 = y_1 + 2y_2$$
$$x_2 = -y_1 + 4y_2 \qquad\qquad (i)$$
$$x_3 = 2y_1 + 3y_2$$

$$y_1 = z_1 + 2z_2 + z_3 + 2z_4$$
$$y_2 = 3z_1 - z_2 + 4z_3 - 5z_4 \qquad\qquad (ii)$$

express x_1, x_2, and x_3 in terms of z_1, z_2, z_3, and z_4.

Solution: We could of course solve the problem by directly sub-stituting the z's in system (ii) for the y's in system (i); however, the solution is easier to obtain using matrix multiplication. System (i) is equivalent to the matrix equation

$$\begin{bmatrix} x_1 \\ x_2 \\ x_3 \end{bmatrix} = \begin{bmatrix} 1 & 2 \\ -1 & 4 \\ 2 & 3 \end{bmatrix} \begin{bmatrix} y_1 \\ y_2 \end{bmatrix}$$

System (ii) is equivalent to the matrix equation

$$\begin{bmatrix} y_1 \\ y_2 \end{bmatrix} = \begin{bmatrix} 1 & 2 & 1 & 2 \\ 3 & -1 & 4 & -5 \end{bmatrix} \begin{bmatrix} z_1 \\ z_2 \\ z_3 \\ z_4 \end{bmatrix}$$

So

$$\begin{bmatrix} x_1 \\ x_2 \\ x_3 \end{bmatrix} = \left(\begin{bmatrix} 1 & 2 \\ -1 & 4 \\ 2 & 3 \end{bmatrix} \begin{bmatrix} 1 & 2 & 1 & 2 \\ 3 & -1 & 4 & -5 \end{bmatrix} \right) \begin{bmatrix} z_1 \\ z_2 \\ z_3 \\ z_4 \end{bmatrix} = \begin{bmatrix} 7 & 0 & 9 & -8 \\ 11 & -6 & 15 & -22 \\ 11 & 1 & 14 & -11 \end{bmatrix} \begin{bmatrix} z_1 \\ z_2 \\ z_3 \\ z_4 \end{bmatrix}$$

Hence

$$x_1 = 7z_1 + 9z_3 - 8z_4$$
$$x_2 = 11z_1 - 6z_2 + 15z_3 - 22z_4$$
$$x_3 = 11z_1 + z_2 + 14z_3 - 11z_4$$

We close this section with a definition that is really just a convenient notation.

Definition 3.17. If A is an $m \times n$ matrix, then A_i denotes the ith row of A and $A^{(j)}$ denotes the jth column of A.

Example 3.19.

If $A = \begin{bmatrix} 1 & 3 & 2 & 0 \\ -1 & 4 & 2 & 5 \\ 7 & 8 & 1 & 4 \end{bmatrix}$, then $A_3 = [7 \quad 8 \quad 1 \quad 4]$ and $A^{(4)} = \begin{bmatrix} 0 \\ 5 \\ 4 \end{bmatrix}$.

Exercise 3.5

In Problems 1 and 2, write each of the given linear systems as equivalent matrix equations.

1. (a) $x_1 + 3x_2 - x_3 + 5x_4 = 3$
$\qquad\qquad x_1 - 4x_3 = 1$
$\qquad x_1 + 7x_2 + x_3 + 2x_4 = 8$

(b) $3x + 2y - 4z = 9$
$\qquad x + 3y - 2z = 4$

2. (a) $3y_1 + 2y_2 - 4y_3 = 1$
$\qquad 8y_1 + 2y_3 = 4$
$\qquad 2y_1 - 3y_2 = 5$
$\qquad y_1 + 3y_2 - 5y_3 = 7$

(b) $2x + 3y + \quad w = 4$
$\qquad 4x - \quad y + 3w = 2$
$\qquad x + 4y - \quad w = 1$
$\qquad 3y - 4w = 6$

In Problems 3 and 4, write each of the matrix equations as equivalent systems of linear equations.

3. (a) $\begin{bmatrix} 2 & 3 & 1 & 0 \\ -1 & 0 & 3 & 4 \\ 2 & 1 & 5 & 4 \end{bmatrix} \begin{bmatrix} x_1 \\ x_2 \\ x_3 \\ x_4 \end{bmatrix} = \begin{bmatrix} 5 \\ 8 \\ 3 \end{bmatrix}$

(b) $\begin{bmatrix} 2 & 3 \\ 4 & 7 \end{bmatrix} \begin{bmatrix} x \\ y \end{bmatrix} = \begin{bmatrix} -1 \\ 6 \end{bmatrix}$

4. (a) $\begin{bmatrix} 1 & 0 & -2 & 3 & 4 \\ 2 & 1 & 3 & 0 & 2 \end{bmatrix} \begin{bmatrix} y_1 \\ y_2 \\ y_3 \\ y_4 \\ y_5 \end{bmatrix} = \begin{bmatrix} 4 \\ 3 \end{bmatrix}$

(b) $\begin{bmatrix} 1 & 3 & 2 \\ 2 & 1 & 0 \\ -1 & 4 & 5 \end{bmatrix} \begin{bmatrix} x \\ y \\ z \end{bmatrix} = \begin{bmatrix} 4 \\ -1 \\ 8 \end{bmatrix}$

In Problems 5 and 6, from the given systems of linear equations express the x's in terms of the z's.

5. $x_1 = 2y_1 + 5y_2 - 3y_3$
$x_2 = -2y_1 + 3y_2 + y_3$
$x_3 = y_1 + 4y_2 - y_3$

$y_1 = 3z_1 + 2z_2$
$y_2 = -z_1 - z_2$
$y_3 = 4z_1 - 3z_2$

6. $x_1 = y_1 + y_2 + y_3 + 3y_4$
$x_2 = 3y_1 - 2y_2 + 5y_3 - 4y_4$
$x_3 = 2y_3 - y_4$
$x_4 = 2y_1 + y_2 - 3y_3$

$y_1 = z_1 + 3z_2 + 2z_3$
$y_2 = 2z_1 - z_2 - z_3$
$y_3 = 3z_1 - 4z_3$
$y_4 = z_1 + 5z_2 + z_3$

7. What is the size (the number of rows and the number of columns) of each product of matrices that makes sense?

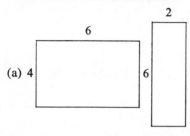

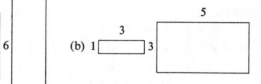

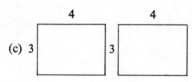

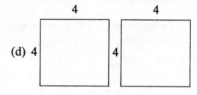

8. $A = \begin{bmatrix} 1 & 2 & 0 \\ 3 & -2 & 1 \end{bmatrix}$, $B = \begin{bmatrix} 1 & 0 & -1 & 2 \\ 3 & 2 & 0 & -3 \\ -1 & 4 & 1 & 0 \end{bmatrix}$, $C = \begin{bmatrix} 2 & 3 \\ 1 & 2 \\ 0 & 4 \\ 1 & -1 \end{bmatrix}$

(a) Find $(AB)C$.
(b) Find $A(BC)$.

(c) Find B_2 and $B^{(4)}$.

9. Find the products AX, AY, $A(2X + 3Y)$, and $2(AX) + 3(AY)$, where

$$A = \begin{bmatrix} 1 & -2 & 0 \\ 3 & -4 & 1 \\ 2 & 1 & -3 \end{bmatrix}, \quad X = \begin{bmatrix} -2 \\ 3 \\ 1 \end{bmatrix}, \quad Y = \begin{bmatrix} 0 \\ 2 \\ -3 \end{bmatrix}$$

10. (a) Find the product AB, where

$$A = \begin{bmatrix} 5 & 2 & -4 \\ 2 & 8 & 2 \\ -4 & 2 & 5 \end{bmatrix} \text{ and } B = \begin{bmatrix} -4 & 2 & -4 \\ 2 & -1 & 2 \\ -4 & 2 & -4 \end{bmatrix}$$

(b) For real numbers we have the theorem: if $ab = 0$, then $a = 0$ or $b = 0$. In view of part (a), what can you say about a corresponding statement for matrices?

11. (a) Find the products AB and AC, where $A = \begin{bmatrix} 1 & 2 \\ 3 & 6 \end{bmatrix}$, $B = \begin{bmatrix} 2 & 1 \\ 3 & 4 \end{bmatrix}$, $C = \begin{bmatrix} 8 & 7 \\ 0 & 1 \end{bmatrix}$.

   (b) In view of part (a), what can you say about a cancellation property for matrix products?

12. (a) Calculate AB and BA if $A = \begin{bmatrix} 1 & 2 \\ 3 & 4 \end{bmatrix}$ and $B = \begin{bmatrix} 0 & 2 \\ 1 & 3 \end{bmatrix}$.
   Is $AB = BA$?

   (b) If $A = \begin{bmatrix} 1 & 2 & 3 \\ 4 & 5 & 6 \end{bmatrix}$ and $B = \begin{bmatrix} 1 & 2 & 0 \\ -1 & 0 & 1 \\ 1 & 1 & 0 \end{bmatrix}$ find AB. Can you calculate BA?

13. If A and B are conformable for multiplication, and if $AB = C = [c_{ij}]$, show that $c_{ij} = A_i \cdot (B^{(j)})^T$. (The dot denotes the dot product of coordinate vectors.)

14. If $A = [a_{ij}]_{(m,n)}$ and B is a $1 \times n$ row vector, show that $AB^T = [A_1 \cdot B, A_2 \cdot B, \ldots, A_m \cdot B]^T$.

15. If $A = [a_{ij}]_{(m,n)}$ and $B = [b_{ij}]_{(n,r)}$, show that
   (a) $(AB)^{(k)} = AB^{(k)}$.
   (b) The ij entry of AB is (the single entry in) $A_i B^{(j)}$.
   (c) $(AB)_i = A_i B$.

Applications of Matrix Multiplication

Example 3.20 Analysis of Food Prices. Suppose that one wishes to compare the total cost of certain food items. The following table, which can be thought of as a matrix, lists the cost in cents per pound of items at three supermarkets.

	ground beef	bread	potatoes	apples	coffee	
store 1	70	40	13	30	330	
store 2	85	38	10	28	310	$= A$
store 3	75	42	12	30	325	

If a purchase is made of 5 lb of ground beef, 3 lb of bread, 10 lb of potatoes, 4 lb of apples, and 2 lb of coffee, we can represent the quantities purchased with the matrix

$$B = \begin{bmatrix} 5 \\ 3 \\ 10 \\ 4 \\ 2 \end{bmatrix}$$

The total cost is given by the matrix product

$$AB = \begin{bmatrix} 70 & 40 & 13 & 30 & 330 \\ 85 & 38 & 10 & 28 & 310 \\ 75 & 42 & 12 & 30 & 325 \end{bmatrix} \begin{bmatrix} 5 \\ 3 \\ 10 \\ 4 \\ 2 \end{bmatrix} = \begin{bmatrix} 1380 \\ 1371 \\ 1391 \end{bmatrix}$$

We see that the total cost at store 2 is 9 cents less than at store 1 and 20 cents less than at store 3. Of course, the problem can be solved without matrices, but the matrices did provide a convenient compact way to state and solve the problem.

The following examples will make use of a fact that we prove in the next section: matrix multiplication is associative. For square matrices we also use A^n to mean A multiplied by itself n times. The reader should note that $A^{(3)}$ means the third column of A, whereas for square matrices A^3 means AAA.

Example 3.21 Interest Compounded Annually. Suppose that we wish to calculate the amount accrued at the end of n years if $100 is invested and compounded annually at interest rates of 5, 6, and 7%. If P dollars is invested for 1 year at a rate r, then the amount at the end of the year is

$$\text{amount} = P + rP = (1 + r)P$$

The matrix product

$$AB = \begin{bmatrix} 1.05 & 0 & 0 \\ 0 & 1.06 & 0 \\ 0 & 0 & 1.07 \end{bmatrix} \begin{bmatrix} 100 \\ 100 \\ 100 \end{bmatrix} = \begin{bmatrix} 105 \\ 106 \\ 107 \end{bmatrix}$$

gives the amount at the end of the first year if $100 is invested at 5, 6, and 7% respectively. At the end of the second year the amount is given by

$$\begin{bmatrix} 1.05 & 0 & 0 \\ 0 & 1.06 & 0 \\ 0 & 0 & 1.07 \end{bmatrix} \begin{bmatrix} 105 \\ 106 \\ 107 \end{bmatrix} = \begin{bmatrix} 1.05 & 0 & 0 \\ 0 & 1.06 & 0 \\ 0 & 0 & 1.07 \end{bmatrix} \left(\begin{bmatrix} 1.05 & 0 & 0 \\ 0 & 1.06 & 0 \\ 0 & 0 & 1.07 \end{bmatrix} \begin{bmatrix} 100 \\ 100 \\ 100 \end{bmatrix} \right)$$

$$= \left(\begin{bmatrix} 1.05 & 0 & 0 \\ 0 & 1.06 & 0 \\ 0 & 0 & 1.07 \end{bmatrix} \begin{bmatrix} 1.05 & 0 & 0 \\ 0 & 1.06 & 0 \\ 0 & 0 & 1.07 \end{bmatrix} \right) \begin{bmatrix} 100 \\ 100 \\ 100 \end{bmatrix}$$

$$= A^2 B$$

In general, the amount at the end of n years is given by $A^n B$. The process of computing AB, $A^2 B = A(AB)$, ..., $A^n B$ is an iterative process that can be programmed on a computer.

Example 3.22   A Markov Chain Example. In a *Markov chain process* there are n states $s_1, s_2, s_3, \ldots, s_n$. At any time the process is in one and only one of the given states, and the process moves successively from one state to another. Also, the probability that the process is in a particular state depends only on the previous state that the process was in. If p_{ij} denotes the *transition probability* that the process moves from state j to state i, then $0 \leq p_{ij} \leq 1$ and $p_{1j} + p_{2j} + \cdots + p_{nj} = 1$. (It is a certainty that the process will move from state j to some state.) The matrix

$$
P = \begin{bmatrix} p_{11} & p_{12} & \cdots & p_{1n} \\ p_{21} & p_{22} & \cdots & p_{2n} \\ \cdot & \cdot & & \cdot \\ \cdot & \cdot & & \cdot \\ \cdot & \cdot & & \cdot \\ p_{n1} & p_{n2} & \cdots & p_{nn} \end{bmatrix}
$$

is called a *transition matrix*. A Markov chain process is determined by specifying the transition matrix and an initial starting state.

As an illustration, consider the working condition of a certain machine. We shall assume that the machine is always in one of three states:

State 1. Broken beyond repair (B).
State 2. In need of adjustment (N).
State 3. Working properly (W).

The transition matrix

$$
\begin{array}{ccc} \text{B} & \text{N} & \text{W} \end{array}
$$
$$
P = \begin{bmatrix} 1 & \frac{1}{2} & 0 \\ 0 & \frac{1}{4} & \frac{1}{2} \\ 0 & \frac{1}{4} & \frac{1}{2} \end{bmatrix} \begin{matrix} \text{B} \\ \text{N} \\ \text{W} \end{matrix}
$$

gives us the various transition probabilities that the machine being in one state will be in another state one time period later. For instance, $p_{11} = 1$ means that it is certain that a machine which is broken beyond repair will be in the same condition one time period later. $p_{23} = \frac{1}{2}$ means that there is a 50% chance that the machine working properly will be in need of adjustment one time period later.

Let's suppose that the machine is working properly initially and that we wish to find the various probabilities that the machine will be in state 1, 2, or 3 after n time periods. Let

$$
X_k = \begin{bmatrix} x_{1,k} \\ x_{2,k} \\ x_{3,k} \end{bmatrix},
$$

where $x_{j,k}$ is the probability that the machine will be in state j after k time periods. Then

$$X_0 = \begin{bmatrix} 0 \\ 0 \\ 1 \end{bmatrix}$$

and we wish to find X_n. Consider the matrix product

$$\begin{bmatrix} p_{11} & p_{12} & p_{13} \\ p_{21} & p_{22} & p_{23} \\ p_{31} & p_{32} & p_{33} \end{bmatrix} \begin{bmatrix} x_{1,k} \\ x_{2,k} \\ x_{3,k} \end{bmatrix} = \begin{bmatrix} y_1 \\ y_2 \\ y_3 \end{bmatrix}$$

The product is a column vector in which, for example, the second entry is

$$y_2 = p_{21}x_{1,k} + p_{22}x_{2,k} + p_{23}x_{3,k}$$

$p_{2j}x_{j,k}$ is the probability that the machine will move from state j to state 2 at the end of the kth time period for $j = 1, 2$, and 3. Therefore, y_2 is the probability that the machine will move from state 1, 2, or 3 to state 2 at the end of the kth time period; that is, $y_2 = x_{2,k+1}$. Similarly, $y_i = x_{i,k+1}$ for $i = 1$ and 3. As an illustration,

$$PX_0 = X_1 = \begin{bmatrix} 0 + 0 + 0 \\ 0 + 0 + \frac{1}{2} \\ 0 + 0 + \frac{1}{2} \end{bmatrix} = \begin{bmatrix} 0 \\ \frac{1}{2} \\ \frac{1}{2} \end{bmatrix}$$

$$y_2 = 0 + 0 + \tfrac{1}{2}$$

probability that the machine moves from working properly to needs adjustment at the end of the first period

probability that the machine moves from needs adjustment to needs adjustment at the end of the first period

probability that the machine moves from broken beyond repair to needs adjustment at the end of the first period

probability that the machine moves from the initial state to needs adjustment at the end of the first time period

We also have $X_2 = PX_1 = P(PX_0) = (PP)X_0 = P^2X_0$, and $X_3 = PX_2 = P(P^2X_0) = (PP^2)X_0 = P^3X_0$. In general, $X_n = P^nX_0$. We see that we can calculate the entries of X_n, which give us the probabilities that the machine will be in state 1, 2, or 3 after n time periods, by calculating P^n. The reader who is familiar with BASIC or one of the other computer programming languages may wish to write a program to calculate various powers of P. In Chapter 7 we shall use *eigenvectors* and *eigenvalues* to calculate P^n. Those interested in pursuing the theory of Markov chains further may consult,

J. Kemeny et al., *Finite Mathematical Structures* (Englewood Cliffs, N.J.: Prentice-Hall, Inc., 1959).

Exercise 3.5

16. The following table gives the cost in cents of 1 can of vegetables at each of three different supermarkets.

$$
\begin{array}{c}
 \\
\text{store 1} \\
\text{store 2} \\
\text{store 3}
\end{array}
\begin{array}{ccc}
\text{peas} & \text{beans} & \text{corn} \\
\left[\begin{array}{ccc}
33 & 25 & 42 \\
34 & 23 & 40 \\
36 & 28 & 35
\end{array} \right]
\end{array}
$$

If a shopper buys 6 cans of peas, 4 cans of beans, and 5 cans of corn, find the total cost to the shopper at each of the stores by performing the appropriate matrix multiplication.

17. In Example 3.21, find the amount at the end of the third and fourth years if \$100 is invested at the rates of 5, 6, and 7%, respectively. Obtain your answer by computing A^3B and A^4B.

18. In Example 3.22, find the probabilities that the machine will be in state 1, 2 or 3
    (a) After two time periods.
    (b) After three time periods.
    (c) After four time periods.

19. Suppose that in a certain work force every man is classified as being in one and only one of the three following states:
    State 1. Professional worker (PW).
    State 2. Unskilled laborer (U).
    State 3. Skilled laborer (S).
    Assume further that every man will have a son and that the transition probability matrix

$$
\begin{array}{c}
\begin{array}{ccc}
\text{PW} & \text{U} & \text{S}
\end{array} \\
\left[\begin{array}{ccc}
0.6 & 0.2 & 0.2 \\
0.1 & 0.6 & 0.3 \\
0.3 & 0.2 & 0.5
\end{array} \right]
\begin{array}{c}
\text{PW} \\
\text{U} \\
\text{S}
\end{array}
\end{array}
$$

gives the probabalities that the next generation will move from one state to another. If 20% of the work force consists of professional workers, 50% of the work force consists of unskilled laborers, and 30% of the work force consists of skilled laborers, find the respective probabilities that the grandson of a man will be a professional, an unskilled laborer, or a skilled laborer.

20. Suppose that an entire population lives in the city or else in one of its suburbs. The states are:
    State 1. Lives in the city (C).
    State 2. Lives in a suburb (S).

Assuming that the conditions of the Markov chain process are satisfied,
(a) Make up what you think would be a reasonable transition matrix.
(b) Assign an initial set of probabilities to obtain X_0, and then compute X_1, X_2, and X_3.

3.6 MATRIX ALGEBRA

The definition of a (real) vector space demands a set $\mathbf{V}$ together with an addition and a scalar multiplication satisfying properties 1A–5A and 6M–10M. We have seen that, if $\mathbf{V} = \mathbf{R}_{m \times n}$ and addition and scalar multiplication are defined according to Definitions 3.12 and 3.14, then $(\mathbf{R}_{m \times n}, +, \text{sm})$ is a vector space.

A vector space may have an additional operation defined on it, and if this new operation (which we call multiplication) satisfies certain properties, which are to be discussed shortly, then the system $(\mathbf{V}, +, \text{sm}, \cdot)$ is called an *algebra* or a *linear algebra*. We shall see that multiplication of matrices has been defined in such a way that the vector space $\mathbf{R}_{n \times n}$ of all square matrices is an algebra.

Definition 3.18. A vector space $(\mathbf{V}, +, \text{sm})$ is called an **algebra (linear algebra)** if and only if $\mathbf{V}$ has a multiplication defined on it that satisfies the following properties. For arbitrary vectors A, B, and C in $\mathbf{V}$ and for an arbitrary scalar r,

11P. AB is a uniquely determined vector in $\mathbf{V}$. (This is the meaning of a binary operation.)

12P. $(AB)C = A(BC)$ (multiplication is associative).

13P. $A(B + C) = AB + AC$ (multiplication on the left distributes over addition).

14P. $(B + C)A = BA + CA$ (multiplication on the right distributes over addition).

15P. $(rA)B = A(rB) = r(AB)$ (we shall say that scalars pull out).

Note 1: On the right side of the equation in property 13P we have adopted the convention that multiplication is performed first and then the addition. For this reason we can write $AB + AC$ in place of the more cumbersome $(AB) + (AC)$.

Note 2: In an algebra it is permissible to write ABC without parentheses because of property 12P.

Before proving that matrix multiplication is associative and distributes over addition, it will be convenient to introduce and discuss the $\sum$ (sigma) notation and some of its properties.

A sum of n numbers $c_1 + c_2 + \cdots + c_n$ may be written in the compact form

$$\sum_{i=1}^{n} c_i$$

The letter i, called the **index** of summation, is a "dummy variable" because it can be replaced by another letter; for instance,

$$\sum_{i=1}^{n} c_i = \sum_{k=1}^{n} c_k$$

We shall use the following properties of $\sum$ notation:

1S. $\displaystyle\sum_{i=1}^{n} (a_i + b_i) = \sum_{i=1}^{n} a_i + \sum_{i=1}^{n} b_i.$

2S. $\displaystyle\sum_{i=1}^{n} ra_i = r \sum_{i=1}^{n} a_i.$

3S. $\displaystyle\sum_{i=1}^{m} \left(\sum_{j=1}^{n} d_{ij} \right) = \sum_{j=1}^{n} \left(\sum_{i=1}^{m} d_{ij} \right).$

Property 1S is a consequence of the associative and commutative properties of addition. Property 2S is an extension of the distributive property of multiplication over addition. To see that property 3S is true, we consider the following array of real numbers:

$$
\begin{array}{ccccc}
d_{11} & d_{12} & d_{13} & \cdots & d_{1n} \\
d_{21} & d_{22} & d_{23} & \cdots & d_{2n} \\
d_{31} & d_{32} & d_{33} & \cdots & d_{3n} \\
\cdot & \cdot & \cdot & & \cdot \\
\cdot & \cdot & \cdot & & \cdot \\
\cdot & \cdot & \cdot & & \cdot \\
d_{m1} & d_{m2} & d_{m3} & \cdots & d_{mn}
\end{array}
$$

There are mn terms in the array. If we add the mn terms by summing across the rows first, we obtain

$$\sum_{j=1}^{n} d_{1j} + \sum_{j=1}^{n} d_{2j} + \sum_{j=1}^{n} d_{3j} + \cdots + \sum_{j=1}^{n} d_{mj} = \sum_{i=1}^{m} \left(\sum_{j=1}^{n} d_{ij} \right)$$

On the other hand, if we first find the sum of each column of terms, we obtain

$$\sum_{i=1}^{m} d_{i1} + \sum_{i=1}^{m} d_{i2} + \sum_{i=1}^{m} d_{i3} + \cdots + \sum_{i=1}^{m} d_{in} = \sum_{j=1}^{n} \left(\sum_{i=1}^{m} d_{ij} \right)$$

Since each double sum is the sum of the mn numbers in the array, we conclude that property 3S is true.

Example 3.23

(a) $\displaystyle\sum_{i=1}^{3} (a_i + b_i) = (a_1 + b_1) + (a_2 + b_2) + (a_3 + b_3)$

$$= (a_1 + a_2 + a_3) + (b_1 + b_2 + b_3) = \sum_{i=1}^{3} a_i + \sum_{i=1}^{3} b_i$$

(b) $\displaystyle\sum_{i=1}^{4} ra_i = ra_1 + ra_2 + ra_3 + ra_4 = r(a_1 + a_2 + a_3 + a_4) = r\sum_{i=1}^{4} a_i$

(c) $\displaystyle\sum_{j=1}^{3}\left(\sum_{i=1}^{2} d_{ij}\right) = \sum_{j=1}^{3}(d_{1j} + d_{2j}) = (d_{11} + d_{21}) + (d_{12} + d_{22}) + (d_{13} + d_{23})$

$\qquad = (d_{11} + d_{12} + d_{13}) + (d_{21} + d_{22} + d_{23})$

$\qquad = \displaystyle\sum_{i=1}^{2}(d_{i1} + d_{i2} + d_{i3}) = \sum_{i=1}^{2}\left(\sum_{j=1}^{3} d_{ij}\right)$

Theorem 3.6. If A, B, and C are matrices that are conformable for multiplication, then $(AB)C = A(BC)$.

Proof: Let A, B, and C be $m \times n$, $n \times r$, and $r \times t$ matrices, respectively. The reader should check the fact that both products $(AB)C$ and $A(BC)$ are $m \times t$ matrices.

Let $D = AB$, $E = DC = (AB)C$, $F = BC$, and $G = AF = A(BC)$. We wish to show that $E = [e_{ij}]_{(m,t)} = G = [g_{ij}]_{(m,t)}$. From the definition of multiplication

$$e_{ij} = \sum_{k=1}^{r} d_{ik}c_{kj} \quad \text{and} \quad d_{ik} = \sum_{s=1}^{n} a_{is}b_{sk}$$

So

$$e_{ij} = \sum_{k=1}^{r} d_{ik}c_{kj} = \sum_{k=1}^{r}\left(\left(\sum_{s=1}^{n} a_{is}b_{sk}\right)c_{kj}\right) = \sum_{k=1}^{r}\left(\sum_{s=1}^{n}(a_{is}b_{sk}c_{kj})\right) \qquad (*)$$

The last equality follows from 2S.

Applying the definition of multiplication to $G = AF$, we get

$$g_{ij} = \sum_{k=1}^{n} a_{ik}f_{kj}.$$

Since $F = BC$, we have

$$f_{kj} = \sum_{p=1}^{r} b_{kp}c_{pj}$$

So

$$g_{ij} = \sum_{k=1}^{n} a_{ik}f_{kj} = \sum_{k=1}^{n}\left(a_{ik}\left(\sum_{p=1}^{r} b_{kp}c_{pj}\right)\right) = \sum_{k=1}^{n}\left(\sum_{p=1}^{r}(a_{ik}b_{kp}c_{pj})\right)$$

Since the indexes are dummy variables, we can write

$$g_{ij} = \sum_{s=1}^{n}\left(\sum_{p=1}^{r}(a_{is}b_{sp}c_{pj})\right) = \sum_{s=1}^{n}\left(\sum_{k=1}^{r}(a_{is}b_{sk}c_{kj})\right) \qquad (**)$$

Comparing (*) and (**) and using property 3S, we see that $e_{ij} = g_{ij}$. Q.E.D.

Theorem 3.7. If B and C are $m \times n$ matrices and A is an $r \times m$ matrix, then $A(B + C) = AB + AC$.

Proof: We first note that both $A(B + C)$ and $AB + AC$ are $r \times n$ matrices. Let $D = B + C$, $E = AD$, $F = AB$, $G = AC$, and $H = F + G$. We wish to show that $E = [e_{ij}]_{(r,n)} = H = [h_{ij}]_{(r,n)}$.

From the definition of multiplication, we obtain

$$e_{ij} = \sum_{k=1}^{m} a_{ik}d_{kj}$$

and from the definition of addition we get

$$d_{kj} = b_{kj} + c_{kj}$$

Substituting, we have

$$e_{ij} = \sum_{k=1}^{m} a_{ik}(b_{kj} + c_{kj}) = \sum_{k=1}^{m} (a_{ik}b_{kj} + a_{ik}c_{kj}) = \sum_{k=1}^{m} a_{ik}b_{kj} + \sum_{k=1}^{m} a_{ik}c_{kj} \quad (*)$$

The last equality follows from property 1S. Applying the definition of addition to H and the definition of multiplication to F and G, we get

$$h_{ij} = f_{ij} + g_{ij} = \sum_{k=1}^{m} a_{ik}b_{kj} + \sum_{k=1}^{m} a_{ik}c_{kj} \qquad (**)$$

Comparing (*) and (**), we see that $e_{ij} = h_{ij}$. Hence $E = H$ and $A(B + C) = AB + AC$. Q.E.D.

Theorem 3.8. If B and C are $m \times n$ matrices and A is an $n \times r$ matrix, then $(B + C)A = BA + CA$.

The proof of this theorem is similar to the proof of Theorem 3.7 and is left as Problem 7, Exercise 3.6.

Theorem 3.9. If r is a scalar, A is an $m \times n$ matrix, and B is an $n \times r$ matrix, then $(rA)B = A(rB) = r(AB)$.

Proof: First we note that each of $(rA)B$, $A(rB)$, and $r(AB)$ is an $m \times r$ matrix. Let $C = AB$, $D = rA$, $E = DB$, and $F = rC$. We will show that $E = [e_{ij}]_{(m,r)} = F = [f_{ij}]_{(m,r)}$. From the definition of multiplication,

$$e_{ij} = \sum_{k=1}^{n} d_{ik}b_{kj}$$

Since $D = rA$, we have $d_{ik} = ra_{ik}$; so

$$e_{ij} = \sum_{k=1}^{n} (ra_{ik})b_{kj} = \sum_{k=1}^{n} r(a_{ik}b_{kj}) = r \sum_{k=1}^{n} a_{ik}b_{kj}$$

The second equality follows because multiplication of real numbers is associative, and the last equality follows from property 2S. Also, since $F = rC$ and $C = AB$, we have

$$f_{ij} = rc_{ij} = r \sum_{k=1}^{n} a_{ik}b_{kj}$$

Comparing the (i, j) entry of E with the (i, j) entry of F, we see that they are identical. Hence $E = F$ and $(rA)B = r(AB)$. The reader is asked to prove that $A(rB) = r(AB)$ in Problem 8, Exercise 3.6. Q.E.D.

The fact that the collection of all square matrices of order n is an algebra now follows from Theorems 3.5–3.9, and we state this result as a corollary.

Corollary 3.3. $(R_{n \times n}, +, \text{sm}, \cdot)$ with addition, scalar multiplication, and multiplication defined as in Definitions 3.12, 3.14, and 3.16 is an algebra.

Example 3.24. (a) Use the result of Theorem 3.9 to simplify the calculation of AB if

$$A = \begin{bmatrix} 9 & 0 & 27 \\ 36 & 18 & 0 \\ -9 & 27 & 45 \end{bmatrix} \quad \text{and} \quad B = \begin{bmatrix} 12 & 8 & 0 & 4 \\ 0 & 20 & -8 & 16 \\ 24 & -4 & 16 & 32 \end{bmatrix}$$

(b) Use the result of Theorem 3.8 to find $BA + CA$ if

$$A = \begin{bmatrix} 1 & 0 & 3 \\ 2 & 1 & 0 \\ 3 & -1 & 2 \end{bmatrix}, \quad B = \begin{bmatrix} 1 & 3 & 2 \\ -1 & 4 & 6 \end{bmatrix}, \quad C = \begin{bmatrix} 4 & -3 & 5 \\ 1 & -4 & 2 \end{bmatrix}$$

Solution: (a) $A = 9C$ and $B = 4D$, where $C = \begin{bmatrix} 1 & 0 & 3 \\ 4 & 2 & 0 \\ -1 & 3 & 5 \end{bmatrix}$ and

$$D = \begin{bmatrix} 3 & 2 & 0 & 1 \\ 0 & 5 & -2 & 4 \\ 6 & -1 & 4 & 8 \end{bmatrix}. \text{ So}$$

$$AB = (9C)(4D) = (36)CD = 36 \begin{bmatrix} 21 & -1 & 12 & 25 \\ 12 & 18 & -4 & 12 \\ 27 & 8 & 14 & 51 \end{bmatrix}$$

(b) $BA + CA = (B + C)A = \begin{bmatrix} 5 & 0 & 7 \\ 0 & 0 & 8 \end{bmatrix} \begin{bmatrix} 1 & 0 & 3 \\ 2 & 1 & 0 \\ 3 & -1 & 2 \end{bmatrix} = \begin{bmatrix} 26 & -7 & 29 \\ 24 & -8 & 16 \end{bmatrix}$

Theorems 3.5–3.9 show that in performing matrix manipulations many of the rules of arithmetic are valid. There are, however, familiar rules that do not hold for matrix multiplication. Problems 10–12, Exercise 3.5, point out some of them. Problem 10 shows that one can have $AB = 0$ with neither $A = 0$ or $B = 0$. Problem 11 shows that **there is no cancellation property for matrix products**; that is, $AB = AC$ and $A \neq 0$ does not imply $B = C$. Problem 12 shows that **matrix multiplication is not commutative**; that is, in general $AB \neq BA$ (even for square matrices). In performing matrix manipulations or in deriving the theory of matrices we must be careful not to use rules of arithmetic that do not hold for matrices. As an illustration, for square matrices of order n, $(A + B)^2 \neq A^2 + 2AB + B^2$. A correct statement is $(A + B)^2 = A^2 + AB + BA + B^2$ (Exercise 3.6, Problem 5).

Prior to Example 3.21 in Section 3.5, we said that we would use A^n to 'mean A multiplied by itself n times; that is,

$$A^n = \underbrace{AA \ldots A}_{n \text{ times}}$$

A more acceptable definition from a logician's point of view is as follows:

Definition 3.19. If A is a square matrix, we define A^n as follows:
(i) $A^1 = A$.
(ii) For any positive integer $n \geq 2$, $A^n = (A^{n-1})A$.

This definition is an example of a **recursive** or **inductive definition**. One thing it does is eliminate the . . . in $A^n = A\,A \ldots A$. Using Definition 3.19 to evaluate A^n for a specific n is an iterative process (see Example 3.25). To find A^3, we first find A^2, to find A^4 we first find A^3, etc. This is the kind of definition that is useful in programming a computer. The computer is directed to loop through the calculations several times until the desired power of A is obtained.

Recursive definitions are also used to prove theorems by *mathematical induction*, a technique of proof that we shall discuss briefly in Section 3.7. The fact that the above recursive definition does indeed define A^n for every positive integer also depends on mathematical induction. The interested reader who wishes to see a proof of this may consult Section 12 of Paul R. Halmos, *Naive Set Theory* (New York: Van Nostrand Reinhold Company, 1961). The next example illustrates how Definition 3.19 can be used to compute powers of A.

Example 3.25. If $A = \begin{bmatrix} 1 & 2 \\ -1 & 0 \end{bmatrix}$, use Definition 3.19 to calculate A^5.

Solution: Using Definition 3.19(ii) several times, we have $A^5 = A^4A$, $A^4 = A^3A$, $A^3 = A^2A$, and $A^2 = A^1A$. By Definition 3.19(i), $A^1 = A$. To calculate A^5, we must first calculate A^4, A^3, and A^2.

$$A^2 = AA = \begin{bmatrix} 1 & 2 \\ -1 & 0 \end{bmatrix}\begin{bmatrix} 1 & 2 \\ -1 & 0 \end{bmatrix} = \begin{bmatrix} -1 & 2 \\ -1 & -2 \end{bmatrix}$$

$$A^3 = A^2A = \begin{bmatrix} -1 & 2 \\ -1 & -2 \end{bmatrix}\begin{bmatrix} 1 & 2 \\ -1 & 0 \end{bmatrix} = \begin{bmatrix} -3 & -2 \\ 1 & -2 \end{bmatrix}$$

$$A^4 = A^3A = \begin{bmatrix} -3 & -2 \\ 1 & -2 \end{bmatrix}\begin{bmatrix} 1 & 2 \\ -1 & 0 \end{bmatrix} = \begin{bmatrix} -1 & -6 \\ 3 & 2 \end{bmatrix}$$

So

$$A^5 = A^4 A = \begin{bmatrix} -1 & -6 \\ 3 & 2 \end{bmatrix}\begin{bmatrix} 1 & 2 \\ -1 & 0 \end{bmatrix} = \begin{bmatrix} 5 & -2 \\ 1 & 6 \end{bmatrix}$$

We conclude this section with the following important theorem concerning the transpose of a product.

Theorem 3.10. If A is an $m \times n$ matrix and B is an $n \times r$ matrix, then $(AB)^T = B^T A^T$. In words, the transpose of a product is the product of the transposes in reverse order.

Proof: First we note that $(AB)^T$ and $B^T A^T$ are both $r \times m$ matrices. Let $A^T = [c_{ij}]_{(n,m)}$ and $B^T = [d_{ij}]_{(r,n)}$, where $c_{ij} = a_{ji}$ and $d_{ij} = b_{ji}$. Then the (i,j) entry of $B^T A^T$ is $\sum_{k=1}^{n} d_{ik}c_{kj}$. Now let $AB = [e_{ij}]_{(m,r)}$, and let $(AB)^T = [f_{ij}]_{(r,m)}$, where $f_{ij} = e_{ji}$. Then

$$f_{ij} = e_{ji} = \sum_{k=1}^{n} a_{jk}b_{ki} = \sum_{k=1}^{n} c_{kj}d_{ik} = \sum_{k=1}^{n} d_{ik}c_{kj}$$

But this is the (i,j) entry of $B^T A^T$.   Q.E.D.

Exercise 3.6

1. (a) Expand $\sum_{k=1}^{4} a_{ik}b_{kj}$. What letter is the index of summation?
   (b) Expand and compare

$$\sum_{s=1}^{2} \left(\sum_{r=1}^{3} a_{rs} \right) \quad \text{and} \quad \sum_{r=1}^{3} \left(\sum_{s=1}^{2} a_{rs} \right)$$

   (c) Expand and compare

$$\sum_{k=1}^{2} a_{ik}(b_{kj} + c_{kj}), \qquad \sum_{k=1}^{2} (a_{ik}b_{kj} + a_{ik}c_{kj})$$

   and

$$\sum_{k=1}^{2} a_{ik}b_{kj} + \sum_{k=1}^{2} a_{ik}c_{kj}$$

2. Expand and compare the following sums:
   (a) $\sum_{k=1}^{2} \left(\left(\sum_{s=1}^{3} a_{is}b_{sk} \right)c_{kj} \right)$ and $\sum_{k=1}^{2} \left(\sum_{s=1}^{3} (a_{is}b_{sk}c_{kj}) \right)$.
   (b) $\sum_{k=1}^{3} \left(a_{ik}\left(\sum_{p=1}^{2} b_{kp}c_{pj} \right) \right)$ and $\sum_{k=1}^{3} \left(\sum_{p=1}^{2} (a_{ik}b_{kp}c_{pj}) \right)$.

3. Use Theorem 3.9 to simplify the calculation of AB if

$$A = \begin{bmatrix} 5 & 20 & 0 \\ 10 & -5 & 25 \\ 0 & 5 & 15 \end{bmatrix} \quad \text{and} \quad B = \begin{bmatrix} 0 & 2 & 6 & 2 \\ 4 & -6 & 8 & 4 \\ -2 & -4 & 2 & 4 \end{bmatrix}$$

4. Use Theorem 3.7 to find $AB + AC$ if

$$A = \begin{bmatrix} 1 & 2 \\ 0 & 1 \\ -1 & 3 \end{bmatrix}, \quad B = \begin{bmatrix} 1 & -7 & -9 \\ 2 & -6 & 3 \end{bmatrix}, \quad C = \begin{bmatrix} 2 & 7 & 9 \\ -4 & 6 & 2 \end{bmatrix}$$

5. Verify that $(A + B)^2 = A^2 + AB + BA + B^2$ for $n \times n$ matrices A and B.

6. Write A^7 in terms of A by using Definition 3.19 several times.

7. Prove Theorem 3.8.

8. Prove that $A(rB) = r(AB)$ in Theorem 3.9.

9. Verify that in an algebra $(rA)(sB) = (rs)AB$.

The *trace* of a square matrix A of order n is defined to be the sum of the elements on the main diagonal. Letting tr (A) denote the trace of A we have

$$\text{tr}\,(A) = \sum_{k=1}^{n} a_{kk}$$

10. Verify that the trace of A has the following properties:
   (a) tr $(A + B) = $ tr $(A) + $ tr (B).
   (b) tr $(rA) = r(\text{tr } A)$
   (c) tr $(AB) = $ tr (BA).

On the collection $\mathbf{R}_{2 \times 2}$ of all 2×2 square matrices, define a product $\circ$ by $A \circ B = AB - BA$. The product AB is the product of Definition 3.16.

11. Verify property 11P for $\circ$ and $\mathbf{R}_{2 \times 2}$; that is, verify that, for any 2×2 matrices A and B, $A \circ B$ is a uniquely determined 2×2 matrix.

12. Prove that property 12P does not hold for $\circ$ by finding three matrices A, B, and C in $\mathbf{R}_{2 \times 2}$ such that $(A \circ B) \circ C \neq A \circ (B \circ C)$.

13. Prove that property 13P holds for $\circ$; that is, $A \circ (B + C) = A \circ B + A \circ C$.

14. Prove that property 14P holds for $\circ$; that is, $(B + C) \circ A = B \circ A + C \circ A$.

15. Prove that property 15P holds for $\circ$; that is, $r(A \circ B) = (rA) \circ B = A \circ (rB)$.

A vector space $(\mathbf{V}, +, \text{sm})$ that has a product defined on it satisfying all the properties of Definition 3.18, with the possible exception of the associative property 12P, is called a **nonassociative algebra**. A **Lie algebra** is a nonassociative algebra in which the product XY satisfies the additional properties $XY = -(YX)$ and $(XY)Z + (YZ)X + (ZX)Y = 0$ for all X, Y, and Z in the algebra (see Exercise 2.4). Lie groups and Lie algebras are named after Marius Sophus Lie because of his research on the subject in the period from 1873 to 1893. Lie algebras have important applications in modern physics.

16. Prove that $A \circ B = -(B \circ A)$.

17. Prove that $(A \circ B) \circ C + (B \circ C) \circ A + (C \circ A) \circ B = 0$. *Hint:* Use the fact that $(AB)C = A(BC)$.

Problems 11–17 and Theorem 3.5 show that $(\mathbf{R}_{2 \times 2}, +, \text{sm}, \circ)$ is a Lie algebra. Another important class of nonassociative algebras is the class of Jordan algebras. Jordan algebras were introduced in the early 1930s by a physicist P. Jordan in connection with quantum mechanics. Much of the mathematical research in Jordan algebras has been done in recent years by such great mathematicians as A. A. Albert (1905–1972). A **Jordan algebra** is a nonassociative algebra that satisfies the identities $XY = YX$ and $(XY)(XX) = X(Y(XX))$ for all X and Y in the algebra.

If we define a product $*$ on $\mathbf{R}_{2 \times 2}$ by $A * B = AB + BA$, the resulting mathematical system $(\mathbf{R}_{2 \times 2}, +, \text{sm}, *)$ is a Jordan algebra, as the following exercises indicate. The proofs depend only on the fact that $\mathbf{R}_{2 \times 2}$ with product AB is an associative algebra and not on any other special properties of $\mathbf{R}_{2 \times 2}$.

18. Prove that property 11P holds for $*$.

19. Prove that $A * B = B * A$.

20. Prove that properties 13P and 14P hold for $*$.

21. Prove that property 15P holds for $*$.

22. Prove that $(A * B) * (A * A) = A * (B * (A * A))$.

23. Show that $*$ is not an associative operation on $\mathbf{R}_{2 \times 2}$.

    An element E in an algebra is called an **idempotent element** if $E \neq 0$ and $E^2 = E$. An element N in an algebra is called **nilpotent** if $N^k = 0$ for some positive integer k.

24. In $(\mathbf{R}_{2 \times 2}, +, \text{sm}, \cdot)$ (a) find an idempotent element $\neq I_2$; (b) find a nonzero nilpotent element.

3.7 NONSINGULAR MATRICES AND ELEMENTARY MATRICES

Although we have already referred to the identity matrix in Section 3.2, we now give a formal definition.

Definition 3.20. The $n \times n$ **identity matrix** or the **identity matrix of order** n is the $n \times n$ matrix with each (i, i) entry equal to 1 and all other entries equal to 0. This matrix is denoted by I or by I_n if we need to stress the size.

The corollary to the next theorem gives the reason for the name *identity* matrix.

Theorem 3.11. If A is an $m \times n$ matrix, then $I_m A = A$ and $A I_n = A$.

Proof: Let $D = I_m A$. Then D is an $m \times n$ matrix. By Problem 13, Exercise 3.5, the (i, j) entry d_{ij} can be viewed as the dot prouct of the ith row of I_m and the transpose of the jth column of A. So

$$d_{ij} = [0\ 0 \ldots 0\ \underset{\text{the } i\text{th entry}}{1}\ 0 \ldots 0] \cdot [a_{1j}\, a_{2j} \ldots a_{mj}] = 1 a_{ij} = a_{ij}$$

Hence $d_{ij} = a_{ij}$ and $I_m A = A$. The proof that $AI_n = A$ is similar and is left as Problem 18, Exercise 3.7. Q.E.D.

We state the following obvious corollary.

Corollary 3.4.   If A is a square matrix of order n and I is the identity matrix of order n, then $AI = IA = A$.

It may be the case that for a given square matrix A there is some matrix B such that $AB = BA = I$. For instance, if

$$A = \begin{bmatrix} 1 & -1 & -1 \\ 0 & 1 & 0 \\ 1 & -2 & 1 \end{bmatrix} \quad \text{and} \quad B = \begin{bmatrix} \frac{1}{2} & \frac{3}{2} & \frac{1}{2} \\ 0 & 1 & 0 \\ -\frac{1}{2} & \frac{1}{2} & \frac{1}{2} \end{bmatrix}$$

the reader can check that $AB = BA = I$.

Definition 3.21.   Let A be any $n \times n$ matrix. An $n \times n$ matrix B having the property that $AB = BA = I_n$ is called a **multiplicative inverse** of the matrix A. A square matrix A that has a multiplicative inverse is called a **nonsingular matrix**. If A has no inverse, it is called **singular**. We shall refer to a multiplicative inverse as simply an inverse.

Example 3.26.   Show that the matrix $A = \begin{bmatrix} 2 & 4 \\ 3 & 6 \end{bmatrix}$ is singular.

Solution: If $B = \begin{bmatrix} b_{11} & b_{12} \\ b_{21} & b_{22} \end{bmatrix}$ is an inverse for A, then $AB = I$ and from the definition of product

$$2b_{11} + 4b_{21} = 1$$
$$3b_{11} + 6b_{21} = 0$$
$$2b_{12} + 4b_{22} = 0$$
$$3b_{12} + 6b_{22} = 1$$

This system is clearly inconsistent. Hence the b's do not exist and A has no inverse.

One can easily show, as we do in the proof of the next theorem, that if a square matrix has an inverse it has only one.

Theorem 3.12.   If A is an $n \times n$ nonsingular matrix and if B and C are inverses of A, then $B = C$.

Proof: The proof depends essentially on the fact that matrix multiplication is associative. We consider the product BAC and associate both ways. $(BA)C = I_n C = C$. On the other hand, $B(AC) = BI_n = B$. Since matrix multiplication is associative, we have $C = B$. Q.E.D.

The unique multiplicative inverse of a nonsingular matrix A is denoted by A^{-1}.

An application of the inverse of a matrix is its use in solving a system of n linear equations in n unknowns, as the next theorem shows. The hypothesis of the theorem refers to the system (3.1) in Section 3.2, where m and n are equal.

Theorem 3.13. If a system of n linear equations and n unknowns has a nonsingular coefficient matrix A then the system has a unique solution given by $X = A^{-1}B$, where B is the column of constants.

Proof: The system of n linear equations in n unknowns is equivalent to a matrix equation $AX = B$, where A is the nonsingular coefficient matrix. Multiplying each side of this equation on the left by A^{-1}, we get $A^{-1}(AX) = A^{-1}B$. But $A^{-1}(AX) = (A^{-1}A)X = IX = X$. Hence $X = A^{-1}B$. Q.E.D.

Example 3.27. Solve the system of linear equations

$$
\begin{aligned}
x_1 - x_2 - x_3 &= b_1 \\
x_2 &= b_2 \\
x_1 - 2x_2 + x_3 &= b_3
\end{aligned}
\quad \text{where} \quad
\begin{bmatrix} b_1 \\ b_2 \\ b_3 \end{bmatrix}
= \text{(a)} \begin{bmatrix} 4 \\ 3 \\ 7 \end{bmatrix}
\text{(b)} \begin{bmatrix} 2 \\ 1 \\ 3 \end{bmatrix}
\text{(c)} \begin{bmatrix} 8 \\ 1 \\ 4 \end{bmatrix}
$$

Solution: The equivalent matrix equation is

$$
\begin{bmatrix} 1 & -1 & -1 \\ 0 & 1 & 0 \\ 1 & -2 & 1 \end{bmatrix}
\begin{bmatrix} x_1 \\ x_2 \\ x_3 \end{bmatrix}
=
\begin{bmatrix} b_1 \\ b_2 \\ b_3 \end{bmatrix}
$$

The inverse of the coefficient matrix is

$$
\begin{bmatrix} \frac{1}{2} & \frac{3}{2} & \frac{1}{2} \\ 0 & 1 & 0 \\ -\frac{1}{2} & \frac{1}{2} & \frac{1}{2} \end{bmatrix}
$$

which was given in an example following Corollary 3.4 of this section. We apply Theorem 3.13. For computational purposes we write the three columns of constants side by side as a 3×3 matrix. We get

$$
\begin{bmatrix} \frac{1}{2} & \frac{3}{2} & \frac{1}{2} \\ 0 & 1 & 0 \\ -\frac{1}{2} & \frac{1}{2} & \frac{1}{2} \end{bmatrix}
\begin{bmatrix} 4 & 2 & 8 \\ 3 & 1 & 1 \\ 7 & 3 & 4 \end{bmatrix}
=
\begin{bmatrix} 10 & 4 & \frac{15}{2} \\ 3 & 1 & 1 \\ 3 & 1 & -\frac{3}{2} \end{bmatrix}
$$

Therefore, the solutions are

$$
\begin{bmatrix} x_1 \\ x_2 \\ x_3 \end{bmatrix}
= \text{(a)} \begin{bmatrix} 10 \\ 3 \\ 3 \end{bmatrix},
\text{(b)} \begin{bmatrix} 4 \\ 1 \\ 1 \end{bmatrix},
\text{(c)} \begin{bmatrix} \frac{15}{2} \\ 1 \\ -\frac{3}{2} \end{bmatrix}
$$

If two matrices of the same order have inverses, then so does their product as the next theorem indicates.

Theorem 3.14. If A and B are nonsingular matrices of order n, then AB is nonsingular and $(AB)^{-1} = B^{-1}A^{-1}$.

Proof: A^{-1} and B^{-1} exist by hypothesis. We see that

$$(AB)(B^{-1}A^{-1}) = A(BB^{-1})A^{-1} = A(IA^{-1}) = AA^{-1} = I$$

Also,

$$(B^{-1}A^{-1})(AB) = B^{-1}(A^{-1}A)B = B^{-1}(IB) = B^{-1}B = I$$

Hence $B^{-1}A^{-1}$ is an inverse for AB. But the inverse of a matrix is unique if one exists. Hence $(AB)^{-1} = B^{-1}A^{-1}$. Q.E.D.

The result of Theorem 3.14 extends to a product of k nonsingular matrices. We could prove this by repeating the process used in the proof of Theorem 3.14. Instead we choose to give a somewhat neater proof by *mathematical induction*, which is valid because of the following theorem.

Principle of Mathematical Induction (PMI). Let $S(n)$ be a statement that is true or false for every positive integer n. (a) If $S(1)$ is true, and (b) if $S(k)$ is true implies $S(k + 1)$ is true, then $S(n)$ is true for all positive integers n.

We give only a brief discussion of why this theorem works. Suppose that $S(n)$ is a statement for which (a) and (b) have been established. $S(2)$ is true because (a) $S(1)$ is true, and (b) $S(1)$ is true implies $S(1 + 1) = S(2)$ is true. Having established that $S(2)$ is true, we see that $S(3)$ is true because $S(2)$ is true, and (b) $S(2)$ is true implies $S(2 + 1) = S(3)$ is true. Intuitively, we see that $S(n)$ must be true for all positive integers. We now use the principle of mathematical induction to prove the following theorem.

Theorem 3.15. If $A_1, A_2, \ldots, A_q$ are nonsingular matrices of order n, then the product $A_1 A_2 \ldots A_q$ is nonsingular, and

$$(A_1 A_2 \ldots A_q)^{-1} = A_q^{-1} A_{q-1}^{-1} \ldots A_1^{-1}$$

Proof: Let $S(k)$ be the statement of the theorem for $q = k$. $S(1)$ is the trivially true statement: if A_1 is a nonsingular matrix of order n, then A_1 is nonsingular and $A_1^{-1} = A_1^{-1}$. Assume that $S(k)$ is true (this is called the *induction hypothesis*). We wish to show that $S(k + 1)$ is true. Let $A_1, A_2, \ldots, A_k, A_{k+1}$ be nonsingular matrices of order n. Let $B = A_1 A_2 \ldots A_k$. Then $A_1 A_2 \ldots A_k A_{k+1} = B A_{k+1}$. By the induction hypothesis, B is nonsingular, and $B^{-1} = A_k^{-1} A_{k-1}^{-1} \ldots A_1^{-1}$. From Theorem 3.14, we have $(B A_{k+1})^{-1} = A_{k+1}^{-1} B^{-1} = A_{k+1}^{-1} A_k^{-1} \ldots A_1^{-1}$. Hence $S(k + 1)$ is true. By the **PMI**, it follows that Theorem 3.15 is true for all positive integers q. Q.E.D.

Note: In the statement of Theorem 3.15 and in its proof when we wrote $A_1 A_2 \ldots A_k$ without parentheses, we were assuming a generalized associative property for multiplication. This result as well as results like a generalized distributive property $A(B_1 + B_2 + \cdots + B_k) = AB_1 + AB_2 + \cdots + AB_k$ could be proved by mathematical induction. We choose instead to assume these results without proof, although some induction proofs are called for in the problems.

Example 3.28. If

$$A = \begin{bmatrix} 1 & 2 \\ 0 & 1 \end{bmatrix}, \quad B = \begin{bmatrix} 2 & 5 \\ 3 & 7 \end{bmatrix}, \quad C = \begin{bmatrix} 3 & 2 \\ 4 & 3 \end{bmatrix}$$

$$A^{-1} = \begin{bmatrix} 1 & -2 \\ 0 & 1 \end{bmatrix}, \quad B^{-1} = \begin{bmatrix} -7 & 5 \\ 3 & -2 \end{bmatrix}, \quad C^{-1} = \begin{bmatrix} 3 & -2 \\ -4 & 3 \end{bmatrix}$$

find $(ABC)^{-1}$.

Solution

$$ABC = (AB)C = \begin{bmatrix} 8 & 19 \\ 3 & 7 \end{bmatrix} \begin{bmatrix} 3 & 2 \\ 4 & 3 \end{bmatrix} = \begin{bmatrix} 100 & 73 \\ 37 & 27 \end{bmatrix}$$

By Theorem 3.15,

$$(ABC)^{-1} = C^{-1}B^{-1}A^{-1} = \begin{bmatrix} -27 & 19 \\ 37 & -26 \end{bmatrix} \begin{bmatrix} 1 & -2 \\ 0 & 1 \end{bmatrix} = \begin{bmatrix} -27 & 73 \\ 37 & -100 \end{bmatrix}$$

As a partial check, we find that

$$\begin{bmatrix} 100 & 73 \\ 37 & 27 \end{bmatrix} \begin{bmatrix} -27 & 73 \\ 37 & -100 \end{bmatrix} = \begin{bmatrix} 1 & 0 \\ 0 & 1 \end{bmatrix}$$

The reader may wonder how the inverses of A, B, and C were found. A technique for finding the inverse of a matrix, when it exists, uses the Gauss–Jordan reduction method. We shall see that a square matrix has an inverse if and only if it is row equivalent to the identity matrix. To derive this result, which we do in the next section, it is necessary to introduce elementary matrices.

Definition 3.22. An **elementary matrix of order** n is an $n \times n$ matrix obtained by applying a single elementary row operation to I_n.

We use the following notation to denote the three types of elementary matrices:

1. F_{st} denotes the elementary matrix obtained by applying the elementary row operation R_{st} to the identity matrix.
2. $F_s(k)$, $k \neq 0$, denotes the elementary matrix obtained by applying the elementary row operation kR_s to the identity matrix.
3. $F_{st}(k)$ denotes the elementary matrix obtained by applying the elementary row operation $kR_s + R_t$ to the identity matrix.

Example 3.29. We list an elementary matrix of each type.

$$F_{13} = \begin{bmatrix} 0 & 0 & 1 \\ 0 & 1 & 0 \\ 1 & 0 & 0 \end{bmatrix} \text{ because } \begin{bmatrix} 1 & 0 & 0 \\ 0 & 1 & 0 \\ 0 & 0 & 1 \end{bmatrix} \xrightarrow{R_{13}} \begin{bmatrix} 0 & 0 & 1 \\ 0 & 1 & 0 \\ 1 & 0 & 0 \end{bmatrix}$$

$$F_2(k) = \begin{bmatrix} 1 & 0 & 0 \\ 0 & k & 0 \\ 0 & 0 & 1 \end{bmatrix} \text{ because } \begin{bmatrix} 1 & 0 & 0 \\ 0 & 1 & 0 \\ 0 & 0 & 1 \end{bmatrix} \xrightarrow{kR_2} \begin{bmatrix} 1 & 0 & 0 \\ 0 & k & 0 \\ 0 & 0 & 1 \end{bmatrix}$$

$$F_{23}(k) = \begin{bmatrix} 1 & 0 & 0 \\ 0 & 1 & 0 \\ 0 & k & 1 \end{bmatrix} \text{ because } \begin{bmatrix} 1 & 0 & 0 \\ 0 & 1 & 0 \\ 0 & 0 & 1 \end{bmatrix} \xrightarrow{kR_2 + R_3} \begin{bmatrix} 1 & 0 & 0 \\ 0 & 1 & 0 \\ 0 & k & 1 \end{bmatrix}$$

Example 3.30. The effect of multiplying a matrix on the left by each of the three types of elementary matrices is illustrated. Let

$$A = \begin{bmatrix} 2 & 3 & 1 & 4 \\ -4 & 5 & 6 & 9 \\ 0 & 7 & 8 & 2 \end{bmatrix}$$

(a)

$$F_{13}A = \begin{bmatrix} 0 & 0 & 1 \\ 0 & 1 & 0 \\ 1 & 0 & 0 \end{bmatrix} \begin{bmatrix} 2 & 3 & 1 & 4 \\ -4 & 5 & 6 & 9 \\ 0 & 7 & 8 & 2 \end{bmatrix} = \begin{bmatrix} 0 & 7 & 8 & 2 \\ -4 & 5 & 6 & 9 \\ 2 & 3 & 1 & 4 \end{bmatrix}$$

We see that multiplication on the left of A by F_{13} is equivalent to applying the elementary row operation R_{13} on A.

(b)

$$F_2(k)A = \begin{bmatrix} 1 & 0 & 0 \\ 0 & k & 0 \\ 0 & 0 & 1 \end{bmatrix} \begin{bmatrix} 2 & 3 & 1 & 4 \\ -4 & 5 & 6 & 9 \\ 0 & 7 & 8 & 2 \end{bmatrix} = \begin{bmatrix} 2 & 3 & 1 & 4 \\ -4k & 5k & 6k & 9k \\ 0 & 7 & 8 & 2 \end{bmatrix}$$

Multiplication on the left of A by $F_2(k)$ is equivalent to applying the elementary row operation kR_2 on A.

(c)

$$F_{23}(k)A = \begin{bmatrix} 1 & 0 & 0 \\ 0 & 1 & 0 \\ 0 & k & 1 \end{bmatrix} \begin{bmatrix} 2 & 3 & 1 & 4 \\ -4 & 5 & 6 & 9 \\ 0 & 7 & 8 & 2 \end{bmatrix} = \begin{bmatrix} 2 & 3 & 1 & 4 \\ -4 & 5 & 6 & 9 \\ -4k & 5k+7 & 6k+8 & 9k+2 \end{bmatrix}$$

Multiplication on the left of A by $F_{23}(k)$ is equivalent to applying the elementary row operation $kR_2 + R_3$ on A.

Theorem 3.16. Let A be an $n \times p$ matrix. The effect of applying the elementary row operations R_{st}, kR_s, $k \neq 0$, and $kR_s + R_t$ on A is obtained by multiplying A on the left by F_{st}, $F_s(k)$, and $F_{st}(k)$, respectively, where each of the elementary matrices is an $n \times n$ matrix.

Proof: Let $F_{st}(k)A = C$. Then C is an $n \times p$ matrix and $c_{ij} = = F_i \cdot (A^{(j)})^T$, where F_i is the ith row of $F_{st}(k)$ and $(A^{(j)})^T$ is the transpose of the jth column of A. For $i \neq t$, we have $F_i = [0\,0\,\ldots\,0\,1\,0\,\ldots\,0] = I_i =$ the
$$\overbrace{}^{i\text{th entry}}$$
ith row of the identity matrix of order n. Also, $F_t = kI_s + I_t$. So

$$c_{ij} = F_i \cdot (A^{(j)})^T = 1a_{ij} = a_{ij} \quad \text{for} \quad i \neq t$$

and

$$c_{tj} = F_t \cdot (A^{(j)})^T = (kI_s + I_t) \cdot (A^{(j)})^T$$
$$= (kI_s) \cdot (A^{(j)})^T + I_t \cdot (A^{(j)})^T = ka_{sj} + 1a_{tj} = ka_{sj} + a_{tj}$$

(property $4I$ of the dot product, Section 2.2). Allowing j to run from 1 to p, we see that the ith row of C is

$$C_i = [a_{i1}\,a_{i2}\,\ldots\,a_{ip}] = A_i \quad \text{for} \quad i \neq t$$

and

$$C_t = [ka_{s1} + a_{t1}\,ka_{s2} + a_{t2}\,\cdots\,ka_{sp} + a_{tp}] = kA_s + A_t$$

Hence C is the matrix obtained by applying the elementary row operation $kR_s + R_t$ on A.

We leave the proof that multiplying A on the left by F_{st} has the effect of interchanging rows s and t of A, and that multiplying A on the left by $F_s(k)$ has the effect of multiplying the sth row of A by k as Problems 19 and 20, Exercise 3.7. Q.E.D.

Next we note that the elementary matrices listed in Example 3.29 have inverses. The example illustrates the general situation with respect to inverses of elementary matrices.

Example 3.31. Find the inverses for the 3×3 matrices F_{13}, $F_2(k)$, $k \neq 0$, and $F_{23}(k)$.
(a) The inverse of F_{13} is F_{13} because

$$\begin{bmatrix} 0 & 0 & 1 \\ 0 & 1 & 0 \\ 1 & 0 & 0 \end{bmatrix} \begin{bmatrix} 0 & 0 & 1 \\ 0 & 1 & 0 \\ 1 & 0 & 0 \end{bmatrix} = \begin{bmatrix} 1 & 0 & 0 \\ 0 & 1 & 0 \\ 0 & 0 & 1 \end{bmatrix}$$

(b) $(F_2(k))^{-1} = F_2(1/k)$ because

$$\begin{bmatrix} 1 & 0 & 0 \\ 0 & 1/k & 0 \\ 0 & 0 & 1 \end{bmatrix} \begin{bmatrix} 1 & 0 & 0 \\ 0 & k & 0 \\ 0 & 0 & 1 \end{bmatrix} = I = \begin{bmatrix} 1 & 0 & 0 \\ 0 & k & 0 \\ 0 & 0 & 1 \end{bmatrix} \begin{bmatrix} 1 & 0 & 0 \\ 0 & 1/k & 0 \\ 0 & 0 & 1 \end{bmatrix}$$

(c) $(F_{23}(k))^{-1} = F_{23}(-k)$ because

$$\begin{bmatrix} 1 & 0 & 0 \\ 0 & 1 & 0 \\ 0 & -k & 1 \end{bmatrix} \begin{bmatrix} 1 & 0 & 0 \\ 0 & 1 & 0 \\ 0 & k & 1 \end{bmatrix} = I = \begin{bmatrix} 1 & 0 & 0 \\ 0 & 1 & 0 \\ 0 & k & 1 \end{bmatrix} \begin{bmatrix} 1 & 0 & 0 \\ 0 & 1 & 0 \\ 0 & -k & 1 \end{bmatrix}$$

Theorem 3.17. Each of the elementary matrices of order n has an inverse that is itself an elementary matrix of the same type. In fact,

(a) $F_{st}^{-1} = F_{st}$.

(b) $(F_s(k))^{-1} = F_s(1/k)$, $k \neq 0$.

(c) $(F_{st}(k))^{-1} = F_{st}(-k)$.

Proof: We give the proof for part (c). Consider the product $F_{st}(-k)F_{st}(k)$. The effect of multiplying $F_{st}(k)$ on the left by $F_{st}(-k)$ is to apply the elementary row operation $(-k)R_s + R_t$ on $F_{st}(k)$ (Theorem 3.16). For $i \neq t$, the ith row of $F_{st}(k)$ is $I_i =$ the ith row of the identity matrix of order n. The tth row of $F_{st}(k)$ is $kI_s + I_t$. Therefore, applying the elementary row operation $(-k)R_s + R_t$ on $F_{st}(k)$ yields a matrix whose tth row is $-kI_s + (kI_s + I_t) = I_t$. Every row i other than the tth row of $F_{st}(k)$ is left unchanged and is I_i. We have shown that every row i, including $i = t$, of the product matrix is I_i. Hence $F_{st}(-k)F_{st}(k) = I$. Similarly, we prove that $F_{st}(k)F_{st}(-k) = I$. The effect of multiplying $F_{st}(-k)$ on the left by $F_{st}(k)$ is to apply the elementary row operation $kR_s + R_t$ on $F_{st}(-k)$. The ith row of $F_{st}(-k)$ is I_i for $i \neq t$. The tth row is $-kI_s + I_t$. So, applying the elementary row operation $kR_s + R_t$ on $F_{st}(-k)$ yields a matrix whose tth row is $kI_s + (-kI_s + I_t) = I_t$. Every row $i \neq t$ is left unchanged and is I_i. Hence $F_{st}(-k)F_{st}(k) = I$.

The reader is asked to give similar proofs for parts (a) and (b) in Problems 21 and 22, Exercise 3.7. Q.E.D.

Exercise 3.7

1. Given that

$$\begin{bmatrix} 1 & 3 & 2 \\ 15 & 2 & 0 \\ 4 & 2 & 1 \end{bmatrix}^{-1} = \begin{bmatrix} 2 & 1 & -4 \\ -15 & -7 & 30 \\ 22 & 10 & -43 \end{bmatrix}$$

solve each of the linear systems

$$x_1 + 3x_2 + 2x_3 = b_1$$
$$15x_1 + 2x_2 = b_2$$
$$4x_1 + 2x_2 + x_3 = b_3$$

if $\begin{bmatrix} b_1 \\ b_2 \\ b_3 \end{bmatrix} =$ (a) $\begin{bmatrix} 2 \\ 2 \\ 1 \end{bmatrix}$; (b) $\begin{bmatrix} 6 \\ 1 \\ 3 \end{bmatrix}$; (c) $\begin{bmatrix} 10 \\ 1 \\ 5 \end{bmatrix}$; (d) $\begin{bmatrix} 1 \\ -2 \\ 0 \end{bmatrix}$.

2. Given that

$$\begin{bmatrix} 2 & 5 \\ 1 & 3 \end{bmatrix}^{-1} = \begin{bmatrix} 3 & -5 \\ -1 & 2 \end{bmatrix}$$

solve each of the systems

$$2x_1 + 5x_2 = b_1$$

$$x_1 + 3x_2 = b_2$$

if $\begin{bmatrix} b_1 \\ b_2 \end{bmatrix} =$ (a) $\begin{bmatrix} 1 \\ 3 \end{bmatrix}$; (b) $\begin{bmatrix} 2 \\ 5 \end{bmatrix}$; (c) $\begin{bmatrix} 4 \\ 6 \end{bmatrix}$; (d) $\begin{bmatrix} 8 \\ 5 \end{bmatrix}$; (e) $\begin{bmatrix} 6 \\ 2 \end{bmatrix}$;

3. Find each of the following 4×4 elementary matrices: (a) $F_{23}(\frac{1}{2})$; (b) F_{14}; (c) $F_2(-3)$.

4. Find each of the following 3×3 elementary matrices: (a) $F_{31}(-7)$; (b) F_{12}; (c) $F_2(-3)$.

5. If $A = \begin{bmatrix} 1 & 2 & 2 \\ 3 & 0 & 4 \\ 2 & 1 & 3 \\ 1 & -1 & 5 \end{bmatrix}$ find (a) $F_{23}(\frac{1}{2})A$; (b) $F_{14}A$; (c) $F_3(4)A$.

6. If $A = \begin{bmatrix} 3 & 1 & 4 & 5 \\ -1 & 0 & 2 & 8 \\ 2 & 1 & 3 & 4 \end{bmatrix}$ find (a) $F_{31}(-7)A$; (b) $F_{12}A$; (c) $F_2(-3)A$.

7. Find the inverse of each of the following 4×4 elementary matrices: (a) $F_{23}(\frac{1}{2})$; (b) F_{14}; (c) $F_3(4)$.

8. Find the inverse of each of the following 3×3 elementary matrices: (a) $F_{31}(-7)$; (b) F_{12}; (c) $F_2(-3)$.

9. If $A = F_{23}(4)$, $B = F_{21}$, and $C = F_3(-2)$ are 3×3 elementary matrices, find (a) ABC; (b) $(ABC)^{-1}$. Check your answer.

10. If $A = F_{13}(-3)$, $B = F_{23}$, and $C = F_1(\frac{1}{3})$ are 3×3 matrices, find (a) ABC; (b) $(ABC)^{-1}$. Check your answer.

11. *Prove:* If A is nonsingular and if $AB = AC$, then $B = C$.

12. If A is nonsingular, prove that A^{-1} is nonsingular and that $(A^{-1})^{-1} = A$.

13. *Prove:* If A is a nonsingular $n \times n$ matrix and B is an $n \times p$ matrix, and if $AB = 0_{(n,p)}$, then $B = 0_{(n,p)}$.

14. If A is an $m \times n$ matrix and each B_i is an $n \times p$ matrix, use mathematical induction to prove that $A(B_1 + B_2 + \cdots + B_q) = AB_1 + AB_2 + \cdots + AB_q$ for every positive integer q. You may assume a generalized associative law for addition of matrices.

In Problems 15–17, use the recursive definition of powers given in Definition 3.19.

15. If A and B are matrices of order n, and if $AB = BA$, use the **PMI** to prove that $(AB)^q = A^q B^q$ for all positive integers q.

16. If A is an $n \times n$ matrix, use the **PMI** to prove that $A^p A^q = A^{p+q}$ for all positive integers p and q. *Hint:* Let $S(k)$ be the statement $A^p A^k = A^{p+k}$ for every positive integer p.

17. If A is an $n \times n$ matrix, use the **PMI** and the result of Problem 16 to prove that $(A^p)^q = A^{pq}$ for all positive integers p and q. *Hint:* Let $S(k)$ be the statement $(A^p)^k = A^{pk}$ for every positive integer p.

18. In Theorem 3.11 prove that $AI_n = A$.

19. Prove Theorem 3.16 for $F_s(k)$, $k \neq 0$.

20. Prove Theorem 3.16 for F_{st}.

21. Prove Theorem 3.17(a).

22. Prove Theorem 3.17(b).

3.8 THE INVERSE OF A MATRIX VIA ELEMENTARY MATRICES

Our main goals in this section are to find necessary and sufficient conditions for a matrix to have a multiplicative inverse, and to develop a technique for finding the inverse of any specific matrix whenever the inverse exists.

We begin by proving that every matrix A can be transformed to a reduced row echelon form by multiplying on the left of A by a finite sequence of elementary matrices or, equivalently (Theorem 3.16), by applying a finite sequence of row operations to A. We already have a technique for doing this, the Gauss–Jordan reduction process. However, to develop the theory we need to prove the result for an arbitrary matrix. We do so by mathematical induction. We list the proof as optional, since the conclusion is rather obvious from our previous work.

Theorem 3.18. If A is an $n \times r$ matrix, then there are elementary matrices $F_1, F_2, \ldots, F_q$ such that $(F_q F_{q-1} \ldots F_2 F_1)A = E$, where E is a reduced row echelon matrix.

Proof: (Optional) Let A be any $n \times r$ matrix. The proof is by induction on the number of rows of A. The plan then is to show that (a) the theorem is true if A has one row, and (b) if the theorem is true for all matrices having k rows, then it is also true for all matrices having $k + 1$ rows. The **PMI** will then assure us that Theorem 3.18 is true.

If A has 1 row, there are two possibilities; either A is the zero matrix, in which case A is already a reduced row echelon matrix and $F_{11}A = E$, or A has a nonzero entry. In this case, let j be the first column (from left to right) of A that has a nonzero entry. Then $F_1(1/a_{1j})A$ is a reduced row echelon matrix and the theorem is true for $q = 1$.

Next we assume that the theorem is true for all matrices having k rows. Let A be a matrix with $k + 1$ rows. If $A = \mathbf{0}$, then it is a reduced row echelon matrix. If $A \neq 0$, then A has at least one nonzero entry. Let t be the first column that has a nonzero entry and let a_{st} be a nonzero entry.

$$A = \begin{array}{c} \\ \\ \\ \\ s \\ \\ \\ \\ \end{array} \begin{bmatrix} 0 & \cdots & 0 & * & \cdots & * \\ 0 & \cdots & 0 & * & \cdots & * \\ & & & & & \\ & & & & & \\ 0 & & 0 & a_{st} & & * \\ & & & & & \\ & & & & & \\ 0 & \cdots & 0 & * & \cdots & * \end{bmatrix}$$

The matrix $F_{1s}F_s(1/a_{st})A$ has all zeros in each of the first $t - 1$ columns, and the entry in the first row and tth column is a 1.

$$F_{1s}F_s\left(\frac{1}{a_{st}}\right)A = \begin{bmatrix} 0 & \cdots & 0 & 1 & * & \cdots & * \\ 0 & \cdots & 0 & * & * & \cdots & * \\ & & & & & & \\ & & & & & & \\ 0 & \cdots & 0 & * & * & \cdots & * \end{bmatrix}$$

The next step is to sweep out the tth column by performing appropriate elementary row operations of the form $c_j R_1 + R_j$, $j = 2, 3, \ldots, n$. The effect of these elementary row operations is obtained by multiplying on the left by the corresponding elementary matrices $F_{1j}(c_j)$, $j = 2, 3, \ldots, n$. We obtain

$$B = F_{n+1}F_n \cdots F_2 F_1 A = \begin{bmatrix} 0 & \cdots & 0 & 1 & * & * & \cdots & * \\ 0 & \cdots & 0 & 0 & & & & \\ & & & & & & C & \\ & & & & & & & \\ 0 & \cdots & 0 & 0 & & & & \end{bmatrix}$$

The matrix C has k rows, and by our induction hypothesis C can be transformed to a reduced row echelon form by multiplying on the left by the appropriate elementary matrices of order k or, what amounts to the same thing, by performing the appropriate elementary row operations on C. Multiplying on the left of B by the elementary matrices of order $k + 1$

corresponding to the elementary row operations that reduce the matrix C, we obtain

$$D = F_p \cdots F_2 F_1 A = \begin{bmatrix} 0 & \cdots & 0 & 1 & * & * & \cdots & * \\ 0 & \cdots & 0 & 0 & \boxed{\begin{array}{c} F \\ \text{reduced} \\ \text{row ech-} \\ \text{form} \end{array}} \\ \vdots & & \vdots & \vdots & \\ 0 & \cdots & 0 & 0 \end{bmatrix}$$

Next the nonzero entries in the first row of D that lie above a leading 1 in the matrix F are changed to zeros by multiplying on the left by appropriate elementary matrices of the form $F_{i1}(d_j)$, where j is the column number and i is the row number of D in which the leading 1 appears. Finally, we have

$$F_u \ldots F_2 F_1 A = E$$

a reduced row echelon matrix. The theorem now follows from the principle of mathematical induction. Q.E.D.

Since multiplying on the left of A by an elementary matrix is equivalent to applying an elementary row operation on A, we can restate Theorem 3.18 in terms of row operations. This restatement, which is the next corollary, was used without proof in solving the general system of linear equations in Section 3.2.

Corollary 3.5.   Any $n \times r$ matrix A is row equivalent to a reduced row echelon matrix E.

Note: The reduced row echelon matrix in Theorem 3.18 and Corollary 3.5 is unique, although we do not prove it here.

We are now in a position to prove the following theorem, which gives necessary and sufficient conditions for a square matrix to have an inverse.

Theorem 3.19.   Let A be any $n \times n$ matrix. The following properties are equivalent.
(a) A is nonsingular.
(b) There exist elementary matrices $F_1, F_2, \ldots, F_k$ of order n such that $(F_k \ldots F_2 F_1)A = I_n$.
(c) A is row equivalent to the identity matrix I_n.
(d) A is a product of elementary matrices.

Proof: Suppose that A is nonsingular. Then A^{-1} exists. Let $P = F_k \ldots F_2 F_1$ be a product of elementary matrices such that $PA = E$ and E is a reduced row echelon matrix (Theorem 3.18). By Theorem 3.17, each elementary matrix F_i is nonsingular, and so by Theorem 3.15 P^{-1} exists. Therefore, E is nonsingular and $E^{-1} = A^{-1}P^{-1}$. We claim that $E = I_n$. Since E is an $n \times n$ reduced row echelon matrix, either E has n leading 1's and is the identity matrix, or else E has less than n leading 1's and consequently the last row of E is a row of zeros. If the latter is the case then the last row of EE^{-1} is also a row of zeros. But $EE^{-1} = I_n$ and the (n, n) entry is a 1. Thus we are forced to conclude that $E = I_n$ and

$$PA = (F_k \ldots F_2 F_1)A = I_n$$

We have proved that (a) $\Rightarrow$ (b) (read "(a) implies (b)"). From Definition 3.11 and the fact that multiplying on the left by an elementary matrix is equivalent to applying an elementary row operation, we see that (b) $\Leftrightarrow$ (c) (read "(b) is equivalent to (c)" or "(b) if and only if (c)"). Now assuming that (c) holds, we have $(F_k \ldots F_2 F_1)A = I_n$. Each elementary matrix F_i has an inverse F_i^{-1} that is itself an elementary matrix by Theorem 3.17. Thus

$$(F_1^{-1}F_2^{-1} \ldots F_k^{-1})(F_k \ldots F_2 F_1)A = F_1^{-1}F_2^{-1} \ldots F_k^{-1}I_n$$

and

$$A = F_1^{-1}F_2^{-1} \ldots F_k^{-1}$$

which is a product of elementary matrices. Hence (c) $\Rightarrow$ (d).

Now suppose that (d) holds and that A is a product of elementary matrices. Each elementary matrix is nonsingular and, by Theorem 3.15, A is nonsingular. Hence (d) $\Rightarrow$ (a). We have proved that (a) $\Rightarrow$ (b), (b) $\Rightarrow$ (c), (c) $\Rightarrow$ (d), and (d) $\Rightarrow$ (a). Consequently, the four properties are equivalent. Q.E.D.

Theorem 3.19 states that every nonsingular matrix A is a product of elementary matrices. This factorization of A into a product of elementary matrices is not unique, as the next example shows.

Example 3.32. If $A = \begin{bmatrix} 2 & -5 \\ -1 & 3 \end{bmatrix}$, then $A^{-1} = \begin{bmatrix} 3 & 5 \\ 1 & 2 \end{bmatrix}$ and A is nonsingular. The reader is asked to check that both of the following factorizations of A are correct.

$$A = F_2(-1)F_{21}(2)F_{12}(-3)F_{12} = \begin{bmatrix} 1 & 0 \\ 0 & -1 \end{bmatrix}\begin{bmatrix} 1 & 2 \\ 0 & 1 \end{bmatrix}\begin{bmatrix} 1 & 0 \\ -3 & 1 \end{bmatrix}\begin{bmatrix} 0 & 1 \\ 1 & 0 \end{bmatrix}$$

$$A = F_{12}F_1(-1)F_{12}(2)F_{21}(-3) = \begin{bmatrix} 0 & 1 \\ 1 & 0 \end{bmatrix}\begin{bmatrix} -1 & 0 \\ 0 & 1 \end{bmatrix}\begin{bmatrix} 1 & 0 \\ 2 & 1 \end{bmatrix}\begin{bmatrix} 1 & -3 \\ 0 & 1 \end{bmatrix}$$

Before illustrating a technique for finding A^{-1} for a specific matrix A, we shall prove that, if B is a left or a right inverse for a square matrix A, then B is an inverse for A.

Theorem 3.20. If A and B are matrices of order n and if B is a left inverse for A in the sense that $BA = I$, then it is also true that $AB = I$. If B is assumed to be only a right inverse of A in the sense that $AB = I$, then $BA = I$.

Proof: Suppose that $BA = I$. By Theorem 3.18 there is a matrix Q, which is a product of elementary matrices, such that $QB = E$ and E is a reduced row echelon matrix. We claim that $E = I$. As in the proof of Theorem 3.19, since E is an $n \times n$ matrix, E has n leading 1's and is the identity matrix, or else E has fewer than n leading 1's and the last row of E is a row of zeros. If the latter is the case, then for any matrix C with n rows EC is a matrix in which the last row is also a row of zeros. The product of elementary matrices Q is nonsingular. Therefore, $QB = E$ implies that $B = Q^{-1}E$. Multiplying on the right by A, we get $BA = Q^{-1}EA$ or $I = Q^{-1}EA$. So $Q = EA$. This means that the last row of Q is a row of zeros. But then $QQ^{-1} = I$ is a matrix with the last row a row of zeros. This is a contradiction, since the last row of I has a 1 in the (n, n) entry. We must conclude that $E = I$. We now have $QB = I$ and $Q = EA = IA = A$. Hence $AB = I$. This proves that B is the inverse of A.

Now suppose that B is a right inverse of A; that is, $AB = I$. Then A is a left inverse for B, and by the proof just concluded $A = B^{-1}$. Thus $AB = BA = I$. Q.E.D.

We shall now describe **a technique for determining whether or not a square matrix is nonsingular.** If the matrix is nonsingular, the method will also find the inverse for us.

Suppose that we have a square matrix A. We apply a sequence of elementary row operations $R_1, R_2, \ldots, R_k$ that transform A to a reduced row echelon matrix E. Let $F_1, F_2, \ldots, F_k$ be the elementary matrices that correspond to $R_1, R_2, \ldots, R_k$, respectively. Then $(F_k \ldots F_2 F_1)A = E$, where E is a reduced row echelon matrix. A is nonsingular if and only if $E = I$, and if $E = I$, then $A^{-1} = F_k \ldots F_2 F_1$. A technique for transforming A to E and at the same time finding $F_k \ldots F_2 F_1$ is as follows:

1. Form the $n \times 2n$ matrix $[A \mid I]$. We then apply the elementary row operations $R_1, R_2, \ldots, R_k$ as indicated.

2. $[A \mid I] \xrightarrow{R_1} [F_1 A \mid F_1 I] = [F_1 A \mid F_1]$.

3. $[F_1 A \mid F_1] \xrightarrow{R_2} [F_2 F_1 A \mid F_2 F_1]$.

.

.

.

k $+$ **1.** $[F_{k-1} \ldots F_2 F_1 A \mid F_{k-1} \ldots F_2 F_1] \xrightarrow{R_k} [F_k \ldots F_2 F_1 A \mid F_k \ldots F_2 F_1] = [E = PA \mid P]$. Here $P = F_k \ldots F_2 F_1$.

If $E = I$, then $PA = I$. By Theorem 3.20, $A^{-1} = P = F_k \ldots F_2 F_1$.

Example 3.33. Use the technique described above to find the inverse of

$$A = \begin{bmatrix} 1 & 2 & 0 \\ 3 & 4 & 2 \\ 1 & 1 & -1 \end{bmatrix}$$

if it has one.

Solution: We form the 3×6 matrix $[A \vdots I]$ and proceed to transform A to reduced row echelon form.

$$\begin{bmatrix} 1 & 2 & 0 \vdots & 1 & 0 & 0 \\ 3 & 4 & 2 \vdots & 0 & 1 & 0 \\ 1 & 1 & -1 \vdots & 0 & 0 & 1 \end{bmatrix} \xrightarrow[-R_1+R_3]{-3R_1+R_2} \begin{bmatrix} 1 & 2 & 0 \vdots & 1 & 0 & 0 \\ 0 & -2 & 2 \vdots & -3 & 1 & 0 \\ 0 & -1 & -1 \vdots & -1 & 0 & 1 \end{bmatrix}$$

$$\xrightarrow[-\tfrac{1}{2}R_2]{R_{23}} \begin{bmatrix} 1 & 2 & 0 \vdots & 1 & 0 & 0 \\ 0 & 1 & 1 \vdots & 1 & 0 & -1 \\ 0 & -2 & 2 \vdots & -3 & 1 & 0 \end{bmatrix}$$

$$\xrightarrow[2R_2+R_3]{-2R_2+R_1} \begin{bmatrix} 1 & 0 & -2 \vdots & -1 & 0 & 2 \\ 0 & 1 & 1 \vdots & 1 & 0 & -1 \\ 0 & 0 & 4 \vdots & -1 & 1 & -2 \end{bmatrix}$$

$$\xrightarrow[2R_3+R_1]{\substack{(1/4)R_3 \\ -R_3+R_2}} \begin{bmatrix} 1 & 0 & 0 \vdots & -\tfrac{3}{2} & \tfrac{1}{2} & 1 \\ 0 & 1 & 0 \vdots & \tfrac{5}{4} & -\tfrac{1}{4} & -\tfrac{1}{2} \\ 0 & 0 & 1 \vdots & -\tfrac{1}{4} & \tfrac{1}{4} & -\tfrac{1}{2} \end{bmatrix}$$

Since $E = I$,

$$A^{-1} = \begin{bmatrix} -\tfrac{3}{2} & \tfrac{1}{2} & 1 \\ \tfrac{5}{4} & -\tfrac{1}{4} & -\tfrac{1}{2} \\ -\tfrac{1}{4} & \tfrac{1}{4} & -\tfrac{1}{2} \end{bmatrix}$$

As a check we multiply

$$\begin{bmatrix} -\tfrac{3}{2} & \tfrac{1}{2} & 1 \\ \tfrac{5}{4} & -\tfrac{1}{4} & -\tfrac{1}{2} \\ -\tfrac{1}{4} & \tfrac{1}{4} & -\tfrac{1}{2} \end{bmatrix} \begin{bmatrix} 1 & 2 & 0 \\ 3 & 4 & 2 \\ 1 & 1 & -1 \end{bmatrix} = \begin{bmatrix} 1 & 0 & 0 \\ 0 & 1 & 0 \\ 0 & 0 & 1 \end{bmatrix}$$

This check shows that the matrix in question is a left inverse for A. Then applying Theorem 3.20 we see that the matrix is the inverse of A.

Example 3.34. (a) Find the inverse of the nonsingular matrix

$$A = \begin{bmatrix} 2 & 0 & 0 \\ 4 & 3 & 0 \\ 6 & 2 & 1 \end{bmatrix}$$

(b) Write A^{-1} as a product of elementary matrices.
(c) Write A as a product of elementary matrices.

Solution

(a)
$$\begin{bmatrix} 2 & 0 & 0 & \vdots & 1 & 0 & 0 \\ 4 & 3 & 0 & \vdots & 0 & 1 & 0 \\ 6 & 2 & 1 & \vdots & 0 & 0 & 1 \end{bmatrix} \xrightarrow[\substack{-4R_1+R_2 \\ -6R_1+R_3}]{(1/2)R_1} \begin{bmatrix} 1 & 0 & 0 & \vdots & \frac{1}{2} & 0 & 0 \\ 0 & 3 & 0 & \vdots & -2 & 1 & 0 \\ 0 & 2 & 1 & \vdots & -3 & 0 & 1 \end{bmatrix}$$

$$\xrightarrow[\substack{-2R_2+R_3}]{(1/3)R_2} \begin{bmatrix} 1 & 0 & 0 & \vdots & \frac{1}{2} & 0 & 0 \\ 0 & 1 & 0 & \vdots & -\frac{2}{3} & \frac{1}{3} & 0 \\ 0 & 0 & 1 & \vdots & -\frac{5}{3} & -\frac{2}{3} & 1 \end{bmatrix}$$

So

$$A^{-1} = \begin{bmatrix} \frac{1}{2} & 0 & 0 \\ -\frac{2}{3} & \frac{1}{3} & 0 \\ -\frac{5}{3} & -\frac{2}{3} & 1 \end{bmatrix}$$

(b) $A^{-1} = F_k \ldots F_2 F_1$, where $F_1, F_2, \ldots, F_k$ are the elementary matrices corresponding to the sequence of elementary row operations $R_1, R_2, \ldots, R_k$, which were applied to transform A to the reduced row echelon matrix I. So

$$A^{-1} = F_{23}(-2)F_2(\tfrac{1}{3})F_{13}(-6)F_{12}(-4)F_1(\tfrac{1}{2})$$

(c) $A = (A^{-1})^{-1} = F_1(2)F_{12}(4)F_{13}(6)F_2(3)F_{23}(2)$.

Note: As was noted in Example 3.32, the factorization of a non-singular matrix into a product of elementary matrices is not unique. The reason for this is that there is more than one sequence of elementary row operations that can be used to transform a matrix to reduced row echelon form.

Example 3.35. Determine if

$$A = \begin{bmatrix} 1 & 2 & 3 \\ 4 & 8 & 0 \\ -2 & -4 & 1 \end{bmatrix}$$

is nonsingular. If A^{-1} exists, find it.

Solution

$$\begin{bmatrix} 1 & 2 & 3 & \vdots & 1 & 0 & 0 \\ 4 & 8 & 0 & \vdots & 0 & 1 & 0 \\ -2 & -4 & 1 & \vdots & 0 & 0 & 1 \end{bmatrix} \xrightarrow[\substack{2R_1+R_3}]{-4R_1+R_2} \begin{bmatrix} 1 & 2 & 3 & \vdots & 1 & 0 & 0 \\ 0 & 0 & -12 & \vdots & -4 & 1 & 0 \\ 0 & 0 & 7 & \vdots & 2 & 0 & 1 \end{bmatrix}$$

$$\xrightarrow[\substack{-7R_2+R_3}]{(-1/12)R_2} \begin{bmatrix} 1 & 2 & 3 & \vdots & 1 & 0 & 0 \\ 0 & 0 & 1 & \vdots & \frac{1}{3} & -\frac{1}{12} & 0 \\ 0 & 0 & 0 & \vdots & -\frac{1}{3} & \frac{7}{12} & 1 \end{bmatrix}$$

$$\xrightarrow[]{-3R_2+R_1} \begin{bmatrix} 1 & 2 & 0 & \vdots & 0 & \frac{1}{4} & 0 \\ 0 & 0 & 1 & \vdots & \frac{1}{3} & -\frac{1}{12} & 0 \\ 0 & 0 & 0 & \vdots & -\frac{1}{3} & \frac{7}{12} & 1 \end{bmatrix}$$

The reduced row echelon form of A is

$$E = \begin{bmatrix} 1 & 2 & 0 \\ 0 & 0 & 1 \\ 0 & 0 & 0 \end{bmatrix} \neq I$$

Therefore, A^{-1} does not exist.

Exercise 3.8

In Problems 1–6, determine if the matrices are nonsingular by transforming $[A \mid I]$ to $[PA = E \mid P]$. Keep track of the elementary operations used, and if A is nonsingular, write both A and A^{-1} as products of elementary matrices.

1. $A = \begin{bmatrix} 1 & 1 & 4 \\ 1 & 2 & 1 \\ 0 & 0 & 1 \end{bmatrix}$ **2.** $A = \begin{bmatrix} 1 & 1 \\ 2 & 1 \end{bmatrix}$ **3.** $A = \begin{bmatrix} 3 & 3 & 4 \\ 1 & 2 & 1 \\ 1 & 6 & 0 \end{bmatrix}$

4. $A = \begin{bmatrix} 1 & 0 \\ 4 & 2 \end{bmatrix}$ **5.** $A = \begin{bmatrix} 1 & 0 & 1 & 0 \\ 0 & 1 & 0 & 1 \\ 2 & 3 & 4 & 0 \\ 0 & 0 & 0 & 1 \end{bmatrix}$ **6.** $A = \begin{bmatrix} 1 & 0 & 0 & 0 \\ 0 & 2 & 2 & 0 \\ 0 & 0 & 3 & 0 \\ 0 & 0 & 0 & 4 \end{bmatrix}$

7. Find A^{-1} if

$$A = \begin{bmatrix} 1 & 0 & 0 & 0 & 0 \\ 0 & 1 & 0 & 0 & 0 \\ 2 & 3 & 1 & 0 & 0 \\ 4 & 2 & 3 & 1 & 0 \\ -2 & 1 & 4 & 2 & 1 \end{bmatrix}$$

Check your answer.

8. Solve the systems $AX = B$ if A is the matrix in Problem 1 and (a) $B = \begin{bmatrix} 3 \\ 2 \\ 4 \end{bmatrix}$;

(b) $B = \begin{bmatrix} 2 \\ 1 \\ 0 \end{bmatrix}$; (c) $B = \begin{bmatrix} -1 \\ 3 \\ 2 \end{bmatrix}$.

9. Write $A = \begin{bmatrix} 3 & 1 \\ 11 & 4 \end{bmatrix}$ as a product of elementary matrices in two different ways.

10. (a) Show that $A = \begin{bmatrix} 4 & 0 & 1 \\ 2 & 3 & 6 \\ 6 & -3 & -4 \end{bmatrix}$ has no inverse.

(b) Find a row echelon matrix E that is row equivalent to A, and write E as $E = F_k \ldots F_2 F_1 A$ for appropriate elementary matrices $F_1, F_2, \ldots, F_k$.

11. Let $A = \begin{bmatrix} a & b \\ c & d \end{bmatrix}$ and assume that $ad - bc \neq 0$.

(a) If $a \neq 0$, use the technique $[A \mid I] \rightarrow [I \mid A^{-1}]$ to find A^{-1}. Give a formula for A^{-1} in terms of a, b, c, and d.

(b) Verify that the formula holds even if $a = 0$.

(c) Show that A^{-1} exists if and only if $ad - bc \neq 0$.

12. Prove that an $n \times n$ matrix is nonsingular if and only if the only $n \times 1$ column vector satisfying the matrix equation $AX = 0_{(n, 1)}$ is $X = 0$. *Hint:* Use Theorem 3.4 of Section 3.2.

13. *Prove:* If A and B are $n \times n$ matrices and AB is nonsingular, then both A and B are nonsingular. *Hint:* Apply Problem 12 to the equation $BX = 0$.

14. If A is an $n \times n$ matrix and one row of A is a multiple of another row of A, show that A is singular.

15. *Prove:* If A and B are $n \times n$ matrices and if A is nonsingular and B is singular, then AB and BA are singular.

16. Let A be the $n \times n$ coefficient matrix of a system of n linear equations in n unknowns. *Prove:* The system of n linear equations in n unknowns has a unique solution if and only if A is nonsingular.

17. Use mathematical induction to prove that, if a square matrix A is nonsingular, then $(A^n)^{-1} = (A^{-1})^n$ for all positive integers n.

4 Determinants

We have already mentioned that the determinant of a 2×2 matrix $A = \begin{bmatrix} a & b \\ c & d \end{bmatrix}$ is the real number, $\det A = ad - bc$. We wish to define the determinant of A for any $n \times n$ matrix A. Our motivation for studying determinants comes primarily from the topics to be covered in Chapter 7. Two of the main properties that we will need at that time are (1) $\det A$ can be calculated from the entries of A in a reasonably simple manner, and (2) $\det A \neq 0$ if and only if A is nonsingular.

4.1 THE DETERMINANT FUNCTION

In this section we define the determinant of an $n \times n$ matrix. There are several ways that this can be done. The definition that we give will enable us to derive rather easily a procedure for calculating determinants that is not unlike the Gauss–Jordan reduction process. Some of the theory of determinants involves rather cumbersome and difficult processes, which we feel may not be too enlightening. Our approach will be to assume without proof those theorems that we feel fall into this category.

We have defined vectors by properties that they possess; that is, vectors are objects that belong to a mathematical system satisfying properties 1A–5A and 6M–10M. Likewise, we shall define the determinant of an $n \times n$ matrix by properties that it must possess. Before giving the definition, we illustrate the properties and the notation to be used with 2×2 determinants.

138

Example 4.1. If $A = \begin{bmatrix} a_{11} & a_{12} \\ a_{21} & a_{22} \end{bmatrix}$, we shall write det $A = \det(A_1, A_2)$, where A_1 and A_2 are the first and second rows of A, respectively. Show that det $A = a_{11}a_{22} - a_{12}a_{21}$ has the following properties.

1d. $\det(A_1, A_2) = 0$, if $A_1 = A_2$.
2d. $\det(rA_1, A_2) = r \det A = \det(A_1, rA_2)$.
3d. $\det(A_1 + B, A_2) = \det(A_1, A_2) + \det(B, A_2)$ and $\det(A_1, A_2 + B) = \det(A_1, A_2) + \det(A_1, B)$, where $B = [b_1 \quad b_2]$.
4d. $\det \begin{bmatrix} 1 & 0 \\ 0 & 1 \end{bmatrix} = 1$.

Verification:

1d. $\det(A_1, A_1) = \det \begin{bmatrix} a_{11} & a_{12} \\ a_{11} & a_{12} \end{bmatrix} = a_{11}a_{12} - a_{12}a_{11} = 0$

2d. $\det(rA_1, A_2) = \det \begin{bmatrix} ra_{11} & ra_{12} \\ a_{21} & a_{22} \end{bmatrix} = (ra_{11})a_{22} - (ra_{12})a_{21}$

$\qquad\qquad\quad = r(a_{11}a_{22} - a_{12}a_{21}) = r \det A$

$\quad \det(A_1, rA_2) = \det \begin{bmatrix} a_{11} & a_{12} \\ ra_{21} & ra_{22} \end{bmatrix} = a_{11}(ra_{22}) - a_{12}(ra_{21})$

$\qquad\qquad\quad = r(a_{11}a_{22} - a_{12}a_{21}) = r \det A$

3d. $\det(A_1 + B, A_2) = \det \begin{bmatrix} a_{11} + b_1 & a_{12} + b_2 \\ a_{21} & a_{22} \end{bmatrix}$

$\qquad\qquad\quad = (a_{11} + b_1)a_{22} - (a_{12} + b_2)a_{21}$

$\qquad\qquad\quad = a_{11}a_{22} + b_1 a_{22} - (a_{12}a_{21} + b_2 a_{21})$

$\qquad\qquad\quad = (a_{11}a_{22} - a_{12}a_{21}) + (b_1 a_{22} - b_2 a_{21})$

$\qquad\qquad\quad = \det \begin{bmatrix} a_{11} & a_{12} \\ a_{21} & a_{22} \end{bmatrix} + \det \begin{bmatrix} b_1 & b_2 \\ a_{21} & a_{22} \end{bmatrix}$

$\qquad\qquad\quad = \det(A_1, A_2) + \det(B, A_2)$

$\quad \det(A_1, A_2 + B) = \det \begin{bmatrix} a_{11} & a_{12} \\ a_{21} + b_1 & a_{22} + b_2 \end{bmatrix}$

$\qquad\qquad\quad = a_{11}(a_{22} + b_2) - a_{12}(a_{21} + b_1)$

$\qquad\qquad\quad = (a_{11}a_{22} - a_{12}a_{21}) + (a_{11}b_2 - a_{12}b_1)$

$\qquad\qquad\quad = \det \begin{bmatrix} a_{11} & a_{12} \\ a_{21} & a_{22} \end{bmatrix} + \det \begin{bmatrix} a_{11} & a_{12} \\ b_1 & b_2 \end{bmatrix}$

$\qquad\qquad\quad = \det(A_1, A_2) + \det(A_1, B)$

4d. $\det \begin{bmatrix} 1 & 0 \\ 0 & 1 \end{bmatrix} = 1(1) - 0(0) = 1$

We take the properties illustrated in Example 4.1 for the determinant of a 2×2 matrix as the defining properties for the determinant of an

$n \times n$ matrix. The notation in Example 4.1 is also the notation we use in the general definition.

For any $n \times n$ matrix A with rows $A_1, A_2, \ldots, A_n$, we write $\det A = \det(A_1, \ldots, A_i, \ldots, A_n)$ and read "determinant of A."

Definition 4.1. For a fixed positive integer n, any function that assigns to each $n \times n$ matrix A a real number, denoted by $\det A = \det(A_1, \ldots, A_i, \ldots, A_n)$, and which has the properties **1d–4d** listed below, is called a **determinant** or an **n × n determinant**.

1d. $\det(A_1, \ldots, A_i, \ldots, A_j, \ldots, A_n) = 0$ if $A_i = A_j$ and $i \neq j$. In words, the determinant of a matrix having two identical rows is zero.

2d. Let B be the matrix obtained from A by multiplying the ith row by a scalar r. $\det B = \det(A_1, \ldots, rA_i, \ldots, A_n) = r \det A$. In words, if any row of A is multiplied by a scalar r, then the determinant of the new matrix B is r times the determinant of A.

3d. $\det(A_1, \ldots, A_i + A_i', \ldots, A_n) = \det(A_1, \ldots, A_i, \ldots, A_n) + \det(A_1, \ldots, A_i', \ldots, A_n)$, where A_i' is any $1 \times n$ row matrix and i is any positive integer from 1 to n.

4d. $\det I_n = 1$, where I_n is the identity matrix of order n.

Definition 4.1 does not tell us that there is a function satisfying properties 1d–4d, but rather if there is such a function then we call it a determinant. Example 4.1 illustrated the existence of a function satisfying Definition 4.1 for 2×2 matrices. We prove that it is unique in the next example.

Example 4.2. Show that the function defined on $\mathbf{R}_{2 \times 2}$ by

$$\det \begin{bmatrix} a_{11} & a_{12} \\ a_{21} & a_{22} \end{bmatrix} = a_{11}a_{22} - a_{12}a_{21}$$

is the only function satisfying properties 1d–4d.

Proof: Suppose that D is a function defined on $\mathbf{R}_{2 \times 2}$ that satisfies 1d–4d. For any 2×2 matrix A, we claim that $D(A) = a_{11}a_{22} - a_{12}a_{21}$. To verify this claim, we write

$$A = \begin{bmatrix} a_{11} & a_{12} \\ a_{21} & a_{22} \end{bmatrix} = \begin{bmatrix} A_1 \\ A_2 \end{bmatrix}$$

where $A_1 = a_{11}[1 \quad 0] + a_{12}[0 \quad 1]$ and $A_2 = a_{21}[1 \quad 0] + a_{22}[0 \quad 1]$. If we let $E_1 = [1 \quad 0]$ and $E_2 = [0 \quad 1]$, we can write

$$A = \begin{bmatrix} a_{11}E_1 + a_{12}E_2 \\ a_{21}E_1 + a_{22}E_2 \end{bmatrix}$$

Then (the notation $\stackrel{3d}{=}$ to be used below means "equal because of property 3d")

$$D(A) = D(A_1, A_2) = D(A_1, a_{21}E_1 + a_{22}E_2)$$

$$\stackrel{3d}{=} D(A_1, a_{21}E_1) + D(A_1, a_{22}E_2)$$

$$\stackrel{2d}{=} a_{21}D(A_1, E_1) + a_{22}D(A_1, E_2)$$

$$= a_{21}D(a_{11}E_1 + a_{12}E_2, E_1) + a_{22}D(a_{11}E_1 + a_{12}E_2, E_2)$$

$$\stackrel{3d}{=} [a_{21}D(a_{11}E_1, E_1) + a_{21}D(a_{12}E_2, E_1)]$$
$$\quad + [a_{22}D(a_{11}E_1, E_2) + a_{22}D(a_{12}E_2, E_2)]$$

$$\stackrel{2d}{=} [a_{21}a_{11}D(E_1, E_1) + a_{21}a_{12}D(E_2, E_1)]$$
$$\quad + [a_{22}a_{11}D(E_1, E_2) + a_{22}a_{12}D(E_2, E_2)]$$

$$\stackrel{1d}{=} 0 + a_{21}a_{12}D(E_2, E_1) + a_{22}a_{11}D(E_1, E_2) + 0$$

$$\stackrel{4d}{=} a_{21}a_{12}D(E_2, E_1) + a_{22}a_{11}$$

We will have our result if we can show that $D(E_2, E_1) = -1$. To this end consider that

$$D(E_1 + E_2, E_1 + E_2) = D(E_1, E_1) + D(E_1, E_2) + D(E_2, E_1) + D(E_2, E_2)$$

Here we have used property 3d several times. This last expression is $0 + 1 + D(E_2, E_1) + 0$ by properties 1d and 4d. But $D(E_1 + E_2, E_1 + E_2) = 0$ by property 1d. So $1 + D(E_2, E_1) = 0$, and $D(E_2, E_1) = -1$. Hence

$$D(A) = a_{21}a_{12}(-1) + a_{22}a_{11} = a_{11}a_{22} - a_{12}a_{21} = \det A \qquad \text{Q.E.D.}$$

It is a fact that there is precisely one function defined on $\mathbf{R}_{n \times n}$ that satisfies 1d–4d. The uniqueness is stated in the next theorem.

Theorem 4.1. If D and D' are functions defined on the collection $\mathbf{R}_{n \times n}$ of all $n \times n$ matrices, and if D and D' satisfy each of the properties 1d–4d, then $D(A) = D'(A)$ for all matrices A of order n.

The proof of Theorem 4.1 is by mathematical induction and follows basically the ideas in the proof of Example 4.2. However, the general proof requires an extensive discussion of permutations, which we wish to avoid. The proof of this theorem and that of Theorem 4.2, which gives a function satisfying properties 1d–4d, can be found in a number of texts. One that uses the same notation and includes the proofs is Nathan Divinsky, *Linear Algebra* (Palo Alto, California: Page Ficklin Publications, 1975).

To give an expression for the value of det A, we need to discuss submatrices. As we shall see shortly, it is possible to express the determinant of an $n \times n$ matrix in terms of determinants of $(n - 1) \times (n - 1)$ matrices.

Definition 4.2. A **submatrix** of an $m \times n$ matrix A is a matrix obtained from A by removing one or more of its rows and/or one or more of its columns.

We shall be concerned with submatrices obtained from an $n \times n$ matrix by deleting one row and one column. If A is an $n \times n$ matrix, we shall use the notation A_{rs} to denote the submatrix of A obtained by deleting the rth row and sth column of A. From

remove
$\downarrow$

$$
A = \begin{bmatrix}
a_{11} & \cdots & a_{1s} & \cdots & a_{1n} \\
\vdots & & \vdots & & \vdots \\
a_{r1} & \cdots & a_{rs} & \cdots & a_{rn} \\
\vdots & & \vdots & & \vdots \\
a_{n1} & \cdots & a_{ns} & \cdots & a_{nn}
\end{bmatrix} \leftarrow \text{remove}
$$

we obtain

$$
A_{rs} = \begin{bmatrix}
a_{11} & \cdots & a_{1,s-1} & a_{1,s+1} & \cdots & a_{1n} \\
\vdots & & \vdots & \vdots & & \vdots \\
a_{r-1,1} & \cdots & a_{r-1,s-1} & a_{r-1,s+1} & \cdots & a_{r-1,n} \\
a_{r+1,1} & \cdots & a_{r+1,s-1} & a_{r+1,s+1} & \cdots & a_{r+1,n} \\
\vdots & & \vdots & \vdots & & \vdots \\
a_{n1} & \cdots & a_{n,s-1} & a_{n,s+1} & \cdots & a_{nn}
\end{bmatrix}
$$

Example 4.3. If

$$
A = \begin{bmatrix}
1 & 2 & -1 & 4 \\
3 & 0 & 4 & 5 \\
6 & 2 & 8 & 7 \\
-3 & 9 & -7 & 0
\end{bmatrix}
$$

the submatrix obtained by deleting the first row and third column of A is

$$
A_{13} = \begin{bmatrix}
3 & 0 & 5 \\
6 & 2 & 7 \\
-3 & 9 & 0
\end{bmatrix}
$$

The submatrix obtained by deleting the second row and second column of A is

$$A_{22} = \begin{bmatrix} 1 & -1 & 4 \\ 6 & 8 & 7 \\ -3 & -7 & 0 \end{bmatrix}$$

A simple way to obtain, say A_{12}, from A is to draw lines through the first row and second column of A and then write out the array that is left:

$$\begin{bmatrix} 1 & 2 & -1 & 4 \\ 3 & 0 & 4 & 5 \\ 6 & 2 & 8 & 7 \\ -3 & 9 & -7 & 0 \end{bmatrix} \quad \text{so} \quad A_{12} = \begin{bmatrix} 3 & 4 & 5 \\ 6 & 8 & 7 \\ -3 & -7 & 0 \end{bmatrix}$$

Definition 4.3. For any $n \times n$ matrix A with $n \geq 2$, the **minor** belonging to the (i, j) entry a_{ij} of A is $\det A_{ij} = M_{ij}$. The **cofactor** of a_{ij} is $C_{ij} = (-1)^{i+j} M_{ij}$.

We are now ready to state a theorem that gives several formulas for the determinant of an $n \times n$ matrix. As was mentioned earlier, a proof can be found in *Linear Algebra* by Nathan Divinsky.

Theorem 4.2

(a) The function defined on $\mathbf{R}_{1 \times 1}$ by $\det [a] = a$ satisfies properties 1d–4d.
(b) A function defined on $\mathbf{R}_{n \times n}$, $n \geq 2$, satisfying properties 1d–4d exists. It is given by any one of the $2n$ formulas

$$\det A = a_{i1}C_{i1} + a_{i2}C_{i2} + \cdots + a_{in}C_{in} \qquad \text{for any fixed } i, \ i = 1, 2, \ldots, n \ (\text{expansion along the } i\text{th row of } A)$$

$$\text{or} \quad \det A = a_{1j}C_{1j} + a_{2j}C_{2j} + \cdots + a_{nj}C_{nj} \qquad \text{for any fixed } j, \ j = 1, 2, \ldots, n \ (\text{expansion along the } j\text{th column of } A)$$

We illustrate Theorem 4.2 with some examples. In addition to $\det A$, another common notation for the determinant of A is to write $|A|$ or, in case the entries are displayed, to write vertical lines around the array. We shall use the two notations interchangeably.

Example 4.4. (a) Find $\det \begin{bmatrix} 1 & 2 & -1 \\ 0 & 3 & 1 \\ 2 & 1 & 4 \end{bmatrix}$.

(b) Find $\det \begin{bmatrix} a_{11} & a_{12} & a_{13} \\ a_{21} & a_{22} & a_{23} \\ a_{31} & a_{32} & a_{33} \end{bmatrix}$.

Solution: (a) Expanding along the first column, we obtain

$$1C_{11} + 0C_{21} + 2C_{31} = C_{11} + 2C_{31}, \qquad C_{11} = +M_{11} = \begin{vmatrix} 3 & 1 \\ 1 & 4 \end{vmatrix} = 11$$

$$C_{31} = +M_{31} = \begin{vmatrix} 2 & -1 \\ 3 & 1 \end{vmatrix} = 5$$

Therefore,

$$\begin{vmatrix} 1 & 2 & -1 \\ 0 & 3 & 1 \\ 2 & 1 & 4 \end{vmatrix} = 11 + 2(5) = 21$$

As a check, we shall find the determinant by expanding along the third row of the matrix.

$$2C_{31} + 1C_{32} + 4C_{33} = 2(-1)^4 \begin{vmatrix} 2 & -1 \\ 3 & 1 \end{vmatrix} + (-1)^5 \begin{vmatrix} 1 & -1 \\ 0 & 1 \end{vmatrix}$$

$$+ 4(-1)^6 \begin{vmatrix} 1 & 2 \\ 0 & 3 \end{vmatrix}$$

$$= 2(5) + (-1)(1) + 4(3) = 21$$

(b) We expand along the first row and get

$$a_{11}(-1)^{1+1}M_{11} + a_{12}(-1)^{1+2}M_{12} + a_{13}(-1)^{1+3}M_{13}$$

$$= a_{11} \begin{vmatrix} a_{22} & a_{23} \\ a_{32} & a_{33} \end{vmatrix} - a_{12} \begin{vmatrix} a_{21} & a_{23} \\ a_{31} & a_{33} \end{vmatrix} + a_{13} \begin{vmatrix} a_{21} & a_{22} \\ a_{31} & a_{32} \end{vmatrix}$$

$$= a_{11}a_{22}a_{33} - a_{11}a_{23}a_{32} - a_{12}a_{21}a_{33} + a_{12}a_{23}a_{31}$$
$$+ a_{13}a_{21}a_{32} - a_{13}a_{22}a_{31}$$

Next we illustrate two schematic devices for evaluating the determinant of a 3×3 matrix.

Device 1

Step 1: Write down A and then write down the first two columns of A as illustrated.

$$\begin{array}{ccc|cc} a_{11} & a_{12} & a_{13} & a_{11} & a_{12} \\ a_{21} & a_{22} & a_{23} & a_{21} & a_{22} \\ a_{31} & a_{32} & a_{33} & a_{31} & a_{32} \end{array}$$

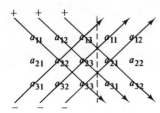

Step 2: Compute the products as indicated by the arrows. The products corresponding to the arrows pointing down will have plus signs attached to them. The products corresponding to arrows pointing up will have negative signs attached to them.

Step 3: Sum the resulting products, using the appropriate signs as determined in step 2.

Device 2

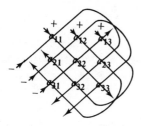

Warning: THE SCHEMATIC DEVICES WORK ONLY FOR 3×3 DETERMINANTS AND CANNOT BE USED TO EVALUATE 4×4 OR HIGHER DETERMINANTS.

Example 4.5. Calculate $\begin{vmatrix} 1 & 2 & -1 \\ 3 & 2 & 1 \\ 0 & 1 & 4 \end{vmatrix}$.

Solution

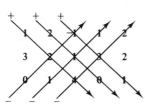

$8 + 0 - 3 - 0 - 1 - 24 = -20$ $8 + 0 - 3 - 0 - 24 - 1 = -20$

Example 4.6. Calculate $A = \begin{vmatrix} 1 & 3 & 4 & -2 \\ 4 & 3 & 0 & -1 \\ 2 & 4 & 1 & 3 \\ -2 & 1 & 0 & 4 \end{vmatrix}$.

Solution: Expanding along column 3, we obtain

$$\det A = 4(-1)^{1+3} \begin{vmatrix} 4 & 3 & -1 \\ 2 & 4 & 3 \\ -2 & 1 & 4 \end{vmatrix} + 0(-1)^{2+3} \begin{vmatrix} 1 & 3 & -2 \\ 2 & 4 & 3 \\ -2 & 1 & 4 \end{vmatrix}$$

$$+ 1(-1)^{3+3} \begin{vmatrix} 1 & 3 & -2 \\ 4 & 3 & -1 \\ -2 & 1 & 4 \end{vmatrix}$$

$$+ (-1)^{3+4}(0) \begin{vmatrix} 1 & 3 & -2 \\ 4 & 3 & -1 \\ 2 & 4 & 3 \end{vmatrix}$$

$$= 4(64 - 18 - 2 - 8 - 24 - 12) + 0$$
$$+ 1(12 + 6 - 8 - 12 - 48 + 1) + 0$$
$$= 4(0) + 1(-49) = -49$$

Note that because column 3 had two zeros the calculation reduced from one of calculating four 3×3 determinants to that of calculating two 3×3 determinants.

We close this section by remarking that many authors use an alternative definition of determinant that involves permutations. We give a brief introduction to this alternative. The reader may have noticed that

$$\det [a_{ij}]_{(3,3)} = a_{11}a_{22}a_{33} - a_{11}a_{23}a_{32} - a_{12}a_{21}a_{33}$$
$$+ a_{12}a_{23}a_{31} + a_{13}a_{21}a_{32} - a_{13}a_{22}a_{31}$$

is a sum of products where each term contains precisely one entry from each row and column. In summation notation we may write

$$\det [a_{ij}]_{(3,3)} = \sum \pm a_{1j_1}a_{2j_2}a_{3j_3}$$

where $(j_1 j_2 j_3)$ ranges over all possible **permutations** (rearrangements) of the numbers 1, 2, and 3. The sign in front of a term is plus if $(j_1 j_2 j_3)$ is an even permutation and minus if it is an odd permutation. The permutation is **even** if it takes an even number of inversions $(i\ k) \rightarrow (k\ i)$ to convert $(j_1 j_2 j_3)$ to the natural order (1 2 3), and it is **odd** if it takes an odd number of inversions. For example, (2 3 1) is even because it can be converted to the natural order (1 2 3) with 2 inversions: $(2\ 3\ 1) \xrightarrow{(2\ 1) \rightarrow (1\ 2)} (1\ 3\ 2) \xrightarrow{(3\ 2) \rightarrow (2\ 3)} (1\ 2\ 3)$. Thus the term $a_{12}a_{23}a_{31}$ is preceded by a plus sign. (3 2 1) is odd because it can be converted to the natural order (1 2 3) in one inversion: $(3\ 2\ 1) \xrightarrow{(3\ 1) \rightarrow (1\ 3)} (1\ 2\ 3)$. Therefore, the term $a_{13}a_{22}a_{31}$ is preceded by a minus sign. An alternative definition for $\det [a_{ij}]_{(n,n)}$ may be given by extending these notions. In summation notation

$$\det [a_{ij}]_{(n,n)} = \sum \pm a_{1j_1}a_{2j_2} \cdots a_{nj_n}$$

where $(j_1 j_2 \ldots j_n)$ ranges over all possible permutations of the numbers 1, 2, ..., n, and the sign of the term is plus if the permutation is even and minus if the permutation is odd.

Exercise 4.1

1. Evaluate $\det \begin{bmatrix} a & b \\ c & d \end{bmatrix}$ by
   (a) Expanding along the first row.
   (b) Expanding along the second row.
   (c) Expanding along the first column.
   (d) Expanding along the second column.

2. (a) Calculate det $\begin{bmatrix} 1 & -2 & 3 \\ 0 & 1 & 2 \\ 1 & 4 & -1 \end{bmatrix}$ by expanding along the third column.

   (b) Check your answer in Problem 2(a) by expanding along the second row.

   (c) Check your answer in Problem 2(a) by using one of the two schematic devices that were given.

3. Express det $\begin{bmatrix} 0 & x-1 & 1 \\ x & 1 & 1 \\ 0 & x+1 & 2 \end{bmatrix}$ as a polynomial in x.

4. Calculate det $\begin{bmatrix} 1 & 2 & -1 & 1 \\ 3 & 0 & 1 & 2 \\ 1 & -1 & 2 & 1 \\ 1 & 0 & 3 & -2 \end{bmatrix}$.

5. A square matrix in which all the entries above the principal diagonal are zero is called a **lower triangular matrix**.

   (a) Determine the determinant of the lower triangular matrix

   $$A = \begin{bmatrix} 1 & 0 & 0 \\ 2 & 3 & 0 \\ 4 & 5 & 6 \end{bmatrix}.$$

   (b) Show that the determinant of the general lower triangular matrix $\begin{bmatrix} a & 0 & 0 \\ b & c & 0 \\ d & e & f \end{bmatrix}$ of order 3 is the product of the elements on the main diagonal.

   (c) Show that the result in Problem 5(b) can be generalized to lower triangular matrices of order n. *Hint:* Continue to expand along the first row of each of the submatrices.

6. A square matrix in which all the entries below the main diagonal are zero is called an **upper triangular matrix**.

   (a) Determine the determinant of the upper triangular matrix $\begin{bmatrix} 1 & 2 & 3 \\ 0 & 4 & 5 \\ 0 & 0 & 6 \end{bmatrix}.$

   (b) Show that the determinant of the general upper triangular matrix $\begin{bmatrix} a & b & c \\ 0 & d & e \\ 0 & 0 & f \end{bmatrix}$ of order 3 is the product of the elements on the main diagonal.

   (c) Show that the result in Problem 6(b) can be generalized to upper triangular matrices of order n.

7. If $A = \begin{bmatrix} 1 & 2 & -1 \\ 3 & 2 & 1 \\ 0 & 1 & 4 \end{bmatrix}$, evaluate $|A - \lambda I|$, where λ is a scalar and I is the 3 × 3 identity matrix.

8. If $A = \begin{bmatrix} a_{11} & a_{12} & a_{13} \\ a_{21} & a_{22} & a_{23} \\ a_{31} & a_{32} & a_{33} \end{bmatrix}$, evaluate $|A^T|$. How does $|A^T|$ compare with $|A|$?

9. Use Theorem 4.2 to verify that $\det A = \det (A^T)$ for any matrix A of order n. *Hint:* The rows of A are the columns of A^T.

10. Compute the determinant of $A = \begin{bmatrix} 1 & 0 & 0 & 2 & 3 \\ 0 & 1 & -1 & 2 & 0 \\ 0 & 2 & 0 & 1 & -2 \\ 2 & 1 & 1 & 0 & 0 \\ 0 & 1 & 0 & 1 & 0 \end{bmatrix}$.

11. Use Device 1 to show that $\begin{vmatrix} \vec{i} & \vec{j} & \vec{k} \\ a_1 & a_2 & a_3 \\ b_1 & b_2 & b_3 \end{vmatrix} = A \times B$, where $A = (a_1, a_2, a_3)$ and $B = (b_1, b_2, b_3)$.

Note: The reader may have correctly observed that this is not a determinant because the entries in the first row are vectors and not real numbers. It is however, a convenient notation for remembering $A \times B$.

12. (a) If $A = \begin{bmatrix} 1 & 0 & 0 & 0 \\ 2 & 3 & 4 & 1 \\ 2 & 6 & 2 & 1 \\ 1 & 4 & 2 & 3 \end{bmatrix}$, evaluate $\det A$ by expanding along a row that

requires the evaluation of only one 3×3 determinant.

(b) Sum the products indicated by the arrows. Is your answer equal to $\det A$?

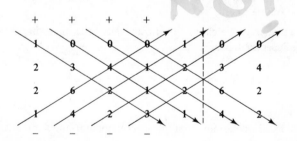

13. Prove that the function on $\mathbf{R}_{1 \times 1}$ defined by $\det [a] = a$ satisfies properties 1d–4d [Theorem 4.2(a)].

14. Evaluate $\begin{vmatrix} 1 & 2 & 32 & 14 \\ 6 & 8 & 19 & 27 \\ 37 & 49 & 82 & 53 \\ 6 & 8 & 19 & 27 \end{vmatrix}$.

15. (a) Use property $2d$ to verify that

$$\det \begin{bmatrix} 6 & 9 & 6 \\ 5 & 0 & 10 \\ 4 & 8 & -8 \end{bmatrix} = 60 \det \begin{bmatrix} 2 & 3 & 2 \\ 1 & 0 & 2 \\ 1 & 2 & -2 \end{bmatrix}$$

(b) Check the result in part (a) by evaluating both determinants.

16. Use property 2d to find det $(F_s(k)A)$ in terms of det A.

17. Prove that for any $n \times n$ matrix A, det $(rA) = r^n$ det A.

18. Prove that for any 3×3 matrix A, det $(A_1, A_2, A_3) = 0$ if $A_3 = aA_1 + bA_2$.

In Problems 19 and 20, there are 24 terms of the form $\pm(a_{1j_1}a_{2j_2}a_{3j_3}a_{4j_4})$ in det $[a_{ij}]_{(4,4)}$. Using the remarks at the end of this section, determine if a plus or a minus sign should precede the given term.

19. (a) $a_{13}a_{22}a_{31}a_{44}$.   (b) $a_{14}a_{23}a_{31}a_{42}$.

     (c) $a_{11}a_{23}a_{34}a_{42}$.   (d) $a_{12}a_{23}a_{34}a_{41}$.

20. (a) $a_{13}a_{21}a_{34}a_{42}$.   (b) $a_{14}a_{22}a_{31}a_{43}$.

     (c) $a_{12}a_{24}a_{31}a_{43}$.   (d) $a_{11}a_{24}a_{32}a_{43}$.

4.2 ADDITIONAL PROPERTIES OF THE DETERMINANT FUNCTION

In Section 4.1 we described a method for calculating the determinant of any $n \times n$ matrix, that is, expansion by cofactors. This method however involves long computations even for n as small as 4. We have seen that unless there are zero entries the calculation of a 3×3 determinant by this method requires the evaluation of $3! = 6$ terms, where each term is a product of three factors. In general, to calculate an $n \times n$ determinant requires the evaluation of $n!$ terms with each term containing n factors. For $n = 12$, this means there are 479,001,600 terms to evaluate, each term consisting of 12 factors. In this section we derive a process that makes the task of evaluating determinants more pleasant.

If a square matrix has enough zero entries, its determinant can easily be calculated. For example, we have seen in Problems 5 and 6 of Exercise 4.1 that a triangular matrix (either upper or lower) has a determinant equal to the product of the elements on the main diagonal. Now the reduced row echelon form (also the row echelon form) of a matrix is an upper triangular matrix, and we know that any matrix can be transformed to a reduced row echelon matrix by applying a sequence of elementary row operations. The question naturally arises, what effect does applying an elementary row operation have on the determinant of a square matrix? The answer is given in the next theorem.

Theorem 4.3. (a) If two rows of an $n \times n$ matrix A are interchanged, then the sign of the determinant is changed.

(b) If all the entries of a row of an $n \times n$ matrix A are multiplied by a scalar k, then the determinant of the resulting matrix is k times the determinant of A.

(c) If the entries of one row of an $n \times n$ matrix are multiplied by a scalar k and added to the corresponding entries of another row, then the determinant is left unchanged.

In terms of the elementary matrices, we can express Theorem 4.3 as follows:

(a) $\det(F_{rs}A) = -\det A$ if $r \neq s$.
(b) $\det(F_s(k)A) = k(\det A)$.
(c) $\det(F_{rs}(k)A) = \det A$.

Note: Theorem 4.3(b) is true even if $k = 0$. As a consequence, we have that *the determinant of a square matrix with a row of zeros is* 0.

Proof: We begin by proving Theorem 4.3(a). Suppose that B is obtained from A by interchanging the rth and sth rows of A and consider

$$\det(A_1, \ldots, A_{r-1}, A_r + A_s, A_{r+1}, \ldots, A_{s-1}, A_r + A_s, A_{s+1}, \ldots, A_n)$$

By property 1d, this determinant is 0 because the rth row is equal to the sth row. Just as we did in the proof of Example 4.2 we use property 3d several times to obtain

$$0 \overset{3d}{=} \det(A_1, \ldots, A_{r-1}, A_r, A_{r+1}, \ldots, A_{s-1}, A_r, A_{s+1}, \ldots, A_n)$$
$$+ \det(A_1, \ldots, A_{r-1}, A_r, A_{r+1}, \ldots, A_{s-1}, A_s, A_{s+1}, \ldots, A_n)$$
$$+ \det(A_1, \ldots, A_{r-1}, A_s, A_{r+1}, \ldots, A_{s-1}, A_r, A_{s+1}, \ldots, A_n)$$
$$+ \det(A_1, \ldots, A_{r-1}, A_s, A_{r+1}, \ldots, A_{s-1}, A_s, A_{s+1}, \ldots, A_n)$$
$$\overset{1d}{=} 0 + \det A + \det B + 0 = \det A + \det B$$

Hence $\det B = -(\det A)$. This completes the proof of part (a). Next we prove part (c).

Let $B = F_{rs}(k)A$, which is the matrix obtained by multiplying the entries of the rth row of A by k and adding the results to the corresponding entries of the sth row of A.

$$\det B = \det(A_1, \ldots, A_r, \ldots, A_{s-1}, kA_r + A_s, A_{s+1}, \ldots, A_n)$$
$$\overset{3d}{=} \det(A_1, \ldots, A_r, \ldots, A_{s-1}, kA_r, A_{s+1}, \ldots, A_n)$$
$$+ \det(A_1, \ldots, A_r, \ldots, A_{s-1}, A_s, A_{s+1}, \ldots, A_n)$$
$$\overset{2d}{=} k \det(A_1, \ldots, A_r, \ldots, A_{s-1}, A_r, A_{s+1}, \ldots, A_n) + \det A$$
$$\overset{1d}{=} 0 + \det A = \det A$$

Hence $\det B = \det A$.

Since Theorem 4.3(b) is just a restatement of property 2d of Definition 4.1, we can conclude that parts (a), (b), and (c) are true. Q.E.D.

If we let $A = I$, where I is the $n \times n$ identity matrix, we obtain

$$\det F_{rs} = \det (F_{rs}I) = -(\det I) = -1$$

$$\det F_s(k) = \det (F_s(k)I) = k(\det I) = k$$

$$\det F_{rs}(k) = \det(F_{rs}(k)I) = \det I = 1$$

Thus we have the following corollary.

Corollary 4.1.   The elementary matrices have the following determinants:

(a) $\det F_{rs} = -1$ if $r \neq s$.
(b) $\det F_s(k) = k$ if $k \neq 0$.
(c) $\det F_{rs}(k) = 1$.

Since $\det F_{rs}A = -\det A = \det F_{rs} \det A$

$$\det F_s(k)A = k \det A = \det F_s(k) \det A$$

$$\det F_{rs}(k)A = \det A = \det F_{rs}(k) \det A$$

we also have the following corollary:

Corollary 4.2.   If F is an $n \times n$ elementary matrix and if A is any $n \times n$ matrix, then $\det FA = \det F \det A$.

By mathematical induction we can extend this result to $\det (F_1F_2 \ldots F_pA)$ $= \det F_1 \det F_2 \ldots \det F_p \det A$ for any elementary matrices $F_1, F_2, \ldots, F_p$ and for any matrix A of order n (Problem 16, Exercise 4.2).

Note: In practice we often wish to reverse the process in Theorem 4.3(b); that is, instead of multiplying a row of a matrix by some number, we may wish to factor a number out. For example, if $B = \begin{bmatrix} 32 & 64 \\ 1 & 3 \end{bmatrix}$, then

$$\det B = \begin{vmatrix} 32 & 64 \\ 1 & 3 \end{vmatrix} = 32 \begin{vmatrix} 1 & 2 \\ 1 & 3 \end{vmatrix} = 32\,(\det A)$$

where $A = \begin{bmatrix} 1 & 2 \\ 1 & 3 \end{bmatrix}$.

Example 4.7.   Use Theorem 4.3 to evaluate $\det A$ if

$$A = \begin{bmatrix} 2 & 3 & 2 & 1 \\ 1 & 4 & -6 & 1 \\ 2 & 1 & 7 & 1 \\ -1 & 0 & 1 & 2 \end{bmatrix}.$$

Solution: We use vertical lines to denote the determinant and keep track of $|A|$ as we reduce the matrix.

$$
\begin{vmatrix} 2 & 3 & 2 & 1 \\ 1 & 4 & -6 & 1 \\ 2 & 1 & 7 & 1 \\ -1 & 0 & 1 & 2 \end{vmatrix} = (-1) \begin{vmatrix} -1 & 0 & 1 & 2 \\ 1 & 4 & -6 & 1 \\ 2 & 1 & 7 & 1 \\ 2 & 3 & 2 & 1 \end{vmatrix} = (-1)(-1) \begin{vmatrix} 1 & 0 & -1 & -2 \\ 1 & 4 & -6 & 1 \\ 2 & 1 & 7 & 1 \\ 2 & 3 & 2 & 1 \end{vmatrix}
$$

$$
= \begin{vmatrix} 1 & 0 & -1 & -2 \\ 0 & 4 & -5 & 3 \\ 0 & 1 & 9 & 5 \\ 0 & 3 & 4 & 5 \end{vmatrix} = (-1) \begin{vmatrix} 1 & 0 & -1 & -2 \\ 0 & 1 & 9 & 5 \\ 0 & 4 & -5 & 3 \\ 0 & 3 & 4 & 5 \end{vmatrix}
$$

$$
= (-1) \begin{vmatrix} 1 & 0 & -1 & -2 \\ 0 & 1 & 9 & 5 \\ 0 & 0 & -41 & -17 \\ 0 & 0 & -23 & -10 \end{vmatrix} = (-1)(-1) \begin{vmatrix} 1 & 0 & -1 & -2 \\ 0 & 1 & 9 & 5 \\ 0 & 0 & 5 & 3 \\ 0 & 0 & 23 & 10 \end{vmatrix}
$$

$$
= 5 \begin{vmatrix} 1 & 0 & -1 & -2 \\ 0 & 1 & 9 & 5 \\ 0 & 0 & 1 & \frac{3}{5} \\ 0 & 0 & 23 & 10 \end{vmatrix} = 5 \begin{vmatrix} 1 & 0 & -1 & -2 \\ 0 & 1 & 9 & 5 \\ 0 & 0 & 1 & \frac{3}{5} \\ 0 & 0 & 0 & -\frac{19}{5} \end{vmatrix}^{*}
$$

$$
= 5(-\tfrac{19}{5}) \begin{vmatrix} 1 & 0 & -1 & -2 \\ 0 & 1 & 9 & 5 \\ 0 & 0 & 1 & \frac{3}{5} \\ 0 & 0 & 0 & 1 \end{vmatrix} = -19 \begin{vmatrix} 1 & 0 & 0 & 0 \\ 0 & 1 & 0 & 0 \\ 0 & 0 & 1 & 0 \\ 0 & 0 & 0 & 1 \end{vmatrix} = -19
$$

Notice that the determinant marked by an asterisk is the determinant of an upper triangular matrix. We could have stopped there and concluded that $\det A = 5(-\frac{19}{5}) = -19$ (Problem 6, Exercise 4.1).

Example 4.8. If $A = \begin{bmatrix} 1 & 0 & 3 & -3 \\ 2 & 2 & -1 & -4 \\ 3 & 1 & 2 & -8 \\ -1 & 4 & 1 & 7 \end{bmatrix}$, evaluate $\det A$ by transforming A to reduced row echelon form.

Solution

$$
|A| = \begin{vmatrix} 1 & 0 & 3 & -3 \\ 0 & 2 & -7 & 2 \\ 0 & 1 & -7 & 1 \\ 0 & 4 & 4 & 4 \end{vmatrix} = (-1) \begin{vmatrix} 1 & 0 & 3 & -3 \\ 0 & 1 & -7 & 1 \\ 0 & 2 & -7 & 2 \\ 0 & 4 & 4 & 4 \end{vmatrix}
$$

$$= (-1) \begin{vmatrix} 1 & 0 & 3 & -3 \\ 0 & 1 & -7 & 1 \\ 0 & 0 & 7 & 0 \\ 0 & 0 & 32 & 0 \end{vmatrix}$$

$$= -7 \begin{vmatrix} 1 & 0 & 3 & -3 \\ 0 & 1 & -7 & 1 \\ 0 & 0 & 1 & 0 \\ 0 & 0 & 32 & 0 \end{vmatrix} = -7 \begin{vmatrix} 1 & 0 & 0 & -3 \\ 0 & 1 & 0 & 1 \\ 0 & 0 & 1 & 0 \\ 0 & 0 & 0 & 0 \end{vmatrix}$$

$$= 0$$

Examples 4.7 and 4.8 illustrate the two possibilities for the reduced row echelon form of a square matrix. Either it is the identity matrix or it has at least one row of zeros. It is this fact that enables us to prove the following important theorem.

Theorem 4.4.   An $n \times n$ matrix A is nonsingular if and only if the determinant of A is not zero.

Proof: According to Theorem 3.18, there are elementary matrices $F_1, F_2, \ldots, F_p$ such that $F_p \ldots F_2 F_1 A = E$, where E is a reduced row echelon matrix. From Corollary 4.2 we have

$$\det (F_p \ldots F_2 F_1 A) = \det F_p \ldots \det F_2 \det F_1 \det A = \det E$$

Now suppose that A is nonsingular. Then

$$E = I \quad \text{and} \quad \det F_p \ldots \det F_1 \det A = \det I = 1.$$

Therefore, $\det A \neq 0$.

Conversely, if A is singular, then since A is an $n \times n$ matrix the last row of E must be a row of zeros. We then have $\det E = 0$. So

$$\det F_p \ldots \det F_1 \det A = \det E = 0.$$

But $\det F_i \neq 0$ by Corollary 4.1. Hence $\det A = 0$.

We have shown that if A is nonsingular then $\det A \neq 0$, and if A is singular then $\det A = 0$. Q.E.D.

We close this section by proving that the determinant of a product is the product of the determinants.

Theorem 4.5.   If A and B are $n \times n$ matrices, then $\det AB = \det A \det B$.

Proof: If AB is nonsingular, then both A and B are nonsingular by Problem 13, Exercise 3.8. According to Theorem 3.19, we can write $A = F_k \ldots F_2 F_1$, where each F_i is an elementary matrix. Then

$$\det AB = \det(F_k \ldots F_2 F_1 B) = \det F_k \ldots \det F_2 \det F_1 \det B$$

by Corollary 4.2 (Problem 16, Exercise 4.2). Also by Corollary 4.2, we have

$$\det(F_k \ldots F_2 F_1) = \det F_k \ldots \det F_2 \det F_1$$

(the A in Corollary 4.2 can be any matrix including F_1). So

$$\det AB = \det(F_k \ldots F_2 F_1) \det B = \det A \det B$$

We have shown that the theorem is true if AB is nonsingular. On the other hand, if AB is singular, then either A or B must be singular, because if A^{-1} and B^{-1} exist then $B^{-1}A^{-1} = (AB)^{-1}$, which is a contradiction. We can now apply Theorem 4.4 and conclude that $\det AB = 0$ and $\det A = 0$ or $\det B = 0$. Hence $\det AB = \det A \det B$. Q.E.D.

Exercise 4.2

In Problems 1 and 2, check the result of Corollary 4.1 by using the cofactor expansion along an appropriate column or row to evaluate the determinants of the given elementary matrices.

1. Each elementary matrix is of order 4.

(a) F_{24}; (b) $F_3(\frac{2}{8})$; (c) $F_{13}(5)$.

2. Each elementary matrix is of order 3.

(a) F_{13}; (b) $F_1(17)$; (c) $F_{23}(7)$.

In Problems 3–6, evaluate $\det A$ by transforming A to reduced row echelon form, as in Examples 4.7 and 4.8.

3. (a) $A = \begin{bmatrix} 1 & 2 & -1 & 1 \\ 3 & 0 & 1 & 2 \\ 1 & -1 & 2 & 1 \\ 1 & 0 & 3 & -2 \end{bmatrix}$; (b) $A = \begin{bmatrix} 1 & 2 & 3 & 4 \\ 0 & 1 & 2 & -1 \\ 0 & 0 & 1 & 2 \\ 2 & -3 & 1 & 2 \end{bmatrix}$;

(c) $A = \begin{bmatrix} 3 & 2 & 1 & 7 \\ 1 & 1 & 4 & 6 \\ 4 & -3 & 5 & 11 \\ 2 & 0 & 2 & 6 \end{bmatrix}$.

4. (a) $A = \begin{bmatrix} 1 & 2 & -1 & 0 \\ 0 & 2 & 1 & 1 \\ -1 & 0 & 2 & 1 \\ 1 & 2 & -4 & 2 \end{bmatrix}$; (b) $A = \begin{bmatrix} 1 & -1 & 0 & 2 & 4 \\ 2 & 0 & 1 & -1 & 2 \\ 1 & 1 & -1 & 0 & 1 \\ 0 & 0 & 1 & -1 & 2 \\ 1 & 0 & 0 & 2 & 6 \end{bmatrix}$;

(c) $A = \begin{bmatrix} 2 & 4 & 6 \\ 3 & 6 & 9 \\ 1 & 2 & 3 \end{bmatrix}$.

5. (a) $A = \begin{bmatrix} 1 & 2 & 4 \\ 3 & 6 & 5 \\ 4 & 3 & 2 \end{bmatrix}$;

(b) $A = \begin{bmatrix} 3 & -2 & 4 \\ 5 & 2 & 3 \\ 2 & 6 & 8 \end{bmatrix}$;

(c) $A = \begin{bmatrix} -1 & 3 & 5 & 2 \\ 2 & 0 & 3 & 7 \\ -3 & 2 & 1 & 1 \\ 0 & 4 & 6 & 3 \end{bmatrix}$.

6. (a) $A = \begin{bmatrix} 14 & 21 & 7 \\ 17 & 34 & 51 \\ 11 & -22 & 33 \end{bmatrix}$;

(b) $A = \begin{bmatrix} 2 & -3 & 1 & 3 \\ 1 & 5 & 4 & 8 \\ 3 & 4 & -2 & 13 \\ 0 & 2 & 3 & 2 \end{bmatrix}$;

(c) $A = \begin{bmatrix} 1 & -2 & 3 \\ 0 & 1 & 2 \\ 1 & 4 & -4 \end{bmatrix}$.

In Problems 7 and 8, use Theorem 4.5 to evaluate det AB. Check your answer by finding the product AB and then evaluating $|AB|$.

7. $A = \begin{bmatrix} 3 & 4 & 2 \\ 1 & 2 & 0 \\ 2 & 5 & -2 \end{bmatrix}$, $B = \begin{bmatrix} 1 & 3 & -1 \\ 2 & 4 & 3 \\ 0 & 2 & 1 \end{bmatrix}$.

8. $A = \begin{bmatrix} 2 & 1 & 0 & 3 \\ -1 & 3 & 4 & 1 \\ 0 & 2 & 3 & 1 \\ 1 & 4 & 0 & -1 \end{bmatrix}$, $B = \begin{bmatrix} -1 & 2 & 3 & 0 \\ 0 & 2 & 4 & 1 \\ 1 & 3 & 0 & 5 \\ 2 & 0 & 1 & 1 \end{bmatrix}$.

9. If $A = \begin{bmatrix} 1 & 3 & 2 \\ 2 & 4 & 4 \\ 3 & 0 & 1 \end{bmatrix}$,

(a) Find A^{-1}.

(b) Find det A and det A^{-1}.

10. *Prove:* If A is nonsingular, then det $(A^{-1}) = 1/\text{det } A$.

11. Use row operations to show that

$$\begin{vmatrix} 1 & 1 & 1 \\ a & b & c \\ a^2 & b^2 & c^2 \end{vmatrix} = (b - a)(c - a)(c - b)$$

This determinant is called a *Vandermonde determinant*.

12. If det $\begin{bmatrix} a & b & c \\ d & e & f \\ g & h & i \end{bmatrix} = 7$, find   (a) det $\begin{bmatrix} a & b & c \\ -d & -e & -f \\ a+g & b+h & c+i \end{bmatrix}$;

(b) det $\begin{bmatrix} d & e & f \\ a & b & c \\ 3g & 3h & 3i \end{bmatrix}$;   (c) det $\begin{bmatrix} g & h & i \\ a & b & c \\ d & e & f \end{bmatrix}$.

13. *Prove:* If A is an $n \times n$ idempotent matrix ($A^2 = A$ and $A \neq 0$), then det A = 0 or det $A = 1$.

14. *Prove:* If A is an $n \times n$ matrix, det $rA = r^n$ det A.

15. *Prove:* The equation of the line through the points $P = (a, b)$ and $Q = (c, d)$ is given by det $\begin{bmatrix} 1 & 1 & 1 \\ x & a & c \\ y & b & d \end{bmatrix} = 0$.

16. Use mathematical induction to prove the following extension of Corollary 4.2. For any positive integer p, if $F_1, F_2, \ldots, F_p$ are $n \times n$ elementary matrices and A is any $n \times n$ matrix, then det $(F_1 F_2 \ldots F_p A) = $ det F_1 det $F_2 \ldots$ det F_p det A.

17. Use mathematical induction to prove the following extension of Theorem 4.5. For any positive integer k, if $A_1, A_2, \ldots, A_k$ are $n \times n$ matrices, then det $(A_1 A_2 \ldots A_k) = $ det A_1 det $A_2 \ldots$ det A_k.

18. *Prove:* A system of n linear equations in n unknowns with coefficient matrix A has a unique solution if and only if det $A \neq 0$.

†4.3 MATRIX INVERSION VIA THE ADJOINT MATRIX; CRAMER'S RULE

We have already seen how to find the inverse of a nonsingular matrix by transforming the matrix to reduced row echelon form. From a computational point of view, this is the preferred method. A second method of inverting a matrix, which is of some interest from a theoretical and a historical point of view, will be discussed in this section. The method depends on the expansion of a determinant by cofactors. To develop the method, our first step will be to prove the following lemma.

Lemma 4.1. If A is an $n \times n$ matrix and $C_{ij} = (-1)^{i+j} |A_{ij}|$ is the cofactor of the (ij) entry of A, then

$$a_{i1} C_{k1} + a_{i2} C_{k2} + \ldots + a_{in} C_{kn} = 0, \quad \text{if } k \neq i$$

In words, if you expand along one row of A but use the corresponding cofactors of a different row, the resulting sum is zero.

Proof: Let $B = [b_{ij}]_{(n,n)}$ be the same matrix as A except that the kth row of B is the ith row of A. Thus $B_k = B_i = A_i$ and

$$\det B = \det (B_1, \ldots, B_i, \ldots, B_k, \ldots, B_n)$$
$$= \det (A_1, \ldots, A_i, \ldots, A_i, \ldots, A_n) = 0$$

---

†This section is optional, for the results are not used in other sections of the text.

Now if we expand det B along the kth row we obtain

$$0 = \det B = b_{k1}(-1)^{k+1}|B_{k1}| + b_{k2}(-1)^{k+2}|B_{k2}|$$
$$+ \ldots + b_{kn}(-1)^{k+n}|B_{kn}|$$
$$= a_{i1}(-1)^{k+1}|A_{k1}| + a_{i2}(-1)^{k+2}|A_{k2}|$$
$$+ \ldots + a_{in}(-1)^{k+n}|A_{kn}|$$
$$= a_{i1}C_{k1} + a_{i2}C_{k2} + \ldots + a_{in}C_{kn} \qquad \text{Q.E.D.}$$

The next step on our way to deriving a formula for A^{-1} is the following definition.

Definition 4.4. If A is any $n \times n$ matrix, the **matrix of cofactors** of A is the $n \times n$ matrix $\tilde{A}$ ($\tilde{A}$ is read "A tilde"), where the $(i\,j)$ entry of $\tilde{A}$ is the cofactor $C_{ij} = (-1)^{i+j}|A_{ij}|$ of a_{ij}. The **adjoint** (also **adjugate**) of A is the transpose of the matrix of cofactors of A. We write adj $A = \tilde{A}^T$.

Theorem 4.6. If A is an $n \times n$ matrix, then $A(\text{adj } A) = (\det A)I_n$.

Proof: Let adj $A = B$ and $D = AB$. The $(i\,j)$ entry of D is

$$d_{ij} = a_{i1}b_{1j} + a_{i2}b_{2j} + \ldots + a_{in}b_{nj}$$
$$= a_{i1}C_{j1} + a_{i2}C_{j2} + \ldots + a_{in}C_{jn}$$

We see that
$$d_{ii} = a_{i1}C_{i1} + a_{i2}C_{i2} + \ldots + a_{in}C_{in} = \det A$$

and by Lemma 4.1
$$d_{ij} = 0, \quad \text{if } i \neq j$$

Therefore, $A(\text{adj } A) = (\det A)I_n$.   Q.E.D.

As a corollary, we have the following formula for the inverse of a nonsingular matrix.

Corollary 4.3. If A is a nonsingular matrix of order n, then

$$A^{-1} = \frac{1}{|A|}(\text{adj } A).$$

Proof: Suppose that A is nonsingular. According to Theorem 4.4, $|A| \neq 0$. From Theorem 4.6, $1/|A|(A(\text{adj } A)) = I$. Since scalars pull out (property 15P), we have $A(1/|A|(\text{adj } A)) = I$. It now follows from Theorem 3.20 that $A^{-1} = 1/|A|(\text{adj } A)$.   Q.E.D.

Example 4.9. If $A = \begin{bmatrix} 1 & 2 & 3 \\ 0 & -2 & 1 \\ 4 & 2 & 1 \end{bmatrix}$, find adj A and $A(\text{adj } A)$. Use the result to find det A. Find A^{-1}.

Solution

$$C_{11} = + \begin{vmatrix} -2 & 1 \\ 2 & 1 \end{vmatrix} = -4, \quad C_{12} = - \begin{vmatrix} 0 & 1 \\ 4 & 1 \end{vmatrix} = 4, \quad C_{13} = + \begin{vmatrix} 0 & -2 \\ 4 & 2 \end{vmatrix} = 8$$

$$C_{21} = - \begin{vmatrix} 2 & 3 \\ 2 & 1 \end{vmatrix} = 4, \quad C_{22} = + \begin{vmatrix} 1 & 3 \\ 4 & 1 \end{vmatrix} = -11, \quad C_{23} = - \begin{vmatrix} 1 & 2 \\ 4 & 2 \end{vmatrix} = 6$$

$$C_{31} = + \begin{vmatrix} 2 & 3 \\ -2 & 1 \end{vmatrix} = 8, \quad C_{32} = - \begin{vmatrix} 1 & 3 \\ 0 & 1 \end{vmatrix} = -1, \quad C_{33} = + \begin{vmatrix} 1 & 2 \\ 0 & -2 \end{vmatrix} = -2$$

So

$$\tilde{A} = \begin{bmatrix} -4 & 4 & 8 \\ 4 & -11 & 6 \\ 8 & -1 & -2 \end{bmatrix} \quad \text{and} \quad \text{adj } A = \begin{bmatrix} -4 & 4 & 8 \\ 4 & -11 & -1 \\ 8 & 6 & -2 \end{bmatrix}$$

We find that

$$A(\text{adj } A) = \begin{bmatrix} 28 & 0 & 0 \\ 0 & 28 & 0 \\ 0 & 0 & 28 \end{bmatrix} = (\det A)I$$

$$\det A = 28$$

$$A^{-1} = \tfrac{1}{28} \begin{bmatrix} -4 & 4 & 8 \\ 4 & -11 & -1 \\ 8 & 6 & -2 \end{bmatrix}$$

Note: Because of the large number of calculations involved in finding adj A, the generally preferred way to find A^{-1} is to use the Gauss–Jordan reduction method.

Historically, both matrices and determinants have their roots in the study of linear equations. In 1750 the work of Gabriel Cramer (1704–1752) in solving a system of n linear equations in n unknowns led to the rudiments of the theory of determinants. The next theorem is named in his honor.

Theorem 4.7 (Cramer's Rule). If the coefficient matrix A of the system of n linear equations and n unknowns

$$a_{11}x_1 + a_{12}x_2 + \cdots + a_{1n}x_n = b_1$$
$$a_{21}x_2 + a_{22}x_2 + \cdots + a_{2n}x_n = b_2$$
$$\cdots\cdots\cdots\cdots$$
$$a_{n1}x_1 + a_{n2}x_2 + \cdots + a_{nn}x_n = b_n$$

is nonsingular, then the system has a unique solution given by

$$x_j = \frac{\det B_j}{\det A}, \quad j = 1, 2, \ldots, n$$

where B_j is the $n \times n$ matrix obtained from A by replacing the jth column of A with the column of constants on the right side of the equations.

Proof: Let $AX = B$ be the matrix equation equivalent to the given system. According to Theorem 3.13, the unique solution is $X = A^{-1}B$. By Corollary 4.3, we have

$$X = \left(\frac{1}{|A|}(\text{adj } A) \right) B = \frac{1}{|A|}((\text{adj } A)B)$$

Let $H = \text{adj } A$ and $D = HB$. D is an $n \times 1$ column matrix, and the jth entry is

$$d_j = h_{j1}b_1 + h_{j2}b_2 + \cdots + h_{jn}b_n$$

$$= b_1 C_{1j} + b_2 C_{2j} + \cdots + b_n C_{nj}$$

$$= \det \begin{bmatrix} a_{11}a_{12} \cdots a_{1,j-1}b_1 a_{1,j+1} \cdots a_{1n} \\ a_{21}a_{22} \cdots a_{2,j-1}b_2 a_{2,j+1} \cdots a_{2n} \\ \cdot \quad \cdot \quad\quad \cdot \quad \cdot \quad \cdot \quad\quad \cdot \\ \cdot \quad \cdot \quad\quad \cdot \quad \cdot \quad \cdot \quad\quad \cdot \\ a_{n1} a_{n2} \cdots a_{n,j-1}b_n a_{n,j+1} \cdots a_{nn} \end{bmatrix}$$

$$= \det B_j$$

The next to last equality is clear if one expands along the jth column of the matrix B_j. We now have

$$x_j = \frac{1}{|A|} d_j = \frac{\det B_j}{\det A}, \qquad j = 1, 2, \ldots, n \qquad\qquad \text{Q. E. D.}$$

Example 4.10. Use Cramer's rule to solve the system

$$x_1 + 2x_2 + x_3 = 1$$
$$2x_1 + 0x_2 + x_3 = 2$$
$$-x_1 + 1x_2 + 2x_3 = 4$$

Solution

$$A = \begin{bmatrix} 1 & 2 & 1 \\ 2 & 0 & 1 \\ -1 & 1 & 2 \end{bmatrix}, \quad B_1 = \begin{bmatrix} 1 & 2 & 1 \\ 2 & 0 & 1 \\ 4 & 1 & 2 \end{bmatrix}, \quad B_2 = \begin{bmatrix} 1 & 1 & 1 \\ 2 & 2 & 1 \\ -1 & 4 & 2 \end{bmatrix},$$

$$B_3 = \begin{bmatrix} 1 & 2 & 1 \\ 2 & 0 & 2 \\ -1 & 1 & 4 \end{bmatrix}$$

So

$$x_1 = \frac{|B_1|}{|A|} = \frac{1}{-9}, \quad x_2 = \frac{|B_2|}{|A|} = \frac{5}{-9}, \quad x_3 = \frac{|B_3|}{|A|} = \frac{-20}{-9} = \frac{20}{9}$$

Check

$$X = \begin{bmatrix} -\frac{1}{9} \\ -\frac{5}{9} \\ \frac{20}{9} \end{bmatrix} = \frac{1}{9} \begin{bmatrix} -1 \\ -5 \\ 20 \end{bmatrix}$$

$$AX = \frac{1}{9} \begin{bmatrix} 1 & 2 & 1 \\ 2 & 0 & 1 \\ -1 & 1 & 2 \end{bmatrix} \begin{bmatrix} -1 \\ -5 \\ 20 \end{bmatrix} = \frac{1}{9} \begin{bmatrix} 9 \\ 18 \\ 36 \end{bmatrix} = \begin{bmatrix} 1 \\ 2 \\ 4 \end{bmatrix}$$

We close this section with an example that gives a geometric interpretation for 2×2 and 3×3 determinants.

Example 4.11. (a) Show that if $A = \begin{bmatrix} a & b \\ c & d \end{bmatrix}$, then $|\det A|$ is the area of the parallelogram with edges (sides) determined by (a, b) and (c, d).

(b) Show that if $F = \begin{bmatrix} a_1 & a_2 & a_3 \\ b_1 & b_2 & b_3 \\ c_1 & c_2 & c_3 \end{bmatrix}$, then $|\det F|$ is the volume of the parallelpiped with edges determined by $A = (a_1, a_2, a_3)$, $B = (b_1, b_2, b_3)$, and $C = (c_1, c_2, c_3)$.

Solution: (a) In 3-space, let $\bar{u} = a\bar{i} + b\bar{j} + 0\bar{k}$ and $\bar{v} = c\bar{i} + d\bar{j} + 0\bar{k}$. The area of the parallelogram determined by $\bar{u}$ and $\bar{v}$ is $\|\bar{u} \times \bar{v}\|$ (see Section 2.4), and

$$\|\bar{u} \times \bar{v}\| = \left\| \left(0, 0, \begin{vmatrix} a & b \\ c & d \end{vmatrix}\right) \right\| = |\det A|$$

Since in 2-space the vectors $\bar{u}$ and $\bar{v}$ correspond to the points (a, b) and (c, d), we have the desired conclusion.

(b) Consider the parallelpiped determined by $A = (a_1, a_2, a_3)$, $B = (b_1, b_2, b_3)$, and $C = (c_1, c_2, c_3)$.

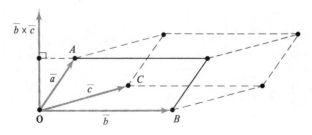

The area of the base of the parallelpiped is $\|\bar{b} \times \bar{c}\|$, and the height is the length of the vector projection of $\bar{a}$ along $\bar{b} \times \bar{c}$ (see Section 2.3). Thus the volume of the parallelpiped is

$$V = \|\bar{b} \times \bar{c}\| \left(\left\| \frac{\bar{a} \cdot (\bar{b} \times \bar{c})}{(\bar{b} \times \bar{c}) \cdot (\bar{b} \times \bar{c})} \, \bar{b} \times \bar{c} \right\| \right) = \|\bar{b} \times \bar{c}\| \left(\frac{|\bar{a} \cdot (\bar{b} \times \bar{c})|}{\|\bar{b} \times \bar{c}\|^2} \|\bar{b} \times \bar{c}\| \right)$$

$$= |\bar{a} \cdot (\bar{b} \times \bar{c})|$$

Now

$$\bar{a} \cdot (\bar{b} \times \bar{c}) = (a_1, a_2, a_3) \cdot \left(\begin{vmatrix} b_2 & b_3 \\ c_2 & c_3 \end{vmatrix}, \; -\begin{vmatrix} b_1 & b_3 \\ c_1 & c_3 \end{vmatrix}, \; \begin{vmatrix} b_1 & b_2 \\ c_1 & c_2 \end{vmatrix} \right)$$

$$= a_1 \begin{vmatrix} b_2 & b_3 \\ c_2 & c_3 \end{vmatrix} - a_2 \begin{vmatrix} b_1 & b_3 \\ c_1 & c_3 \end{vmatrix} + a_3 \begin{vmatrix} b_1 & b_2 \\ c_1 & c_2 \end{vmatrix} = \begin{vmatrix} a_1 & a_2 & a_3 \\ b_1 & b_2 & b_3 \\ c_1 & c_2 & c_3 \end{vmatrix}$$

Hence the volume of the parallelpiped is $|\det F|$.

Exercise 4.3

1. Verify Lemma 4.1 for the matrix $A = \begin{bmatrix} 1 & 3 & 2 \\ -1 & 2 & 0 \\ 2 & 4 & 1 \end{bmatrix}$ if $i = 1$ and $k = 3$.

2. Verify Lemma 4.1 for the matrix $A = \begin{bmatrix} 1 & 2 & 4 & 3 \\ -1 & 0 & 2 & 1 \\ 2 & 4 & 1 & 3 \\ 2 & 1 & -2 & 0 \end{bmatrix}$ if $i = 2$ and $k = 4$.

3. If $A = \begin{bmatrix} 3 & 2 & -1 \\ 0 & 4 & -3 \\ 1 & -2 & 2 \end{bmatrix}$, find (a) the matrix of cofactors $\tilde{A}$; (b) adj A; (c) $A(\text{adj } A)$; (d) $\det A$; (e) A^{-1} if it exists.

4. If $A = \begin{bmatrix} 1 & 0 & 2 & -3 \\ 2 & -1 & 4 & 0 \\ 3 & 2 & 1 & -3 \\ 0 & -1 & 3 & -2 \end{bmatrix}$ find (a) the matrix of cofactors $\tilde{A}$; (b) adj A; (c) $A(\text{adj } A)$; (d) $\det A$; (e) A^{-1} if it exists.

In Problems 5 and 6, determine if A^{-1} exists by computing $|A|$. If A^{-1} does exist, compute it by finding adj A.

5. $A = \begin{bmatrix} 1 & 1 \\ 2 & 1 \end{bmatrix}$.

6. $A = \begin{bmatrix} 1 & 0 & 0 & 0 \\ 0 & 2 & 2 & 0 \\ 0 & 0 & 3 & 0 \\ 0 & 0 & 0 & 4 \end{bmatrix}$.

7. Use Cramer's rule (if it applies) to solve the system

$$x_1 + 2x_2 + 3x_3 = -1$$
$$2x_1 + x_2 - 4x_3 = 9$$
$$x_1 - x_2 + 2x_3 = -2$$

8. Use Cramer's rule (if it applies) to solve the system

$$x_1 + 0x_2 + x_3 - x_4 = -4$$
$$2x_1 + x_2 - x_3 + x_4 = 8$$
$$-x_1 + 2x_2 + 0x_3 - 2x_4 = -5$$
$$x_1 + 0x_2 + 2x_3 + 2x_4 = 3$$

9. Assuming that the coefficient matrix is nonsingular, use Cramer's rule to write out the unique solution to the system

$$ax + by = e$$
$$cx + dy = f$$

10. Use determinants to find
   (a) The area of the parallelogram with edges determined by $(3, 4)$ and $(2, 3)$.
   (b) The volume of the parallelpiped with edges determined by $(2, 3, 1)$, $(1, 3, 2)$, and $(4, -1, 0)$.

11. Use determinants to find
   (a) The area of the rectangle with edges determined by $(2, 3)$ and $(6, -4)$. Check your answer with the formula area = length × width.
   (b) The volume of a rectangular parallelpiped with edges determined by $(2, 2, 0)$, $(4, -4, 1)$, and $(2, -2, -16)$. Check your answer with the formula V = length × width × height.

12. If $A = \begin{bmatrix} a & b \\ c & d \end{bmatrix}$, show that adj (adj A) = A.

13. *Prove:* If A is an $n \times n$ matrix and det $A = 0$, then det (adj A) = 0.

14. (a) If A and B are nonsingular $n \times n$ matrices, prove that adj (AB) = (adj B)(adj A). *Hint:* Show that (adj AB)(AB) = (adj B)(adj A)AB.
   (b) Verify part (a) for the matrices $A = \begin{bmatrix} 1 & 2 \\ 2 & 3 \end{bmatrix}$ and $B = \begin{bmatrix} 4 & 7 \\ 3 & 5 \end{bmatrix}$.

15. If A is an $n \times n$ matrix, prove that det (adj A) = (det $A)^{n-1}$.

5 Abstract Vector Spaces

5.1 DEFINITION AND EXAMPLES OF VECTOR SPACES

In the previous chapters we have seen several examples of mathematical systems called vector spaces. To be classified as a vector space, the mathematical system must consist of three things, a collection of objects $\mathbf{V}$ (called vectors), a binary operation defined on $\mathbf{V}$ and satisfying properties 1A–5A, and a scalar multiplication satisfying properties 6M–10M. In our work the scalars will be real numbers or, on occasion, complex numbers. In a more abstract setting the only requirement on the scalars is that they be elements of a mathematical system called a *field*. Although we have already given the definition for a vector space, we repeat that definition here.

Definition 5.1. A (real) **vector space** is a set $\mathbf{V}$ together with a binary operation $+$ and a scalar multiplication that satisfies the following properties.

Properties of Addition
1A. For any X and Y in $\mathbf{V}$, $X + Y$ is a uniquely determined element in $\mathbf{V}$. (This is the meaning of the statement "$+$ is a **binary operation**.")
2A. For any X and Y in $\mathbf{V}$, $X + Y = Y + X$. (Addition of vectors is **commutative**.)
3A. For any X, Y, and Z in $\mathbf{V}$, $(X + Y) + Z = X + (Y + Z)$. (Addition of vectors is **associative**.)
4A. There is a vector $\mathbf{0}$ in $\mathbf{V}$ such that for any X in $\mathbf{V}$, $X + \mathbf{0} = \mathbf{0} + X = X$. ($\mathbf{0}$ is an **identity element** for the operation called addition.)

5A. For any vector X in $\mathbf{V}$ there is a vector Y in $\mathbf{V}$ such that $X + Y = Y + X = 0$. (Y is an **additive inverse** for the vector X.)

Properties of Scalar Multiplication

6M. For any real number r and any vector X in $\mathbf{V}$, rX is a uniquely determined vector in $\mathbf{V}$. (This is the meaning of **scalar multiplication**.)

7M. For any real numbers r and s and any vector X in $\mathbf{V}$, $(rs)X = r(sX) = s(rX)$.

8M. For any real numbers r and s and any vector X in $\mathbf{V}$, $(r + s)X = rX + sX$.

9M. For any real number r and vectors X and Y in $\mathbf{V}$, $r(X + Y) = rX + rY$.

10M. For any X in $\mathbf{V}$, $1X = X$, where 1 is the real number 1.

We now list several examples of vector spaces. Examples (a)–(d) have already been encountered in Chapters 1–3.

Example 5.1.   Some Vector Spaces

(a) $\mathbf{V}$ is the collection of all line vectors in a plane (or in space) with addition and scalar multiplication defined as in Sections 1.1 and 1.2.

(b) $\mathbf{V} = \mathbf{R_n}$ is the collection of all n-tuples of real numbers with addition and scalar multiplication as defined in Definitions 1.5 and 1.6. Note that each positive integer n determines a vector space, and thus we have infinitely many vector spaces.

(c) $\mathbf{V}$ is the collection of all solutions to a homogeneous system of m linear equations in n unknowns. The elements of $\mathbf{V}$ are n-tuples. Addition and scalar multiplication are defined as in part (b). (See Problem 13, Exercise 3.2.)

(d) $\mathbf{V} = \mathbf{R_{m \times n}}$ is the collection of all $m \times n$ matrices with real number entries. Addition and scalar multiplication are defined as in Definitions 3.12 and 3.14.

(e) $\mathbf{V} = \mathbf{F}$ is the collection of all real-valued functions defined on $\mathbf{R}$ (the set of real numbers). For functions $f: \mathbf{R} \longrightarrow \mathbf{R}$ and $g: \mathbf{R} \longrightarrow \mathbf{R}$ in $\mathbf{F}$, addition is defined by $f + g: \mathbf{R} \longrightarrow \mathbf{R}$, where $(f + g)(x) = f(x) + g(x)$ for all x in $\mathbf{R}$. Multiplication of a function f by a scalar r is defined by $rf: \mathbf{R} \longrightarrow \mathbf{R}$, where $(rf)(x) = r(f(x))$ for all x in $\mathbf{R}$. (The reader who needs a brief review of functions and function notation may wish to read Section 6.1.)

(f) $\mathbf{V} = \mathbf{P} \subset \mathbf{F}$ is the set of all polynomial functions. Addition and scalar multiplication are defined as in part (e).

(g) $\mathbf{V} = \mathbf{P_n} \subset \mathbf{F}$ is the set of all polynomial functions of degree less than n. Addition and scalar multiplication are defined as in part (e).

(h) $\mathbf{V}$ is the set of all complex numbers. For complex numbers $a + bi$ and $c + di$ ($i = \sqrt{-1}$), addition is defined by $(a + bi) + (c + di) = (a + c)$

$+ (b + d)i$. We define multiplication of a complex number by a scalar (real number) by $r(a + bi) = ra + (rb)i$.

The reader who is familiar with elementary calculus may also wish to consider the following examples. In each example addition and scalar multiplication are defined as in Example 5.1(e). However, the functions are defined on a closed interval $[a, b]$ rather than on **R**, as is the case in Example 5.1(e).

(i) **V** = **C** is the collection of all continuous real valued functions defined on a closed interval $[a, b]$.

(j) **V** is the collection of all differentiable real-valued functions defined on a closed interval $[a, b]$.

(k) **V** is the collection of all integrable real-valued functions defined on a closed interval $[a, b]$.

(l) **V** is the collection of all real-valued functions f defined on a closed interval $[a, b]$ that satisfy the differential equation $3f' - f = 0$.

We verify that Examples 5.1(e) and (f) are vector spaces and leave the remaining verifications to the reader.

Verification of Example 5.1(e):   We first remark that, in the definition of $f + g$ given by $(f + g)(x) = f(x) + g(x)$, the plus sign on the left side of the equation denotes the *sum of functions* that is being defined, and the plus sign on the right side of the equation is the *sum of real numbers*. Second, we note that two functions g and h are equal if and only if (1) they are defined on the same set (domain) and (2) for every x in the common domain, $g(x) = h(x)$.

Now let f and g be functions in **F**. It is clear from the definition of addition that $f + g$ is in **F** and is uniquely determined by f and g. So property 1A holds.

Since

$$(f + g)(x) = f(x) + g(x) = g(x) + f(x) = (g + f)(x)$$

it follows that $f + g = g + f$ and property 2A holds.

For functions f, g, and h in **F**, we have

$$((f + g) + h)(x) = (f + g)(x) + h(x) = (f(x) + g(x)) + h(x)$$
$$= f(x) + (g(x) + h(x)) = f(x) + ((g + h)(x))$$
$$= (f + (g + h))(x)$$

(The reader is asked to supply reasons for each of the equalities in Problem 2, Exercise 5.1.) Therefore, $(f + g) + h = f + (g + h)$ and property 3A holds.

Define a function **0**: **R** $\rightarrow$ **R** by $\mathbf{0}(x) = 0$ for all x in **R**. Then **0** is in **F** and, for any f in **F**,

$$(f + \mathbf{0})(x) = f(x) + \mathbf{0}(x) = f(x) + 0 = f(x)$$

So $f + \mathbf{0} = f$. By property 2A we have $f + \mathbf{0} = \mathbf{0} + f = f$ and property 4A holds. We call **0** the **zero function**.

For any function f in $\mathbf{F}$, the function $(-1)f$ is in $\mathbf{F}$ and

$$(f + (-1)f)(x) = f(x) + ((-1)f)(x) = f(x) + (-1)f(x) = 0$$

Since this holds for every x in $\mathbf{R}$, we have $f + (-1)f = 0$. From property 2A we get $(-1)f + f = f + (-1)f = 0$. Thus property 5A holds.

Let f be any function in $\mathbf{F}$. From the definition of multiplication of a function by a scalar, it is clear that rf is in $\mathbf{F}$ and is uniquely determined by r and f. So property 6M holds.

For any scalars r and s and for any f in $\mathbf{F}$,

$$((rs)f)(x) = (rs)(f(x)) = r(sf(x)) = r((sf)(x)) = (r(sf))(x)$$

So $(rs)f = r(sf)$. Also,

$$((rs)f)(x) = (rs)(f(x)) = s(rf(x)) = s((rf)(x)) = (s(rf))(x)$$

Thus $(rs)f = s(rf)$ and property 7M holds.

For any scalars r and s and for any f in $\mathbf{F}$,

$$((r + s)f)(x) = (r + s)(f(x)) = r(f(x)) + s(f(x))$$
$$= (rf)(x) + (sf)(x) = (rf + sf)(x)$$

Therefore, $(r + s)f = rf + sf$ and property 8M holds. For f and g in $\mathbf{F}$ and any scalar r,

$$(r(f + g))(x) = r((f + g)(x)) = r(f(x) + g(x)) = r(f(x)) + r(g(x))$$
$$= (rf)(x) + (rg)(x) = (rf + rg)(x)$$

So $r(f + g) = rf + rg$, and property 9M holds.

For any f in $\mathbf{F}$, $(1f)(x) = 1(f(x)) = f(x)$. Therefore, $1f = f$ and property 10M holds. Q.E.D.

Verification of 5.1(f): If f is a polynomial function of degree n, we shall write $f: a_0 + a_1x + \cdots + a_nx^n$ and read this as "f **is the function defined by** $f(x) = a_0 + a_1x + \cdots + a_nx^n$.

Suppose that $f: a_0 + a_1x + \cdots + a_nx^n$, $g: b_0 + b_1x + \cdots + b_mx^m$, and without loss of generality suppose that $n \geq m \geq 0$. Then

$$f + g: (a_0 + b_0) + (a_1 + b_1)x + \cdots + (a_m + b_m)x^m + a_{m+1}x^{m+1}$$
$$+ \cdots + a_nx^n$$

So $f + g$ is in $\mathbf{P}$ and property 1A holds. Properties 2A and 3A hold for *all* functions in $\mathbf{F}$. In particular, these properties hold for functions in $\mathbf{P}$, and so properties 2A and 3A are satisfied. The function $\mathbf{0}$ in $\mathbf{F}$ is also an identity for $\mathbf{P}$, and thus property 4A holds. (We shall say that all constant functions including $\mathbf{0}$ are polynomial functions of degree 0.) If $f: a_0 + a_1x + \cdots + a_nx^n$, then $(-1)f: (-a_0) + (-a_1)x + \cdots + (-a_n)x^n$ is a polynomial function and $(-1)f$ is an additive inverse for f by the proof of Example 5.1(e). So property 5A holds. If r is a scalar, then $rf: (ra_0) + (ra_1)x + \cdots + (ra_n)x^n$

is a polynomial function, and property 6M is satisfied. Properties 7M–9M hold for *all* functions in **F**. In particular, they hold for functions in **P**, and thus 7M–9M hold. Finally, if f is a polynomial function, then so is $1f$, and $1f = f$. Q.E.D.

Next we show that the additive identity of a vector space is unique and that the additive inverse of a vector is unique. We could have taken these properties as requirements of a vector space in Definition 5.1; however, the proofs are quite simple. We also list several simple but important properties of vectors, the proofs of which are left as problems in Exercise 5.1.

Theorem 5.1.   (a) If **0** and **0′** are elements in a vector space **V** and each satisfies property 4A, then $\mathbf{0} = \mathbf{0}'$.

(b) If X is in a vector space **V** and if Y and Y' are additive inverses of X in the sense of property 5A, then $Y = Y'$.

(c) In a vector space **V**, the scalar 0 times a vector X is the identity element in **V**. In symbols, $0X = \mathbf{0}$.

(d) In a vector space **V**, the scalar -1 times a vector X is the additive inverse of X. In symbols, $(-1)X = -X$.

(e) In a vector space **V**, a scalar a times the identity element in **V** is the identity element. In symbols, $a\mathbf{0} = \mathbf{0}$.

(f) In a vector space **V**, if $rX = \mathbf{0}$ and $r \neq 0$, then $X = \mathbf{0}$.

Proof: (a) Suppose that **0** and **0′** are identities for a vector space *V*. Since **0** is an identity for **V**, we have $\mathbf{0} + \mathbf{0}' = \mathbf{0}'$. But **0′** is also an identity for **V**; therefore, $\mathbf{0} + \mathbf{0}' = \mathbf{0}$. Hence $\mathbf{0} = \mathbf{0}'$.

(b) Suppose that Y and Y' are additive inverses for a vector X. We have $(Y + X) + Y' = \mathbf{0} + Y' = Y'$. Also, $Y + (X + Y') = Y + \mathbf{0} = Y$. Since addition is associative, we have $Y = Y'$. The proofs of parts (c)–(f) are called for in Problems 12–15, Exercise 5.1.   Q.E.D.

The unique additive identity in a vector space will always be denoted by **0** and be called the **zero vector**. Also, the unique additive inverse of a vector X will be denoted by $-X$.

We close this section by defining subtraction in a vector space.

Definition 5.2.   If X and Y are vectors in a vector space **V**, we define **subtraction** by $X - Y = X + (-Y)$.

Exercise 5.1

In each of Problems 1–8 you are given a set together with a binary operation (addition) and a scalar multiplication. Determine which systems are vector spaces. For those that are not, list all the properties 1A–5A and 6M–10M that fail to hold.

1. V is the set of all ordered pairs (x, y) of real numbers. Addition is defined by $(x, y) + (v, w) = (x + 2v, y + 2w)$, and scalar multiplication is defined by $r(x, y) = (rx, ry)$.

2. The set V is as in Problem 1. Addition is defined by $(x, y) + (v, w) = (x + v, y + w)$, and scalar multiplication is defined by $r(x, y) = (0, 0)$.

3. The set V is as in Problem 1. Addition is defined by $(x, y) + (v, w) = (x + v, y + w)$, and scalar multiplication is defined by $r(x, y) = (ry, rx)$.

4. The set V is as in Problem 1. Addition is defined by $(x, y) + (v, w) = (x, y)$, and scalar multiplication is defined by $r(x, y) = (rx, ry)$.

5. V is the set of all 2×2 matrices with real number entries. Addition is the usual matrix addition as defined in Definition 3.12 and scalar multiplication is defined by

$$r\begin{bmatrix} a & b \\ c & d \end{bmatrix} = \begin{bmatrix} 0 & 0 \\ 0 & 0 \end{bmatrix}$$

6. The set V is as in Problem 5. Addition is defined by

$$\begin{bmatrix} a & b \\ c & d \end{bmatrix} + \begin{bmatrix} e & f \\ g & h \end{bmatrix} = \begin{bmatrix} e & f \\ g & h \end{bmatrix}$$

and scalar multiplication is as defined in Definition 3.14.

7. The set V is as in Problem 5. Addition is defined as in Definition 3.12 and scalar multiplication is defined by

$$r\begin{bmatrix} a & b \\ c & d \end{bmatrix} = \begin{bmatrix} 0 & rb \\ rc & 0 \end{bmatrix}$$

8. The set V is as in Problem 5. Addition is defined by

$$\begin{bmatrix} a & b \\ c & d \end{bmatrix} + \begin{bmatrix} e & f \\ g & h \end{bmatrix} = \begin{bmatrix} a + e & 0 \\ 0 & d + h \end{bmatrix}$$

and scalar multiplication is defined by

$$r\begin{bmatrix} a & b \\ c & d \end{bmatrix} = \begin{bmatrix} ra & rb \\ rc & rd \end{bmatrix}$$

9. Show that Example 5.1(g) is a vector space for $n = 3$. Use the notation for polynomial functions that was used to verify that Example 5.1(f) is a vector space.

10. Supply reasons for each of the equalities in the string(s) of equalities given in the verification of Example 5.1(e), property (a) 3A; (b) 5A; (c) 7M; (d) 8M; (e) 9M.

11. Verify that Example 5.1(h) is a vector space. You may assume the familiar properties of the complex numbers.

12. Prove Theorem 5.1(c). *Hint:* Write $0X$ as $(0 + 0)X$. Then use property 8M.

13. Prove Theorem 5.1(d). *Hint:* Add $(-1)X$ to $1X$ and use property 8M.

14. Prove Theorem 5.1(e). *Hint:* Write 0 as $0 + 0$ and use property 9M.

15. Prove Theorem 5.1(f).

16. Verify that in a vector space **V**
(a) $(-r)X = -(rX)$. *Hint:* Show that $rX + (-r)X = 0$ and use the fact that the additive inverse of rX is unique.
(b) $-(rX) = r(-X)$.
(c) $(r - s)X = rX - sX$.

In Problems 17 and 18, you may assume a generalized associative law for addition of vectors. The r's are scalars and the X's are vectors in a vector space **V**.

17. Use mathematical induction to prove that $r(X_1 + X_2 + \cdots + X_n) = rX_1 + rX_2 + \cdots + rX_n$ for any positive integer n.

18. (a) Use mathematical induction to prove that $(r_1 + r_2 + \cdots + r_n)X = r_1X + r_2X + \cdots + r_nX$.
(b) Show that $\underbrace{X + X + \cdots + X}_{n \text{ times}} = nX$.

Problems 19–22 require a knowledge of elementary calculus.

19. Verify that Example 5.1(i) is a vector space.

20. Verify that Example 5.1(j) is a vector space.

21. Verify that Example 5.1(k) is a vector space.

22. Verify that Example 5.1(l) is a vector space.

5.2 SUBSPACES

The collection of vectors in the vector space of Example 5.1(c) is a subcollection of $\mathbf{R_n}$ in Example 5.1(b). Also, the addition and scalar multiplication in Example 5.1(c) are defined as in Example 5.1(b). For these reasons we call the vector space of Example 5.1(c) a subspace of the vector space of coordinate vectors $\mathbf{R_n}$. For similar reasons, each of the vector spaces $\mathbf{P_n}$ in Example 5.1(g) is a subspace of the vector space $\mathbf{P}$ in Example 5.1(f).

According to Definition 5.1, a vector space is a set **V** together with an operation addition and a scalar multiplication that must satisfy properties 1A–5A and 6M–10M. Although for convenience we have been and will continue to refer to the set **V** as the vector space, the reader should keep in mind that **V** is only a part of the mathematical system $(\mathbf{V}, +, \text{sm})$, which is the vector space.

Definition 5.3. A nonempty subset **W** of a vector space **V** is a **subspace** of **V** if and only if **W**, together with the same addition and scalar multiplication† as that defined on **V**, is itself a vector space.

†Technically, the addition and the scalar multiplication on **W** are the addition and the scalar multiplication on **V** *restricted* to **W**.

On the surface it appears that to determine if a subset **W** is a subspace one should check if the 10 properties 1A–5A and 6M–10M hold. In fact, as the next theorem shows, it is necessary to check only properties 1A and 6M, provided that the set **W** is not empty.

Theorem 5.2. If (**V**, +, sm) is a vector space and if **W** is a nonempty subset of **V**, then **W** is a subspace of **V** provided that (i) for all X and Y in **W**, $X + Y$ is in **W**, and (ii) for all r in **R** and for all X in **W**, rX is in **W**. Conversely, if **W** is a subspace of **V**, it is clear that (i) and (ii) hold.

Proof: By hypothesis (i), for any vectors X and Y in **W**, $X + Y$ is in **W**. Furthermore, $X + Y$ is uniquely determined in **W** because the sum is uniquely determined in **V**. Therefore, property 1A holds.

For any vectors X and Y in **W**, the vectors X and Y are in **V**, and so $X + Y = Y + X$. Thus property 2A holds for **W**. We say that **W** *inherits* property 2A from **V**. Similarly, **W** inherits property 3A from **V**.

W is not empty, so we can select a vector X_1 from **W**. According to hypothesis (ii), $(-1)X_1$ must be in **W**. By Theorem 5.1(d), we have $(-1)X_1 = -X_1$, which is the additive inverse of X_1. Using hypothesis (i), we see that the sum of any two vectors in **W** is again in **W** and, in particular, $X_1 + (-1)X_1 = 0$ is in **W**. Since $X + 0 = 0 + X = X$ for any vector X in **V**, it is surely true that $X + 0 = X = 0 + X$ for all vectors X in **W**. Hence property 4A holds for **W**.

For any vector X in **W**, we have, by hypothesis (ii), that $(-1)X = -X$ is in **W**. Thus property 5A holds.

By hypothesis (ii), for any vector X in **W** and any scalar r in **R**, we have that rX is in **W**. The vector rX is uniquely determined in **V**, and therefore it is uniquely determined in **W**. So property 6M holds. Properties 7M–10M are all properties that **W** inherits from **V**. Hence **W** is a subspace of **V**. Q.E.D.

Note: In the proof of property 4A we not only proved that **W** has a zero vector, but also that the *zero for the subspace* **W** *is the same as the zero for the superspace* **V**.

The proof that the set of solutions to a homogeneous system of m linear equations in n unknowns is a vector space was called for in Problem 13, Exercise 3.2. We shall illustrate the use of Theorem 5.2 to show that this space is a subspace of $\mathbf{R_n}$.

Example 5.2. Prove that the collection of all solutions to a homogeneous system of m linear equations in n unknowns is a subspace of $\mathbf{R_n}$.

Solution: Let S denote the set of solutions to the system

$$a_{11}x_1 + a_{12}x_2 + \cdots + a_{1n}x_n = 0$$
$$a_{21}x_1 + a_{22}x_2 + \cdots + a_{2n}x_n = 0$$
$$\cdots\cdots\cdots\cdots$$
$$a_{m1}x_1 + a_{m2}x_2 + \cdots + a_{mn}x_n = 0$$

First we note that S is not empty because $0 = (0, 0, \ldots, 0)$ is a solution. Now let $B = (b_1, b_2, \ldots, b_n)$ and $C = (c_1, c_2, \ldots, c_n)$ be any solutions. Then

$$a_{i1}(b_1 + c_1) + a_{i2}(b_2 + c_2) + \cdots + a_{in}(b_n + c_n)$$
$$= (a_{i1}b_1 + a_{i2}b_2 + \cdots + a_{in}b_n) + (a_{i1}c_1 + a_{i2}c_2 + \cdots + a_{in}c_n)$$
$$= 0 + 0 = 0, \quad \text{for } i = 1, 2, \ldots, m$$

Therefore, $B + C$ is a solution and Theorem 5.2(i) holds. For a solution B and a scalar r, we have

$$a_{i1}(rb_1) + a_{i2}(rb_2) + \cdots + a_{in}(rb_n) = r(a_{i1}b_1 + a_{i2}b_2 + \cdots + a_{in}b_n)$$
$$= r0 = 0, \quad \text{for } i = 1, 2, \ldots, m$$

Thus rB is a solution and Theorem 5.2(ii) is satisfied. Hence S is a subspace of $\mathbf{R_n}$.

Example 5.3. Show that the set of all solutions to the system

$$3x_1 + 2x_2 + (-4)x_3 = 0$$
$$2x_1 + 3x_2 + (-2)x_3 = 2$$

is not a subspace of $\mathbf{R_3}$.

Solution: $0 = (0, 0, 0)$ is not a solution to this system. Since the zero of a subspace is the same as the zero for the superspace, this is sufficient to verify that the set of solutions to the given system is not a subspace of $\mathbf{R_3}$. We also note that $(4, 0, 3)$ is a solution and thus the solution set is not empty. According to Theorem 5.2, either property 1A or 6M must fail. It is clear that 6M fails, because, for example, $0(4, 0, 3)$ is not a solution.

Example 5.4. Describe both algebraically and geometrically the "smallest" subspace of $\mathbf{R_3}$ that contains the vector $(1, 0, 3)$.

Solution: It is easy to show that the collection $\mathbf{W}$ of all scalar multiples of $(1, 0, 3)$ is a subspace of $\mathbf{R_3}$. For, if $r(1, 0, 3)$ and $s(1, 0, 3)$ are in $\mathbf{W}$, then $r(1, 0, 3) + s(1, 0, 3) = (r + s)(1, 0, 3)$ is in $\mathbf{W}$. Also, if $r(1, 0, 3)$ is in $\mathbf{W}$ and s is in $\mathbf{R}$, then $s(r(1, 0, 3)) = (sr)(1, 0, 3)$, which is in $\mathbf{W}$. The vector $(1, 0, 3) = 1(1, 0, 3)$ is in $\mathbf{W}$ and, by Theorem 5.2, $\mathbf{W}$ is a subspace of $\mathbf{R_3}$. $\mathbf{W}$ is the smallest such subspace in the sense that any subspace that contains $(1, 0, 3)$ must contain each of the elements of $\mathbf{W}$. This is clear because a subspace of $\mathbf{R_3}$ containing $(1, 0, 3)$ must satisfy property 6M.

The geometric description of $\mathbf{W}$ is that it is the line in 3-space that passes through $(0, 0, 0)$ and $(1, 0, 3)$.

Example 5.5. Describe both algebraically and geometrically the "smallest" subspace of $\mathbf{R_3}$ that contains the vectors $(1, 2, -4)$ and $(2, 1, 0)$.

Solution: It is clear from Theorem 5.2 that any subspace that contains $(1, 2, -4)$ and $(2, 1, 0)$ must also contain $r(1, 2, -4)$ and $s(2, 1, 0)$ and, consequently, it must contain $r(1, 2, -4) + s(2, 1, 0)$ for all scalars r and s. On the other hand, we can show that

$$\mathbf{W} = \{r(1, 2, -4) + s(2, 1, 0) \mid r \text{ and } s \text{ are in } \mathbf{R}\}$$

is a subspace of $\mathbf{R}_3$. For, let

$$X = r_1(1, 2, -4) + s_1(2, 1, 0)$$

$$Y = r_2(1, 2, -4) + s_2(2, 1, 0) \in \mathbf{W}$$

Then

$$X + Y = (r_1 + r_2)(1, 2, -4) + (s_1 + s_2)(2, 1, 0) \in \mathbf{W}$$

Also,

$$tX = t(r_1(1, 2, -4) + s_1(2, 1, 0)) = tr_1(1, 2, -4) + ts_1(2, 1, 0) \in \mathbf{W}$$

Hence $\mathbf{W}$ is a subspace of $\mathbf{R}_3$. Geometrically, $\mathbf{W}$ is the plane through the three points $(0, 0, 0)$, $(1, 2, -4)$ and $(2, 1, 0)$.

Examples 5.4 and 5.5 have an important generalization. Before we state the generalization, we need to make a comment on notation and state a definition. The comment is in regard to parentheses or the lack thereof. If X, Y, and Z are vectors in a vector space $\mathbf{V}$, we usually write the sum of the three vectors as $X + Y + Z$. No parentheses are needed because, according to property 3A, $(X + Y) + Z = X + (Y + Z)$. A consequence of property 3A, which probably seems clear to the reader, is that for n vectors $X_1, X_2, \ldots,$ X_n in $\mathbf{V}$ we can write the sum $X_1 + X_2 + \cdots + X_n$ without parentheses. No matter how one inserts parentheses, the same vector is obtained. We are now ready to state the following definition.

Definition 5.4. Let $X_1, X_2, \ldots, X_n$ be vectors in a vector space $\mathbf{V}$. A vector X in $\mathbf{V}$ is a **linear combination** of $X_1, X_2, \ldots, X_n$ if and only if $X = r_1 X_1 + r_2 X_2 + \cdots + r_n X_n$ for some scalars $r_1, r_2, \ldots, r_n$.

Note that if $n = 1$ a linear combination becomes a scalar multiple. Next we state and prove a theorem that is a generalization of Examples 5.4 and 5.5.

Theorem 5.3. If $X_1, X_2, \ldots, X_n$ are vectors in a vector space $\mathbf{V}$ then

$$\mathbf{W} = \{a_1 X_1 + a_2 X_2 + \cdots + a_n X_n \mid a_i \in \mathbf{R}, i = 1, 2, \ldots, n\}$$

is a subspace of $\mathbf{V}$ that contains each of the vectors $X_1, X_2, \ldots, X_n$. $\mathbf{W}$ is the smallest subspace of $\mathbf{V}$ that contains these vectors in the sense that, if $\mathbf{U}$ is a subspace of $\mathbf{V}$ and $X_i \in \mathbf{U}$ for $i = 1, 2, \ldots, n$, then $\mathbf{W} \subseteq \mathbf{U}$.

Proof: We can write

$$X_i = 0X_1 + \cdots + 0X_{i-1} + 1X_i + 0X_{i+1} + \cdots + 0X_n$$

Therefore, each of the vectors $X_1, X_2, \ldots, X_n$ is in $\mathbf{W}$ and $\mathbf{W} \neq \varnothing$. Now let

Y and Z be vectors in $\mathbf{W}$. Then

$$Y = r_1 X_1 + r_2 X_2 + \cdots + r_n X_n$$
$$Z = s_1 X_1 + s_2 X_2 + \cdots + s_n X_n$$

for certain scalars $r_1, r_2, \ldots, r_n$ and $s_1, s_2, \ldots, s_n$. Because of generalizations of properties 2A and 3A, which we think are obvious to the reader (but may be difficult to give proofs for), we can write

$$Y + Z = (r_1 X_1 + r_2 X_2 + \cdots + r_n X_n) + (s_1 X_1 + s_2 X_2 + \cdots + s_n X_n)$$
$$= (r_1 X_1 + s_1 X_1) + (r_2 X_2 + s_2 X_2) + \cdots + (r_n X_n + s_n X_n)$$

Because of property 8M, this last vector is equal to

$$(r_1 + s_1) X_1 + (r_2 + s_2) X_2 + \cdots + (r_n + s_n) X_n$$

which is a vector in $\mathbf{W}$. So $\mathbf{W}$ is closed under addition.

Now suppose that Y is as above and s is an arbitrary scalar. Using a generalization of property 9M, we have

$$sY = s(r_1 X_1 + r_2 X_2 + \cdots + r_n X_n)$$
$$= s(r_1 X_1) + s(r_2 X_2) + \cdots + s(r_n X_n)$$

By property 7M, this vector is equal to

$$(sr_1) X_1 + (sr_2) X_2 + \cdots + (sr_n) X_n$$

which is in $\mathbf{W}$. We have proved that $\mathbf{W}$ is a subspace of $\mathbf{V}$, which contains the vectors $X_1, X_2, \ldots, X_n$.

Now let $\mathbf{U}$ be a subspace of $\mathbf{V}$, which contains $X_1, X_2, \ldots, X_n$, and let $Y = r_1 X_1 + r_2 X_2 + \cdots + r_n X_n$ be an arbitrary vector in $\mathbf{W}$. For $i = 1, 2, \ldots, n$, we must have $r_i X_i \in \mathbf{U}$ because property 6M holds in the vector space $\mathbf{U}$. Using property 1A repeatedly (or mathematical induction), we see that $Y = r_1 X_1 + r_2 X_2 + \cdots + r_n X_n$ must be in $\mathbf{U}$ and hence $\mathbf{W} \subseteq \mathbf{U}$. Q.E.D.

Definition 5.5. If $\{X_1, X_2, \ldots, X_n\}$ is a set of vectors in a vector space $\mathbf{V}$ the subspace of all linear combinations of these vectors is called the **subspace generated (spanned)** by $\{X_1, X_2, \ldots, X_n\}$ and is denoted by $\langle X_1, X_2, \ldots, X_n \rangle$. We also say that the subspace $\langle X_1, X_2, \ldots, X_n \rangle$ is **generated (spanned) by the vectors** $X_1, X_2, \ldots, X_n$.

Example 5.6. Determine if $(4, 6)$ is in $\mathbf{W} = \langle (1, 1), (3, 2) \rangle$.

Solution: If $(4, 6) \in \mathbf{W}$, then there must be scalars r and s such that $(4, 6) = r(1, 1) + s(3, 2) = (r + 3s, r + 2s)$. It is now clear that the problem is equivalent to determining if the system

$$r + 3s = 4$$
$$r + 2s = 6$$

has a solution. The fact that the coefficient matrix has a nonzero determinant tells us that there is a (unique) solution. The solution is $(r, s) = (10, -2)$. So $(4, 6)$ is in $\mathbf{W}$, and in fact $(4, 6) = 10(1, 1) + (-2)(3, 2)$.

Example 5.7. Determine the subspace $\mathbf{W}$ of $\mathbf{P_4}$ that is spanned by $\{f_1 : x^2 + 1, f_2 : 2x - 1, f_3 : 3\}$.

Solution: $\mathbf{W}$ is the set of all linear combinations of f_1, f_2, and f_3. Therefore $f \in \mathbf{W}$ if and only if there are scalars a_1, a_2, and a_3 such that

$$f = a_1 f_1 + a_2 f_2 + a_3 f_3 : a_1(x^2 + 1) + a_2(2x - 1) + a_3(3)$$
$$= a_1 x^2 + 2a_2 x + (a_1 - a_2 + 3a_3)$$

Since a_1, a_2, and a_3 are completely arbitrary, f is any polynomial function of the form $f : ax^2 + bx + c$. Therefore, $\mathbf{W} = \mathbf{P_3}$.

Example 5.8. Determine the subspace $\mathbf{W}$ of $\mathbf{R}_{2 \times 4}$ spanned by

$$A = \begin{bmatrix} 1 & 0 & 0 & 2 \\ 0 & 0 & 3 & 0 \end{bmatrix} \quad \text{and} \quad B = \begin{bmatrix} 0 & 4 & 1 & 0 \\ 0 & 0 & 1 & 0 \end{bmatrix}$$

Solution: $\mathbf{W} = \langle A, B \rangle = \{aA + bB \mid a, b \in \mathbf{R}\}$.

$$aA + bB = \begin{bmatrix} a & 0 & 0 & 2a \\ 0 & 0 & 3a & 0 \end{bmatrix} + \begin{bmatrix} 0 & 4b & b & 0 \\ 0 & 0 & b & 0 \end{bmatrix} = \begin{bmatrix} a & 4b & b & 2a \\ 0 & 0 & 3a+b & 0 \end{bmatrix}$$

So

$$W = \left\{ \begin{bmatrix} a & 4b & b & 2a \\ 0 & 0 & 3a+b & 0 \end{bmatrix} \,\middle|\, a, b \in \mathbf{R} \right\}$$

To find a specific element in $\mathbf{W}$, assign values to a and b. For example, if $a = 3$ and $b = 5$, we get $\begin{bmatrix} 3 & 20 & 5 & 6 \\ 0 & 0 & 14 & 0 \end{bmatrix}$. What values should we assign to a and b in order to obtain A? to obtain B?

If $\mathbf{U}$ and $\mathbf{W}$ are subspaces of a vector space $\mathbf{V}$, it is easy to show that the *intersection* $\mathbf{U} \cap \mathbf{W}$ *is a subspace* of $\mathbf{V}$ (Problem 18, Exercise 5.2). However, the *union of subspaces* $\mathbf{U}$ *and* $\mathbf{W}$ *is in general not a subspace* because the sum $X + Y$ of a vector X in $\mathbf{U}$ and a vector Y in $\mathbf{W}$ need not be in $\mathbf{U} \cup \mathbf{W}$ (see Problem 19, Exercise 5.2).

Definition 5.6. If $\mathbf{U}$ and $\mathbf{W}$ are arbitrary subspaces of a vector space $\mathbf{V}$, we define the **sum of the subspaces** to be

$$\mathbf{U} + \mathbf{W} = \{X + Y \mid X \in \mathbf{U} \text{ and } Y \in \mathbf{W}\}$$

The sum of subspaces $\mathbf{U} + \mathbf{W}$ defined above is easily shown to be a subspace of $\mathbf{V}$ that contains every vector in $\mathbf{U} \cup \mathbf{W}$. In fact, $\mathbf{U} + \mathbf{W}$ is the **smallest subspace of $\mathbf{V}$ that contains $\mathbf{U} \cup \mathbf{W}$** in the sense that, if $\mathbf{S}$ is a subspace of **V** such that **U** ∪ **W** ⊆ **S**, then **U** + **W** ⊆ **S**. The reader is asked to verify these remarks in Problem 20, Exercise 5.2. We shall see that the sum **U** + **W** plays an important part in the theory of vector spaces and of linear transformations, which is the topic of Chapter 6.

Exercise 5.2

1. If $X_1 = (-2, 1, -3)$ and $X_2 = (4, 0, 6)$,
   (a) Write $(10, 3, 15)$ as a linear combination of the vectors X_1 and X_2.
   (b) Determine if $(-4, 5, -6)$ is a linear combination of the vectors X_1 and X_2.

2. If $\mathbf{W} = \langle (1, 2), (3, -1) \rangle$,
   (a) Find a vector in **W** different from $(1, 2)$ and $(3, -1)$.
   (b) How many vectors are there in **W**?
   (c) How many vectors are there in $\{(1, 2), (3, -1)\}$?

3. If $\mathbf{W} = \langle (3, 2, 0), (1, 0, 3) \rangle$,
   (a) Find a vector in **W** different from $(3, 2, 0)$ and $(1, 0, 3)$.
   (b) Describe **W** geometrically.
   (c) Determine whether or not $(3, 4, 1)$ is in **W**.

4. If $X_1 = f_1: x + 1$, $X_2 = f_2: x^2 + 1$, and $X_3 = f_3: 7$ are vectors (polynomial functions) in **P**, and $X = f: 2x^2 + 3x + 33$,
   (a) Write f as a linear combination of the vectors X_1, X_2, and X_3.
   (b) Describe the space spanned by $\{X_1, X_2, X_3\}$.

5. Describe the smallest subspace of $\mathbf{R}_{2 \times 3}$ that contains

$$A = \begin{bmatrix} 1 & 0 & 0 \\ 0 & 0 & 0 \end{bmatrix}, \quad B = \begin{bmatrix} 0 & 0 & 1 \\ 0 & 0 & 0 \end{bmatrix}, \quad C = \begin{bmatrix} 0 & 0 & 0 \\ 0 & 1 & 0 \end{bmatrix}$$

6. Describe the subspace of $\mathbf{R}_{2 \times 3}$ spanned by $A = \begin{bmatrix} 1 & 1 & 0 \\ 0 & 0 & 0 \end{bmatrix}$ and $B = \begin{bmatrix} 0 & 0 & 0 \\ 0 & 1 & 1 \end{bmatrix}$.

7. Determine whether or not the given subset is a subspace of $\mathbf{R}_3$. If it is a subspace, prove it. If it is not, give a reason why it is not.
   (a) The set of all vectors of the form $(x, y, 0)$.
   (b) The set of all vectors (x, y, z) satisfying $z = 3x + y$.
   (c) The set of all vectors of the form $(x, 3, z)$.
   (d) $\{(0, 0, 0)\}$.
   (e) The set of all vectors of the form $(x, 2x, 3x)$.
   (f) The set of all vectors (x, y, z) satisfying $x + 3y + z = 2$.

8. Determine whether or not the given subset is a subspace of $\mathbf{R}_3$. If it is a subspace, prove it. If it is not, give a reason why it is not.
   (a) $\{(x, 0, y) \mid x, y \in \mathbf{R}\}$.
   (b) $\{(x, 2, 0) \mid x \in \mathbf{R}\}$.
   (c) $\{(x, y, z) \mid x + y = 4z\}$.
   (d) $\{(x, y, z) \mid x + y - 3 = 2z\}$.

9. Use Theorem 5.2 to verify that $\mathbf{P}_3$ is a subspace of **P**.

10. If **0** is the zero vector in a vector space **V**, show that {**0**} is a subspace of **V**.

11. Let **W** be a nonempty subset of a vector space **V**. *Prove:* **W** is a subspace of **V** if and only if **W** is closed under linear combinations; that is, if X and Y are in **W** and if r and s are scalars, then $rX + sY$ is in **W**.

12. Show that the vectors $E_1 = (1, 0, 0)$ and $E_2 = (0, 1, 0)$ span the space **W** = {$(a, b, 0) \,|\, a, b \in \mathbf{R}$}.

13. Decide which of the following sets of vectors spans $\mathbf{R}_3$.
    (a) {$E_1 = (1, 0, 0)$, $E_2 = (0, 1, 0)$, $E_3 = (0, 0, 1)$}.
    (b) {$X_1 = (1, 1, 0)$, $X_2 = (0, 0, 1)$}.
    (c) {$X_1 = (1, -1, 0)$, $X_2 = (0, 2, 1)$, $X_3 = (2, 4, 3)$}.

14. Decide which of the following subsets of $\mathbf{P}_4$ is a subspace of $\mathbf{P}_4$.
    (a) **W** = {$f \in \mathbf{P}_4 \,|\, f(2) = 0$}.
    (b) The subset of all polynomial functions of degree 3 together with the zero function.
    (c) The subset of all polynomial functions for which the constant term of the polynomial is 0.

15. Decide which of the following subsets of $\mathbf{R}_{2 \times 3}$ is a subspace of $\mathbf{R}_{2 \times 3}$.
    (a) $\left\{ \begin{bmatrix} a & b & a \\ 0 & a+b & b \end{bmatrix} \middle| a, b \in \mathbf{R} \right\}$.
    (b) $\left\{ \begin{bmatrix} a & b & c \\ b+c & 0 & a+b \end{bmatrix} \middle| a, b \in \mathbf{R} \right\}$.

16. Determine if the set **W** = {$(1, 2, 0)$, $(4, -2, 3)$, $(5, 0, 3)$, $(-1, -2, 0)$, $(-4, 2, -3)$, $(-5, 0, -3)$, $(0, 0, 0)$} is a subspace of $\mathbf{R}_3$. Verify your answer.

17. In the proof of Theorem 5.2, verify that property 4A holds by considering $0X_1$, where X_1 is a vector in **W**.

18. If **U** and **W** are subspaces of **V**, prove that $\mathbf{U} \cap \mathbf{W}$ is a subspace of **V**.

19. If **U** = $\langle (1, 0, 2) \rangle$ and **W** = $\langle (1, 0, 0) \rangle$, show that $\mathbf{U} \cup \mathbf{W}$ is not a subspace of $\mathbf{R}_3$.

20. If **U** and **W** are arbitrary subspaces of a vector space **V**,
    (a) Prove that $\mathbf{U} \cup \mathbf{W} \subseteq \mathbf{U} + \mathbf{W}$.
    (b) Prove that $\mathbf{U} + \mathbf{W}$ is a subspace of **V**.
    (c) If **S** is a subspace of **V** and $\mathbf{U} \cup \mathbf{W} \subseteq \mathbf{S}$, prove that $\mathbf{U} + \mathbf{W} \subseteq \mathbf{S}$.

Problems 21 and 22 require a knowledge of elementary calculus.

21. Show that the set of all functions f that satisfy the differential equation $f''(x) + 5f(x) = 0$ is a subspace of **F** in Example 5.1(e).

22. Assuming results from elementary calculus, verify that
    (a) The space **D** of differentiable functions in Example 5.1(j) is a subspace of the space **C** of continuous functions in Example 5.1(i).
    (b) The space **C** of continuous functions is a subspace of the space of integrable functions in Example 5.1(k).

5.3 LINEAR INDEPENDENCE AND LINEAR DEPENDENCE

In a vector space $\mathbf{V}$, it is clear that by taking each of the scalars to be zero one can write the zero vector as a linear combination of any finite set of vectors. A question that one might ask is, for a given finite set of vectors is it possible to write the zero vector as a linear combination of the vectors if some of the scalars are not zero? The answer is that it depends on the particular set that we have.

Consider the sets of vectors $A = \{(1, 2, 0), (3, 2, 1), (7, 6, 2)\}$ and $B = \{(1, 2, 0), (3, 2, 1), (1, 0, 0)\}$. For the vectors in A, we see that $-1(1, 2, 0) + (-2)(3, 2, 1) + 1(7, 6, 2) = (0, 0, 0)$. On the other hand, for the vectors in B we can easily show that, if $a(1, 2, 0) + b(3, 2, 1) + c(1, 0, 0) = (0, 0, 0)$, then $a = b = c = 0$.

This distinction in the way in which the zero vector can be expressed as a linear combination of a given set of vectors is extremely important. The set of vectors A is a linearly dependent set, and the set of vectors B is a linearly independent set according to the following definitions.

Definition 5.7. A nonempty set $\{X_1, X_2, \ldots, X_n\}$ of (distinct) vectors in a vector space $\mathbf{V}$ is **linearly dependent** if and only if there are scalars $r_1, r_2, \ldots, r_n$ not all zero such that $r_1 X_1 + r_2 X_2 + \cdots + r_n X_n = \mathbf{0}$.
The negation of linearly dependent is linearly independent.

Definition 5.8. A nonempty set $\{X_1, X_2, \ldots, X_n\}$ of vectors in a vector space $\mathbf{V}$ is **linearly independent** if and only if, for all scalars $r_1, r_2, \ldots, r_n$, if $r_1 X_1 + r_2 X_2 + \cdots + r_n X_n = \mathbf{0}$, then $r_1 = r_2 = \cdots = r_n = 0$.

Example 5.9. Determine if the given sets of coordinate vectors are linearly independent or linearly dependent.
(a) $C = \{X_1 = (1, 2, 6), X_2 = (2, 4, 12)\}$.
(b) $D = \{X_1 = (1, 2, 0, 1), X_2 = (2, 4, -1, 0), X_3 = (0, 0, 1, 0)\}$.

Solution: (a) Since $X_2 = 2X_1$, we have $2X_1 + (-1)X_2 = \mathbf{0}$. Therefore, C is a linearly dependent set.

(b) Since there is no obvious dependency relationship as there was in Example 5.9(a), we solve the equation $aX_1 + bX_2 + cX_3 = \mathbf{0}$ for a, b, and c. Using the definitions of addition and scalar multiplication in $\mathbf{R}_3$, we have

$$(a + 2b + 0c, 2a + 4b + 0c, 0a + (-1)b + c, a + 0b + 0c) = (0, 0, 0, 0)$$

which is equivalent to the system

$$a + 2b + 0c = 0$$
$$2a + 4b + 0c = 0$$
$$0a - b + c = 0$$
$$a + 0b + 0c = 0$$

We see that the only solution is $a = 0, b = 0$, and $c = 0$. Hence D is a linearly independent set.

Example 5.10. Determine if the given subsets of **P** are linearly independent or linearly dependent.
(a) $A = \{f_1: x^2 + 1, f_2: 2x - 1, f_3: 3\}$.
(b) $B = \{f_1: x^2 + x, f_2: 3x^2 + 4x, f_3: 2x\}$.

Solution: (a) We solve the equation $af_1 + bf_2 + cf_3 = 0$ for a, b, and c. Call the function on the left side of the equation f. Then

$$f: a(x^2 + 1) + b(2x - 1) + c(3) = 0: 0$$

So

$$ax^2 + 2bx + (a - b + 3c) = 0$$

for every real number x. The only way that this is possible is for $a = 0, 2b = 0$ and $a - b + 3c = 0$, because, from algebra, a polynomial equation of degree 2 (or less) that is not identically zero can have at most two real roots. Thus $a = b = c = 0$, and the set A is linearly independent.

(b) As in part (a), we solve the equation $af_1 + bf_2 + cf_3 = 0$ for a, b, and c. If f is the function on the left side of the equation, we have

$$f: a(x^2 + x) + b(3x^2 + 4x) + c(2x) = 0: 0$$

Therefore,

$$(a + 3b)x^2 + (a + 4b + 2c)x = 0$$

for every real number x. As in part (a), this is possible only if

$$a + 3b = 0$$
$$a + 4b + 2c = 0$$

The coefficient matrix of this system reduces to $\begin{bmatrix} 1 & 0 & -6 \\ 0 & 1 & 2 \end{bmatrix}$. Thus the solutions are $a = 6c$ and $b = -2c$. Setting $c = 1$, for example, we have $f = 6f_1 + (-2)f_2 + 1f_3 = 0$. Hence B is a linearly dependent set.

Although the definitions of linearly dependent and linearly independent refer to sets of vectors, it will be convenient to also say that **vectors are linearly independent or dependent.** For instance, we shall say that the vectors X_1 and X_2 in Example 5.9(a) are linearly dependent, whereas the vectors X_1, X_2, and X_3 in Example 5.9(b) are linearly independent. Also, we shall sometimes drop the word "linearly" and simply refer to a set of vectors as being independent or dependent.

Example 5.11. (a) Give a geometric interpretation to the statement, the nonzero vectors X_1 and X_2 are linearly dependent vectors in $\mathbf{R}_3$.

(b) Give a geometric interpretation to the statement, the nonzero vectors X_1, X_2, and X_3 are linearly dependent vectors in $\mathbf{R_3}$.

Solution: (a) If X_1 and X_2 are linearly dependent, then there are scalars a and b not both zero such that $aX_1 + bX_2 = \mathbf{0} = (0, 0, 0)$. Let's say that $a \neq 0$. Then $X_1 = -(b/a)X_2$ (also $b \neq 0$ since $X_1 \neq \mathbf{0}$). Consequently, the points X_1, X_2, and $\mathbf{0}$ are collinear. The position vectors corresponding to X_1 and X_2 are parallel.

(b) If X_1, X_2, and X_3 are nonzero linearly dependent vectors in $\mathbf{R_3}$, then there are scalars a, b, and c not all zero such that $aX_1 + bX_2 + cX_3 = \mathbf{0}$. Let's say that $a \neq 0$. Then $X_1 = -(b/a)X_2 + -(c/a)X_3$. There are two possible situations. If X_2 and X_3 are linearly dependent, then they are scalar multiples of each other [see part (a)]; consequently, X_1 is a scalar multiple of X_2 (and of X_3). Thus X_1, X_2, and X_3 are points on a line through the origin. On the other hand, if X_2 and X_3 are linearly independent, then $\mathbf{0}$, X_2, and X_3 are three noncollinear points. These three noncollinear points determine a plane, and since X_1 is a linear combination of X_2 and X_3, it follows that X_1 is in this plane.

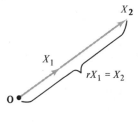

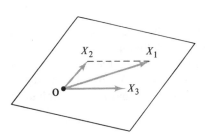

X_1 and X_2 are linearly dependent.

X_2 and X_3 are linearly independent, but X_1, X_2, and X_3 are linearly dependent.

Example 5.12. Show that the vector $X_4 = (-6, 6, -1, -1)$ is in the space $\langle X_1 = (2, 3, 1, 0), X_2 = (4, 0, 1, 2), X_3 = (0, 1, 0, 1) \rangle$, and consequently the set $\{X_1, X_2, X_3, X_4\}$ is linearly dependent.

Solution: $X_4 = 1X_1 + -2X_2 + 3X_3$ and so $X_4 \in \langle X_1, X_2, X_3 \rangle$. Thus

$$0 = (0, 0, 0, 0) = -X_4 + X_4 = (-1)X_1 + 2X_2 + (-3)X_3 + 1X_4$$

and $\{X_1, X_2, X_3, X_4\}$ is a linearly dependent set.

A generalization of Example 5.12 is stated in the next theorem.

Theorem 5.4. If $X_1, X_2, \ldots, X_{k-1}, X_k$ are distinct vectors in a vector space $\mathbf{V}$, and if $X_k \in \langle X_1, X_2, \ldots, X_{k-1} \rangle$, then $\{X_1, X_2, \ldots, X_{k-1}, X_k\}$ is a linearly dependent set.

Proof: Suppose that $X_k \in \langle X_1, X_2, \ldots, X_{k-1} \rangle$. Then

$$X_k = r_1 X_1 + r_2 X_2 + \cdots + r_{k-1} X_{k-1}$$

for some scalars $r_1, r_2, \ldots, r_{k-1}$. Adding $-(r_1 X_1 + r_2 X_2 + \cdots + r_{k-1} X_{k-1})$ to both sides, we obtain

$$-(r_1 X_1 + r_2 X_2 + \cdots + r_{k-1} X_{k-1}) + X_k = 0$$

Therefore,

$$(-r_1)X_1 + (-r_2)X_2 + \cdots + (-r_{k-1})X_{k-1} + 1X_k = 0$$

(see Problem 16, Exercise 5.3). Not all of the scalars $-r_1, -r_2, \ldots, -r_{k-1}, 1$ are zero. Hence $X_1, X_2, \ldots, X_k$ are linearly dependent.   Q.E.D.

The converse is not true. For example, $\{(1, 1, 0), (2, 2, 0), (1, 0, 0)\}$ is a linearly dependent set but $(1, 0, 0) \notin \langle (1, 1, 0), (2, 2, 0) \rangle$. However, the following "near converse" is true.

Theorem 5.5. If $A = \{X_1, X_2, \ldots, X_k\}$ is a set of linearly dependent vectors in $\mathbf{V}$ and $X_1 \neq 0$, then there is a vector X_j in the set with $j \geq 2$ such that X_j is a linear combination of the preceding vectors ($X_j \in \langle X_1, X_2, \ldots, X_{j-1} \rangle$).

Proof: Since A is linearly dependent, there are scalars $a_1, a_2, \ldots, a_k$ not all zero such that $a_1 X_1 + a_2 X_2 + \cdots + a_k X_k = 0$. Let a_j be the nonzero coefficient having the largest subscript. Then $a_{j+1} = \cdots = a_k = 0$ and $a_1 X_1 + \cdots + a_j X_j = 0$. We also note that $j > 1$ since $a_1 X_1 = 0$ and $X_1 \neq 0$ (by hypothesis) implies that $a_1 = 0$. Solving for X_j, we have

$$X_j = \frac{-a_1}{a_j} X_1 + \cdots + \frac{-a_{j-1}}{a_j} X_{j-1}$$

and thus $X_j \in \langle X_1, \ldots, X_{j-1} \rangle$ and $j \geq 2$.   Q.E.D.

Note: The hypothesis that the vector $X_1 \neq 0$ is essential to obtain the conclusion. For example, if $A = \{X_1, X_2\}$, where $X_1 = (0, 0, 0)$ and $X_2 = (1, 2, 3)$, then A is a linearly dependent set, but X_2 is not in the space $\langle X_1 \rangle = \{0\}$.

As a consequence of Theorem 5.5, it is possible to replace a linearly dependent set of generators for a vector space by a linearly independent subset that still generates the same space. Before proving the theorem, we look at an example to illustrate the idea of the proof.

Example 5.13.   Let $\mathbf{W}$ be the subspace of $\mathbf{R}_3$ generated by the vectors $X_1 = (1, 2, 1)$, $X_2 = (3, 6, 3)$, $X_3 = (4, 2, 1)$, and $X_4 = (11, 4, 2)$. Show that the set $A = \{X_1, X_2, X_3, X_4\}$ is a linearly dependent set, and that there is a linearly independent subset of A that generates the same space $\mathbf{W}$ as that generated by A.

Solution: Set A is a linearly dependent set because $(-1)X_1 + 0X_2 + 3X_3 + (-1)X_4 = 0$. According to Theorem 5.5, some X_j is a linear combination of the preceding ones. Scanning the vectors from left to right, we find the first such vector, which is $X_2 = 3X_1$. We discard the vector X_2 and claim that $W = \langle X_1, X_3, X_4 \rangle$. The fact that $\langle X_1, X_3, X_4 \rangle \subseteq W$ is clear, because any linear combination $aX_1 + bX_3 + cX_4$ in $\langle X_1, X_3, X_4 \rangle$ can be written as $aX_1 + 0X_2 + bX_3 + cX_4$, which is in W. On the other hand, suppose that $X = aX_1 + bX_2 + cX_3 + dX_4$ is in W. Since $X_2 = 3X_1$, we can write

$$X = aX_1 + b(3X_1) + cX_3 + dX_4$$
$$= (a + 3b)X_1 + cX_3 + dX_4$$

which is in $\langle X_1, X_3, X_4 \rangle$. We have shown that $W \subseteq \langle X_1, X_3, X_4 \rangle$ and that $\langle X_1, X_3, X_4 \rangle \subseteq W$. Thus $W = \langle X_1, X_3, X_4 \rangle$. Next we check to see if X_1, X_3, and X_4 are linearly dependent. We see that $(-1)X_1 + 3X_3 + (-1)X_4 = 0$. Therefore, by Theorem 5.5, some vector in $\{X_1, X_3, X_4\}$ is a linear combination of the preceding vectors. Scanning the vectors again from left to right, we find that the first such vector is $X_4 = (-1)X_1 + 3X_3$. We discard X_4 and claim that $\langle X_1, X_3, X_4 \rangle = \langle X_1, X_3 \rangle$. The fact that $\langle X_1, X_3 \rangle \subseteq \langle X_1, X_3, X_4 \rangle$ is obvious, since $aX_1 + bX_3 = aX_1 + bX_3 + 0X_4$. Now suppose that $aX_1 + bX_3 + cX_4$ is an arbitrary vector in the space $\langle X_1, X_3, X_4 \rangle$. Then

$$aX_1 + bX_3 + cX_4 = aX_1 + bX_3 + c(-1X_1 + 3X_3)$$
$$= (a - c)X_1 + (b + 3c)X_3$$

which is in $\langle X_1, X_3 \rangle$. So $\langle X_1, X_3, X_4 \rangle \subseteq \langle X_1, X_3 \rangle$ and consequently $\langle X_1, X_3 \rangle = \langle X_1, X_3, X_4 \rangle = \langle X_1, X_2, X_3, X_4 \rangle = W$. The vectors X_1 and X_3 are linearly independent.

What we have done is replace the linearly dependent set of generators $\{X_1, X_2, X_3, X_4\}$ with the linearly independent set $\{X_1, X_3\}$, which generates the same space W. We generalize Example 5.13 in the next theorem. The proof of the theorem imitates the solution to Example 5.13 and may be omitted.

Theorem 5.6.   If $A = \{X_1, X_2, \ldots, X_k\}$ is a set of linearly dependent generators for a vector space V and $X_1 \neq 0$, then there exists a linearly independent subset of A that generates the same space V as that generated by A.

Proof: (Optional) Since A is linearly dependent and $X_1 \neq 0$, we know that there are at least two vectors in A, and according to Theorem 5.5, there is some $j \geq 2$ such that X_j is a linear combination of the preceding vectors. Let

$$A_1 = \{X_1, \ldots, X_{j-1}, X_{j+1}, \ldots, X_k\}$$

Obviously, the space generated by A_1 is a subspace of the space V generated by A. Let

$$X = a_1X_1 + a_2X_2 + \cdots + a_kX_k$$

be an arbitrary vector in $\mathbf{V}$. Since X_j is a linear combination of the preceding vectors, we can write

$$X = a_1 X_1 + \cdots + a_{j-1} X_{j-1} + (b_1 X_1 + \cdots + b_{j-1} X_{j-1})$$
$$+ a_{j+1} X_{j+1} + \cdots + a_k X_k$$
$$= (a_1 + b_1) X_1 + \cdots + (a_{j-1} + b_{j-1}) X_{j-1} + a_{j+1} X_{j+1} + \cdots + a_k X_k$$

Thus X is a linear combination of the vectors in A_1, and consequently the space generated by A_1 is equal to the space generated by A. If A_1 is linearly independent, we have established the theorem. Otherwise, we can repeat the above process on A_1, obtaining a set A_2 having $k - 2$ vectors that again generates $\mathbf{V}$. Continuing, we must eventually obtain a set A_m with $m < k$ such that A_m generates $\mathbf{V}$ and A_m is linearly independent. This is so because at each step we throw out a vector, and yet the remaining set still generates $\mathbf{V}$. This is possible only if we throw out fewer than k vectors. Hence the process terminates with some $m < k$. Q.E.D.

An alternative but useful test to determine if a set of vectors is linearly independent is given in the next theorem.

Theorem 5.7. A set of vectors $B = \{X_1, X_2, \ldots, X_k\}$ in a vector space $\mathbf{V}$ is linearly independent if and only if every vector $X \in \langle X_1, X_2, \ldots, X_k \rangle$ can be expressed in a unique way (the scalar coefficients are uniquely determined) as a linear combination of $X_1, X_2, \ldots, X_k$.

Proof: Suppose that the set B is linearly independent. Let X be a vector such that

$$X = a_1 X_1 + a_2 X_2 + \cdots + a_k X_k = b_1 X_1 + b_2 X_2 + \cdots + b_k X_k$$

Then

$$0 = X - X = (a_1 X_1 + a_2 X_2 + \cdots + a_k X_k)$$
$$- (b_1 X_1 + b_2 X_2 + \cdots + b_k X_k)$$
$$= (a_1 - b_1) X_1 + (a_2 - b_2) X_2 + \cdots + (a_k - b_k) X_k$$

Since B is linearly independent, we have

$$a_1 - b_1 = 0, \, a_2 - b_2 = 0, \ldots, a_k - b_k = 0$$

and thus

$$a_1 = b_1, \, a_2 = b_2, \ldots, a_k = b_k$$

Conversely, suppose that every vector that can be written as a linear combination of the vectors in B has uniquely determined coefficients. Suppose that $a_1 X_1 + a_2 X_2 + \cdots + a_k X_k = 0$. Now $0 = 0 X_1 + 0 X_2 + \cdots + 0 X_k$. Since the coefficients are uniquely determined, we must have $a_1 = 0$, $a_2 = 0$, $\ldots, a_k = 0$. Q.E.D.

We close this section with a remark concerning Definition 5.7. If we had allowed repetitions when listing vectors in a set, then the set $\{X_1, X_2\}$ with $X_1 = X_2 \neq 0$ is linearly dependent because $1X_1 + (-1)X_2 = 0$. However, $\{X_1, X_2\} = \{X_1\}$ and $\{X_1\}$ is linearly independent (Exercise 5.3, Problem 13). We used the word "distinct" in Definition 5.7 in order to avoid this potential problem.

Exercise 5.3

In Problems 1–6, determine which sets of vectors are linearly independent and which sets are linearly dependent. If the vectors are linearly dependent, write **0** as a linear combination of the vectors in a nontrivial (not all the scalars are 0) way. The vector spaces are among those listed in Example 5.1.

1. V = R₃.
   (a) $A = \{X_1 = (1, 2, 4), X_2 = (3, 4, -16), X_3 = (3, 5, -2)\}$.
   (b) $A = \{X_1 = (1, 0, 2), X_2 = (3, 2, 1), X_3 = (2, 0, 4)\}$.
   (c) $A = \{X_1 = (1, 3, -1), X_2 = (0, 3, 2), X_3 = (1, 1, 0)\}$.

2. V = R₃.
   (a) $A = \{X_1 = (1, 2, 3), X_2 = (3, 6, 10)\}$.
   (b) $A = \{X_1 = (0, 1, -2), X_2 = (3, 4, -7), X_3 = (-6, -5, -8)\}$.
   (c) $A = \{X_1 = (1, 0, 0), X_2 = (0, 1, 0), X_3 = (0, 0, 1)\}$.

✓ **3. V = P.**
   (a) $A = \{f_1: x^3 + 3x, f_2: 3x, f_3: 2x^3 + 4x\}$.
   (b) $A = \{f_1: x + 3, f_2: 2x + 5, f_3: x^2 + 1\}$.
   (c) $A = \{f_1: x + 3, f_2: 3x + 4, f_3: 2x - 1\}$.
   (d) $A = \{f_1: 1, f_2: x, f_3: x^2, \ldots, f_n: x^{n-1}\}$.

4. V = R₂ₓ₃.
   (a) $A = \left\{ X_1 = \begin{bmatrix} 1 & 2 & 0 \\ 3 & 0 & 0 \end{bmatrix}, X_2 = \begin{bmatrix} 2 & 0 & 0 \\ 4 & 0 & 0 \end{bmatrix}, X_3 = \begin{bmatrix} 0 & 1 & 0 \\ 0 & 0 & 0 \end{bmatrix} \right\}$.

   (b) $A = \left\{ X_1 = \begin{bmatrix} 1 & 0 & 0 \\ 2 & 3 & 1 \end{bmatrix}, X_2 = \begin{bmatrix} 0 & 0 & 0 \\ 2 & 0 & 2 \end{bmatrix}, X_3 = \begin{bmatrix} 0 & 0 & 0 \\ 0 & 4 & 0 \end{bmatrix} \right\}$.

   (c) $A = \left\{ \begin{array}{l} X_1 = \begin{bmatrix} 1 & 0 & 0 \\ 0 & 0 & 0 \end{bmatrix}, X_2 = \begin{bmatrix} 0 & 1 & 0 \\ 0 & 0 & 0 \end{bmatrix}, X_3 = \begin{bmatrix} 0 & 0 & 1 \\ 0 & 0 & 0 \end{bmatrix} \\ X_4 = \begin{bmatrix} 0 & 0 & 0 \\ 1 & 0 & 0 \end{bmatrix}, X_5 = \begin{bmatrix} 0 & 0 & 0 \\ 0 & 1 & 0 \end{bmatrix}, X_6 = \begin{bmatrix} 0 & 0 & 0 \\ 0 & 0 & 1 \end{bmatrix} \end{array} \right\}$.

5. V is the vector space of complex numbers.
   (a) $A = \{X_1 = 2 + 3i, X_2 = 4 - i\}$.
   (b) $A = \{X_1 = 2i, X_2 = 2\}$.
   (c) $A = \{X_1 = 5, X_2 = 13\}$.
   (d) $A = \{X_1 = 1, X_2 = i\}$.

6. $V = R_4$.

(a) $A = \{X_1 = (1, 0, 2, 1), X_2 = (-2, 3, -4, 1), X_3 = (0, 1, 0, 1)\}$.

(b) $A = \{X_1 = (1, 0, 0, 0), X_2 = (0, 1, 0, 0), X_3 = (0, 0, 1, 0), X_4 = (0, 0, 0, 1)\}$.

7. Determine if $(0, 18, 6) \in \langle (2, 3, 1), (1, 6, 2) \rangle$.

✓ **8.** Show that the vectors $(1, 1, 1)$ and $(0, 2, 0)$ do not generate R_3.

In Problems 9–12, reduce the given set of generators (for the space spanned by the given set) to a set of linearly independent generators by using the method of Example 5.13 and of the proof of Theorem 5.6. Label the unlabeled vectors in each of the given sets as $X_1, X_2, \ldots, X_k$ for the appropriate k.

9. $A = \{(1, -2, -3), (0, 1, 2), (2, -1, 0), (5, -4, -3)\}$.

✓ **10.** $A = \{(1, -3, 2, 4), (2, -6, 4, 8), (2, 3, 0, 1), (2, 12, -4, -6)\}$.

11. $A = \{f_1 : 2x^2 + 1, f_2 : x^2 + 4, f_3 : 3x^2 + 6, f_4 : x^2\}$.

12. $A = \left\{ \begin{bmatrix} 1 & 0 \\ 0 & 0 \end{bmatrix}, \begin{bmatrix} 1 & 1 \\ 0 & 0 \end{bmatrix}, \begin{bmatrix} 1 & 1 \\ 1 & 0 \end{bmatrix}, \begin{bmatrix} 1 & 1 \\ 1 & 1 \end{bmatrix}, \begin{bmatrix} 2 & 0 \\ 0 & 2 \end{bmatrix} \right\}$.

13. If X is a vector in a vector space V, verify that $\{X\}$ is a linearly independent set if and only if $X \neq 0$.

14. *Prove:* If $X = (a, b)$ and $Y = (c, d)$ are nonzero vectors in R_2, then $\{X, Y\}$ is linearly independent if and only if $\det \begin{bmatrix} a & b \\ c & d \end{bmatrix} \neq 0$.

15. *Prove:* If $X_1, X_2, \ldots, X_k$ are linearly independent vectors in a vector space V, and if Y is a vector in V such that Y is not in $\langle X_1, X_2, \ldots, X_k \rangle$, then $\{X_1, X_2, \ldots, X_k, Y\}$ is a linearly independent set.

16. Use mathematical induction to verify that if $X_1, X_2, \ldots, X_k$ are vectors in a vector space V then

$$-(r_1 X_1 + r_2 X_2 + \cdots + r_k X_k) = (-r_1) X_1 + (-r_2) X_2 + \cdots + (-r_k) X_k$$

5.4 BASIS AND DIMENSION

In this section we give a precise definition of the term *dimension* of a vector space. Any vector space that is generated by a finite number of vectors is called a finite-dimensional vector space. Other than mentioning that there are infinite-dimensional vector spaces, and giving some examples, we shall not be concerned with infinite-dimensional vector spaces.

All of us have an intuitive idea of what dimension is. We think of a line as being one dimensional, a plane as being two dimensional, and space as being three dimensional. Our definition of dimension will agree with this intuitive notion but will be more general. We begin by defining a basis for a vector space.

Definition 5.9. If B is a set of vectors in a vector space V, then B is a **basis** for V if and only if

(i) The space V is generated by B.
(ii) B is a linearly independent set.

Example 5.14. (a) Show that each of the sets $B_1 = \{(1, 0), (0, 1)\}$ and $B_2 = \{(2, 3), (1, 1)\}$ is a basis for $\mathbf{R}_2$.

(b) Show that each of the sets $C_1 = \{f_1 : x^2, f_2 : x, f_3 : 1\}$ and $C_2 = \{g_1 : x^2, g_2 : 3x + 4, g_3 : 4\}$ is a basis for $\mathbf{P}_3$.

Solution: (a) Any vector (a, b) in $\mathbf{R}_2$ can be written as $a(1, 0) + b(0, 1)$. Therefore, B_1 generates $\mathbf{R}_2$. If $(0, 0) = a(1, 0) + b(0, 1)$, it is clear that $a = b = 0$, and B_1 is linearly independent.

To determine if B_2 generates $\mathbf{R}_2$, we take an arbitrary vector (a, b) and ask the question, are there scalars x and y such that $x(2, 3) + y(1, 1) = (a, b)$? This vector equation is equivalent to the linear system

$$2x + y = a$$
$$3x + y = b$$

Since

$$\det \begin{bmatrix} 2 & 1 \\ 3 & 1 \end{bmatrix} = -1 \neq 0$$

we know that there is a (unique) *solution*, and hence B_2 generates $\mathbf{R}_2$. (The actual solution is $x = b - a$ and $y = 3a - 2b$.) Also, $\det \begin{bmatrix} 2 & 1 \\ 3 & 1 \end{bmatrix} \neq 0$ tells us that any vector $X = (a, b)$ is expressible as a *unique* linear combination of $X_1 = (2, 3)$ and $X_2 = (1, 1)$. Thus, by Theorem 5.7, B_2 is linearly independent.

(b) Let f be a polynomial function in $\mathbf{P}_3$. Then $f: ax^2 + bx + c$ and $f = af_1 + bf_2 + cf_3$. Thus C_1 generates $\mathbf{P}_3$. If $af_1 + bf_2 + cf_3 = \mathbf{0}$, then $ax^2 + bx + c = 0$ for all x in $\mathbf{R}$. This is possible only if $a = b = c = 0$. Hence C_1 is a basis.

Again, let $f: ax^2 + bx + c$ be an arbitrary function in $\mathbf{P}_3$. To determine if C_2 generates $\mathbf{P}_3$, we find out if there are scalars r_1, r_2, and r_3 such that

$$r_1 g_1 + r_2 g_2 + r_3 g_3 : r_1 x^2 + r_2(3x + 4) + r_3(4) = f: ax^2 + bx + c$$

If so, then we must have

$$r_1 x^2 + 3r_2 x + 4r_2 + 4r_3 = ax^2 + bx + c$$

or

$$(r_1 - a)x^2 + (3r_2 - b)x + (4r_2 + 4r_3 - c) = 0$$

for every x. Since a polynomial equation of degree 2 or less that is not identically zero can have at most two real solutions, it follows that the coefficients must all be zero and

$$r_1 - a = 0, \qquad 3r_2 - b = 0, \qquad 4r_2 + 4r_3 - c = 0$$

Thus $r_1 = a$, $r_2 = b/3$, and $r_3 = c/4 - b/3$. Checking, we see that

$$ax^2 + bx + c = a(x^2) + \frac{b}{3}(3x + 4) + \left(\frac{c}{4} - \frac{b}{3}\right)(4)$$

Thus C_2 generates $\mathbf{P}_3$.

Next we check to see if C_2 is linearly independent. In this case, we solve for r_1, r_2, and r_3 when $f = 0$. Therefore,

$$a = b = c = 0 \quad \text{and} \quad r_1 x^2 + 3r_2 x + (4r_2 + 4r_3) = 0$$

for every x. As before, each of the coefficients must be zero, and we have

$$r_1 = 0, \qquad 3r_2 = 0, \qquad 4r_2 + 4r_3 = 0$$

Hence $r_1 = r_2 = r_2 = 0$ and C_2 is linearly independent.

Example 5.14 illustrates the fact that a vector space can have more than one basis. In the example, each of the bases for $\mathbf{R}_2$ has two elements, and each of the bases for $\mathbf{P}_3$ has three elements. We shall see that, in general, if a vector space has a basis of n vectors, then any other basis for the space must have n vectors. The next theorem, the Steinitz replacement theorem (after Ernst Steinitz, 1871–1928) leads to this result.

Theorem 5.8 (Steinitz Replacement Theorem). If $\{Y_1, Y_2, \ldots, Y_r\}$ is a linearly independent set of vectors in a vector space $\mathbf{W}$, and if $\mathbf{W} = \langle X_1, X_2, \ldots, X_n \rangle$, then
(i) $r \leq n$; that is, the number of linearly independent vectors in $\mathbf{W}$ does not exceed the number of vectors in a generating set for $\mathbf{W}$.
(ii) If $r < n$, then r of the X's can be replaced by the linearly independent Y's so that $\mathbf{W} = \langle Y_1, Y_2, \ldots, Y_r, X_{i_1}, X_{i_2}, \ldots, X_{i_{n-r}} \rangle$.
(iii) If $r = n$, then $\langle Y_1, Y_2, \ldots, Y_r \rangle = \mathbf{W}$.
We illustrate both the theorem and a method of proof in the next example. The proof of Theorem 5.8 is left as an exercise and is outlined in Problem 20, Exercise 5.4.

Example 5.15. Let $Y_1 = (4, 2, 0)$, $Y_2 = (2, -1, 4)$, $X_1 = (1, 0, 1)$, $X_2 = (2, 0, 2)$, $X_3 = (2, 1, 0)$, $X_4 = (1, -1, 3)$, and $X_5 = (4, -1, 6)$. Let $\mathbf{W} = \langle X_1, X_2, X_3, X_4, X_5 \rangle$. (a) Show that Y_1 and Y_2 are linearly independent vectors in $\mathbf{W}$, and thus the hypothesis of Theorem 5.8 is satisfied.
   (b) Show that $\mathbf{W} = \langle Y_1, Y_2, X_{i_1}, X_{i_2}, X_{i_3} \rangle$ for the appropriate X vectors, and thus conclusion (ii) of Theorem 5.8 holds.

Solution: (a) Since $Y_1 = 2X_3$ and $Y_2 = X_1 + X_4$, it follows that Y_1 and Y_2 are in $\mathbf{W}$. The set $\{Y_1, Y_2\}$ is linearly independent since neither vector is a scalar multiple of the other.
   (b) Consider the vectors $Y_1, X_1, X_2, X_3, X_4, X_5$, which span $\mathbf{W}$ and are linearly dependent, because Y_1 is a linear combination of the X's. One of the X's must be a linear combination of the preceding vectors; we see for example that $X_2 = 0Y_1 + 2X_1$. Discarding X_2, we are left with vectors Y_1, X_1, X_3, X_4, X_5, which still span $\mathbf{W}$. This is so since any vector Z in $\mathbf{W}$ may be written as

$$Z = a_1 X_1 + a_2 X_2 + a_3 X_3 + a_4 X_4 + a_5 X_5$$
$$= a_1 X_1 + a_2(0Y_1 + 2X_1) + a_3 X_3 + a_4 X_4 + a_5 X_5$$
$$= 0Y_1 + (a_1 + 2a_2)X_1 + a_3 X_3 + a_4 X_4 + a_5 X_5$$

Now consider $Y_1, Y_2, X_1, X_3, X_4, X_5$. These vectors are dependent because $Y_2 \in \langle Y_1, X_1, X_3, X_4, X_5 \rangle$. Thus some vector must be a linear combination of the preceding ones. It cannot be a Y, because the Y's are linearly independent. We see that $X_5 = 3X_1 + X_4$. Discarding X_5, we are left with the vectors Y_1, Y_2, X_1, X_3, X_4, which still span W. We have

$$W = \langle Y_1, Y_2, X_{i_1}, X_{i_2}, X_{i_3} \rangle$$

where $i_1 = 1$, $i_2 = 3$, and $i_3 = 4$.

Corollary 5.1. If a vector space V has a basis $B = \{X_1, X_2, \ldots, X_n\}$ consisting of n vectors, then the number of vectors in any other basis for V is also n.

Proof: Let $B_2 = \{Y_1, Y_2, \ldots, Y_k\}$ be a basis for V. Since B_2 is linearly independent and B generates V, we have $k \leq n$ by the Steinitz replacement theorem. From the hypothesis that B_2 is a basis for V, we have $V = \langle Y_1, Y_2, \ldots, Y_k \rangle$. Also, B is linearly independent. Applying the Steinitz replacement theorem, where the roles of the X's and the Y's have been reversed, we get $n \leq k$. Hence $k = n$. Q.E.D.

Corollary 5.1 allows us to define the dimension of a vector space in terms of the number of vectors in a basis.

Definition 5.10. The **dimension** of a vector space $V \neq \{0\}$ is n if and only if V has a basis containing n vectors. In the trivial case that $V = \{0\}$, we shall say that V has dimension 0.

Example 5.16. Show that the dimension of R_n is n.

Solution: Consider the set $B = \{E_1, E_2, \ldots, E_n\}$, where E_i is the vector $(0, \ldots, 0, 1, 0, \ldots, 0)$ having a 1 as the ith entry and all other entries 0.

Any vector $A = (a_1, a_2, \ldots, a_n)$ in R_n can be written as $A = a_1 E_1 + a_2 E_2 + \cdots + a_n E_n$. Therefore, B generates R_n.

If $b_1 E_1 + b_2 E_2 + \cdots + b_n E_n = 0$, then $(b_1, b_2, \ldots, b_n) = (0, 0, \ldots, 0)$. Hence $b_1 = b_2 = \cdots = b_n = 0$, and B is a linearly independent set. So R_n has a basis containing n elements and the dimension of R_n is n.

The basis $B = \{E_1, E_2, \ldots, E_n\}$ is called the **natural** or **standard basis** for R_n.

For $n = 1$ we can identify the vector (a_1) with the real number a_1, and the vectors in R_1 can be thought of as real numbers. Scalar multiplication becomes real-number multiplication, and vector addition becomes real-number addition. The set $\{a\}$, where a is any nonzero number, is a basis for R_1. The natural basis for R_1 is $\{1\}$.

Note: We shall use E_i to denote the ith vector in the natural basis for the space R_n, for every n. For example, in R_3, $E_2 = (0, 1, 0)$, but in R_4, $E_2 = (0, 1, 0, 0)$.

Example 5.17. Give a geometric description of $\mathbf{R}_1$, $\mathbf{R}_2$, and $\mathbf{R}_3$. Identify an arbitrary vector X in each of the spaces by picturing it as a linear combination of the standard basis.

Solution

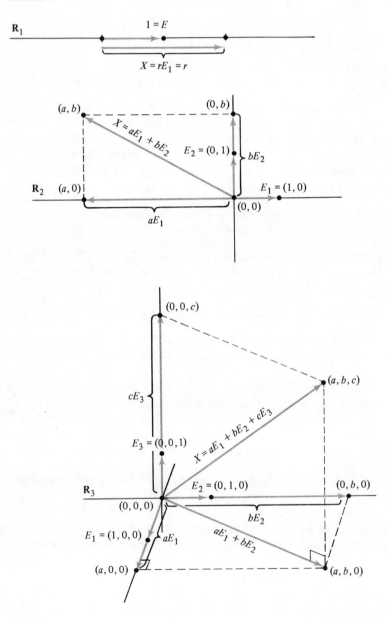

Example 5.18. Give a geometric description of R_2 using the basis $B = \{X_1 = (1, -1), X_2 = (2, 4)\}$.

Solution: Since B is a basis, every vector in R_2 is a linear combination of X_1 and X_2. Generating R_2 by this basis is geometrically equivalent to coordinatizing the plane by choosing the axes as the lines determined by the points $(0, 0)$, $(1, -1)$, and $(2, 4)$. Note that $(a, b) = cX_1 + dX_2$, where $c = (2a - b)/3$ and $d = (a + b)/6$.

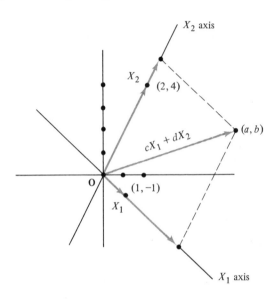

As was mentioned at the beginning of this section, a vector space is called a **finite-dimensional vector space** if and only if it is generated by a finite number of vectors. Because of Theorem 5.6, this is equivalent to saying that the space has a finite basis (or the space is the trivial space $\{0\}$). An **infinite-dimensional vector space** is a vector space that is not finite dimensional.

Example 5.19. (a) Show that the space **P** of all polynomial functions is infinite dimensional.

(b) Show that the space P_n of all polynomial functions of degree less than n has dimension n.

Solution: (a) Let S be any finite subset of **P**, and let $\deg f$ denote the degree of the polynomial function f. Since there are only a finite number of functions in S, there is a function f_1 in S having maximum degree m; that is, if g is in S, then $\deg g \leq m$. Now let h be any function in the space spanned by S. Then h is a sum of scalar multiples of polynomial functions each having degree $\leq m$. Thus the degree of h cannot exceed m. But **P** has polynomial functions of degree $m + 1$ and larger. Therefore, S does not generate $\mathbf{P}$. Since S was arbitrary, we have shown that $\mathbf{P}$ has no finite set of generators, and hence $\mathbf{P}$ is infinite dimensional.

(b) Consider the set

$$B = \{f_1: x^{n-1}, f_2: x^{n-2}, \ldots, f_{n-1}: x, f_n: 1\}$$

If $f: a_1 x^{n-1} + a_2 x^{n-2} + \cdots + a_{n-1} x + a_n 1$ is an arbitrary function in $\mathbf{P_n}$, then

$$f = a_1 f_1 + a_2 f_2 + \cdots + a_{n-1} f_{n-1} + a_n f_n$$

and B generates $\mathbf{P_n}$. Now suppose that $b_1, b_2, \ldots, b_n$ are scalars such that

$$b_1 f_1 + b_2 f_2 + \cdots + b_n f_n = 0$$

Then

$$b_1 x^{n-1} + b_2 x^{n-2} + \cdots + b_{n-1} x + b_n = 0$$

for every real number x. If not all of the b's are zero, then we have a polynomial equation of degree at most $n - 1$, and thus there can be at most $n - 1$ real solutions. Since *every* real number is a solution we must have

$$b_1 = b_2 = \cdots = b_n = 0$$

Hence B is a linearly independent set and is a basis for $\mathbf{P_n}$.

Exercise 5.4

In Problems 1–3, show that each of the sets of vectors is a basis for the given vector space. You may label each of the unlabeled vectors in a set as $X_1, X_2, \ldots$.

1. $V = \mathbf{R}_2$.
  (a) $A = \{(3, 4), (2, -1)\}$.
  (b) $B = \{(1, 0), (1, 1)\}$.

2. $V = \mathbf{P}_3$.
  (a) $A = \{f_1: 1, f_2: x + 1, f_3: x^2 + x + 1\}$.
  (b) $B = \{f_1: x^2, f_2: 1 - x, f_3: 1 + x\}$.

3. $V = \mathbf{R}_{2 \times 2}$.
  (a) $A = \left\{ \begin{bmatrix} 1 & 0 \\ 0 & 0 \end{bmatrix}, \begin{bmatrix} 0 & 1 \\ 0 & 0 \end{bmatrix}, \begin{bmatrix} 0 & 0 \\ 1 & 0 \end{bmatrix}, \begin{bmatrix} 0 & 0 \\ 0 & 1 \end{bmatrix} \right\}$.

  (b) $B = \left\{ \begin{bmatrix} 1 & 0 \\ 0 & 0 \end{bmatrix}, \begin{bmatrix} 1 & 1 \\ 0 & 0 \end{bmatrix}, \begin{bmatrix} 1 & 1 \\ 1 & 0 \end{bmatrix}, \begin{bmatrix} 1 & 1 \\ 1 & 1 \end{bmatrix} \right\}$.

In Problems 4–8, you are given a set G of generators for a space $\mathbf{W}$ and a set L of linearly independent vectors.

  (a) Show that the vectors in L are in the space $\mathbf{W}$ generated by G.
  (b) Follow the solution to Example 5.15 and replace some of the X's in G by the Y's in L to obtain a new generating set for $\mathbf{W}$ that contains the linearly independent Y's.

4. $G = \{X_1 = (1, 2), X_2 = (3, 6), X_3 = (1, 0)\}$ and
   $L = \{Y_1 = (1, 0), Y_2 = (1, 1)\}$.

5. $G = \{X_1 = (1, 2, 0), X_2 = (0, 0, 1), X_3 = (1, 2, 2)\}$ and
   $L = \{Y_1 = (2, 4, 0), Y_2 = (3, 6, 1)\}$.

6. $G = \{f_1: 1, f_2: x + 1, f_3: 2x + 3, f_4: 4, f_5: x\}$ and
   $L = \{g_1: x + 2, g_2: x - 3\}$.

7. $G = \left\{ X_1 = \begin{bmatrix} 1 & 1 \\ 0 & 0 \end{bmatrix}, X_2 = \begin{bmatrix} 2 & 2 \\ 0 & 0 \end{bmatrix}, X_3 = \begin{bmatrix} 4 & 0 \\ 0 & 0 \end{bmatrix}, X_4 = \begin{bmatrix} 0 & 4 \\ 0 & 0 \end{bmatrix} \right\}$ and

   $L = \left\{ Y_1 = \begin{bmatrix} 1 & 0 \\ 0 & 0 \end{bmatrix}, Y_2 = \begin{bmatrix} 2 & 1 \\ 0 & 0 \end{bmatrix} \right\}$.

8. $G = \{X_1 = (1, 1, 0), X_2 = (1, 1, 1), X_3 = (0, 1, 1), X_4 = (0, 1, 0), X_5 = (1, 0, 1\}$
   and $L = \{Y_1 = (1, 0, 0), Y_2 = (2, 3, 0)\}$.

9. Verify that $\mathbf{R}_{m \times n}$ has dimension mn. Look for a "natural" basis.

   For Problems 10 and 11, see and follow Example 5.18.

10. Give a geometric description of $\mathbf{R}_2$ using the basis $\{X_1 = (1, 2), X_2 = (3, -1)\}$.

11. Give a geometric description of $\mathbf{R}_3$ using the basis $\{X_1 = (1, 0, 0), X_2 = (1, 1, 0),$
    $X_3 = (1, 1, 1)\}$.

12. (a) If $\vec{a}$ and $\vec{b}$ are nonzero position vectors in a plane relative to some fixed
        origin O, state a necessary and sufficient condition, in terms of parallel
        vectors, that $\{\vec{a}, \vec{b}\}$ be a basis for the space of all position vectors in the given
        plane.
    (b) If $\vec{a}$ and $\vec{b}$ are nonzero position vectors in a plane relative to some fixed
        origin O, state a necessary and sufficient condition, in terms of the points
        at the heads of $\vec{a}$ and $\vec{b}$, that $\{\vec{a}, \vec{b}\}$ be a basis for the space of all position
        vectors in the given plane.
    (c) If $\vec{a}, \vec{b}$, and $\vec{c}$ are nonzero position vectors in 3-space relative to a fixed
        origin O, find conditions under which $\{\vec{a}, \vec{b}, \vec{c}\}$ will be a basis for 3-space.

13. What is the dimension of the vector space of Example 5.1(h)? Find a basis for
    this space and verify that you have a basis.

An **infinite set** B of vectors in a vector space $\mathbf{V}$ is **linearly independent** if and only if
every finite nonempty subset of B is linearly independent.

14. If $B = \{f_1: 1, f_2: x, f_3: x^2, \ldots, f_n: x^{n-1}, \ldots\}$, (a) show that B is a linearly
    independent set; (b) show that B generates $\mathbf{P}$.
    Remark: The vector spaces of Example 5.1(e), (f), (i), (j), and (k) are all
    infinite dimensional.

15. *Prove:* If $A = \{Y, X_1, X_2, \ldots, X_n\}$ is a set of distinct vectors in a vector space
    $\mathbf{V}$, and if $Y \in \langle X_1, X_2, \ldots, X_n \rangle$, then A is a linearly dependent set.

16. *Prove:* If A is a linearly independent set of vectors in a vector space $\mathbf{V}$ and if
    $Y \in A$, then $Y \neq 0$.

17. *Prove:* If $\{Y_1, Y_2, \ldots, Y_k\}$ is a linearly independent set of vectors in $\mathbf{V}$, then
    $\{Y_1, Y_2, \ldots, Y_j\}$ is a linearly independent set of vectors for any j, $1 \leq j \leq k$.

(It should be clear in fact that any proper subset of a linearly independent set of vectors is linearly independent. Can you prove it?)

18. For a given set of vectors $B = \{X_1, X_2, \ldots, X_n\}$ in V prove that B is a basis for V if and only if
    (i) Every vector in V is a linear combination of the vectors in B, and
    (ii) If $a_1 X_1 + a_2 X_2 + \cdots + a_n X_n = b_1 X_1 + b_2 X_2 + \cdots + b_n X_n$, then $a_1 = b_1, a_2 = b_2, \ldots, a_n = b_n$.

19. In the vector space V of Example 5.1(l), use your knowledge of calculus to solve the differential equation $3f' - f = 0$. Determine a basis for V.

20. Use the outline below to prove the Steinitz replacement theorem.
    (a) Consider the vectors $Y_1, X_1, X_2, \ldots, X_n$, which span W. Show that some X_i can be removed and that the vectors $Y_1, X_1, \ldots, X_{i-1}, X_{i+1}, \ldots, X_n$ still span W.
    (b) Consider the vectors $Y_1, Y_2, X_1, \ldots, X_{i-1}, X_{i+1}, \ldots, X_n$. Show that some X_j can be removed and that the vectors $Y_1, Y_2, X_1, \ldots, X_{i-1}, X_{i+1}, \ldots, X_{j-1}, X_{j+1}, \ldots, X_n$ still span W.
    (c) Continue the above process of inserting a Y vector and removing an X vector. Since for every Y vector put in we are able to remove an X vector, it follows that there are at least r X's and $r \leq n$. After r steps, the Y's are exhausted and we conclude that $W = \langle Y_1, Y_2, \ldots, Y_r, X_{i_1}, \ldots, X_{i_{r-n}} \rangle$ if $r < n$. If $r = n$, we have $W = \langle Y_1, Y_2, \ldots, Y_r \rangle$.

5.5 SOME DIMENSION THEOREMS

By its definition, a basis for a vector space is required to *generate* the space, *and* it must be a *linearly independent* set. We shall see as a consequence of Theorems 5.6 and 5.8, however, that in a vector space that is known to have dimension n, if a set of n vectors satisfies one of these properties then it must satisfy the other property. We shall also see that in a finite-dimensional vector space a basis can be characterized as a minimal set of generating vectors or as a maximal set of linearly independent vectors.

Although the next result is just a restatement of conclusion (iii) of the Steinitz replacement theorem, we state it as a corollary.

Corollary 5.2. If V has dimension n and L is a linearly independent set of n vectors in V, then L is a basis for V.

As an illustration, consider the set $L = \{(1, 3), (6, 4)\}$. Neither vector in L is a scalar multiple of the other, and therefore the set is linearly independent. Since the dimension of R_2 is 2 and L is a linearly independent set containing two vectors, Corollary 5.2 assures us that L is a basis for R_2.

Theorem 5.9. If the dimension of V is n and S is a set of n vectors that spans V, then S is a basis for V.

Proof: If S were a linearly dependent set, we could apply Theorem 5.6 to "reduce" S to a linearly independent subset that still generates V. This linearly independent subset would be a basis for V consisting of fewer than n vectors. Since this is impossible, S must be a linearly independent set. Q.E.D.

Example 5.20. (a) Determine if the set of vectors $S = \{X = (2, 0, 0), Y = (2, 1, 0), Z = (2, 1, 1)\}$ is a basis for $\mathbf{R_3}$.
(b) Determine if the set $S = \{f_1 : x + 4, f_2 : x - 3\}$ is a basis for $\mathbf{P_2}$.

Solution: (a) To determine if S generates $\mathbf{R_3}$, we solve the vector equation $(a, b, c) = rX + sY + tZ$ for r, s, and t. This is equivalent to solving the system

$$2r + 2s + 2t = a$$
$$s + t = b$$
$$t = c$$

The solution is $(r, s, t) = ((a - 2b)/2, b - c, c)$. Therefore,

$$(a, b, c) = \frac{a - 2b}{2} X + (b - c)Y + cZ$$

and S generates $\mathbf{R_3}$. By Theorem 5.9, S is a basis for $\mathbf{R_3}$.
(b) An arbitrary function g in $\mathbf{P_2}$ is of the form $g : ax + b$ (a and b can be any real numbers including 0). We solve the vector (function) equation

$$rf_1 + sf_2 : r(x + 4) + s(x - 3) = g : ax + b$$

for r and s. We have $ax + b = (r + s)x + (4r - 3s)$ for every real number x. Therefore,

$$r + s = a$$
$$4r - 3s = b$$

Solving this system, we obtain

$$(r, s) = \left(\frac{b + 3a}{7}, \frac{4a - b}{7}\right)$$

So

$$\frac{b + 3a}{7} f_1 + \frac{4a - b}{7} f_2 = g$$

By Theorem 5.9, S is a basis for $\mathbf{P_2}$.

Theorem 5.10. If V has dimension n, then (a) no subset of V with fewer than n vectors can generate V, and (b) no subset of V with more than n vectors (finite or infinite) can be linearly independent.

Proof: Suppose that S is a subset of V having k vectors and $k < n$. Since the dimension of V is n, there is a basis B of n linearly independent vectors. If S were a generating set for **V**, one would conclude from the Steinitz replacement theorem that $n \leq k$. Since $k < n$, it follows that S cannot generate **V**, and we have established conclusion (a).

Now let L be a subset of **V**, and suppose that L has k or more vectors and $k > n$. Let $A = \{X_1, X_2, \ldots, X_k\}$ be a subset of L, and suppose that L is linearly independent. We wish to show that A is linearly independent. If L is infinite, then A is linearly independent by the definition of an infinite linearly independent set. If L is finite and has m vectors, we order and label them $X_1, \ldots, X_k, X_{k+1}, \ldots, X_m$, where the first k vectors are the vectors in A. Now suppose that

$$a_1 X_1 + a_2 X_2 + \cdots + a_k X_k = 0$$

Then

$$a_1 X_1 + \cdots + a_k X_k + 0 X_{k+1} + \cdots + 0 X_m = 0$$

Since L is linearly independent, $a_1 = a_2 = \cdots = a_k = 0$, and thus A is linearly independent. The dimension of **V** is n, and therefore **V** has a subset of n generators. By the Steinitz replacement theorem, the number of linearly independent vectors k in A cannot exceed the number of generators n. But $k \leq n$ is a contradiction. Hence L cannot be linearly independent.   Q.E.D.

Example 5.21. $S = \{(1, 0, 4), (2, 6, 3)\}$ cannot generate **R₃** because any generating set for **R₃** must have at least three vectors. Of course, a set of vectors in **R₃** having more than three vectors may or may not generate **R₃**; for example, $S = \{(1, 0, 0), (0, 1, 0), (0, 0, 1), (1, 2, 3)\}$ generates **R₃**, but $S = \{(1, 0, 0), (2, 0, 0), (3, 0, 0), (1, 2, 3)\}$ does not generate **R₃**.

Example 5.22.   (a) Determine if the set $L = \{(1, 3), (4, 7), (21, 57)\}$ is linearly independent or dependent.

(b) Find a basis for **R₃** that contains the vectors $X = (1, 2, 1)$ and $Y = (1, 3, 0)$.

Solution: (a) Since **R₂** has dimension 2 and L has more than two vectors, it follows from Theorem 5.10(b) that L is linearly dependent.

(b) The vectors X and Y are linearly independent because neither vector is a scalar multiple of the other. The vectors E_1, E_2, and E_3 (natural basis) generate **R₃**. By the Steinitz replacement theorem, two of the E vectors can be replaced by X and Y to obtain another basis. Following the procedure outlined in Example 5.15, we first consider Y, E_1, E_2, E_3.

$$E_2 = \tfrac{1}{3} Y + (-\tfrac{1}{3}) E_1$$

and so E_2 can be eliminated, and Y, E_1, E_3 still generate **R₃**. Next consider X, Y, E_1, E_3.

$$E_3 = 1X + (-\tfrac{2}{3}) Y + (-\tfrac{1}{3}) E_1$$

Therefore, E_3 can be discarded, and $\{X, Y, E_1\}$ is a basis for $\mathbf{R}_3$. Another way of looking at the situation is to think of the set $\{X, Y\}$ as being **extended** to a basis.

It probably seems obvious to the reader that a subspace of an n-dimensional vector space has dimension not greater than n. This is the case, and we give a proof that indicates how one can actually find a basis for the subspace.

Theorem 5.11. If $\mathbf{U}$ is a subspace of an n-dimensional vector space $\mathbf{V}$, then $\mathbf{U}$ has dimension less than or equal to n.

Proof: It will be convenient to denote the dimension of a space $\mathbf{V}$ by dim $\mathbf{V}$. If $\mathbf{U} = \{\mathbf{0}\}$, then dim $\mathbf{U} = 0$. If $\mathbf{U} \neq \{\mathbf{0}\}$, we can obtain a basis for $\mathbf{U}$ in the following way. Select a vector X_1 from $\mathbf{U}$ such that $X_1 \neq \mathbf{0}$. $\{X_1\}$ is a linearly independent set. If $\mathbf{U} = \langle X_1 \rangle$, we have found a basis. If not, there is a vector X_2 in $\mathbf{U}$, which is not in $\langle X_1 \rangle$. By Theorem 5.5, $\{X_1, X_2\}$ must be a linearly independent set (otherwise, $X_2 \in \langle X_1 \rangle$). If $\mathbf{U} = \langle X_1, X_2 \rangle$, then we have a basis for $\mathbf{U}$. If not, we select a vector X_3, which is in $\mathbf{U}$ but not in $\langle X_1, X_2 \rangle$. Again by Theorem 5.5, $\{X_1, X_2, X_3\}$ must be linearly independent. Continuing this process, eventually we obtain a linearly independent set of vectors $\{X_1, X_2, \ldots, X_k\}$ in $\mathbf{U}$ for which $\{X_1, X_2, \ldots, X_k, X\}$ is linearly dependent for any X in $\mathbf{U}$ and $k \leq n$. This is true because $\mathbf{U} \subseteq \mathbf{V}$, and $\mathbf{V}$ cannot have more than n linearly independent vectors. It follows that $\{X_1, X_2, \ldots, X_k\}$ is a basis for $\mathbf{U}$ and dim $\mathbf{U} = k \leq n$. Q.E.D.

An important result concerning the dimension of the subspace $\mathbf{U} + \mathbf{W}$ (see Definition 5.6) is the following theorem.

Theorem 5.12. If $\mathbf{U}$ and $\mathbf{W}$ are subspaces of a finite-dimensional vector space $\mathbf{V}$, then $\dim(\mathbf{U} + \mathbf{W}) = \dim \mathbf{U} + \dim \mathbf{W} - \dim(\mathbf{U} \cap \mathbf{W})$.

Theorem 5.12 and a method of proof are illustrated in the next example. The actual proof is left as an exercise and is outlined in Problem 21, Exercise 5.5.

Example 5.23. Let $\mathbf{U} = \langle (1, 0, 0, 0), (0, 1, 0, 0) \rangle$ and $\mathbf{W} = \langle (1, 1, 0, 0), (0, 0, 1, 1) \rangle$. Then $\mathbf{U}$ and $\mathbf{W}$ are both subspaces of $\mathbf{V} = \mathbf{R}_4$. (a) Find a basis B_1 for $\mathbf{U} \cap \mathbf{W}$.

(b) Extend B_1 to a basis B_2 for $\mathbf{U}$ and extend B_1 to a basis B_3 for $\mathbf{W}$.

(c) Show that the union $B_4 = B_2 \cup B_3$ is a basis for $\mathbf{U} + \mathbf{W}$, and conclude that $\dim(\mathbf{U} + \mathbf{W}) = \dim \mathbf{U} + \dim \mathbf{W} - \dim(\mathbf{U} \cap \mathbf{W})$.

Solution: (a) A vector X is in $\mathbf{U} \cap \mathbf{W}$ if and only if X is in both $\mathbf{U}$ and $\mathbf{W}$. Thus $X = (a, b, 0, 0)$ and $X = (c, c, d, d)$ for real numbers $a, b, c,$ and d. Solving, we see that $a = c$, $b = c$, and $d = 0$. So $X = (c, c, 0, 0)$ and $B_1 = \{X_1 = (1, 1, 0, 0)\}$ is a basis for $\mathbf{U} \cap \mathbf{W}$.

(b) $U \cap W$ is a subspace of U, and thus B_1 can be extended to a basis for U. For example, $B_2 = \{X_1 = (1, 1, 0, 0), Y_2 = (0, 1, 0, 0)\}$ is a basis for U. Also, $U \cap W$ is a subspace of W and B_1 can be extended to $B_3 = \{X_1 = (1, 1, 0, 0), Z_2 = (0, 0, 1, 1)\}$, which is a basis for W.

(c) Consider $B_4 = B_2 \cup B_3 = \{X_1, Y_2, Z_2\}$. Every vector in $U + W$ is of the form $Y = A + B$, where A is in U and B is in W. So $Y = A + B = (aX_1 + bY_2) + (cX_1 + dZ_2) = (a + c)X_1 + bY_2 + dZ_2$ and B_4 generates $U + W$. Next we show that B_4 is linearly independent. Suppose that

$$a_1 X_1 + b_2 Y_2 + c_2 Z_2 = 0 \tag{1}$$

We let $C = c_2 Z_2$, which is a vector in W. From equation (1) we have $C = -a_1 X_1 + -b_2 Y_2$ and we see that C is a vector in U. Hence C is a vector in $U \cap W$ and we must have $C = d_1 X_1$ for some d_1. We can now write

$$0 = C - C = d_1 X_1 - (-a_1 X_1 - b_2 Y_2) = (a_1 + d_1)X_1 + b_2 Y_2$$

The set B_2 is linearly independent, and therefore b_2 (as well as $a_1 + d_1$) must be zero. From equation (1) we now have $a_1 X_1 + c_2 Z_2 = 0$. Since B_3 is a linearly independent set, we must have $a_1 = c_2 = 0$, and thus $a_1 = b_2 = c_2 = 0$. Consequently, B_4 is a linearly independent set. The number of elements in B_4 is $\#B_4 = \#B_2 + \#B_3 - \#B_1$, and we conclude that $\dim(U + W) = \dim U + \dim W - \dim(U \cap W)$.

Example 5.24.   If $U = \langle(1, 2, 1), (0, 1, 2)\rangle$ and $W = \langle(1, 0, 0), (0, 1, 0)\rangle$, (a) find a basis for $U \cap W$.

(b) Determine the dimension of $U + W$.

(c) Describe U, W, $U \cap W$, and $U + W$ geometrically.

Solution: (a) $X \in U \cap W$ means that

$$X = a(1, 2, 1) + b(0, 1, 2) = c(1, 0, 0) + d(0, 1, 0)$$

for some scalars a, b, c, and d. So

$$(a, 2a + b, a + 2b) = (c, d, 0)$$

and

$$a - c = 0$$
$$2a + b - d = 0$$
$$a + 2b = 0$$

We solve this system for a, b, c, and d.

$$\begin{bmatrix} 1 & 0 & -1 & 0 \\ 2 & 1 & 0 & -1 \\ 1 & 2 & 0 & 0 \end{bmatrix} \xrightarrow[-R_1 + R_3]{-2R_1 + R_2} \begin{bmatrix} 1 & 0 & -1 & 0 \\ 0 & 1 & 2 & -1 \\ 0 & 2 & 1 & 0 \end{bmatrix} \xrightarrow{-2R_2 + R_3} \begin{bmatrix} 1 & 0 & -1 & 0 \\ 0 & 1 & 2 & -1 \\ 0 & 0 & -3 & 2 \end{bmatrix}$$

$$\xrightarrow[\substack{-2R_3 + R_2 \\ R_3 + R_1}]{(-1/3)R_3} \begin{bmatrix} 1 & 0 & 0 & -\frac{2}{3} \\ 0 & 1 & 0 & \frac{1}{3} \\ 0 & 0 & 1 & -\frac{2}{3} \end{bmatrix}$$

Therefore,

$$a = \tfrac{2}{3}d$$
$$b = -\tfrac{1}{3}d$$
$$c = \tfrac{2}{3}d$$

Since $X = (c, d, 0)$, we have

$$X = (\tfrac{2}{3}d, d, 0) = \tfrac{1}{3}d(2, 3, 0)$$

Hence $U \cap W = \langle(2, 3, 0)\rangle$.

(b) $\dim(U + W) = \dim U + \dim W - \dim(U \cap W) = 2 + 2 - 1 = 3$.

(c) U is the plane determined by the three points $(0, 0, 0)$, $(1, 2, 1)$, and $(0, 1, 2)$. W is the plane determined by the three points $(0, 0, 0)$, $(1, 0, 0)$, and $(0, 1, 0)$. $U \cap W$ is the line determined by $(0, 0, 0)$ and $(2, 3, 0)$. $U + W$ is all of 3-space.

In the special case that $U \cap W = \{0\}$, we see that $\dim(U + W) = \dim U + \dim W$. In this case, $U + W$ is called a **direct sum**, and we write $U \oplus W$. Repeating, use of the symbol $U \oplus W$ means that (1) $U + W = U \oplus W$ and (2) $U \cap W = \{0\}$.

Exercise 5.5

In Problems 1–4, use Corollary 5.2 or Theorem 5.9 to determine which sets are bases for the given vector space. If a set is not a basis, give at least one reason why it is not.

1. For $V = R_2$.
   (a) $B = \{(1, 2), (3, 4)\}$.
   (b) $B = \{(3, 2), (6, 4)\}$.
   (c) $B = \{(3, 2), (4, 7), (\sqrt{2}, \pi)\}$.
   (d) $B = \{(3, 2)\}$.

✓ 2. For $V = R_3$.
   (a) $B = \{(1, 2, 3), (0, 1, 2), (3, 4, 6)\}$.
   (b) $B = \{(1, 2, 3), (8, -6, 1), (4, 3, 0), (1, 6, 2)\}$.

3. For $V = P_4$.
   (a) $B = \{f_1: x + 1, f_2: x^2 + 1, f_3: 3x^2 + x\}$.
   (b) $B = \{f_1: 1, f_2: x, f_3: x^2 + x, f_4: x^3 + x^2 + x\}$.
   (c) $B = \{f_1: 3, f_2: x + 1, f_3: x^2 + 3, f_4: x^3, f_5: 2x + 1\}$.
   (d) $B = \{f_1: x, f_2: x^2 + 2, f_3: x^3 + x, f_4: x^2 - 4\}$.

4. For $V = P_3$.
   (a) $B = \{f_1: 2, f_2: 7, f_3: x^2, f_4: x^3\}$.
   (b) $B = \{f_1: x^2, f_2: x^2 + 1, f_3: x\}$.
   (c) $B = \{f_1: 1, f_2: x + 1, f_3: x^2\}$.

√ 5. Explain why each of the following sets is or is not a basis for the given vector space.

(a) $V = R_2$, $A = \{(1, 3), (7, 9), (4, 11)\}$.

(b) $V = R_3$, $A = \{(1, 2, 0), (3, 4, 6)\}$.

(c) $V = P_4$, $A = \{f_1: x^3, f_2: x^2 + 1, f_3: x + 4, f_4: x - 3, f_5: 3\}$.

(d) $V = R_{2 \times 2}$, $A = \left\{ \begin{bmatrix} 1 & 0 \\ 0 & 0 \end{bmatrix}, \begin{bmatrix} 0 & 1 \\ 1 & 1 \end{bmatrix}, \begin{bmatrix} 0 & 0 \\ 1 & 1 \end{bmatrix} \right\}$.

In Problems 6–8, following Example 5.22(b), extend the given set A to a basis for the indicated vector space.

6. $V = P_3$, $A = \{f_1: x + 1, f_2: 2x\}$.

√ 7. $V = R_4$, $A = \{X = (2, 1, 0, 0), \ Y = (0, 1, 0, 1)\}$.

8. $V = R_{2 \times 2}$, $A = \left\{ \begin{bmatrix} 1 & 0 \\ 0 & 0 \end{bmatrix}, \begin{bmatrix} 1 & 1 \\ 0 & 0 \end{bmatrix} \right\}$.

9. If $U = \langle (1, 2, 0), (0, 2, 1) \rangle$ and $W = \langle (0, 0, 2), (0, 1, 0) \rangle$,

(a) Find a basis for $U \cap W$.

(b) Determine the dimension of $U + W$.

(c) Describe U, W, $U \cap W$, and $U + W$ geometrically.

10. If $U = \langle f_1: x + 1, f_2: x^2 \rangle$ and $W = \langle g_1: x^2 + 1, g_2: x \rangle$,

(a) Find a basis for $U \cap W$.

(b) Determine the dimension of $U + W$.

11. Verify that $R_3 = U \oplus W$, where $U = \{(a, b, 0) \,|\, a, b \in R\}$ and $W = \{(0, 0, c) \,|\, c \in R\}$.

12. If $U = \{(x, y, z) \,|\, 2x + 3y + z = 0\}$ and $W = \{(x, y, z) \,|\, x + 2y - z = 0\}$,

(a) Find a basis for $U \cap W$.

(b) Determine dim $U + W$.

(c) Describe U, W, $U \cap W$, $U + W$ geometrically.

13. If $S = \{(x, y, z) \,|\, x + 2y + 2z = 0\}$, (a) Why is S a subspace of R_3? (b) Find a basis for S.

√ 14. Find a basis for the space of all solutions to the system

$$x + 2y + 3z + \ 4w = 0$$
$$4x - 7y - 3z - \ 2w = 0$$
$$2x - \ y + \ z + \ 2w = 0$$
$$4x + 3y + 7z + 10w = 0$$

15. Show that $W = \left\{ \begin{bmatrix} a & b \\ c & d \end{bmatrix} \middle| b = c \right\}$ is a subspace of $R_{2 \times 2}$. Find the dimension of W.

16. *Prove:* If $S = \{X_1, X_2, \ldots, X_k\}$ is a set of nonzero vectors that generates a vector space V, then there is a subset of S that is a basis for V.

17. *Prove:* If V is a finite-dimensional vector space and U is a subspace of V, then there is a subspace W of V such that $V = U \oplus W$.

18. Find subspaces U, W_1, and W_2 of R_3 such that $U \oplus W_1 = U \oplus W_2$ but $W_1 \neq W_2$.

19. *Prove:* If V is a finite-dimensional vector space and W is a subspace of V such that dim W = dim V, then $V = W$.

20. *Prove:* If $B = \{X_1, X_2, \ldots, X_t\}$ is a linearly independent subset of vectors in an n-dimensional vector space V, and if $t < n$, then B can be extended to a basis for V; that is, there are vectors $Y_{t+1}, \ldots, Y_n$ in V such that $C = \{X_1, \ldots, X_t, Y_{t+1}, \ldots, Y_n\}$ is a basis for V. *Hint:* Start with a basis of Y's for V and use the Steinitz replacement theorem.

21. Use the outline below to prove Theorem 5.12. The outline follows Example 5.23.

    (a) Show that $U \cap W$ is a subspace of each of the vector spaces U and W. Let dim $(U \cap W) = r$, dim $U = s$, and dim $W = t$. Let $B_1 = \{X_1, X_2, \ldots, X_r\}$ be a basis for $U \cap W$.

    (b) Extend B_1 to a basis B_2 for U, and also extend B_1 to a basis B_3 for W. Let $Y_{r+1}, \ldots, Y_s$ and $Z_{r+1}, \ldots, Z_t$ be the additional vectors in B_2 and B_3, respectively.

    (c) Show that $B_4 = \{X_1, \ldots, X_r, Y_{r+1}, \ldots, Y_s, Z_{r+1}, \ldots, Z_t\} = B_2 \cup B_3$ is a basis for $U + W$.

5.6 SYSTEMS OF LINEAR EQUATIONS REVISITED

In this section we shall discuss a method for determining a basis for and hence the dimension of the solution space to an arbitrary homogeneous system of linear equations. We shall also show how the solutions to a nonhomogeneous system of linear equations ($AX = B$) can be expressed in terms of the solutions of the associated homogeneous system of linear equations ($AX = 0$).

We begin with an example to illustrate the method used to find a basis for the solution space of a homogeneous system.

Example 5.25. Find a basis for the solution space of the system

$$x_1 + 3x_2 + 0x_3 + 2x_4 + 0x_5 = 0$$
$$2x_1 + 6x_2 + x_3 + 5x_4 + 0x_5 = 0$$
$$3x_1 + 9x_2 + 0x_3 + 6x_4 + x_5 = 0$$
$$x_1 + 3x_2 + x_3 + 3x_4 + 0x_5 = 0$$

Solution: First we find the solution space by taking the coefficient matrix of the system and transforming it to reduced row echelon form.

$$
\begin{bmatrix} 1 & 3 & 0 & 2 & 0 \\ 2 & 6 & 1 & 5 & 0 \\ 3 & 9 & 0 & 6 & 1 \\ 1 & 3 & 1 & 3 & 0 \end{bmatrix}
\xrightarrow[\substack{-3R_1+R_3 \\ -R_1+R_4}]{-2R_1+R_2}
\begin{bmatrix} 1 & 3 & 0 & 2 & 0 \\ 0 & 0 & 1 & 1 & 0 \\ 0 & 0 & 0 & 0 & 1 \\ 0 & 0 & 1 & 1 & 0 \end{bmatrix}
\xrightarrow{-1R_2+R_4}
\begin{bmatrix} 1 & 3 & 0 & 2 & 0 \\ 0 & 0 & 1 & 1 & 0 \\ 0 & 0 & 0 & 0 & 1 \\ 0 & 0 & 0 & 0 & 0 \end{bmatrix}
$$

The equivalent system corresponding to the reduced row echelon matrix is

$$x_1 + 3x_2 + 2x_4 = 0$$

$$x_3 + x_4 = 0$$

$$x_5 = 0$$

Solving for the basic variables (the variables corresponding to the leading 1's) in terms of the parameters (nonbasic variables), we get

$$x_1 = -3x_2 - 2x_4$$

$$x_3 = -x_4$$

$$x_5 = 0$$

We see that an arbitrary solution is given by

$$x_1 = -3x_2 - 2x_4$$

$$x_2 = 1x_2 + 0x_4$$

$$x_3 = 0x_2 - 1x_4$$

$$x_4 = 0x_2 + 1x_4$$

$$x_5 = 0x_2 + 0x_4$$

This can be written as an equivalent vector equation $X = x_2 Y_1 + x_4 Y_2$, where

$$X = \begin{bmatrix} x_1 \\ x_2 \\ x_3 \\ x_4 \\ x_5 \end{bmatrix}, \quad Y_1 = \begin{bmatrix} -3 \\ 1 \\ 0 \\ 0 \\ 0 \end{bmatrix}, \quad Y_2 = \begin{bmatrix} -2 \\ 0 \\ -1 \\ 1 \\ 0 \end{bmatrix}$$

The vectors Y_1 and Y_2 are solutions to the given system (take $x_2 = 1$ and $x_4 = 0$ to obtain Y_1, and let $x_2 = 0$ and $x_4 = 1$ to obtain Y_2), and they clearly generate the solution space. Furthermore, Y_1 and Y_2 are linearly independent since solving $a Y_1 + b Y_2 = 0$ immediately yields $a = b = 0$. Hence the solution space is the two-dimensional subspace $W = \langle Y_1, Y_2 \rangle$ of $\mathbf{R}_5$.† Note that the dimension of the solution space is $n - r$, where n is the number of variables and r is the row rank (number of leading 1's) of the reduced row echelon matrix.

---

†Technically, $\mathbf{R}_5$ is a vector space of row vectors, whereas the solution space is a vector space of column vectors. As vector spaces, the space $\mathbf{R}_n$ and the space of $n \times 1$ column vectors are essentially the same, and we shall interpret the elements of $\mathbf{R}_n$ as column vectors when it is convenient to do so.

Example 5.26.   Find a basis for the solution space to the system

$$x_1 + 3x_2 + 0x_3 + 2x_4 - x_5 + 5x_6 + x_7 = 0$$
$$-x_1 - 3x_2 + x_3 + x_4 + 5x_5 - 2x_6 - x_7 = 0$$
$$-2x_1 - 6x_2 + 2x_3 + 2x_4 + 10x_5 - 4x_6 - x_7 = 0$$
$$x_1 + 3x_2 - x_3 - x_4 - 5x_5 + 2x_6 + 3x_7 = 0$$
$$x_1 + 3x_2 + x_3 + 5x_4 + 3x_5 + 8x_6 + 4x_7 = 0$$

Solution: Applying elementary row operations to the coefficient matrix A, we obtain the reduced row echelon matrix

$$E = \begin{matrix} x_1 & x_2 & x_3 & x_4 & x_5 & x_6 & x_7 \end{matrix}$$

$$E = \begin{bmatrix} 1 & 3 & 0 & 2 & -1 & 5 & 0 \\ 0 & 0 & 1 & 3 & 4 & 3 & 0 \\ 0 & 0 & 0 & 0 & 0 & 0 & 1 \\ 0 & 0 & 0 & 0 & 0 & 0 & 0 \\ 0 & 0 & 0 & 0 & 0 & 0 & 0 \end{bmatrix}$$

We can obtain a basis for the solution space directly from the matrix E as follows.

The basic variables are x_1, x_3, and x_7, and the remaining variables are parameters. As in Example 5.25, we see that the solutions are of the form $X = x_2 Y_1 + x_4 Y_2 + x_5 Y_3 + x_6 Y_4$. We wish to determine the basis vectors Y_1, Y_2, Y_3, and Y_4 directly from the matrix E. The second entry of Y_1 is 1, and the second entry of each of the other Y's is 0 (this gives us $x_2 = 1x_2$). Similarly, the fourth entry of Y_2 is 1 and the fourth entry of each of the other Y's is 0; the fifth entry of Y_3 is 1 and all other fifth entries are 0; the sixth entry of Y_4 is 1 and all other sixth entries are 0. We obtain

$$[Y_1 \,|\, Y_2 \,|\, Y_3 \,|\, Y_4] = \begin{bmatrix} - & - & - & - \\ 1 & 0 & 0 & 0 \\ - & - & - & - \\ 0 & 1 & 0 & 0 \\ 0 & 0 & 1 & 0 \\ 0 & 0 & 0 & 1 \\ - & - & - & - \end{bmatrix}$$

The remaining entries in the vectors Y_1, Y_2, Y_3, and Y_4 are the negatives of the entries of the submatrix

$$F = \begin{matrix} x_2 & x_4 & x_5 & x_6 \\ \downarrow & \downarrow & \downarrow & \downarrow \end{matrix}$$

$$F = \begin{bmatrix} 3 & 2 & -1 & 5 \\ 0 & 3 & 4 & 3 \\ 0 & 0 & 0 & 0 \end{bmatrix}$$

F is obtained from E by deleting the zero rows of E and by deleting the columns of E that correspond to the basic variables. We then have

$$[Y_1 \mid Y_2 \mid Y_3 \mid Y_4] = \begin{bmatrix} -3 & -2 & 1 & -5 \\ 1 & 0 & 0 & 0 \\ 0 & -3 & -4 & -3 \\ 0 & 1 & 0 & 0 \\ 0 & 0 & 1 & 0 \\ 0 & 0 & 0 & 1 \\ 0 & 0 & 0 & 0 \end{bmatrix}$$

The solution space is the four-dimensional subspace of $\mathbf{R}_7$ generated by the basis $\{Y_1, Y_2, Y_3, Y_4\}$. An easy way to check the answer is to compute the product AY, where A is the coefficient matrix of the given system and Y is the 7×4 matrix $Y = [Y_1 \mid Y_2 \mid Y_3 \mid Y_4]$. If the work is correct, the product AY should be the 5×4 zero matrix.

The next theorem generalizes the conclusion of Examples 5.25 and 5.26. Although the proof is omitted, a proof can be constructed along the lines of the above examples.

Theorem 5.13. If a homogeneous system of m linear equations in n unknowns has an equivalent matrix equation $AX = 0$, then the dimension of the solution space is $n - r$, where r is the row rank of the reduced row echelon form of the coefficient matrix.

The nonhomogeneous system

$$a_{11}x_1 + a_{12}x_2 + \cdots + a_{1n}x_n = b_1$$
$$a_{21}x_1 + a_{22}x_2 + \cdots + a_{2n}x_n = b_2 \tag{N}$$
$$\cdots$$
$$a_{m1}x_1 + a_{m2}x_2 + \cdots + a_{mn}x_n = b_n$$

has the equivalent matrix equation $AX = B$, where A is the coefficient matrix of the system and B is the column of constants and $B \neq 0$. The system

$$a_{11}x_1 + a_{12}x_2 + \cdots + a_{1n}x_n = 0$$
$$a_{21}x_1 + a_{22}x_2 + \cdots + a_{2n}x_n = 0 \tag{H}$$
$$\cdots$$
$$a_{m1}x_1 + a_{m2}x_2 + \cdots + a_{mn}x_n = 0$$

which has the equivalent matrix equation $AX = 0$ is called the **associated homogeneous system** of the system (N). Although the solution set to (N) is not a vector space (0 is not a solution), the next theorem shows that its solution set is closely related to the vector space of solutions of its associated homogeneous system of linear equations.

Theorem 5.14. If there is a particular solution C_0 to the nonhomogeneous system (N), then Y is a solution to (N) if and only if $Y = C_0 + Z$ for some Z in the solution space of the associated homogeneous system (H).

Proof: Assume that a solution C_0 to the system (N) exists, and let Z be a solution to the associated homogeneous system (H). Then $A(C_0 + Z) = AC_0 + AZ = B + 0 = B$. Therefore, $C_0 + Z$ is a solution to (N). Conversely, if Y is a solution to (N), then $A(Y - C_0) = AY - AC_0 = B - B = 0$. So $Y - C_0$ is a solution to (H). But $Y = C_0 + (Y - C_0)$. Therefore, Y is of the form $C_0 + Z$ for some Z in the solution space of (H).   Q.E.D.

Example 5.27. Solve the system

$$x + 3y + 2z = 12$$
$$-x - 3y - z = -7$$

Use Theorem 5.14 to interpret the solution set geometrically.

Solution: We transform the augmented matrix to its reduced row echelon form.

$$\begin{bmatrix} 1 & 3 & 2 & \vdots & 12 \\ -1 & -3 & -1 & \vdots & -7 \end{bmatrix} \xrightarrow{R_1 + R_2} \begin{bmatrix} 1 & 3 & 2 & \vdots & 12 \\ 0 & 0 & 1 & \vdots & 5 \end{bmatrix} \xrightarrow{-2R_2 + R_1} \begin{bmatrix} 1 & 3 & 0 & \vdots & 2 \\ 0 & 0 & 1 & \vdots & 5 \end{bmatrix}$$

Observe that the reduced row echelon form of the coefficient matrix is $E = \begin{bmatrix} 1 & 3 & 0 \\ 0 & 0 & 1 \end{bmatrix}$. Using E to solve the associated homogeneous system, we see that the basic variables are x and z, the solution space has dimension 1, $\begin{bmatrix} x \\ y \\ z \end{bmatrix}$

$= y \begin{bmatrix} -3 \\ 1 \\ 0 \end{bmatrix}$ and $\mathbf{W} = \left\langle \begin{bmatrix} -3 \\ 1 \\ 0 \end{bmatrix} \right\rangle$ is the solution space. From the reduced row echelon form $\begin{bmatrix} 1 & 3 & 0 & \vdots & 2 \\ 0 & 0 & 1 & \vdots & 5 \end{bmatrix}$ of the augmented matrix, we see that by setting $y = 0$ we can obtain the particular solution $C_0 = \begin{bmatrix} 2 \\ 0 \\ 5 \end{bmatrix}$. By Theorem 5.14, the solution set S to the given nonhomogeneous system is

$$S = \left\{ \begin{bmatrix} 2 \\ 0 \\ 5 \end{bmatrix} + Y \mid Y \in \mathbf{W} \right\}$$

Geometrically, $\mathbf{W}$ is a line through $\begin{bmatrix} 0 \\ 0 \\ 0 \end{bmatrix}$ and $\begin{bmatrix} -3 \\ 1 \\ 0 \end{bmatrix}$. Therefore, S is a line

through $\begin{bmatrix} 2 \\ 0 \\ 5 \end{bmatrix}$ and parallel to the line through $\begin{bmatrix} 0 \\ 0 \\ 0 \end{bmatrix}$ and $\begin{bmatrix} -3 \\ 1 \\ 0 \end{bmatrix}$ (see Section 2.5).

In general, if C_0 is a solution to a nonhomogeneous system, and $\mathbf{W}$ is the solution space of the associated homogeneous system, one may think of the solution set S of the nonhomogeneous system as the space $\mathbf{W}$ having been translated away from $\mathbf{0}$ to the point C_0.

Exercise 5.6

In each of Problems 1–6, use the method of Example 5.26 to solve the system, find a basis for the solution space, and determine the dimension of the solution space. Also, check your answers by the method suggested in Example 5.26.

1. $x_1 + 3x_2 + \frac{2}{3}x_4 + x_5 = 0$
$\qquad x_3 + \frac{1}{3}x_4 + \frac{1}{2}x_5 = 0$

2. $x_1 + 2x_2 + x_3 - 4x_4 = 0$
$\qquad\quad x_2 - 2x_4 = 0$
$\qquad\qquad\quad x_3 = 0$

3. $\qquad x_1 + 3x_2 + 2x_4 = 0$
$\quad 2x_1 + 6x_2 + x_3 + 8x_4 = 0$

4. $2x_1 + 4x_2 + 6x_3 + 8x_4 + 2x_5 = 0$
$\quad 3x_1 + 6x_2 + 9x_3 + 12x_4 + 5x_6 = 0$

5. $3x_1 + 6x_2 + x_3 + 15x_4 = 0$
$\qquad\qquad\quad x_3 + 3x_4 = 0$
$\quad x_1 + 2x_2 + 4x_4 = 0$

6. $\qquad x_1 + 3x_2 + 2x_3 + 2x_4 = 0$
$\quad -4x_1 - 12x_2 + 8x_3 - 2x_4 = 0$
$\qquad 5x_1 + 15x_2 - 10x_3 + 3x_4 = 0$
$\qquad 3x_1 + 9x_2 - 6x_3 + 2x_4 = 0$

In Problems 7 and 8, find a basis for the solution space $\mathbf{W}$ of the associated homogeneous system. Find a particular solution C_0 for the given system, and express each solution in the form $C_0 + Y$ for some Y in $\mathbf{W}$.

7. $x_1 + 3x_2 + 2x_3 + 2x_4 = \quad 3$
$\quad 2x_1 + 6x_2 + 4x_3 + 8x_4 = -2$
$\quad 3x_1 + 9x_2 + 6x_3 + 7x_4 = \quad 7$

8. $2x_1 - 2x_2 + 4x_3 + 8x_4 + \quad x_5 + 11x_6 = 8$
$\quad x_1 - \quad x_2 + 2x_3 + 7x_4 + 2x_5 + 10x_6 = 3$

9. Solve the system

$$2x + y + 7z = 4$$
$$x - y + 2z = 5$$

As in Example 5.27, use Theorem 5.14 to interpret the solution set geometrically.

10. (a) Solve the system consisting of the single equation $2x + y + 3z = 0$ and find a basis for the solution space.

(b) Give a geometric interpretation to the solution space.

(c) Find a basis of vectors for the solution space having integer coordinates. Interpret this change of basis geometrically.

(d) Give a geometric interpretation to the solution set of the nonhomogeneous system $2x + y + 3z = 5$.

11. Let $X = (x_1, x_2, x_3)$, $Y = (y_1, y_2, y_3)$, and $Z = (z_1, z_2, z_3)$. Give a geometric interpretation of each of the following statements.

(a) X and Y are nonzero linearly dependent vectors.

(b) X and Y are linearly independent vectors.

(c) X, Y, and Z are linearly dependent vectors, but any two of the vectors are linearly independent.

(d) X, Y, and Z are linearly independent vectors.

6
Linear
Transformations
and Matrices

In this chapter we shall study certain functions called *linear transformations.* A connection will be made between linear transformations and matrices. Just as matrices can be added, multiplied together, and multiplied by a scalar, we shall see that similar operations can be defined for linear transformations. In fact, we shall see that the algebra of matrices is essentially the same as the algebra of linear transformations.

6.1 A BRIEF REVIEW OF FUNCTIONS

A basic concept in mathematics is the notion of function. The words *function, transformation,* and *mapping* are often used in mathematics to refer to the same idea. In calculus, one learns that an equation such as $y = 2x^2 + 1$ can be used to define a function that maps or transforms each real number into a real number. If we call the function f, then f maps the number 2 into $2(2)^2 + 1 = 9$, and we write $f(2) = 9$. Similarly, f maps 3 into $2(3)^2 + 1 = 19$, and we write $f(3) = 19$. In this text we shall use the notation $f: \mathbf{R} \to \mathbf{R}$ (read "f is a function from the reals into the reals") to denote the fact that f is a function from the set of real numbers into the set of real numbers. We shall also use $x \xrightarrow{f} 2x^2 + 1$ (read "f maps x into $2x^2 + 1$") to mean the same thing as $f(x) = 2x^2 + 1$ (read "f of x is $2x^2 + 1$").

Definition 6.1. If A and B are nonempty sets, a **transformation (function, mapping)** T with **domain** A and **codomain** B is a correspondence that associates with each element a in A a *unique* element b in B. The notation $T: A \to B$ is used to denote the transformation, and the notation $a \xrightarrow{T} b$ or $T(a) = b$ is used to indicate that b is the unique element in B that is associated with the element a in A. We say that b is the **image** of a, and a is a **preimage** of b. The collection $\{T(a) \mid a \in A\}$ is called the **range** of T. The range of T is a subset of the co-domain B, and if the range of T equals B, we say that T is a transformation **onto** B. If for all elements a_1 and a_2 in the domain of T we have $T(a_1) = T(a_2)$ implies $a_1 = a_2$, then T is a **one-to-one** (1–1) function. Another way of saying this is that T is one to one if each element in the range of T has only one preimage. We call T a **one-to-one correspondence between A and B** if T is both one to one and onto B.

Example 6.1. Let $\mathbf{Z} = \{\ldots, -2, -1, 0, 1, 2, \ldots\}$, which is the set of all integers. Let $T: \mathbf{Z} \rightarrow \mathbf{Z}$ be defined by $T(k) = k^2 - 1$. The domain of T is $\mathbf{Z}$ and the codomain of T is $\mathbf{Z}$. To find the range of T, we let $j = T(k) = k^2 - 1$. Then j is in the range of T if and only if there is a preimage k in $\mathbf{Z}$ such that $k^2 - 1 = j$. Solving for k, we get $k = \pm\sqrt{j + 1}$. Therefore, j is in the range of T if and only if $j + 1$ is a perfect square. For example, 8 is in the range of T because 9 is a perfect square. 3 and -3 are both preimages of 8, and we write $3 \xrightarrow{T} 8$ and $-3 \xrightarrow{T} 8$. This means that T is not 1–1. [$T(3) = T(-3)$ but $3 \neq -3$.]

A more precise but less intuitive way to define a function is to define a function as a set of ordered pairs for which no two distinct ordered pairs have the same first coordinate. For instance, the function T above could be described as $T = \{(k, k^2 - 1) \mid k \text{ is an integer}\}$. The correspondence is given by the ordered pairs, for example, $4 \xrightarrow{T} 15$, because $(4, 15)$ belongs to the set of ordered pairs.

Example 6.2. Consider the function $T: \mathbf{R}_3 \rightarrow \mathbf{R}_2$ defined by $T((x, y, z)) = (x - y, y + z)$. The domain of this function is $\mathbf{R}_3$ and the codomain is $\mathbf{R}_2$. The range is the set of all 2-tuples (a, b) such that $(a, b) = (x - y, y + z)$ for some (x, y, z) in $\mathbf{R}_3$. Solving for x, y, and z, we have

$$x - y = a$$
$$y + z = b$$

Reducing the augmented matrix, we get

$$\begin{bmatrix} 1 & -1 & 0 & | & a \\ 0 & 1 & 1 & | & b \end{bmatrix} \xrightarrow{R_2 + R_1} \begin{bmatrix} 1 & 0 & 1 & | & a + b \\ 0 & 1 & 1 & | & b \end{bmatrix}$$

Therefore,

$$x = -z + a + b$$
$$y = -z + b$$
$$z = z$$

Thus any vector (a, b) in $\mathbf{R}_2$ has infinitely many preimages in $\mathbf{R}_3$, and T is onto $\mathbf{R}_2$. A preimage of $(4, 3)$ is $(6, 2, 1)$; that is, $(6, 2, 1) \xrightarrow{T} (6 - 2, 2 + 1) = (4, 3)$. Another preimage of $(4, 3)$ is $(2, -2, 5)$, since $(2, -2, 5) \xrightarrow{T} (2 - (-2), -2 + 5) = (4, 3)$. T is not 1-1 because each image has (infinitely) many preimages.

A key word in Definition 6.1 is the word "unique." A correspondence that is defined on a set A fails to be a function if there is some $a \in A$ that has more than one image. For example, if we define f on the set $\mathbf{Q}$ of rational numbers by $f(p/q) = p$, we have $f(\frac{2}{4}) = 2$ and $f(\frac{1}{2}) = 1$. f is not a function because the image of $\frac{1}{2}$ is not unique. A correspondence such as this is called a **relation** but not a function. In terms of sets, a relation is just a set of ordered pairs (which may or may not have two different ordered pairs having the same first coordinate).

Functions f and g are **equal** if and only if they are equal as sets of ordered pairs. This is true if and only if they are defined on the same set A (they have the same domain) and for each $x \in A$, $f(x) = g(x)$.

Example 6.3. (a) Determine if the functions f and g are equal if $f = \{(1, 6), (2, 8), (3, 7), (4, 6)\}$ and $g = \{(2, 8), (3, 6), (4, 6), (1, 7)\}$.

(b) Verify that $T = S$ if $T: \mathbf{R}_2 \longrightarrow \mathbf{R}_3$ is defined by $T(1, 0) = (1, 0, 2)$, $T(0, 1) = (3, 1, 0)$, and $T(x, y) = xT(1, 0) + yT(0, 1)$, and $S: \mathbf{R}_2 \longrightarrow \mathbf{R}_3$ is defined by $S(x, y) = (x, y)\begin{bmatrix} 1 & 0 & 2 \\ 3 & 1 & 0 \end{bmatrix}$.

Solution: (a) Each of the functions f and g has domain $\{1, 2, 3, 4\}$. Also, the range of $f = \{6, 8, 7\} = $ the range of g. However, the functions are not equal because they are not the same set of ordered pairs. In terms of images, $f(3) = 7 \neq g(3) = 6$ [also, $f(1) = 6 \neq g(1) = 7$].

(b) Each of the functions T and S has domain $\mathbf{R}_2$. For any (x, y) in $\mathbf{R}_2$,

$$T(x, y) = xT(1, 0) + yT(0, 1) = x(1, 0, 2) + y(3, 1, 0)$$
$$= (x + 3y, y, 2x)$$

Also,

$$S(x, y) = (x, y)\begin{bmatrix} 1 & 0 & 2 \\ 3 & 1 & 0 \end{bmatrix} = (x + 3y, y, 2x)$$

Therefore, $T = S$.

Exercise 6.1

In Problems 1–12, for the given transformations, determine by inspection the domain and codomain, determine the range, determine if the function is *onto* the codomain, and determine if the function is 1–1. The symbols $\mathbf{R}$, $\mathbf{Q}$, $\mathbf{Z}$, and $\mathbf{N}$ are used to represent the set of real numbers, the set of rational numbers, the set of integers, and the set of natural or counting numbers, respectively.

1. $f: \mathbf{Z} \longrightarrow \mathbf{Z}$ defined by $f(k) = 3k + 1$.

2. $f: \mathbf{Z} \longrightarrow \mathbf{Z}$ defined by $f(k) = k^2$.

3. $f: \mathbf{R} \longrightarrow \mathbf{R}$ defined by $f(x) = \sqrt{x^2 + 1}$.

4. $f: \mathbf{N} \longrightarrow \mathbf{Q}$ defined by $f(n) = 1/n$.

5. $f: \mathbf{N} \longrightarrow \mathbf{N}$ defined by $f(n) = k$ if $n = 2k - 1$ for some k in $\mathbf{N}$, and $f(n) = k$ if $n = 2k$ for some k in $\mathbf{N}$.

6. $f: \mathbf{N} \longrightarrow \mathbf{N}$ defined by $f(n) = 2n$.

7. $f: \mathbf{R} \longrightarrow \mathbf{Z}$ defined by $f(x) = [[x]]$. $([[x]]$ is the greatest integer less than or equal to x.)

8. $T: \mathbf{R}_3 \longrightarrow \mathbf{R}_2$ defined by $T(x, y, z) = (x, z)$. [Here we have written $T((x, y, z))$ as $T(x, y, z)$.]

9. $T: \mathbf{R}_2 \longrightarrow \mathbf{R}_2$ defined by $T(x, y) = (x + 2, y - 3)$.

10. $T: \mathbf{R}_3 \longrightarrow \mathbf{R}_2$ defined by

$$T\begin{bmatrix} x \\ y \\ z \end{bmatrix} = \begin{bmatrix} 1 & 0 & 2 \\ 2 & 0 & 4 \end{bmatrix}\begin{bmatrix} x \\ y \\ z \end{bmatrix} = \begin{bmatrix} x + 2z \\ 2x + 4z \end{bmatrix}$$

Here we are thinking of the elements of $\mathbf{R}_3$ and of $\mathbf{R}_2$ as column vectors in order to have matrices that are conformable for multiplication. An alternative would be to use row vectors and write

$$T(x, y, z) = (x, y, z)\begin{bmatrix} 1 & 2 \\ 0 & 0 \\ 2 & 4 \end{bmatrix} = (x + 2z, 2x + 4z)$$

11. $T: \mathbf{R}_{2 \times 2} \longrightarrow \mathbf{R}$ defined by $T\begin{bmatrix} a & b \\ c & d \end{bmatrix} = \det \begin{bmatrix} a & b \\ c & d \end{bmatrix}$.

12. $T: \mathbf{R}_{2 \times 2} \longrightarrow \mathbf{R}$ defined by $T\begin{bmatrix} a & b \\ c & d \end{bmatrix} = a + b + c + d$.

13. (a) Find $f(\sqrt{2})$ and $f(-1.7)$ in Problem 7.
    (b) Find $T(3, 4, 2)$ and $T(-1, 5, 9)$ in Problem 8.
    (c) Find $T\begin{bmatrix} 3 \\ 1 \\ 2 \end{bmatrix}$ and $T\begin{bmatrix} -5 \\ 0 \\ 4 \end{bmatrix}$ in Problem 10.

14. (a) Find $f(-2)$ and $f(3)$ in Problem 3.
    (b) Find $f(6)$ and $f(9)$ in Problem 5.
    (c) Find $T\begin{bmatrix} 3 & 2 \\ 4 & 1 \end{bmatrix}$ and $T\begin{bmatrix} -2 & 5 \\ 2 & 3 \end{bmatrix}$ in Problem 11.

15. Show that the relation f with domain $\mathbf{Q}$ and codomain $\mathbf{Z}$ defined by $f(p/q) = p + q$ for all p/q in $\mathbf{Q}$ is *not* a function.

16. Describe each of the functions in Problems 1–4 as a a set of ordered pairs.

17. Describe each of the functions in Problems 9–11 as a set of ordered pairs.

18. (a) State precisely what it means for a function $f: A \longrightarrow B$ to *not* be onto its codomain.
   (b) State precisely what it means for a function $f: A \longrightarrow B$ to *not* be 1-1.

19. If $f: \mathbf{N} \longrightarrow \mathbf{N}$ is defined by $f(n) = 1 + 2 + \cdots + n$ and $g: \mathbf{N} \longrightarrow \mathbf{N}$ is defined by $g(n) = n(n + 1)/2$,
   (a) Find $f(4)$ and $g(4)$.
   (b) Use mathematical induction to prove that $f = g$.

6.2 DEFINITION AND EXAMPLES OF LINEAR TRANSFORMATIONS

The transformation $T: \mathbf{R_3} \longrightarrow \mathbf{R_2}$ (Problem 10, Exercise 6.1) defined by

$$T\begin{bmatrix} x \\ y \\ z \end{bmatrix} = \begin{bmatrix} 1 & 0 & 2 \\ 2 & 0 & 4 \end{bmatrix}\begin{bmatrix} x \\ y \\ z \end{bmatrix} = \begin{bmatrix} x + 2z \\ 2x + 4z \end{bmatrix}$$

has the nice properties that $T(X + Y) = T(X) + T(Y)$ and $T(aX) = aT(X)$. For example,

$$T\begin{bmatrix} 1 \\ 2 \\ 3 \end{bmatrix} = \begin{bmatrix} 7 \\ 14 \end{bmatrix}, \quad T\begin{bmatrix} -1 \\ 0 \\ 2 \end{bmatrix} = \begin{bmatrix} 3 \\ 6 \end{bmatrix},$$

$$T\left(\begin{bmatrix} 1 \\ 2 \\ 3 \end{bmatrix} + \begin{bmatrix} -1 \\ 0 \\ 2 \end{bmatrix}\right) = T\begin{bmatrix} 0 \\ 2 \\ 5 \end{bmatrix} = \begin{bmatrix} 10 \\ 20 \end{bmatrix} = \begin{bmatrix} 7 \\ 14 \end{bmatrix} + \begin{bmatrix} 3 \\ 6 \end{bmatrix}$$

Also,

$$T\left(4\begin{bmatrix} 1 \\ 2 \\ 3 \end{bmatrix}\right) = T\begin{bmatrix} 4 \\ 8 \\ 12 \end{bmatrix} = \begin{bmatrix} 28 \\ 56 \end{bmatrix} = 4\begin{bmatrix} 7 \\ 14 \end{bmatrix}$$

In words, *the image of a sum is equal to the sum of the images, and the image of a scalar multiple of a vector is equal to the scalar multiple of the image of the vector.* The fact that T has these properties for all X and Y in $\mathbf{R_3}$ follows easily from properties of matrices. If we let

$$A = \begin{bmatrix} 1 & 0 & 2 \\ 2 & 0 & 4 \end{bmatrix}, \quad X = \begin{bmatrix} x_1 \\ x_2 \\ x_3 \end{bmatrix}, \quad Y = \begin{bmatrix} y_1 \\ y_2 \\ y_3 \end{bmatrix}$$

and let a be any scalar, then

$$T(X + Y) = A(X + Y) = AX + AY = T(X) + T(Y)$$

and

$$T(aX) = A(aX) = a(AX) = aT(X)$$

The above illustration is an example of a linear transformation according to the following definition.

Definition 6.2.   A **linear transformation** is a transformation $T: V \to W$ from a vector space V into a vector space W having the additional properties (i) for any vectors X and Y in V, $T(X + Y) = T(X) + T(Y)$, and (ii) for any scalar a and vector X in V, $T(aX) = aT(X)$.

Note that the addition $X + Y$ on the left side of the equation in (i) is addition in V because X and Y are in V. The addition on the right side of the equation is addition in W because $T(X)$ and $T(Y)$ are in W. Similarly, aX is scalar multiplication in V, and $aT(X)$ is scalar multiplication in W.

Example 6.4.   If A is an $m \times n$ matrix and the elements in $\mathbf{R}_m$ and $\mathbf{R}_n$ are considered column vectors (rather than row vectors), then $T: \mathbf{R}_n \to \mathbf{R}_m$ defined by $T(X) = AX$ is a linear transformation.

Verification: The proof is the same as that used in the earlier illustration. If X and Y are $n \times 1$ column vectors, then multiplication on the right of A by both X and Y is defined and

$$T(X + Y) = A(X + Y) = AX + AY = T(X) + T(Y)$$

Also,

$$T(aX) = A(aX) = a(AX) = aT(X)$$

Example 6.5   A Rotation Counterclockwise about the Origin.   Consider a rotation R_θ counterclockwise about the origin through an angle of measure θ radians. It can be shown geometrically (Exercise 6.2, Problem 17) that $R_\theta: \mathbf{R}_2 \to \mathbf{R}_2$ is a linear transformation. In Figure 6.1, $R_\theta(A)$ is denoted by A'. We

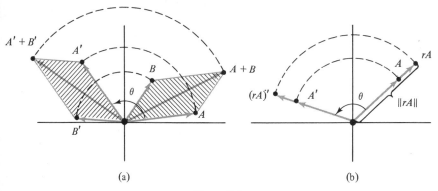

(a)   (b)

Figure 6.1

shall find $R_\theta(x, y)$ by finding $R_\theta(1, 0)$ and $R_\theta(0, 1)$, and using the fact that R_θ is a linear transformation.

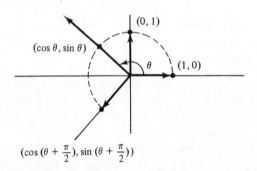

Since a rotation preserves length $(\|A\| = \|R_\theta(A)\|)$, we have $(1, 0) \xrightarrow{R_\theta} (\cos\theta, \sin\theta)$ and

$$(0, 1) \xrightarrow{R_\theta} \left(\cos\left(\theta + \frac{\pi}{2}\right), \sin\left(\theta + \frac{\pi}{2}\right)\right) = (-\sin\theta, \cos\theta)$$

Then

$$
\begin{aligned}
R_\theta(x, y) &= R_\theta((x, 0) + (0, y)) \\
&= R_\theta(x, 0) + R_\theta(0, y) \quad\text{[Definition 6.2(i)]} \\
&= R_\theta(x(1, 0)) + R_\theta(y(0, 1)) \\
&= xR_\theta(1, 0) + yR_\theta(0, 1) \quad\text{[Definition 6.2(ii)]} \\
&= x(\cos\theta, \sin\theta) + y(-\sin\theta, \cos\theta) \\
&= (x\cos\theta - y\sin\theta, \, x\sin\theta + y\cos\theta)
\end{aligned}
$$

Therefore, $R_\theta(x, y) = (x', y')$, where $x' = x\cos\theta - y\sin\theta$ and $y' = x\sin\theta + y\cos\theta$.

The rotation $R_\theta: \mathbf{R_2} \rightarrow \mathbf{R_2}$ can be defined in terms of the matrix

$$C = \begin{bmatrix} \cos\theta & -\sin\theta \\ \sin\theta & \cos\theta \end{bmatrix}$$

If $X = \begin{bmatrix} x \\ y \end{bmatrix}$,

$$R_\theta(X) = CX = \begin{bmatrix} x\cos\theta - y\sin\theta \\ x\sin\theta + y\cos\theta \end{bmatrix}$$

Note that the first column of C is the image of $\begin{bmatrix} 1 \\ 0 \end{bmatrix}$ and the second column of C is the image of $\begin{bmatrix} 0 \\ 1 \end{bmatrix}$.

Example 6.6   A Reflection in the Line y = x. Using the fact that a reflection T (of the plane) in a line preserves both the length of a vector ($\|A\| = \|T(A)\|$) and the magnitude of the angle between two vectors [the positive measure of the angle between A and B is equal to the positive measure of the angle between $T(A)$ and $T(B)$], one can show that the reflection T in the line $y = x$ is a linear transformation (Exercise 6.2, Problem 18). In Figure 6.2, $T(A)$ is denoted by A'.

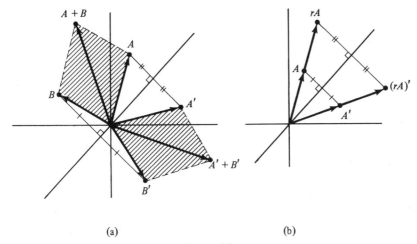

(a)   (b)

Figure 6.2

Another way to show that T is a linear transformation is to note that $T(x, y) = (y, x)$. Using column vectors, this transformation can be accomplished by writing

$$T\begin{bmatrix} x \\ y \end{bmatrix} = \begin{bmatrix} 0 & 1 \\ 1 & 0 \end{bmatrix}\begin{bmatrix} x \\ y \end{bmatrix} = \begin{bmatrix} y \\ x \end{bmatrix}$$

Since T is now defined by a matrix, we know by Example 6.4 that T is a linear transformation.

Example 6.7   The Derivative. (This example requires a knowledge of calculus.) Let $\mathbf{W}$ be the vector space of all real-valued functions on $[a, b]$, and let $\mathbf{V}$ be the subspace of all functions in $\mathbf{W}$ that are differentiable [see Example 5.1(j)]. Consider the mapping $D: \mathbf{V} \rightarrow \mathbf{W}$ defined by $D(f) = f'$ where f' is the derivative of f. From calculus, $D(f + g) = (f + g)' = f' + g' = D(f) + D(g)$. Also, $D(af) = (af)' = a(f') = aD(f)$. Therefore, D is a linear transformation.

Example 6.8. Show that the transformation $T: \mathbf{R}_3 \rightarrow \mathbf{R}_3$ defined by $T(x, y, z) = (x - y, 0, y + z)$ is a linear transformation.

Solution: We must show that properties (i) and (ii) of Definition 6.2 hold. Let $X_1 = (x_1, y_1, z_1)$ and $X_2 = (x_2, y_2, z_2)$ be arbitrary elements of $\mathbf{R}_3$. Then

$$T(X_1 + X_2) = T(x_1 + x_2, y_1 + y_2, z_1 + z_2)$$
$$= ((x_1 + x_2) - (y_1 + y_2), 0, (y_1 + y_2) + (z_1 + z_2))$$
$$= (x_1 - y_1, 0, y_1 + z_1) + (x_2 - y_2, 0, y_2 + z_2)$$
$$= T(X_1) + T(X_2)$$

Also,

$$T(rX_1) = T(rx_1, ry_1, rz_1) = (rx_1 - ry_1, 0, ry_1 + rz_1)$$
$$= r(x_1 - y_1, 0, y_1 + z_1) = rT(X_1)$$

Hence T is a linear transformation.

Example 6.9.   Show that the following mappings are *not* linear transformations.
(a) A translation $T: \mathbf{R}_2 \rightarrow \mathbf{R}_2$ defined by $T(x, y) = (x + a, y + b)$, where (a, b) is a fixed point in $\mathbf{R}_2$ and $(a, b) \neq (0, 0)$.
(b) $T: \mathbf{R} \rightarrow \mathbf{R}$ defined by $T(x) = \sin x$.
(c) $T: \mathbf{R}_3 \rightarrow \mathbf{R}_2$ defined by $T(x, y, z) = (x^2, 0)$.
(d) $T: \mathbf{R} \rightarrow \mathbf{R}$ defined by $T(x) = |x|$.

Solution: (a) If $X_1 = (x_1, y_1)$ and $X_2 = (x_2, y_2)$, then

$$T(X_1 + X_2) = T(x_1 + x_2, y_1 + y_2) = (x_1 + x_2 + a, y_1 + y_2 + b)$$

But

$$T(X_1) + T(X_2) = (x_1 + a, y_1 + b) + (x_2 + a, y_2 + b)$$
$$= (x_1 + x_2 + 2a, y_1 + y_2 + 2b)$$

Since $(a, b) \neq (0, 0)$, we see that $T(X_1 + X_2) \neq T(X_1) + T(X_2)$.

(b) Let A and B be real numbers. Then $T(A + B) = \sin(A + B) = \sin A \cos B + \sin B \cos A$. But $T(A) + T(B) = \sin A + \sin B$. If $A = \pi/4$ and $B = \pi/4$, then $T(A + B) = 1$ and $T(A) + T(B) = \sqrt{2} \neq 1$. Therefore, T is not a linear transformation.

(c) $T(3(2, 1, 1)) = T(6, 3, 3) = (36, 0)$ and $3T(2, 1, 1) = 3(4, 0) = (12, 0) \neq (36, 0)$. Therefore, T is not a linear transformation.

(d) $T(4 + (-2)) = 2$, but $T(4) + T(-2) = 6$. Therefore, T is not a linear transformation.

Exercise 6.2

In Problems 1–9, determine which of the transformations are linear transformations. If the transformation is linear, verify it. If it is not linear, show that at least one of the two properties of Definition 6.2 does not hold.

1. $T: \mathbf{R}_2 \longrightarrow \mathbf{R}_2$ defined by (a) $T(x, y) = 7(x, y)$; (b) $T(x, y) = k(x, y)$, k a constant.

2. $T: \mathbf{R}_2 \longrightarrow \mathbf{R}_2$ defined by $T(x, y) = (x + 1, y + 3)$.

3. $T: \mathbf{R}_3 \longrightarrow \mathbf{R}_2$ defined by $T(x, y, z) = (x + y, x + z)$.

4. $T: \mathbf{R}_2 \longrightarrow \mathbf{R}_2$ defined by $T(x, y)$ is the reflection of (x, y) in the line $x = 0$ (the Y axis).

5. $T: \mathbf{R}_2 \longrightarrow \mathbf{R}_2$ defined by $T(x, y) = (x^2, y + x)$.

6. $T: \mathbf{R}_3 \longrightarrow \mathbf{R}_4$ defined by $T(x_1, x_2, x_3) = (x_1, x_2, x_3) \begin{bmatrix} 1 & 2 & 0 & 3 \\ 0 & 1 & -1 & 0 \\ -1 & 3 & 1 & -2 \end{bmatrix}$.

7. $T: \mathbf{R} \longrightarrow \mathbf{R}$ defined by $T(x) = kx$, k a constant.

8. $T: \mathbf{V} \longrightarrow \mathbf{V}$, where $\mathbf{V}$ is the space of complex numbers and $T(a + bi) = a - bi$.

9. $T: \mathbf{P}_2 \longrightarrow \mathbf{P}_3$ is defined by $T(f: ax + b) = g: ax^2 + bx$.

10. If $\mathbf{V}$ and $\mathbf{W}$ are any vector spaces and if $\mathbf{0'}$ is the zero vector of $\mathbf{W}$, verify that the mapping $T: \mathbf{V} \longrightarrow \mathbf{W}$ defined by $T(X) = \mathbf{0'}$ for all X in $\mathbf{V}$ is a linear transformation.

11. If $\mathbf{V}$ is any vector space, verify that the mapping $T: \mathbf{V} \longrightarrow \mathbf{V}$ defined by $T(X) = X$ for all X in $\mathbf{V}$ is a linear transformation.

12. Let A be a fixed vector in the plane $\mathbf{R}_2$. Define $P: \mathbf{R}_2 \longrightarrow \mathbf{R}_2$ by $P(X)$ is the vector projection of X onto A. Show that P is a linear transformation.

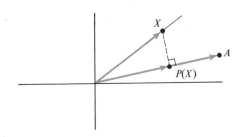

13. Show that the following functions are *not* linear transformations (as some students seem to think).
    (a) $T: \mathbf{R} \longrightarrow \mathbf{R}$ defined by $T(x) = \tan x$.
    (b) $T: \mathbf{R} \longrightarrow \mathbf{R}$ defined by $T(x) = \log x$.

14. If $T: \mathbf{R}_2 \longrightarrow \mathbf{R}_2$ is a linear transformation, and if $T(1, 0) = (3, 4)$ and $T(0, 1) = (-1, 2)$, find $T(x, y)$ (see Example 6.5).

15. If $T: \mathbf{R}_2 \longrightarrow \mathbf{R}_2$ is a linear transformation, and if $T(1, 1) = (2, 0)$ and $T(0, 2) = (3, 1)$, find $T(x, y)$ (see Example 6.5).

16. Given that $T: \mathbf{R}_2 \longrightarrow \mathbf{R}_2$ is a linear transformation and that $T(1, 1) = (3, 6)$ and $T(2, 2) = (6, 12)$, explain why one cannot find $T(x, y)$ without additional information. Compare this problem with Problem 15 (or 14).

17. Use Figure 6.1(a) and (b) to give a geometric argument that R_θ is a linear transformation; that is, show that (a) $(A + B)' = A' + B'$, and (b) $(rA)' = rA'$. *Hint:* Use the fact that a rotation preserves both distance and the angle between vectors. Show that $\|A + B\| = \|A' + B'\|$ and dir $(A' + B') = $ dir $(A + B)'$.

18. Use Figure 6.2(a) and (b) to give a geometric argument that the reflection T in Example 6.6 is a linear transformation; that is, show that
    (a) $(A + B)' = A' + B'$.
    (b) $(rA)' = rA'$.
    (See the hint for Problem 17.)

19. If **V** is the vector space of all integrable real-valued functions defined on a closed interval $[a, b]$ [Example 5.1(k)], show that the transformation $T: \mathbf{V} \longrightarrow \mathbf{R}$ defined by $T(f) = \int_a^b f(x)\, dx$ is a linear transformation. (A linear transformation from a vector space **V** into the real numbers is called a *linear functional.*)

20. *Prove:* A mapping $T: \mathbf{V} \longrightarrow \mathbf{W}$ from a vector space **V** into a vector space **W** is a linear transformation if and only if, for all vectors X and Y in **V** and for all scalars r and s, $T(rX + sY) = rT(X) + sT(Y)$.

6.3 PROPERTIES OF LINEAR TRANSFORMATIONS

In this section we discuss several properties of linear transformations. One of the most important is that a linear transformation is completely determined by what it does to a basis. This means, for example, that a linear transformation with domain $\mathbf{R}_2$ is completely determined by the images of any two nonzero vectors X_1 and X_2 that are not scalar multiples of each other.

If a vector space **V** has a basis of n vectors and if n vectors (not necessarily distinct) are chosen from a vector space **W**, we shall see that there is one and only one linear transformation $T: \mathbf{V} \longrightarrow \mathbf{W}$ that maps the given basis vectors onto the n vectors from **W** in a prescribed order.

The first theorem of this section is a rather obvious extension of the definition of a linear transformation.

Theorem 6.1. If $T: \mathbf{V} \longrightarrow \mathbf{W}$ is a linear transformation, and if $X_1, X_2, \ldots, X_n$ are vectors in **V**, then for any scalars $a_1, a_2, \ldots, a_n$,

$$T(a_1 X_1 + a_2 X_2 + \cdots + a_n X_n) = a_1 T(X_1) + a_2 T(X_2) + \cdots + a_n T(X_n)$$

Proof: Let $X_1, X_2, \ldots, X_n$ be any vectors in V, and let $Y = a_1 X_1 + a_2 X_2 + \cdots + a_n X_n$. Property (i) of Definition 6.2 can be extended by mathematical induction (Exercise 6.3, Problem 15) to obtain

$$T(Y) = T(a_1 X_1 + a_2 X_2 + \cdots + a_n X_n)$$
$$= T(a_1 X_1) + T(a_2 X_2) + \cdots + T(a_n X_n)$$

By property (ii) of Definition 6.2, $T(a_i X_i) = a_i T(X_i)$ for $i = 1, 2, \ldots, n$. Therefore,

$$T(Y) = a_1 T(X_1) + a_2 T(X_2) + \cdots + a_n T(X_n) \qquad \text{Q.E.D.}$$

In Example 6.5, we first found the images of the basis $\{(1, 0), (0, 1)\}$ under R_θ. Combining this information with the fact that R_θ was a linear transformation, we were able to find the image of an arbitrary vector (x, y). The next theorem shows that this procedure works for any linear transformation.

Theorem 6.2. If $T: \mathbf{V} \rightarrow \mathbf{W}$ is a linear transformation, if $B = \{X_1, X_2, \ldots, X_n\}$ is a basis for $\mathbf{V}$, and if $T(X_i) = Y_i$, $i = 1, 2, \ldots, n$, then for any vector $X \in \mathbf{V}$, $T(X)$ is determined and $T(X) = a_1 Y_1 + a_2 Y_2 + \cdots + a_n Y_n$, where $a_1, a_2, \ldots, a_n$ are scalars such that $X = a_1 X_1 + a_2 X_2 + \cdots + a_n X_n$.

Proof: Let $X \in \mathbf{V}$. Since B is a basis for $\mathbf{V}$, there are scalars $a_1, a_2, \ldots, a_n$ such that $X = a_1 X_1 + a_2 X_2 + \cdots + a_n X_n$. Applying Theorem 6.1, we get

$$T(X) = T(a_1 X_1 + a_2 X_2 + \cdots + a_n X_n)$$
$$= a_1 T(X_1) + a_2 T(X_2) + \cdots + a_n T(X_n)$$
$$= a_1 Y_1 + a_2 Y_2 + \cdots + a_n Y_n \qquad \text{Q.E.D.}$$

Example 6.10. If $X_1 = (1, 0, 0)$, $X_2 = (1, 1, 0)$, $X_3 = (1, 1, 1)$, and if $T: \mathbf{R}_3 \rightarrow \mathbf{R}_2$ is a linear transformation such that $T(X_1) = (-1, 0)$, $T(X_2) = (2, 1)$, and $T(X_3) = (3, 3)$, (a) find $T(x, y, z)$; (b) find $T(4, 3, 2)$.

Solution: (a) Let $X = (x, y, z)$. We solve the vector equation $X = a_1 X_1 + a_2 X_2 + a_3 X_3$ for a_1, a_2, and a_3. This vector equation is equivalent to the system

$$a_1 + a_2 + a_3 = x$$
$$a_2 + a_3 = y$$
$$a_3 = z$$

The solutions are

$$a_1 = x - y$$
$$a_2 = y - z$$
$$a_3 = z$$

Therefore,

$$X = (x - y)X_1 + (y - z)X_2 + zX_3$$

We apply Theorem 6.2 to obtain

$$T(x, y, z) = (x - y)(-1, 0) + (y - z)(2, 1) + z(3, 3)$$
$$= (-x + 3y + z, y + 2z)$$

(b) Using the solution to part (a), we have

$$T(4, 3, 2) = (-4 + 9 + 2, 3 + 4)$$
$$= (7, 7)$$

According to Theorem 6.2, a linear transformation is determined by the images of a basis. The next example will show how one can define a linear transformation that maps, in a prescribed manner, a basis of n vectors in one vector space onto a collection of n vectors (or fewer) in another vector space.

Example 6.11. Define a linear transformation $T: \mathbf{R}_3 \to \mathbf{R}_2$ that maps the basis vectors $X_1 = (1, 0, 0)$, $X_2 = (1, 1, 0)$, and $X_3 = (1, 1, 1)$ into $Y_1 = (1, 3)$, $Y_2 = (4, 2)$, and $Y_3 = (3, 0)$, respectively. Show that there is only one such linear transformation, and find $T(x, y, z)$.

Solution: Let $X = (x, y, z)$ be an arbitrary vector in $\mathbf{R}_3$. Since $B = \{X_1, X_2, X_3\}$ is a basis for $\mathbf{R}_3$, there are *unique* scalars a_1, a_2, and a_3 such that $X = a_1 X_1 + a_2 X_2 + a_3 X_3$ (in fact, $a_1 = x - y$, $a_2 = y - z$, and $a_3 = z$). How should we define $T(X)$? If T is to be linear, we must have

$$T(X) = T(a_1 X_1 + a_2 X_2 + a_3 X_3) = a_1 T(X_1) + a_2 T(X_2) + a_3 T(X_3)$$

Also, we must have $T(X_i) = Y_i$ for $i = 1, 2, 3$ if the given condition is to be satisfied. Thus we are led to *define* T by

$$T(X) = a_1 Y_1 + a_2 Y_2 + a_3 Y_3$$

where $X = a_1 X_1 + a_2 X_2 + a_3 X_3$. $T(X)$ is uniquely determined by X because the scalars a_1, a_2, and a_3 are uniquely determined. Also,

$$T(X_1) = T(1X_1 + 0X_2 + 0X_3) = 1 Y_1 + 0 Y_2 + 0 Y_3 = Y_1$$

Similarly, $T(X_2) = Y_2$ and $T(X_3) = Y_3$. We must still check to see if T, as defined above, is linear. To check, we let $X = a_1 X_1 + a_2 X_2 + a_3 X_3$ and $Z = b_1 X_1 + b_2 X_2 + b_3 X_3$ be arbitrary vectors in $\mathbf{R}_3$. Then

$$T(X + Z) = T((a_1 + b_1)X_1 + (a_2 + b_2)X_2 + (a_3 + b_3)X_3)$$
$$= (a_1 + b_1)Y_1 + (a_2 + b_2)Y_2 + (a_3 + b_3)Y_3$$
$$= (a_1 Y_1 + a_2 Y_2 + a_3 Y_3) + (b_1 Y_1 + b_2 Y_2 + b_3 Y_3)$$
$$= T(X) + T(Z)$$

Also,

$$T(rX) = T(ra_1 X_1 + ra_2 X_2 + ra_3 X_3) = ra_1 Y_1 + ra_2 Y_2 + ra_3 Y_3$$
$$= r(a_1 Y_1 + a_2 Y_2 + a_3 Y_3) = rT(X)$$

Hence T is indeed a linear transformation. Now suppose that $S: \mathbf{R}_3 \to \mathbf{R}_2$ is a linear transformation such that $S(X_i) = Y_i$ for $i = 1, 2, 3$. If $X = a_1 X_1 + a_2 X_2 + a_3 X_3$ is an arbitrary vector in $\mathbf{R}_3$, then

$$S(X) = S(a_1 X_1 + a_2 X_2 + a_3 X_3)$$
$$= a_1 S(X_1) + a_2 S(X_2) + a_3 S(X_3)$$
$$= a_1 Y_1 + a_2 Y_2 + a_3 Y_3 = T(X)$$

Thus there is precisely one linear transformation from $\mathbf{R}_3$ that maps the basis vectors X_1, X_2, and X_3 into Y_1, Y_2, and Y_3, respectively. Finally,

$$T(x, y, z) = T((x - y)X_1 + (y - z)X_2 + zX_3)$$
$$= (x - y)(1, 3) + (y - z)(4, 2) + z(3, 0)$$
$$= (x + 3y - z, 3x - y - 2z)$$

The next theorem generalizes Example 6.11. The proof imitates the previous discussion and is left as Problem 19, Exercise 6.3.

Theorem 6.3. If V is a vector space with basis $B = \{X_1, X_2, \ldots, X_n\}$, and if $Y_1, Y_2, \ldots, Y_n$ are vectors in a vector space W (they need not be all different), then there exists one and only one linear transformation $T: V \to W$ such that $T(X_i) = Y_i$ for $i = 1, 2, \ldots, n$.

Theorem 6.3 indicates that in a sense a basis can be mapped anywhere by a linear transformation. However, the images of the basis elements determine where all the other vectors are mapped. The next theorem shows that the zero vector must map into the zero vector (of the codomain), and that the inverse of a vector X must map into the inverse of the image of X.

Theorem 6.4. If $T: V \to W$ is a linear transformation, then (a) the image of the zero vector in V is the zero vector in W, and (b) if $X \xrightarrow{T} Y = T(X)$, then $-X \xrightarrow{T} -Y = -T(X)$.

Proof: Let $\mathbf{0}$ denote the zero in V and let $\mathbf{0}'$ denote the zero in W. Then $T(\mathbf{0}) = T(\mathbf{0}) + \mathbf{0}'$ and $T(\mathbf{0}) = T(\mathbf{0} + \mathbf{0}) = T(\mathbf{0}) + T(\mathbf{0})$. So $T(\mathbf{0}) + \mathbf{0}' = T(\mathbf{0}) + T(\mathbf{0})$. From this equation we obtain, after adding $-T(\mathbf{0})$ to both sides, the result $\mathbf{0}' = T(\mathbf{0})$. This completes the proof of part (a).

Now suppose that $X \xrightarrow{T} Y = T(X)$ and consider $T(-X)$. We have $T(-X) = T((-1)X) = (-1)T(X) = -T(X)$. Hence $-X \xrightarrow{T} -Y = -T(X)$.   Q.E.D.

We close this section with some applications of linear transformations.

Applications

The following examples illustrate how linear transformations (or matrices) can be used to advantage in solving certain problems.

Example 6.12   Business.† A toy manufacturer produces toys that have three parts: wheels, axles, and bodies. The table indicates how many of each of the parts is required to make a particular toy.

†S. R. Searle and W. H. Hausman, *Matrix Algebra for Business and Economics* (New York: Wiley-Interscience/A Division of John Wiley & Sons, 1970), pp. 41–43. Reprinted by permission of John Wiley & Sons, Inc.

	toy 1	toy 2	toy 3
part 1 (wheels)	4	3	2
part 2 (axles)	2	2	1
part 3 (bodies)	1	1	1

Let x_i equal the number of toys of type i that are requested on a sales order for $i = 1, 2,$ or 3. Let y_i equal the number of parts of type i that are required to assemble the toys needed to fill the order.

The mapping $T: \mathbf{R}_3 \longrightarrow \mathbf{R}_3$ that sends the sales order vector

$$X = \begin{bmatrix} x_1 \\ x_2 \\ x_3 \end{bmatrix}$$

into the parts vector

$$Y = \begin{bmatrix} y_1 \\ y_2 \\ y_3 \end{bmatrix}$$

is given by

$$Y = T(X) = \begin{bmatrix} 4 & 3 & 2 \\ 2 & 2 & 1 \\ 1 & 1 & 1 \end{bmatrix} X$$

If the order calls for 8 toys of type 1, 12 toys of type 2, and 10 toys of type 3, then

$$\begin{bmatrix} 4 & 3 & 2 \\ 2 & 2 & 1 \\ 1 & 1 & 1 \end{bmatrix} \begin{bmatrix} 8 \\ 12 \\ 10 \end{bmatrix} = \begin{bmatrix} 88 \\ 50 \\ 30 \end{bmatrix}$$

To fill the order, the company must produce 88 wheels, 50 axles, and 30 bodies.

The problem can be extended by considering the amount of raw material that must be used. Suppose that the toys are made of steel and plastic. Let z_1 be the amount of steel and let z_2 be the amount of plastic required to make y_1 wheels, y_2 axles, and y_3 bodies. The table indicates the amount of material used in the production of the various parts.

Material (lb)	part 1, wheels	part 2, axles	part 3, bodies
material 1 (steel)	0	1	4
material 2 (plastic)	0.6	0	1

The transformation that sends the parts vector Y into the materials vector $Z = \begin{bmatrix} z_1 \\ z_2 \end{bmatrix}$ is given by $S: \mathbf{R}_3 \to \mathbf{R}_2$, where

$$S(Y) = Z = \begin{bmatrix} 0 & 1 & 4 \\ 0.6 & 0 & 1 \end{bmatrix} \begin{bmatrix} y_1 \\ y_2 \\ y_3 \end{bmatrix}.$$

To find the amount of material needed in terms of the sales order vector X, we write

$$S(Y) = S(T(X)) = \begin{bmatrix} 0 & 1 & 4 \\ 0.6 & 0 & 1 \end{bmatrix} \begin{bmatrix} 4 & 3 & 2 \\ 2 & 2 & 1 \\ 1 & 1 & 1 \end{bmatrix} \begin{bmatrix} x_1 \\ x_2 \\ x_3 \end{bmatrix} = \begin{bmatrix} 6 & 6 & 5 \\ 3.4 & 2.8 & 2.2 \end{bmatrix} \begin{bmatrix} x_1 \\ x_2 \\ x_3 \end{bmatrix}$$

To fill the order for $X = \begin{bmatrix} 8 \\ 12 \\ 10 \end{bmatrix}$ toys, the material used would be

$$\begin{bmatrix} 6 & 6 & 5 \\ 3.4 & 2.8 & 2.2 \end{bmatrix} \begin{bmatrix} 8 \\ 12 \\ 10 \end{bmatrix} = \begin{bmatrix} 170 \\ 82.8 \end{bmatrix}$$

or 170 lb of steel and 82.8 lb of plastic. Of course, the same result can be obtained by first calculating the parts vector $\begin{bmatrix} 88 \\ 50 \\ 30 \end{bmatrix}$, and then the material used would be

$$\begin{bmatrix} 0 & 1 & 4 \\ 0.6 & 0 & 1 \end{bmatrix} \begin{bmatrix} 88 \\ 50 \\ 30 \end{bmatrix} = \begin{bmatrix} 170 \\ 82.8 \end{bmatrix}$$

Without using matrices, the parts required would be given by

$$y_1 = 4x_1 + 3x_2 + 2x_3$$
$$y_2 = 2x_1 + 2x_2 + 1x_3$$
$$y_3 = x_1 + x_2 + x_3$$

and the materials by

$$z_1 = 0y_1 + y_2 + 4y_3$$
$$z_2 = 0.6y_1 + 0y_2 + y_3$$

To find z_1 and z_2 in terms of the number of toys ordered, we would substitute the x's for the y's. Contrast this with the matrix solution, which simply requires the multiplication of two matrices.

Example 6.13   Cryptography, the Art of Secret Writing. We wish to send a message, which, if intercepted, cannot be decoded easily. The message is

CONTACT CIA IN SUPERIOR. A first step in setting up a code could be to transform the alphabet into numbers. For example,

Letter	A	B	C	D	E	F	G	H	I	J	K	L	M	N	O	P	Q	R	S	T	U	V	W	X	Y	Z
Image	26	25	24	23	22	21	20	19	18	17	16	15	14	13	12	11	10	9	8	7	6	5	4	3	2	1

Using this code our message becomes 24, 12, 13, 7, 26, 24, 7, 24, 18, 26, 18, 13, 8, 6, 11, 22, 9, 18, 12, 9. This message could be deciphered too easily by experts. To avoid this, we do the following. Construct a 4×5 matrix (the message has 20 letters).

$$
A = \begin{bmatrix} 24 & 26 & 18 & 8 & 9 \\ 12 & 24 & 26 & 6 & 18 \\ 13 & 7 & 18 & 11 & 12 \\ 7 & 24 & 13 & 22 & 9 \end{bmatrix} \longleftrightarrow \begin{bmatrix} C|A|I|S|R \\ O|C|A|U|I \\ N|T|I|P|O \\ T|C|N|E|R \end{bmatrix}
$$

Next choose an invertible 4×4 matrix, for example,

$$
B = \begin{bmatrix} 0 & -1 & 1 & 0 \\ -2 & 0 & 0 & 1 \\ 0 & -2 & 1 & 0 \\ -3 & 0 & 0 & 1 \end{bmatrix}
$$

The message is sent via the matrix BA.

$$
BA = \begin{bmatrix} 1 & -17 & -8 & 5 & -6 \\ -41 & -28 & -23 & 6 & -9 \\ -11 & -41 & -34 & -1 & -24 \\ -65 & -54 & -41 & -2 & -18 \end{bmatrix}
$$

The receiver, who has in his possession the matrix

$$
B^{-1} = \begin{bmatrix} 0 & 1 & 0 & -1 \\ 1 & 0 & -1 & 0 \\ 2 & 0 & -1 & 0 \\ 0 & 3 & 0 & -2 \end{bmatrix}
$$

deciphers the message by calculating

$$
B^{-1}(BA) = \begin{bmatrix} 0 & 1 & 0 & -1 \\ 1 & 0 & -1 & 0 \\ 2 & 0 & -1 & 0 \\ 0 & 3 & 0 & -2 \end{bmatrix} \begin{bmatrix} 1 & -17 & -8 & 5 & -6 \\ -41 & -28 & -23 & 6 & -9 \\ -11 & -41 & -34 & -1 & -24 \\ -65 & -54 & -41 & -2 & -18 \end{bmatrix}
$$

$$= \begin{bmatrix} 24 & 26 & 18 & 8 & 9 \\ 12 & 24 & 26 & 6 & 18 \\ 13 & 7 & 18 & 11 & 12 \\ 7 & 24 & 13 & 22 & 9 \end{bmatrix} \longleftrightarrow \begin{bmatrix} C|A|I|S|R \\ O|C|A|U|I \\ N|T|I|P|O \\ T|C|N|E|R \end{bmatrix}$$

Exercise 6.3

In Problems 1–7 you are given a linear transformation and the images of a basis. Find the required images.

1. $T: \mathbf{R}_3 \longrightarrow \mathbf{R}_2$, $T(1, 0, 0) = (3, 4)$, $T(0, 1, 0) = (1, -2)$, and $T(0, 0, 1) = (-1, 3)$. Find (a) $T(x, y, z)$; (b) $T(1, 6, -2)$.

2. $T: \mathbf{R}_2 \longrightarrow \mathbf{R}_2$, $T(1, 2) = (1, 1)$ and $T(2, 0) = (3, 4)$. Find (a) $T(x, y)$; (b) $T(3, -1)$.

3. $T: \mathbf{R}_3 \longrightarrow \mathbf{R}_4$, $T(1, 0, 0) = (1, 0, 1, -1)$, $T(1, 1, 0) = (2, 1, 3, 0)$, and $T(1, 1, 1) = (0, 0, 0, 0)$. Find (a) $T(x, y, z)$; (b) $T(3, -1, 0)$.

4. $T: \mathbf{R}_3 \longrightarrow \mathbf{R}_2$, $T(1, 0, 0) = (2, 4)$, $T(0, 1, 0) = (1, 2)$, and $T(0, 0, 1) = (3, 6)$. Find (a) $T(x, y, z)$; (b) $T(5, 3, 4)$.

5. $T: \mathbf{P}_4 \longrightarrow \mathbf{P}_3$, $T(f_1: x^3) = g_1: 3x$, $T(f_2: x^2) = g_2: 4x^2$, $T(f_3: x) = g_3: 3$, and $T(f_4: 1) = g_4: 2x$. Find (a) $T(f)$ if $f: ax^3 + bx^2 + cx + d$; (b) $T(f)$ if $f: 3x + 2$.

6. $T: \mathbf{R}_{2 \times 2} \longrightarrow \mathbf{R}_{2 \times 3}$, $T\begin{bmatrix} 1 & 0 \\ 0 & 0 \end{bmatrix} = \begin{bmatrix} 3 & 1 & 0 \\ 2 & 1 & -2 \end{bmatrix}$, $T\begin{bmatrix} 0 & 1 \\ 0 & 0 \end{bmatrix} = \begin{bmatrix} 1 & 0 & 0 \\ 0 & 1 & 0 \end{bmatrix}$, $T\begin{bmatrix} 0 & 0 \\ 1 & 0 \end{bmatrix} = \begin{bmatrix} 0 & 0 & 0 \\ 0 & 0 & 0 \end{bmatrix}$, and $T\begin{bmatrix} 0 & 0 \\ 0 & 1 \end{bmatrix} = \begin{bmatrix} 2 & 0 & 0 \\ 0 & 1 & 2 \end{bmatrix}$. Find (a) $T\begin{bmatrix} a & b \\ c & d \end{bmatrix}$; (b) $T\begin{bmatrix} 1 & 3 \\ 2 & 4 \end{bmatrix}$.

7. $V = \{a + bi \,|\, a, b \in \mathbf{R}\}$. $T: V \longrightarrow V$, $T(1) = 2 + 3i$, and $T(i) = 3 - 4i$. Find (a) $T(a + bi)$; (b) $T(6 - 3i)$.

In Problems 8–14, you are given a vector space V, a basis B of n vectors in V, a vector space W, and vectors $Y_1, Y_2, \ldots, Y_n$ in W. Determine a linear transformation $T: V \longrightarrow W$ such that $T(X_i) = Y_i$, $i = 1, 2, \ldots, n$. If X is an arbitrary vector in V, find $T(X)$.

8. $V = \mathbf{R}_3$, $B = \{X_1 = (1, 0, 0), X_2 = (0, 1, 0), X_3 = (0, 0, 1)\}$, $W = \mathbf{R}_2$, and $Y_1 = (1, -1)$, $Y_2 = (0, 2)$, and $Y_3 = (0, 0)$.

9. $V = \mathbf{R}_2$, $B = \{X_1 = (1, 2), X_2 = (2, 0)\}$, $W = \mathbf{R}_2$, $Y_1 = (1, 1)$ and $Y_2 = (2, 2)$.

10. $V = \mathbf{R}_3$, $B = \{X_1 = (1, 0, 0), X_2 = (1, 1, 0), X_3 = (1, 1, 1)\}$, $W = \mathbf{R}_3$, $Y_1 = (1, 0, 1)$, $Y_2 = (2, 1, 1)$, and $Y_3 = (1, 0, 1)$.

11. $V = \mathbf{P}_4$, $B = \{f_1: x^3, f_2: x^2, f_3: x, f_4: 1\}$, $W = \mathbf{R}_4$, $Y_1 = (1, 0, 0, 0)$, $Y_2 = (0, 1, 0, 0)$, $Y_3 = (0, 0, 1, 0)$, and $Y_4 = (0, 0, 0, 1)$.

12. $V = \mathbf{R}_2$, $B = \{X_1 = (1, 0), X_2 = (0, 1)\}$, $W = \{a + bi \,|\, a, b \in \mathbf{R}\}$, $Y_1 = 1$ and $Y_2 = i$.

13. $V = R_4$, $B = \{X_1, X_2, X_3, X_4\}$ is the standard basis for R_4, $W = R_{2\times 2}$, $Y_1 = \begin{bmatrix} 1 & 0 \\ 0 & 0 \end{bmatrix}$, $Y_2 = \begin{bmatrix} 0 & 1 \\ 0 & 0 \end{bmatrix}$, $Y_3 = \begin{bmatrix} 0 & 0 \\ 1 & 0 \end{bmatrix}$, and $Y_4 = \begin{bmatrix} 0 & 0 \\ 0 & 1 \end{bmatrix}$.

14. $V = \{a + bi \,|\, a, b \in R\}$, $B = \{X_1 = 1, X_2 = i\}$, $W = P_2$, $Y_1 = f_1 : 1$, and $Y_2 = f_2 : x$.

15. If $T: V \longrightarrow W$ is a linear transformation and if $X_1, X_2, \ldots, X_n$ are vectors in V, use mathematical induction to prove $T(X_1 + X_2 + \cdots + X_n) = T(X_1) + T(X_2) + \cdots + T(X_n)$.

16. If $T: V \longrightarrow W$ is a linear transformation, prove $T(X - Y) = T(X) - T(Y)$.

17. In Example 6.12, suppose that the parts and materials are given by the following tables:

	toy 1	toy 2	toy 3
part 1	6	8	4
part 2	3	4	2
part 3	1	2	1

	part 1	part 2	part 3
material 1	0	1	6
material 2	0.5	0	4

(a) If $X = \begin{bmatrix} x_1 \\ x_2 \\ x_3 \end{bmatrix}$ is the sales order vector and $Y = \begin{bmatrix} y_1 \\ y_2 \\ y_3 \end{bmatrix}$ is the parts vector, and $X \xrightarrow{T} Y$, write $T(X)$ as a matrix product.

(b) If $X = \begin{bmatrix} 30 \\ 20 \\ 25 \end{bmatrix}$, find $Y = T(X)$.

(c) If $Z = \begin{bmatrix} z_1 \\ z_2 \end{bmatrix}$ is the materials vector and $Y \xrightarrow{S} Z$, write Z as a matrix product.

(d) Using matrices, find the amounts of material required to fill the sales order in part (b).

18. Using the letter–number code in Example 6.13, convert the message "The space ship has landed" to a 4×6 matrix A. [There are 21 letters in the message and 24 entries in A. A zero could be placed in the $(2, 6)$ entry to indicate to the receiver that the message ends with the $(1, 6)$ entry.] Send the message via BA (find BA), where B is the invertible matrix in Example 6.13. Decode the message by multiplying on the left by B^{-1}.

19. Prove Theorem 6.3. Follow the solution to Example 6.11.

6.4 THE RANGE AND THE NULL SPACE; ISOMORPHISMS

In this section we study two subsets that are associated with any linear transformation $T: V \longrightarrow W$. One of these, the range of T, is shown to be a subspace of the codomain W. The other, called the *kernel* of T or the *null space* of T (all the vectors that map onto zero), is shown to be a subspace of the domain **V**. We shall prove a dimension theorem that relates the dimension of these two spaces to the dimension of **V**. As a consequence, we shall find necessary and sufficient conditions for a linear transformation to be 1–1 and onto its codomain. Such a linear transformation is called an *isomorphism*.

In our first theorem of this section we show that the range of T is a vector space, and also find a set of generators for it. The range of T is denoted by $\mathbf{R_T}$.

Theorem 6.5.   If $T: \mathbf{V} \longrightarrow \mathbf{W}$ is a linear transformation, then $\mathbf{R_T}$ is a subspace of the codomain **W**. Furthermore, if $\{X_1, X_2, \ldots, X_n\}$ is a basis for **V**, then the vectors $T(X_1), T(X_2), \ldots, T(X_n)$ generate $\mathbf{R_T}$.

Proof: According to Theorem 5.2, it is sufficient to show that: (0) $\mathbf{R_T}$ has something in it, (1) if Y and Z are in $\mathbf{R_T}$, then $Y + Z \in \mathbf{R_T}$, and (2) if a is a scalar and $Y \in \mathbf{R_T}$, then $aY \in \mathbf{R_T}$.

$\mathbf{R_T}$ is not empty because, according to Theorem 6.4(a), the zero of **W** is in $\mathbf{R_T}$. Suppose that Y and Z are in $\mathbf{R_T}$. This means that there are preimages X and U in **V** such that $X \overset{T}{\longrightarrow} Y$ and $U \overset{T}{\longrightarrow} Z$. Since T is a *linear* transformation, we have

$$X + U \overset{T}{\longrightarrow} T(X + U) = T(X) + T(U) = Y + Z$$

Therefore, $X + U$ is a preimage of $Y + Z$, and thus $Y + Z \in \mathbf{R_T}$.

Now suppose that a is a scalar and Y is a vector in $\mathbf{R_T}$ with a preimage X. Since $aX \overset{T}{\longrightarrow} T(aX) = aT(X) = aY$, we see that $aY \in \mathbf{R_T}$. Hence $\mathbf{R_T}$ is a subspace of **W**. Now let $B = \{X_1, X_2, \ldots, X_n\}$ be a basis for **V**, and let Y be any vector in $\mathbf{R_T}$. There must be at least one vector X in **V** such that $T(X) = Y$. Since B is a basis for **V**, there are scalars $a_1, a_2, \ldots, a_n$ such that

$$X = a_1 X_1 + a_2 X_2 + \cdots + a_n X_n$$

By Theorem 6.1,

$$Y = T(X) = a_1 T(X_1) + a_2 T(X_2) + \cdots + a_n T(X_n)$$

Since each of the vectors $T(X_1), T(X_2), \ldots, T(X_n)$ is in $\mathbf{R_T}$, it follows that $\mathbf{R_T}$ is generated by these vectors. Q.E.D.

Applying Theorem 6.5 to the transformation in Example 6.10, we see that $\{(-1, 0), (2, 1), (3, 3)\}$ generates $\mathbf{R_T}$. The linearly independent subset $\{(-1, 0), (2, 1)\}$ is a basis for $\mathbf{R_T}$, and we see that T is *onto* its codomain.

Definition 6.3.   The **rank** of a linear transformation $T: \mathbf{V} \longrightarrow \mathbf{W}$ is the dimension of the range ($\mathbf{R_T}$) of T.

Definition 6.4.   The **null space** or the **kernel** of a linear transformation $T: \mathbf{V} \longrightarrow \mathbf{W}$ is the set $\mathbf{N_T} = \{X \in \mathbf{V} \mid T(X) = \mathbf{0}'\}$. Here $\mathbf{0}'$ denotes the zero of the codomain **W**.

We emphasize that N_T is a *subset of the domain* space **V**. The next theorem states that N_T is a vector space. The dimension of this space is called the **nullity** of T.

Theorem 6.6. If $T: \mathbf{V} \to \mathbf{W}$ is a linear transformation and if $0'$ is the zero of **W**, then $N_T = \{X \in \mathbf{V} \mid X \xrightarrow{T} 0'\}$ is a subspace of **V**.

Proof: Let X and Y be arbitrary vectors in N_T, and let a be any scalar. Since $T(X + Y) = T(X) + T(Y) = 0' + 0' = 0'$, we have $X + Y \in N_T$. Also, $T(aX) = aT(X) = a0' = 0'$ and $aX \in N_T$. The set N_T is not empty because $T(0) = 0'$, and thus $0 \in N_T$. By Theorem 5.2, N_T is a subspace of **V**. Q.E.D.

Example 6.14. Find the null space and the nullity of the linear transformation $T: \mathbf{R}_3 \to \mathbf{R}_2$ of Example 6.10.

Solution: From the solution to Example 6.10, we have $(x, y, z) \xrightarrow{T} (-x + 3y + z, y + 2z)$. Therefore, $(x, y, z) \in N_T$ if and only if

$$-x + 3y + z = 0$$
$$y + 2z = 0$$

The reduced row echelon form of the coefficient matrix of this system is $\begin{bmatrix} 1 & 0 & 5 \\ 0 & 1 & 2 \end{bmatrix}$. Hence

$$N_T = \{(-5z, -2z, z) \mid z \in \mathbf{R}\} = \langle (-5, -2, 1) \rangle$$

and the nullity of T is 1.

A transformation is 1–1 if and only if distinct elements in the domain have different images. Thus it is clear that a 1–1 linear transformation can have only the zero vector in the null space. It is also true, but not obvious, that if $N_T = \{0\}$ then T is 1–1.

Theorem 6.7. A linear transformation $T: \mathbf{V} \to \mathbf{W}$ is 1–1 if and only if $N_T = \{0\}$.

Proof: Suppose that $N_T = \{0\}$ and let X and Y be vectors in the domain **V** such that $T(X) = T(Y)$. Then $T(X) - T(Y) = 0'$ (the zero of **W**). By Problem 16, Exercise 6.3, $T(X) - T(Y) = T(X - Y)$. So $X - Y$ is in the null space of T. But, $N_T = \{0\}$. Thus $X - Y = 0$ and $X = Y$. Hence T is 1–1. As we mentioned prior to stating the theorem, the converse is rather obvious, but we ask the reader to give a formal proof in Exercise 6.4, Problem 8.

Q.E.D.

We now relate the nullity and rank of a linear transformation to the dimension of the domain of the transformation.

Theorem 6.8.   If $T: \mathbf{V} \longrightarrow \mathbf{W}$ is a linear transformation from a finite-dimensional vector space $\mathbf{V}$ into a vector space $\mathbf{W}$, then dim $\mathbf{V} = $ dim $\mathbf{N_T} + $ dim $\mathbf{R_T}$.

Proof: Since $\mathbf{V}$ is finite dimensional, we know (Theorem 5.11) that the subspace $\mathbf{N_T}$ is finite dimensional. Let $B_1 = \{X_1, X_2, \ldots, X_t\}$ be a basis for $\mathbf{N_T}$, and extend B_1 to a basis $B_2 = \{X_1, X_2, \ldots, X_t, Y_{t+1}, \ldots, Y_n\}$ for $\mathbf{V}$ (Exercise 5.5, Problem 20). We shall establish the conclusion of the theorem by proving that $C = \{T(Y_{t+1}), \ldots, T(Y_n)\}$ is a basis for $\mathbf{R_T}$. Applying Theorem 6.5, we see that the vectors

$$T(X_1), \ldots, T(X_t), T(Y_{t+1}), \ldots, T(Y_n)$$

generate $\mathbf{R_T}$. However,

$$T(X_1) = T(X_2) = \cdots = T(X_t) = \mathbf{0}$$

because the X's are all in the null space of T. Consequently, $T(Y_{t+1}), \ldots, T(Y_n)$ generate $\mathbf{R_T}$.

Next we show that these vectors are linearly independent. Suppose that

$$c_{t+1} T(Y_{t+1}) + \cdots + c_n T(Y_n) = \mathbf{0}$$

Using the fact that T is a linear transformation, we get

$$T(c_{t+1} Y_{t+1} + \cdots + c_n Y_n) = \mathbf{0}$$

So

$$Z = c_{t+1} Y_{t+1} + \cdots + c_n Y_n$$

is in $\mathbf{N_T}$. Since B_1 is a basis for $\mathbf{N_T}$, we have

$$Z = a_1 X_1 + a_2 X_2 + \cdots + a_t X_t$$

for certain scalars $a_1, a_2, \ldots, a_t$. Adding $-Z$ to both sides of this equation, we obtain

$$\mathbf{0} = a_1 X_1 + a_2 X_2 + \cdots + a_t X_t + -c_{t+1} Y_{t+1} + \cdots + -c_n Y_n$$

This is a linear combination of the vectors in B_2, which is a basis for $\mathbf{V}$. Therefore, all the scalar coefficients are zero and in particular all the c's are zero. Hence $\{T(Y_{t+1}), \ldots, T(Y_n)\}$ is a linearly independent set of generators for $\mathbf{R_T}$. We now have dim $\mathbf{R_T} = n - t$. Thus

$$\dim \mathbf{V} = n = t + (n - t) = \dim \mathbf{N_T} + \dim \mathbf{R_T} \qquad \text{Q.E.D.}$$

Example 6.15.   If $T: \mathbf{P}_3 \longrightarrow \mathbf{P}_2$ is the linear transformation that maps the basis $B = \{f_1: x^2, f_2: x, f_3: 1\}$ into the elements $g_1: x + 1, g_2: x$, and $g_3: 6$, respectively, find the range of T, the null space of T, and verify the conclusion of Theorem 6.8 for this specific T.

Solution: From Theorem 6.5, $\mathbf{R_T} = \langle g_1: x + 1, g_2: x, g_3: 6 \rangle$. Since $g_3 = 6g_1 - 6g_2$, we have $\mathbf{R_T} = \langle g_1: x + 1, g_2: x \rangle$. The functions g_1 and g_2 are linearly independent, and thus $B = \{g_1, g_2\}$ is a basis for $\mathbf{R_T}$. Next consider N_T. If f: $ax^2 + bx + c$, then

$$T(f) = g: (a + b)x + (a + 6c)$$

So $f \in N_T$ if and only if

$$a + b = 0$$
$$a + 6c = 0$$

Therefore $a = -6c$, $b = 6c$, and f: $(-6c)x^2 + 6cx + c$. It follows that $N_T = \langle h: 6x^2 - 6x - 1 \rangle$. We have

$$\dim R_T = 2, \qquad \dim N_T = 1, \qquad \dim P_3 = \dim N_T + \dim R_T$$

We state the next result as a corollary to Theorems 6.7 and 6.8.

Corollary 6.1. Let T: $V \to W$ be a linear transformation and let $n = \dim V$. Then T is 1–1 if and only if the rank of T is n.

Proof: By Theorem 6.7, T is 1–1 if and only if $\dim N_T = 0$. Using Theorem 6.8, and the hypothesis that $\dim V = n$, we have $\dim N_T = 0$ if and only if the rank of T is n. Q.E.D.

Definition 6.5. A linear transformation T: $V \to W$ is an **isomorphism** if and only if T is 1–1 and onto W. If there is an isomorphism T: $V \to W$, we say that V and W are **isomorphic**.

Theorem 6.9. Let V and W be finite-dimensional vector spaces. V is isomorphic to W if and only if the dimension of V is equal to the dimension of W.

Proof: Suppose that $\dim V = \dim W$. Then each space has a basis of n vectors. Let $B = \{X_1, X_2, \ldots, X_n\}$ and $C = \{Y_1, Y_2, \ldots, Y_n\}$ be bases for V and W, respectively. By Theorem 6.3, we know that there is a linear transformation T: $V \to W$ such that

$$T(X_i) = Y_i, \qquad \text{for } i = 1, 2, \ldots, n$$

According to Theorem 6.5,

$$R_T = \langle Y_1, Y_2, \ldots, Y_n \rangle = W$$

Also, applying Theorem 6.8, we see that $\dim N_T = 0$ and, by Theorem 6.7, T is 1–1. Hence V and W are isomorphic. Conversely, suppose that V is isomorphic to W. Then there exists a linear transformation T: $V \to W$ that is 1–1 and onto W. Since T is onto W, we have $R_T = W$. From Theorem 6.8, we have $\dim V = \dim N_T + \dim W$. Since T is 1–1, $N_T = \{0\}$. Thus

$$\dim V = 0 + \dim W = \dim W \qquad \text{Q.E.D.}$$

The reader has probably already observed the strong similarities between such vector spaces as R_4, P_4, and $R_{2 \times 2}$. Each has dimension 4 and so they are isomorphic. As the proof of Theorem 6.9 indicates, one can obtain an isomorphism between any two of these vector spaces by mapping a basis onto a basis. The bases

$B_1 = \{E_1 = (1, 0, 0, 0), E_2 = (0, 1, 0, 0), E_3 = (0, 0, 1, 0), E_4 = (0, 0, 0, 1)\}$

$B_2 = \{f_1: 1, f_2: x, f_3: x^2, f_4: x^3\}$

$B_3 = \left\{ E_{11} = \begin{bmatrix} 1 & 0 \\ 0 & 0 \end{bmatrix}, E_{12} = \begin{bmatrix} 0 & 1 \\ 0 & 0 \end{bmatrix}, E_{21} = \begin{bmatrix} 0 & 0 \\ 1 & 0 \end{bmatrix}, E_{22} = \begin{bmatrix} 0 & 0 \\ 0 & 1 \end{bmatrix} \right\}$

are regarded as "natural" bases for the spaces R_4, P_4, and $R_{2 \times 2}$, respectively. The linear transformation $T: R_4 \rightarrow P_4$ that sends E_i onto f_i for $i = 1, 2, 3, 4$ will map the vector $X = (a, b, c, d)$ onto the polynomial function f: $a + bx + cx^2 + dx^3$. The intuitive notion that the vector $X = (a, b, c, d)$ acts like the function $T(X) = f: a + bx + cx^2 + dx^3$ is made mathematically precise by the concept of isomorphism. Under the given isomorphism we have, for example,

$$X_1 = (3, 2, 1, 4) \xrightarrow{T} g_1: 3 + 2x + x^2 + 4x^3$$

$$X_2 = (1, 0, 2, 1) \xrightarrow{T} g_2: 1 + 0x + 2x^2 + x^3$$

$$X_3 = X_1 + X_2 = (4, 2, 3, 5) \xrightarrow{T} (g_3: 4 + 2x + 3x^2 + 5x^3) = g_1 + g_2$$

$$3X_1 = (9, 6, 3, 12) \xrightarrow{T} (g_4: 9 + 6x + 3x^2 + 12x^3) = 3g_1$$

Under addition the images g_1 and g_2 act just like the preimages X_1 and X_2, and under scalar multiplication $3g_1$ acts like $3X_1$. The isomorphism indicates that, as vector spaces, R_4 and P_4 are essentially the same.

   Similarly, the mapping $S: R_4 \rightarrow R_{2 \times 2}$ that sends $E_1 \xrightarrow{S} E_{11}$, $E_2 \xrightarrow{S} E_{12}$, $E_3 \xrightarrow{S} E_{21}$, and $E_4 \xrightarrow{S} E_{22}$ will send

$$X = (a, b, c, d) \xrightarrow{S} \begin{bmatrix} a & b \\ c & d \end{bmatrix} = S(X)$$

Intuitively, $S(X)$ acts like X.

   In Section 1.3 we defined addition in R_2 so that, if $P = (x_1, y_1)$ and $Q = (x_2, y_2)$ were the points at the heads of the position vectors $\bar{p}$ and $\bar{q}$, then $P + Q$ would be the point at the head of $\bar{p} + \bar{q}$. Also, rP was defined so that it would be the point at the head of $r\bar{p}$. In other words, we defined addition and scalar multiplication so that the ordered pairs in R_2 would act like position vectors. More precisely, the definition of addition and scalar multiplication in R_2 forced the mapping T, defined by $\bar{p} \xrightarrow{T} P$, to be an isomorphism.

   The mathematical significance of Theorem 6.9 is this. Any two vector spaces of dimension n are essentially the same. Probably the simplest n-dimensional space to study is R_n. This "simple" n-dimensional vector space serves as a prototype or model for any other n-dimensional vector space V, because R_n and V are isomorphic. Any property of R_n that is derived from the vector space axioms can be transferred to the n-dimensional vector space V.

Exercise 6.4

1–7. For each of the transformations in Problems 1–7 of Exercise 6.3, find a set of generators for $\mathbf{R}_T$ and reduce this set of generators to a basis for $\mathbf{R}_T$. Find a basis for $\mathbf{N}_T$ and verify that dim (domain space) = dim $\mathbf{N}_T$ + dim $\mathbf{R}_T$.

8. *Prove:* If $T: \mathbf{V} \longrightarrow \mathbf{W}$ is a 1–1 linear transformation, then $\mathbf{N}_T = \{0\}$.

9. (a) If $T: \mathbf{R}_3 \longrightarrow \mathbf{R}_7$ is a linear transformation, what are the minimum and maximum possible values for the rank of T?
(b) If $T: \mathbf{R}_7 \longrightarrow \mathbf{R}_3$ is a linear transformation, what are the minimum and maximum possible values for the nullity of T?

10. If $T: \mathbf{R}_3 \longrightarrow \mathbf{R}_3$ is defined by $T(x, y, z) = (2x - z, 3x - 2y, x - 2y + z)$,
(a) Determine the null space of T.
(b) Use your answer to part (a) to determine if T is 1–1.
(c) Find a basis for $\mathbf{N}_T$ and use Theorem 6.8 to determine the rank of T.

11. Repeat Problem 10 for the linear transformation $T: \mathbf{R}_3 \longrightarrow \mathbf{R}_3$ defined by $T(x, y, z) = (x + y, z, 2x + 2y)$.

12. (a) Show that it is impossible for a linear transformation $T: \mathbf{R}_3 \longrightarrow \mathbf{R}_5$ to be onto $\mathbf{R}_5$.
(b) Show that it is impossible for a linear transformation $T: \mathbf{R}_5 \longrightarrow \mathbf{R}_3$ to be 1–1.

13. If $\mathbf{V}$ is an n-dimensional vector space with basis $B = \{X_1, X_2, \ldots, X_n\}$, and $T: \mathbf{V} \longrightarrow \mathbf{R}_n$ is the isomorphism that sends $X_i \overset{T}{\longrightarrow} E_i$ for $i = 1, 2, \ldots, n$, find the image of an arbitrary vector $X = a_1 X_1 + a_2 X_2 + \cdots + a_n X_n$.

14. For convenience, we have at times thought of $\mathbf{R}_n$ as the collection of all column vectors, rather than as n-tuples or row vectors. Give a precise mathematical formulation of the following statement: the column vector $\begin{bmatrix} x_1 \\ x_2 \\ \cdot \\ \cdot \\ \cdot \\ x_n \end{bmatrix}$ acts like the n-tuple $(x_1, x_2, \ldots, x_n)$.

15. Verify that $T: \mathbf{R}_6 \longrightarrow \mathbf{R}_{2 \times 3}$ defined by $T(a, b, c, d, e, f) = \begin{bmatrix} a & b & c \\ d & e & f \end{bmatrix}$ is an isomorphism.

16. *Prove:* If $T: \mathbf{V} \longrightarrow \mathbf{V}$ is a linear transformation onto $\mathbf{V}$, and if dim $\mathbf{V} = n$, then T is 1–1.

17. *Prove:* If $T: \mathbf{V} \longrightarrow \mathbf{V}$ is a linear transformation that is 1–1, and if dim $\mathbf{V} = n$, then T is onto $\mathbf{V}$.

18. If $D: \mathbf{P}_3 \longrightarrow \mathbf{P}_2$ and $D(f: a + bx + cx^2) = f': b + 2cx$ (the derivative of f), (a) show that D is a linear transformation; (b) find a basis for the null space of D and for the range of D.

19. If $T: \mathbf{P_3} \longrightarrow \mathbf{P_2}$ is defined by $T(f: a + bx + cx^2) = f'': 2c$ (the second derivative of f), (a) show that T is a linear transformation. Find $\mathbf{R_T}$ and $\mathbf{N_T}$.

20. Consider a homogeneous system of m linear equations and n unknowns with coefficient matrix $A_{(m,n)}$. Define $T: \mathbf{R_n} \longrightarrow \mathbf{R_m}$ by $T(X) = AX$ (X is an $n \times 1$ column vector). Use Theorem 6.6 to prove that the set of all solutions to the given homogeneous system is a subspace of $\mathbf{R_n}$.

6.5 THE MATRIX OF A TRANSFORMATION RELATIVE TO A PAIR OF BASES

In this section we shall show that any linear transformation $T: \mathbf{V} \longrightarrow \mathbf{W}$ from a finite-dimensional vector space $\mathbf{V}$ into a finite-dimensional vector space $\mathbf{W}$ can be represented by a matrix. The matrix representation of T depends on both a basis for $\mathbf{V}$ and a basis for $\mathbf{W}$. However, once we choose a pair of bases and consider them as *ordered bases*, then the matrix representation is uniquely determined. Before proceeding, we need a definition.

Definition 6.6. Let $B = \{X_1, X_2, \ldots, X_n\}$ be an ordered basis for a vector space $\mathbf{V}$. The **coordinate vector of a vector** X **in** $\mathbf{V}$ **relative to the ordered basis** B is the column vector $X = \begin{bmatrix} a_1 \\ a_2 \\ \vdots \\ a_n \end{bmatrix}$, where $a_1, a_2, \ldots, a_n$ are the uniquely determined scalars such that $X = a_1 X_1 + a_2 X_2 + \cdots + a_n X_n$. The scalars $a_1, a_2, \ldots, a_n$ are called the **coordinates** of X **relative to the basis** B. We denote the coordinate vector of X relative to the basis B by $[X]_B$.

Note: The reason for insisting that B be an *ordered* basis is so that the coordinate vector is uniquely determined.

Example 6.16. Let $\mathbf{V}$ be the four-dimensional vector space of all polynomial functions of degree less than 4. The coordinate vector of $f: 3 + 6x - 2x^2 + 4x^3$ relative to the ordered basis

$$B = \{f_1: 1, f_2: x, f_3: x^2, f_4: x^3\}$$

is

$$[f]_B = \begin{bmatrix} 3 \\ 6 \\ -2 \\ 4 \end{bmatrix}$$

If

$$B_2 = \{g_1: x^3, g_2: x^2, g_3: 1, g_4: x\}$$

then

$$[f]_{B_2} = \begin{bmatrix} 4 \\ -2 \\ 3 \\ 6 \end{bmatrix}$$

As sets, B and B_2 are identical, but as ordered sets they are different. Before we define the matrix of a linear transformation, we give an example to motivate the definition.

Example 6.17. Let

$$B_1 = \left\{ X_1 = \begin{bmatrix} 1 \\ 0 \\ 0 \end{bmatrix}, X_2 = \begin{bmatrix} 1 \\ 1 \\ 0 \end{bmatrix}, X_3 = \begin{bmatrix} 1 \\ 1 \\ 1 \end{bmatrix} \right\}$$

and

$$B_2 = \left\{ Y_1 = \begin{bmatrix} 1 \\ 0 \end{bmatrix}, Y_2 = \begin{bmatrix} 1 \\ 1 \end{bmatrix} \right\}$$

be ordered bases for $\mathbf{R}_3$ and $\mathbf{R}_2$, respectively. Let $T: \mathbf{R}_3 \rightarrow \mathbf{R}_2$ be the linear transformation such that

$$T(X_1) = 2Y_1 + 3Y_2, \qquad T(X_2) = 0Y_1 + (-1)Y_2, \qquad T(X_3) = 3Y_1 - 4Y_2$$

Relative to the ordered basis B_2, we have

$$[T(X_1)]_{B_2} = \begin{bmatrix} 2 \\ 3 \end{bmatrix}, \qquad [T(X_2)]_{B_2} = \begin{bmatrix} 0 \\ -1 \end{bmatrix}, \qquad [T(X_3)]_{B_2} = \begin{bmatrix} 3 \\ -4 \end{bmatrix}$$

We form the matrix

$$\begin{bmatrix} 2 & 0 & 3 \\ 3 & -1 & -4 \end{bmatrix} = [[T(X_1)]_{B_2} \mathbin{|} [T(X_2)]_{B_2} \mathbin{|} [T(X_3)]_{B_2}] = A\dagger$$

For any vector X in $\mathbf{R}_3$, we claim that

$$A[X]_{B_1} = [T(X)]_{B_2}$$

To verify this, let $X = x_1 X_1 + x_2 X_2 + x_3 X_3$ be any vector in $\mathbf{R}_3$. Then

$$A[X]_{B_1} = \begin{bmatrix} 2 & 0 & 3 \\ 3 & -1 & -4 \end{bmatrix} \begin{bmatrix} x_1 \\ x_2 \\ x_3 \end{bmatrix} = \begin{bmatrix} 2x_1 + 0x_2 + 3x_3 \\ 3x_1 - 1x_2 - 4x_3 \end{bmatrix}$$

†Technically, the columns of A are just the coordinates of $[T(X_i)]_{B_2}$ and not the coordinate vectors.

Next we write $T(X)$ as a linear combination of the vectors in B_2. Using the fact that T is linear, we get

$$T(X) = x_1 T(X_1) + x_2 T(X_2) + x_3 T(X_3)$$
$$= x_1(2Y_1 + 3Y_2) + x_2(0Y_1 + (-1)Y_2) + x_3(3Y_1 - 4Y_2)$$
$$= (2x_1 + 0x_2 + 3x_3)Y_1 + (3x_1 - 1x_2 - 4x_3)Y_2$$

Thus

$$[T(X)]_{B_2} = \begin{bmatrix} 2x_1 + 0x_2 + 3x_3 \\ 3x_1 - 1x_2 - 4x_3 \end{bmatrix} = A[X]_{B_1}$$

For this reason, we call A the matrix of T relative to the bases B_1 and B_2.

Definition 6.7. Let $T: V \longrightarrow W$ be a linear transformation. Let $B_1 = \{X_1, X_2, \ldots, X_n\}$ and $B_2 = \{Y_1, Y_2, \ldots, Y_m\}$ be bases for V and W, respectively. Let $T(X_j) = a_{1j}Y_1 + a_{2j}Y_2 + \cdots + a_{mj}Y_m$ for $j = 1, 2, \ldots, n$. **The matrix of T relative to the ordered bases B_1 and B_2 is**

$$[T]_{B_1 B_2} = \begin{bmatrix} a_{11} & a_{12} & \cdots & a_{1n} \\ a_{21} & a_{22} & \cdots & a_{2n} \\ \cdot & \cdot & & \cdot \\ \cdot & \cdot & & \cdot \\ \cdot & \cdot & & \cdot \\ a_{m1} & a_{m2} & \cdots & a_{mn} \end{bmatrix} = [[T(X_1)]_{B_2} \vdots [T(X_2)]_{B_2} \vdots \ldots \vdots [T(X_n)]_{B_2}]$$

In words, the matrix of T relative to the bases B_1 and B_2 is the matrix whose jth column is the coordinate vector of $T(X_j)$ relative to the basis B_2.

Note: Whenever we refer to the matrix of a linear transformation relative to a pair of bases, **the bases are assumed to be ordered bases** whether or not it is explicitly stated.

The next theorem is a generalization of Example 6.17.

Theorem 6.10. Let $B_1 = \{X_1, X_2, \ldots, X_n\}$ and $B_2 = \{Y_1, Y_2, \ldots, Y_m\}$ be bases for V and W, respectively, let $T: V \longrightarrow W$ be a linear transformation, and let $[T]_{B_1 B_2} = A = [a_{ij}]_{(m,n)}$. Then for any vector X in V, $A[X]_{B_1} = [T(X)]_{B_2}$.

Proof: (Optional) Since $[T]_{B_1 B_2} = A$, we have $[T(X_j)]_{B_2} = A^{(j)}$ and $T(X_j) = a_{1j}Y_1 + a_{2j}Y_2 + \cdots + a_{mj}Y_m$ for $j = 1, 2, \ldots, n$. Let $X \in V$ and

let $[X]_{B_1} = \begin{bmatrix} x_1 \\ x_2 \\ \cdot \\ \cdot \\ \cdot \\ x_n \end{bmatrix}$. Then $X = x_1 X_1 + x_2 X_2 + \cdots + x_n X_n$, and

$$T(X) = x_1 T(X_1) + x_2 T(X_2) + \cdots + x_n T(X_n)$$

$$= x_1 \sum_{i=1}^{m} a_{i1} Y_i + x_2 \sum_{i=1}^{m} a_{i2} Y_i + \cdots + x_n \sum_{i=1}^{m} a_{in} Y_i$$

$$= \sum_{i=1}^{m} (x_1 a_{i1} Y_i) + \sum_{i=1}^{m} (x_2 a_{i2} Y_i) + \cdots + \sum_{i=1}^{m} (x_n a_{in} Y_i)$$

$$= \sum_{i=1}^{m} (x_1 a_{i1} Y_i + x_2 a_{i2} Y_i + \cdots + x_n a_{in} Y_i)$$

$$= \sum_{i=1}^{m} \left(\sum_{j=1}^{n} x_j a_{ij} Y_i \right)$$

$$= \sum_{j=1}^{n} \left(\sum_{i=1}^{m} x_j a_{ij} Y_i \right)$$

$$= \sum_{j=1}^{n} (x_j a_{1j} Y_1 + x_j a_{2j} Y_2 + \cdots + x_j a_{mj} Y_m)$$

$$= \sum_{j=1}^{n} (x_j a_{1j} Y_1) + \sum_{j=1}^{n} (x_j a_{2j} Y_2) + \cdots + \sum_{j=1}^{n} (x_j a_{mj} Y_m)$$

$$= \left(\sum_{j=1}^{n} x_j a_{1j} \right) Y_1 + \left(\sum_{j=1}^{n} x_j a_{2j} \right) Y_2 + \cdots + \left(\sum_{j=1}^{n} x_j a_{mj} \right) Y_m$$

Therefore,

$$[T(X)]_{B_2} = \begin{bmatrix} \sum_{j=1}^{n} x_j a_{1j} \\ \sum_{j=1}^{n} x_j a_{2j} \\ \cdot \\ \cdot \\ \cdot \\ \sum_{j=1}^{n} x_j a_{mj} \end{bmatrix} = \begin{bmatrix} A_1 \cdot [X]_{B_1}^T \\ A_2 \cdot [X]_{B_1}^T \\ \cdot \\ \cdot \\ \cdot \\ A_m \cdot [X]_{B_1}^T \end{bmatrix} = A[X]_{B_1}$$

Q.E.D.

Note: If $X = \begin{bmatrix} x_1 \\ x_2 \\ \cdot \\ \cdot \\ \cdot \\ x_n \end{bmatrix}$ is a vector in $\mathbf{R_n}$ and if $B = \{E_1, E_2, \ldots, E_n\}$ is the

natural basis for $\mathbf{R_n}$, then $X = x_1 E_1 + x_2 E_2 + \cdots + x_n E_n$. Thus $[X]_B = X$; that is, *the coordinate vector of X relative to the natural basis is identical to the vector X*. Therefore, the matrix of a linear transformation $T: \mathbf{R_n} \rightarrow \mathbf{R_m}$ relative to the natural bases B and B' for $\mathbf{R_n}$ and $\mathbf{R_m}$, respectively, is the matrix $[T]_{BB'} = [T(E_1) | T(E_2) | \ldots | T(E_n)]$. Also, the conclusion to Theorem 6.10, that $A[X]_B = [T(X)]_{B'}$, becomes $AX = T(X)$.

Example 6.18.   Let $T: \mathbf{R}_4 \rightarrow \mathbf{R}_3$ be the linear transformation defined by

$$T(X) = \begin{bmatrix} x + 2y \\ x - 3z + w \\ 2y + 3z + 4w \end{bmatrix}, \qquad \text{where } X = \begin{bmatrix} x \\ y \\ z \\ w \end{bmatrix}$$

If B and B' are the natural bases for $\mathbf{R}_4$ and $\mathbf{R}_3$, respectively, (a) find $A = [T]_{BB'}$; (b) use A to find $T(X)$.

Solution: (a) From the definition of T we see that

$$T(E_1) = \begin{bmatrix} 1 \\ 1 \\ 0 \end{bmatrix}, \, T(E_2) = \begin{bmatrix} 2 \\ 0 \\ 2 \end{bmatrix}, \, T(E_3) = \begin{bmatrix} 0 \\ -3 \\ 3 \end{bmatrix}, \quad T(E_4) = \begin{bmatrix} 0 \\ 1 \\ 4 \end{bmatrix}$$

Since $[T(E_i)]_{B'} = T(E_i)$, we have

$$A = \begin{bmatrix} 1 & 2 & 0 & 0 \\ 1 & 0 & -3 & 1 \\ 0 & 2 & 3 & 4 \end{bmatrix} = [T]_{BB'}$$

(b) Since $[X]_B = X$ and $[T(X)]_{B'} = T(X)$, we have from Theorem 6.10 that

$$T(X) = AX = \begin{bmatrix} 1 & 2 & 0 & 0 \\ 1 & 0 & -3 & 1 \\ 0 & 2 & 3 & 4 \end{bmatrix} \begin{bmatrix} x \\ y \\ z \\ w \end{bmatrix} = \begin{bmatrix} x + 2y \\ x - 3z + w \\ 2y + 3z + 4w \end{bmatrix}$$

which agrees with the given definition of T.

Example 6.19.   Let $T: \mathbf{R}_3 \rightarrow \mathbf{R}_2$ be the linear transformation defined by

$$T \begin{bmatrix} x \\ y \\ z \end{bmatrix} = \begin{bmatrix} -\frac{1}{2}x + -\frac{11}{2}y + \frac{5}{2}z \\ \frac{7}{2}x + -\frac{7}{2}y + \frac{7}{2}z \end{bmatrix}$$

If

$$B_1 = \left\{ X_1 = \begin{bmatrix} 1 \\ 0 \\ 1 \end{bmatrix}, \quad X_2 = \begin{bmatrix} 2 \\ 0 \\ 0 \end{bmatrix}, \quad X_3 = \begin{bmatrix} 0 \\ 1 \\ 1 \end{bmatrix} \right\}$$

and

$$B_2 = \left\{ Y_1 = \begin{bmatrix} 1 \\ 2 \end{bmatrix}, \quad Y_2 = \begin{bmatrix} 0 \\ 3 \end{bmatrix} \right\}$$

find $A = [T]_{B_1 B_2}$.

Solution: The columns of A are $[T(X_1)]_{B_2}$, $[T(X_2)]_{B_2}$, and $[T(X_3)]_{B_2}$. To find the coordinate vectors of $T(X_i)$ relative to B_2, we solve the equations

$$T(X_i) = a_i Y_1 + b_i Y_2 = a_i \begin{bmatrix} 1 \\ 2 \end{bmatrix} + b_i \begin{bmatrix} 0 \\ 3 \end{bmatrix}$$

for a_i and b_i, $i = 1, 2$, and 3. These three equations are equivalent to the matrix equations

$$\begin{bmatrix} 1 & 0 \\ 2 & 3 \end{bmatrix} \begin{bmatrix} a_i \\ b_i \end{bmatrix} = T(X_i)$$

for $i = 1, 2$, and 3. From the definition of T, we obtain

$$T(X_1) = \begin{bmatrix} 2 \\ 7 \end{bmatrix}, \qquad T(X_2) = \begin{bmatrix} -1 \\ 7 \end{bmatrix}, \qquad T(X_3) = \begin{bmatrix} -3 \\ 0 \end{bmatrix}$$

We solve the three matrix equations simultaneously by reducing the matrix

$$[Y_1 Y_2 \,\vert\, T(X_1) T(X_2) T(X_3)] = \begin{bmatrix} 1 & 0 & \vert & 2 & -1 & -3 \\ 2 & 3 & \vert & 7 & 7 & 0 \end{bmatrix}$$

$$\xrightarrow{-2R_1 + R_2} \begin{bmatrix} 1 & 0 & \vert & 2 & -1 & -3 \\ 0 & 3 & \vert & 3 & 9 & 6 \end{bmatrix} \xrightarrow{(1/3)R_2} \begin{bmatrix} 1 & 0 & \vert & 2 & -1 & -3 \\ 0 & 1 & \vert & 1 & 3 & 2 \end{bmatrix}$$

Therefore,

$$[T(X_1)]_{B_2} = \begin{bmatrix} 2 \\ 1 \end{bmatrix}, \qquad [T(X_2)]_{B_2} = \begin{bmatrix} -1 \\ 3 \end{bmatrix}, \qquad [T(X_3)]_{B_2} = \begin{bmatrix} -3 \\ 2 \end{bmatrix}$$

Hence

$$[T]_{B_1 B_2} = \begin{bmatrix} 2 & -1 & -3 \\ 1 & 3 & 2 \end{bmatrix}$$

The method used in Example 6.19 to find $[T]_{B_1 B_2}$ can be generalized. If $T: \mathbf{R}_n \to \mathbf{R}_m$ is a linear transformation, and if $B_1 = \{X_1, X_2, \ldots, X_n\}$ and $B_2 = \{Y_1, Y_2, \ldots, Y_m\}$ are bases for $\mathbf{R}_n$ and $\mathbf{R}_m$, respectively, we can obtain $[T]_{B_1 B_2}$ as follows:

1. Form the $m \times (m + n)$ matrix

$$[Y_1 \ Y_2 \ldots Y_m \,\vert\, T(X_1) T(X_2) \ldots T(X_n)] = [C \,\vert\, D]$$

where $C = [Y_1 \,\vert\, Y_2 \,\vert\, \ldots \,\vert\, Y_m]_{(m,m)}$ and $D = [T(X_1) \,\vert\, T(X_2) \,\vert\, \ldots \,\vert\, T(X_n)]_{(m,n)}$.
2. Reduce the matrix $[C \,\vert\, D] \to [I_m \,\vert\, A]$. Then $A = [T]_{B_1 B_2}$.

Example 6.20.   Let $T: \mathbf{R}_3 \to \mathbf{R}_2$, B_1, B_2, and $A = [T]_{B_1 B_2}$ be the linear transformation, the bases, and the matrix of the linear transformation in Example 6.19. Let $Z_1 = \begin{bmatrix} 1 \\ 2 \\ 3 \end{bmatrix}$ and $Z_2 = \begin{bmatrix} 3 \\ 0 \\ 4 \end{bmatrix}$.

(a) Use A to find $T(Z_1)$ and $T(Z_2)$.

(b) Find the matrix of T relative to the natural bases B and B' for $\mathbf{R}_3$ and $\mathbf{R}_2$, respectively.

(c) Use $[T]_{BB'}$ to find $T(Z_1)$ and $T(Z_2)$, and compare the result with part (a).

Solution: (a) Since A is the matrix of T relative to the bases B_1 and B_2, Theorem 6.10 tells us that $A[Z_i]_{B_1} = [T(Z_i)]_{B_2}$. Thus we need to find the coordinate vectors $[Z_1]_{B_1}$ and $[Z_2]_{B_1}$. The problem is similar to the problem of finding the coordinate vectors of $T(X_1)$, $T(X_2)$, and $T(X_3)$ relative to the basis vectors Y_1 and Y_2, which we solved in Example 6.19 by reducing $[Y_1 Y_2 | T(X_1)T(X_2)T(X_3)]$. We now wish to find the coordinate vectors of Z_1 and Z_2 relative to the basis vectors X_1, X_2, and X_3. Therefore, we reduce the matrix

$$[X_1X_2X_3 | Z_1Z_2] = \begin{bmatrix} 1 & 2 & 0 & | & 1 & 3 \\ 0 & 0 & 1 & | & 2 & 0 \\ 1 & 0 & 1 & | & 3 & 4 \end{bmatrix} \xrightarrow[R_3+R_1]{-R_1+R_3} \begin{bmatrix} 1 & 0 & 1 & | & 3 & 4 \\ 0 & 0 & 1 & | & 2 & 0 \\ 0 & -2 & 1 & | & 2 & 1 \end{bmatrix}$$

$$\xrightarrow[-R_3+R_2]{R_{23}} \begin{bmatrix} 1 & 0 & 1 & | & 3 & 4 \\ 0 & -2 & 0 & | & 0 & 1 \\ 0 & 0 & 1 & | & 2 & 0 \end{bmatrix} \xrightarrow[-(1/2)R_2]{-R_3+R_1} \begin{bmatrix} 1 & 0 & 0 & | & 1 & 4 \\ 0 & 1 & 0 & | & 0 & -\frac{1}{2} \\ 0 & 0 & 1 & | & 2 & 0 \end{bmatrix}$$

$$= [I_3 | [Z_1]_{B_1} [Z_2]_{B_1}]$$

We can find $A[Z_1]_{B_1} = [T(Z_1)]_{B_2}$ and $A[Z_2]_{B_1} = [T(Z_2)]_{B_2}$ as follows.

$$A[[Z_1]_{B_1} | [Z_2]_{B_1}] = \begin{bmatrix} 2 & -1 & -3 \\ 1 & 3 & 2 \end{bmatrix} \begin{bmatrix} 1 & | & 4 \\ 0 & | & -\frac{1}{2} \\ 2 & | & 0 \end{bmatrix} = \begin{bmatrix} -4 & | & \frac{17}{2} \\ 5 & | & \frac{5}{2} \end{bmatrix}$$

Therefore,

$$T(Z_1) = -4Y_1 + 5Y_2 = \begin{bmatrix} -4 \\ 7 \end{bmatrix} \quad \text{and} \quad T(Z_2) = \frac{17}{2}Y_1 + \frac{5}{2}Y_2 = \begin{bmatrix} \frac{17}{2} \\ \frac{49}{2} \end{bmatrix}$$

The reader should check the result by calculating $T(Z_1)$ and $T(Z_2)$ directly from the definition of T.

(b) As the note following the proof of Theorem 6.10 indicates, the matrix of T relative to the natural bases B and B' for $\mathbf{R}_3$ and $\mathbf{R}_2$, respectively, is

$$[T]_{BB'} = [T(E_1)T(E_2)T(E_3)] = \begin{bmatrix} -\frac{1}{2} & -\frac{11}{2} & \frac{5}{2} \\ \frac{7}{2} & -\frac{7}{2} & \frac{7}{2} \end{bmatrix}$$

The reader should note that the method of Example 6.19 for finding the matrix $[T]_{BB'}$ can be used. However, the matrix

$$[C | D] = \begin{bmatrix} 1 & 0 & | & -\frac{1}{2} & -\frac{11}{2} & \frac{5}{2} \\ 0 & 1 & | & \frac{7}{2} & -\frac{7}{2} & \frac{7}{2} \end{bmatrix}$$

is already in the reduced form $[I_2 | [T]_{BB'}]$, and the matrix $[T]_{BB'}$ is obtained immediately.

(c) Again, as the note following the proof of Theorem 6.10 indicates, $T(Z_i) = A'Z_i$, where $A' = [T]_{BB'}$. Therefore,

$$A'[Z_1 | Z_2] = \begin{bmatrix} -\frac{1}{2} & -\frac{11}{2} & \frac{5}{2} \\ \frac{7}{2} & -\frac{7}{2} & \frac{7}{2} \end{bmatrix} \begin{bmatrix} 1 & 3 \\ 2 & 0 \\ 3 & 4 \end{bmatrix} = \begin{bmatrix} -4 & \frac{17}{2} \\ 7 & \frac{49}{2} \end{bmatrix} = [T(Z_1) | T(Z_2)]$$

So

$$T(Z_1) = \begin{bmatrix} -4 \\ 7 \end{bmatrix} \quad \text{and} \quad T(Z_2) = \begin{bmatrix} \frac{17}{2} \\ \frac{49}{2} \end{bmatrix}$$

which checks with the solution to part (a).

Example 6.21. Let $A = \begin{bmatrix} 4 & 2 & 1 \\ 0 & 1 & 3 \end{bmatrix}$. (a) Find the unique transformation $T: \mathbf{R}_3 \to \mathbf{R}_2$ such that the matrix of T relative to the bases

$$B_1 = \left\{ X_1 = \begin{bmatrix} 1 \\ 0 \\ 0 \end{bmatrix}, X_2 = \begin{bmatrix} 1 \\ 1 \\ 0 \end{bmatrix}, X_3 = \begin{bmatrix} 1 \\ 1 \\ 1 \end{bmatrix} \right\}$$

and

$$B_2 = \left\{ Y_1 = \begin{bmatrix} 1 \\ 0 \end{bmatrix}, Y_2 = \begin{bmatrix} 1 \\ 1 \end{bmatrix} \right\}$$

is A.

(b) Find $T \begin{bmatrix} x \\ y \\ z \end{bmatrix}$.

Solution: (a) If $[T]_{B_1 B_2} = A$, then

$$[T(X_1)]_{B_2} = \begin{bmatrix} 4 \\ 0 \end{bmatrix}, \quad [T(X_2)]_{B_2} = \begin{bmatrix} 2 \\ 1 \end{bmatrix}, \quad \text{and} \quad [T(X_3)]_{B_2} = \begin{bmatrix} 1 \\ 3 \end{bmatrix}$$

Therefore,

$$T(X_1) = 4Y_1 + 0Y_2 = \begin{bmatrix} 4 \\ 0 \end{bmatrix}, \quad T(X_2) = 2Y_1 + 1Y_2 = \begin{bmatrix} 3 \\ 1 \end{bmatrix}$$

$$T(X_3) = 1Y_1 + 3Y_2 = \begin{bmatrix} 4 \\ 3 \end{bmatrix}$$

T is unique because a transformation is completely determined by the images of a basis.

(b)

$$X = \begin{bmatrix} x \\ y \\ z \end{bmatrix} = (x - y) \begin{bmatrix} 1 \\ 0 \\ 0 \end{bmatrix} + (y - z) \begin{bmatrix} 1 \\ 1 \\ 0 \end{bmatrix} + z \begin{bmatrix} 1 \\ 1 \\ 1 \end{bmatrix}$$

Therefore,

$$[X]_{B_1} = \begin{bmatrix} x - y \\ y - z \\ z \end{bmatrix}$$

So

$$\begin{bmatrix} 4 & 2 & 1 \\ 0 & 1 & 3 \end{bmatrix} \begin{bmatrix} x - y \\ y - z \\ z \end{bmatrix} = \begin{bmatrix} 4x - 2y - z \\ y + 2z \end{bmatrix} = [T(X)]_{B_2}$$

Hence

$$T(X) = (4x - 2y - z)Y_1 + (y + 2z)Y_2 = \begin{bmatrix} 4x - y + z \\ y + 2z \end{bmatrix}$$

We can check the result by using this expression for $T(X)$ to find the images of the vectors in B_1.

$$T \begin{bmatrix} 1 \\ 0 \\ 0 \end{bmatrix} = \begin{bmatrix} 4 - 0 + 0 \\ 0 + 0 \end{bmatrix} = \begin{bmatrix} 4 \\ 0 \end{bmatrix}, \qquad T \begin{bmatrix} 1 \\ 1 \\ 0 \end{bmatrix} = \begin{bmatrix} 4 - 1 \\ 1 + 0 \end{bmatrix} = \begin{bmatrix} 3 \\ 1 \end{bmatrix},$$

$$T \begin{bmatrix} 1 \\ 1 \\ 1 \end{bmatrix} = \begin{bmatrix} 4 - 1 + 1 \\ 1 + 2 \end{bmatrix} = \begin{bmatrix} 4 \\ 3 \end{bmatrix}$$

Exercise 6.5

1. If $B_1 = \left\{ E_1 = \begin{bmatrix} 1 \\ 0 \\ 0 \end{bmatrix}, \ E_2 = \begin{bmatrix} 0 \\ 1 \\ 0 \end{bmatrix}, \ E_3 = \begin{bmatrix} 0 \\ 0 \\ 1 \end{bmatrix} \right\}$, $B_2 = \{E_3, E_2, E_1\}$, B_3
$= \{E_2, E_3, E_1\}$, and $B_4 = \{E_1, E_3, E_2\}$ are ordered bases for $\mathbf{R}_3$ and $X = \begin{bmatrix} a \\ b \\ c \end{bmatrix}$, find $[X]_{B_1}$, $[X]_{B_2}$, $[X]_{B_3}$, and $[X]_{B_4}$.

2. If $B = \{f_1 : 1, f_2 : x, f_3 : x^2\}$, $C = \{f_2, f_3, f_1\}$, and $D = \{f_3, f_1, f_2\}$ are ordered bases for $\mathbf{P}_3$ and $f : a + bx + cx^2$, find $[f]_B$, $[f]_C$, and $[f]_D$.

3. Let $B_1 = \{X_1, X_2, X_3\}$ and $B_2 = \{Y_1, Y_2, Y_3, Y_4\}$ be bases for vector spaces $\mathbf{V}$ and $\mathbf{W}$, respectively. If $T : \mathbf{V} \longrightarrow \mathbf{W}$ is a linear transformation such that $T(X_1) = 3Y_1 + 2Y_2 - 3Y_3 + 4Y_4$, $T(X_2) = 3Y_2 + 4Y_3 - Y_4$, and $T(X_3) = Y_1 - Y_2 + 2Y_3$, (a) find $A = [T]_{B_1 B_2}$; (b) use A to find $T(X)$ if $X = aX_1 + bX_2 + cX_3$.

4. Let $B_1 = \{X_1, X_2, X_3\}$ and $B_2 = \{Y_1, Y_2\}$ be bases for vector spaces $\mathbf{V}$ and $\mathbf{W}$, respectively. If $T : \mathbf{V} \longrightarrow \mathbf{W}$ is a linear transformation such that $T(X_1) = 3Y_1 - 4Y_2$, $T(X_2) = 7Y_1 + 3Y_2$, and $T(X_3) = -2Y_1 + 5Y_2$, (a) find $A = [T]_{B_1 B_2}$; (b) use A to find $T(X)$ if $X = aX_1 + bX_2 + cX_3$.

5. If $B_1 = \{f_1: x + 1, f_2: 3\}$ and $B_2 = \{g_1: x^2 + 1, g_2: x, g_3: 1\}$, and if

$T: P_2 \longrightarrow P_3$ is a linear transformation such that $[T]_{B_1 B_2} = \begin{bmatrix} 1 & 2 \\ 3 & 0 \\ 0 & -1 \end{bmatrix}$, find

(a) $T(f_1)$; (b) $T(f_2)$; (c) $T(f: a + bx)$.

6. If $T: V \longrightarrow W$ is a linear transformation, if $B_1 = \{X_1, X_2, X_3\}$, and $B_2 = \{Y_1, Y_2\}$ are bases for V and W, respectively, and if $[T]_{B_1 B_2} = \begin{bmatrix} 1 & 3 & 4 \\ 2 & 0 & -1 \end{bmatrix}$, find (a) $T(X_1)$; (b) $T(X_2)$; (c) $T(X_3)$; (d) $T(X)$, if $X = aX_1 + bX_2 + cX_3$.

In each of Problems 7–10 you are given a linear transformation T, an ordered basis B_1 for the domain, an ordered basis B_2 for the codomain, and the images of the elements of B_1. Use the method of Examples 6.19 and 6.20 to do the following:

(a) Find the matrix of the transformation relative to the given bases.
(b) Use the matrix in part (a) to find the image of the given vectors Z_1 and Z_2.

7. $T: R_3 \longrightarrow R_2$, $B_1 = \left\{ X_1 = \begin{bmatrix} 1 \\ 1 \\ 0 \end{bmatrix}, X_2 = \begin{bmatrix} 0 \\ 0 \\ \frac{1}{2} \end{bmatrix}, X_3 = \begin{bmatrix} 3 \\ 0 \\ 0 \end{bmatrix} \right\}$,

$B_2 = \left\{ Y_1 = \begin{bmatrix} 2 \\ 3 \end{bmatrix}, Y_2 = \begin{bmatrix} 1 \\ 0 \end{bmatrix} \right\}$, $T(X_1) = \begin{bmatrix} 1 \\ 2 \end{bmatrix}$, $T(X_2) = \begin{bmatrix} 2 \\ 0 \end{bmatrix}$, $T(X_3) = \begin{bmatrix} 6 \\ 4 \end{bmatrix}$, $Z_1 = \begin{bmatrix} 2 \\ 4 \\ 3 \end{bmatrix}$, $Z_2 = \begin{bmatrix} 4 \\ 6 \\ 2 \end{bmatrix}$.

8. $T: R_2 \longrightarrow R_2$, $B_1 = \left\{ X_1 = \begin{bmatrix} 1 \\ 3 \end{bmatrix}, X_2 = \begin{bmatrix} 0 \\ 2 \end{bmatrix} \right\} = B_2$, $T(X_1) = \begin{bmatrix} 2 \\ 6 \end{bmatrix}$, $T(X_2) = \begin{bmatrix} 2 \\ 0 \end{bmatrix}$, $Z_1 = \begin{bmatrix} 4 \\ 2 \end{bmatrix}$, $Z_2 = \begin{bmatrix} 6 \\ 8 \end{bmatrix}$.

9. $T: R_4 \longrightarrow R_4$, $B_1 = \left\{ X_1 = \begin{bmatrix} 1 \\ 0 \\ 0 \\ 0 \end{bmatrix}, X_2 = \begin{bmatrix} 0 \\ 2 \\ 0 \\ 0 \end{bmatrix}, X_3 = \begin{bmatrix} 0 \\ 0 \\ 1 \\ 1 \end{bmatrix}, X_4 = \begin{bmatrix} 0 \\ 0 \\ 1 \\ 0 \end{bmatrix} \right\} = B_2$,

$T(X_1) = \begin{bmatrix} 2 \\ 3 \\ 4 \\ 2 \end{bmatrix}$, $T(X_2) = \begin{bmatrix} 1 \\ 4 \\ 0 \\ 6 \end{bmatrix}$, $T(X_3) = \begin{bmatrix} 0 \\ 3 \\ 2 \\ 0 \end{bmatrix}$, $T(X_4) = \begin{bmatrix} 3 \\ 0 \\ 2 \\ 1 \end{bmatrix}$, $Z_1 = \begin{bmatrix} 2 \\ 2 \\ 4 \\ 3 \end{bmatrix}$,

$Z_2 = \begin{bmatrix} 4 \\ 0 \\ 1 \\ 1 \end{bmatrix}$.

10. $T: \mathbf{R_3} \longrightarrow \mathbf{R_3}$, $B_1 = \{E_1, E_2, E_3\}$, $B_2 = \{E_2, E_3, E_1\}$, $T(E_1) = \begin{bmatrix} 1 \\ 2 \\ 3 \end{bmatrix}$, $T(E_2) =$

$\begin{bmatrix} 0 \\ 1 \\ 1 \end{bmatrix}$, $T(E_3) = \begin{bmatrix} 2 \\ 1 \\ 0 \end{bmatrix}$, $Z_1 = \begin{bmatrix} 4 \\ 6 \\ 8 \end{bmatrix}$, $Z_2 = \begin{bmatrix} 3 \\ 1 \\ 2 \end{bmatrix}$.

11. If $T: \mathbf{R_3} \longrightarrow \mathbf{R_2}$ is the linear transformation defined by $T\begin{bmatrix} x \\ y \\ z \end{bmatrix} = \begin{bmatrix} 3x + 2y \\ y - 2z \end{bmatrix}$,

if $B_1 = \left\{ X_1 = \begin{bmatrix} 1 \\ 0 \\ 1 \end{bmatrix}, X_2 = \begin{bmatrix} 1 \\ 2 \\ 1 \end{bmatrix}, X_3 = \begin{bmatrix} 0 \\ 1 \\ 2 \end{bmatrix} \right\}$, and

if $B_2 = \left\{ Y_1 = \begin{bmatrix} 1 \\ 2 \end{bmatrix}, Y_2 = \begin{bmatrix} 4 \\ 0 \end{bmatrix} \right\}$,

(a) Find $A = [T]_{B_1 B_2}$.

(b) Use $[T]_{B_1 B_2}$ to find $T(Z_1)$ and $T(Z_2)$ if $Z_1 = \begin{bmatrix} 4 \\ 6 \\ 2 \end{bmatrix}$ and $Z_2 = \begin{bmatrix} 3 \\ 0 \\ 4 \end{bmatrix}$.

(c) Find $[T]_{BB'}$ if B and B' are the natural bases for $\mathbf{R_3}$ and $\mathbf{R_2}$, respectively.

(d) Use $[T]_{BB'}$ to find $T(Z_1)$ and $T(Z_2)$ with Z_i as in part (b).

In Problems 12 and 13, B denotes the natural basis for the domain space, and B' denotes the natural basis for the codomain.

12. If $T: \mathbf{R_2} \longrightarrow \mathbf{R_3}$ and $T\begin{bmatrix} x \\ y \end{bmatrix} = \begin{bmatrix} x + y \\ x \\ 2y \end{bmatrix}$, (a) find $[T]_{BB'}$; (b) use $[T]_{BB'}$ to find

$T\begin{bmatrix} x \\ y \end{bmatrix}$.

13. If $T: \mathbf{R_3} \longrightarrow \mathbf{R_3}$ and $T\begin{bmatrix} x \\ y \\ z \end{bmatrix} = \begin{bmatrix} x + 3y \\ x - 2y + z \\ x + 4y \end{bmatrix}$, (a) find $[T]_{BB'}$; (b) use $[T]_{BB'}$ to

find $T\begin{bmatrix} 4 \\ 6 \\ 1 \end{bmatrix}$.

14. If $T: \mathbf{R_3} \longrightarrow \mathbf{R_2}$ and

$$T\begin{bmatrix} x_1 \\ x_2 \\ x_3 \end{bmatrix} = \begin{bmatrix} a_{11}x_1 + a_{12}x_2 + a_{13}x_3 \\ a_{21}x_1 + a_{22}x_2 + a_{23}x_3 \end{bmatrix}$$

use matrices to prove that T is a linear transformation.

15. If $T: \mathbf{R}_3 \longrightarrow \mathbf{R}_2$ is a linear transformation and if $T \begin{bmatrix} x_1 \\ x_2 \\ x_3 \end{bmatrix} = \begin{bmatrix} y_1 \\ y_2 \end{bmatrix}$, prove that

there are fixed scalars a_{ij}, $i = 1, 2$ and $j = 1, 2, 3$ such that

$$y_1 = a_{11}x_1 + a_{12}x_2 + a_{13}x_3$$
$$y_2 = a_{21}x_1 + a_{22}x_2 + a_{23}x_3$$

Hint: Let $T(E_i) = a_{1i}E_1 + a_{2i}E_2$ for $i = 1, 2, 3$. (This result can be extended to a linear transformation $T: \mathbf{R}_n \longrightarrow \mathbf{R}_m$.)

16. If $\mathbf{V}$ is an n-dimensional vector space, and if $B = \{X_1, X_2, \ldots, X_n\}$ is a basis for $\mathbf{V}$, show that the mapping $T: \mathbf{V} \longrightarrow \mathbf{R}_n$ defined by $T(X) = [X]_B$ is an isomorphism.

6.6 SUM, PRODUCT, AND SCALAR MULTIPLES OF LINEAR TRANSFORMATIONS

In this section we define an addition and a multiplication of linear transformations and multiplication of a linear transformation by a scalar. The operations are defined in such a way that they reflect the corresponding operations on matrices. We begin with a short discussion of the algebra of functions.

The reader has probably had previous experience with real-valued functions of a real variable, that is, functions with domain a subset of the real numbers and codomain the set of real numbers. An example is the function defined by $f(x) = \sqrt{x}$. The domain of a function defined by a rule such as $y = \sqrt{x}$ is understood to be all the real-number replacements for x for which $f(x)$ is a real number. For the above f, the domain is the set of all nonnegative real numbers. As in Example 5.1(e), if $f: D_f \longrightarrow \mathbf{R}$ and $g: D_g \longrightarrow \mathbf{R}$, then $f + g$ is the function defined by $f + g: D_f \cap D_g \longrightarrow \mathbf{R}$, where $(f + g)(x) = f(x) + g(x)$. Multiplication of a function $f: D_f \longrightarrow \mathbf{R}$ by a scalar r is defined to be the function $rf: D_f \longrightarrow \mathbf{R}$, where $(rf)(x) = r(f(x))$. A product called the composite of f and g is defined by $(f \circ g)(x) = f(g(x))$. The domain of this function is the set of all numbers x in the domain of g such that $g(x)$ is in the domain of f. We shall write $(f + g)(x)$ as $f + g(x)$ (read "the value of $f + g$ at x" or "$f + g$ of x"), and $(f \circ g)(x)$ as $f \circ g(x)$ (read "the value of f circle g at x").

Example 6.22. If functions f and g are defined by $f(x) = \sqrt{x}$ and $g(x) = x - 5$, describe the functions $f + g$, rf, $f \circ g$, and $g \circ f$.

Solution: $f + g$ is the function defined by $f + g(x) = f(x) + g(x) = \sqrt{x} + x - 5$. The domain of $f + g$ is the set of numbers that belong to both the domain of f and the domain of g, which is $[0, +\infty)$. The function rf is defined by $rf(x) = r(f(x)) = r\sqrt{x}$. The domain of rf is the domain of f, which is $[0, +\infty)$. $f \circ g$ is the function defined by $f \circ g(x) = f(g(x)) = f(x - 5) = \sqrt{x - 5}$. The domain of $f \circ g$ is the set of all real numbers x such that $x - 5$ is in the domain of f, which is $[5, +\infty)$. $g \circ f(x) = g(f(x)) = g(\sqrt{x}) = \sqrt{x} - 5$. The domain of $g \circ f$ is $[0, +\infty)$. Note that $f \circ g \neq g \circ f$.

The above definitions of $f + g$ and rf depend on the fact that we can add real numbers and multiply a real number by a scalar. We are led to define addition of linear transformations and multiplication of a linear transformation by a scalar in a similar way. After all, we can add two vectors in the same space, and we can multiply a vector by a scalar.

Definition 6.8.   Let $T: \mathbf{V} \longrightarrow \mathbf{W}$ and $S: \mathbf{V} \longrightarrow \mathbf{W}$ be linear transformations. The **sum** $T + S$ is the transformation $T + S: \mathbf{V} \longrightarrow \mathbf{W}$ defined by $T + S(X) = T(X) + S(X)$ for every X in $\mathbf{V}$. If r is a scalar, the **scalar multiple** rT is the transformation $rT: \mathbf{V} \longrightarrow \mathbf{W}$ defined by $rT(X) = r(T(X))$ for all X in $\mathbf{V}$.

The reader should note that in the equation $T + S(X) = T(X) + S(X)$ the addition on the left is the addition of transformations being defined, and the addition on the right is addition of vectors in $\mathbf{W}$. Similarly, in the equation $rT(X) = r(T(X))$ the symbol rT on the left denotes the new function being defined, and $r(T(X))$ is scalar multiplication in $\mathbf{W}$. In Exercise 6.6, Problem 8, the reader is asked to verify that $T + S$ and rT are *linear* transformations.

If T and S are linear transformations with corresponding matrices A and B relative to a pair of bases, we shall show that the matrix of $T + S$ is $A + B$. We also show that the matrix of rT is rA. Before proving the result we illustrate it with an example.

Example 6.23.   Let $T: \mathbf{R}_4 \longrightarrow \mathbf{R}_3$ and $S: \mathbf{R}_4 \longrightarrow \mathbf{R}_3$ be the linear transformations defined by

$$
T(X) = \begin{bmatrix} x + 2y \\ 3y + 4z \\ -2x + 5w \end{bmatrix} \quad \text{and} \quad S(X) = \begin{bmatrix} 2x + y + z + w \\ y + 2z + w \\ 2x - 3y + 4z \end{bmatrix}
$$

where $X = \begin{bmatrix} x \\ y \\ z \\ w \end{bmatrix}$. For the bases,

$$
C_1 = \left\{ X_1 = \begin{bmatrix} 1 \\ 0 \\ 0 \\ 0 \end{bmatrix}, \quad X_2 = \begin{bmatrix} 1 \\ 1 \\ 0 \\ 0 \end{bmatrix}, \quad X_3 = \begin{bmatrix} 1 \\ 1 \\ 1 \\ 0 \end{bmatrix}, \quad X_4 = \begin{bmatrix} 1 \\ 1 \\ 1 \\ 1 \end{bmatrix} \right\}
$$

and

$$C_2 = \left\{ Y_1 = \begin{bmatrix} 1 \\ 0 \\ 0 \end{bmatrix}, \quad Y_2 = \begin{bmatrix} 0 \\ 1 \\ 0 \end{bmatrix}, \quad Y_3 = \begin{bmatrix} 0 \\ 1 \\ 1 \end{bmatrix} \right\}$$

(a) Find $A = [T]_{C_1 C_2}$ and $B = [S]_{C_1 C_2}$.

(b) Find $[T + S]_{C_1 C_2}$ and show that $[T + S]_{C_1 C_2} = A + B$.

(c) Find $[rT]_{C_1 C_2}$ and show that $[rT]_{C_1 C_2} = rA$.

Solution: (a) The columns of A are the coordinate vectors of $T(X_1)$, $T(X_2)$, $T(X_3)$, and $T(X_4)$ relative to the C_2 basis. From the definition of T, we have

$$T(X_1) = \begin{bmatrix} 1 \\ 0 \\ -2 \end{bmatrix} = Y_1 + 2Y_2 - 2Y_3$$

$$T(X_2) = \begin{bmatrix} 3 \\ 3 \\ -2 \end{bmatrix} = 3Y_1 + 5Y_2 - 2Y_3$$

$$T(X_3) = \begin{bmatrix} 3 \\ 7 \\ -2 \end{bmatrix} = 3Y_1 + 9Y_2 - 2Y_3$$

$$T(X_4) = \begin{bmatrix} 3 \\ 7 \\ 3 \end{bmatrix} = 3Y_1 + 4Y_2 + 3Y_3$$

So

$$A = \begin{bmatrix} 1 & 3 & 3 & 3 \\ 2 & 5 & 9 & 4 \\ -2 & -2 & -2 & 3 \end{bmatrix}$$

Similarly, the columns of B are obtained by finding

$$S(X_1) = \begin{bmatrix} 2 \\ 0 \\ 2 \end{bmatrix} = 2Y_1 - 2Y_2 + 2Y_3$$

$$S(X_2) = \begin{bmatrix} 3 \\ 1 \\ -1 \end{bmatrix} = 3Y_1 + 2Y_2 - Y_3$$

$$S(X_3) = \begin{bmatrix} 4 \\ 3 \\ 3 \end{bmatrix} = 4Y_1 + 0Y_2 + 3Y_3$$

$$S(X_4) = \begin{bmatrix} 5 \\ 4 \\ 3 \end{bmatrix} = 5Y_1 + 1Y_2 + 3Y_3$$

We see that

$$B = \begin{bmatrix} 2 & 3 & 4 & 5 \\ -2 & 2 & 0 & 1 \\ 2 & -1 & 3 & 3 \end{bmatrix}$$

(b) To find $[T + S]_{C_1 C_2}$, we need the coordinate vectors of $T + S(X_1)$, $T + S(X_2)$, $T + S(X_3)$, and $T + S(X_4)$ relative to the C_2 basis. But $T + S(X_i) = T(X_i) + S(X_i)$, and this implies that

$$[T + S(X_i)]_{C_2} = [T(X_i)]_{C_2} + [S(X_i)]_{C_2}$$

Thus each column of $[T + S]_{C_1 C_2}$ is the sum of the corresponding columns of $[T]_{C_1 C_2}$ and $[S]_{C_1 C_2}$, and

$$[T + S]_{C_1 C_2} = \begin{bmatrix} 1+2 & 3+3 & 3+4 & 3+5 \\ 2-2 & 5+2 & 9+0 & 4+1 \\ -2+2 & -2-1 & -2+3 & 3+3 \end{bmatrix}$$

$$= \begin{bmatrix} 3 & 6 & 7 & 8 \\ 0 & 7 & 9 & 5 \\ 0 & -3 & 1 & 6 \end{bmatrix} = A + B$$

(c) To find the matrix of rT, we must find the coordinate vectors of $rT(X_i)$ relative to the C_2 basis. But $rT(X_i) = r(T(X_i))$ implies that $[rT(X_i)]_{C_2} = r[T(X_i)]_{C_2}$. Therefore, each column of $[rT]_{C_1 C_2}$ is r times the corresponding column of $[T]_{C_1 C_2}$. Hence

$$[rT]_{C_1 C_2} = \begin{bmatrix} 1r & 3r & 3r & 3r \\ 2r & 5r & 9r & 4r \\ -2r & -2r & -2r & 3r \end{bmatrix} = rA$$

Theorem 6.11. Let $T: V \longrightarrow W$ and $S: V \longrightarrow W$ be linear transformations from an n-dimensional vector space V to an m-dimensional vector space W, and let $C_1 = \{X_1, X_2, \ldots, X_n\}$ and $C_2 = \{Y_1, Y_2, \ldots, Y_m\}$ be bases for V and W, respectively. If $[T]_{C_1 C_2} = A = [a_{ij}]_{(m,n)}$ and $[S]_{C_1 C_2} = B = [b_{ij}]_{(m,n)}$, and r is a scalar, then $[T + S]_{C_1 C_2} = A + B$ and $[rT]_{C_1 C_2} = rA$.

Proof: The coordinate vector of $T(X_j)$ relative to the basis C_2 is

$$A^{(j)} = \begin{bmatrix} a_{1j} \\ a_{2j} \\ \cdot \\ \cdot \\ \cdot \\ a_{mj} \end{bmatrix}. \text{ Also, } B^{(j)} = \begin{bmatrix} b_{1j} \\ b_{2j} \\ \cdot \\ \cdot \\ \cdot \\ b_{mj} \end{bmatrix} \text{ is the coordinate vector of } S(X_j) \text{ relative to}$$

the basis C_2. So

$$T + S(X_j) = T(X_j) + S(X_j)$$
$$= (a_{1j}Y_1 + a_{2j}Y_2 + \cdots + a_{mj}Y_m)$$
$$+ (b_{1j}Y_1 + b_{2j}Y_2 + \cdots + b_{mj}Y_m)$$
$$= (a_{1j} + b_{1j})Y_1 + (a_{2j} + b_{2j})Y_2 + \cdots + (a_{mj} + b_{mj})Y_m$$

Therefore, the coordinate vector of $T + S(X_j)$ is

$$\begin{bmatrix} a_{1j} + b_{1j} \\ a_{2j} + b_{2j} \\ \cdot \\ \cdot \\ \cdot \\ a_{mj} + b_{mj} \end{bmatrix} = A^{(j)} + B^{(j)} = \text{the } j\text{th column of } A + B$$

Since this is true for $j = 1, 2, \ldots, n$, we have that the matrix of $T + S$ relative to the bases C_1 and C_2 is $A + B$. The reader is asked to prove that the matrix of rT is rA in Problem 9 of Exercise 6.6. Q.E.D.

Example 6.24. Let $T: \mathbf{P}_3 \rightarrow \mathbf{P}_2$ be the linear transformation defined by $T(f: a + bx + cx^2) = g: (a - 3c) + (b + 2a)x$, and let $S: \mathbf{P}_3 \rightarrow \mathbf{P}_2$ be the linear transformation defined by $S(f: a + bx + cx^2) = h: (2a + b) + cx$. If $C_1 = \{f_1: 1, f_2: x, f_3: x^2\}$, and $C_2 = \{g_1: 1, g_2: x\}$ are bases for $\mathbf{P}_3$ and $\mathbf{P}_2$, respectively,

(a) Find $[T]_{C_1C_2}$.
(b) Find $[S]_{C_1C_2}$.
(c) Find $[T + S]_{C_1C_2}$ and show that $[T + S]_{C_1C_2} = [T]_{C_1C_2} + [S]_{C_1C_2}$.
(d) Find $[rT]_{C_1C_2}$ and show that $[rT]_{C_1C_2} = r[T]_{C_1C_2}$.
(e) Use $[T + S]_{C_1C_2}$ to calculate $(T + S)(f: a + bx + cx^2)$.

Solution

(a) $[T]_{C_1C_2} = [[T(f_1)]_{c_2} | [T(f_2)]_{c_2} | [T(f_3)]_{c_2}] = \begin{bmatrix} 1 & 0 & -3 \\ 2 & 1 & 0 \end{bmatrix}$.

(b) $[S]_{C_1C_2} = [[S(f_1)]_{c_2} | [S(f_2)]_{c_2} | [S(f_3)]_{c_2}] = \begin{bmatrix} 2 & 1 & 0 \\ 0 & 0 & 1 \end{bmatrix}$.

(c) $[T + S]_{C_1C_2} = [[T + S(f_1)]_{c_2} | [T + S(f_2)]_{c_2} | [T + S(f_3)]_{c_2}]$
$$= \begin{bmatrix} 3 & 1 & -3 \\ 2 & 1 & 1 \end{bmatrix} = \begin{bmatrix} 1 & 0 & -3 \\ 2 & 1 & 0 \end{bmatrix} + \begin{bmatrix} 2 & 1 & 0 \\ 0 & 0 & 1 \end{bmatrix}$$
$$= [T]_{C_1C_2} + [S]_{C_1C_2}$$

(d) $[rT]_{C_1C_2} = [[rT(f_1)]_{c_2} | [rT(f_2)]_{c_2} | [rT(f_3)]_{c_2}]$
$$= \begin{bmatrix} r & 0 & -3r \\ 2r & r & 0 \end{bmatrix} = r\begin{bmatrix} 1 & 0 & -3 \\ 2 & 1 & 0 \end{bmatrix} = r[T]_{C_1C_2}$$

(e) $\begin{bmatrix} 3 & 1 & -3 \\ 2 & 1 & 1 \end{bmatrix} \begin{bmatrix} a \\ b \\ c \end{bmatrix} = \begin{bmatrix} 3a + b - 3c \\ 2a + b + c \end{bmatrix}.$

So

$$T + S(f: a + bx + cx^2) = k: (3a + b - 3c) + (2a + b + c)x$$
$$= T(f) + S(f)$$

Next we define a product or composite of linear transformations that corresponds to the definition of the composite $f \circ g$ for real-valued functions of a real variable. If T and S are linear transformations and if $T(S(X))$ is to make sense, then X would have to be a vector in the domain of S, and $S(X)$ would have to be a vector in the domain of T. To assure that this is true, we shall define $T \circ S$ only for linear transformations for which the codomain of S is equal to the domain of T.

Definition 6.9. If $S: \mathbf{V} \longrightarrow \mathbf{W}$ and $T: \mathbf{W} \longrightarrow \mathbf{U}$ are linear transformations, we define **the composite** $T \circ S$ to be the transformation $T \circ S: \mathbf{V} \longrightarrow \mathbf{U}$, where $(T \circ S)(X) = T(S(X))$ for all X in $\mathbf{V}$. We will often drop the parentheses and write $T \circ S(X)$ instead of $(T \circ S)(X)$.

Theorem 6.12. If $S: \mathbf{V} \longrightarrow \mathbf{W}$ and $T: \mathbf{W} \longrightarrow \mathbf{U}$ are linear transformations, then $T \circ S: \mathbf{V} \longrightarrow \mathbf{U}$ is a *linear* transformation.

Proof: We must show that

$$T \circ S(X + Y) = T \circ S(X) + T \circ S(Y)$$

and

$$T \circ S(rX) = r(T \circ S(X))$$

To this end, let X and Y be vectors in $\mathbf{V}$ and let r be a scalar. Then

$$
\begin{aligned}
T \circ S(X + Y) &= T(S(X + Y)) && \text{(definition of } T \circ S) \\
&= T(S(X) + S(Y)) && (S \text{ is linear}) \\
&= T(S(X)) + T(S(Y)) && (T \text{ is linear}) \\
&= T \circ S(X) + T \circ S(Y) && \text{(definition of } T \circ S)
\end{aligned}
$$

Therefore,

$$T \circ S(X + Y) = T \circ S(X) + T \circ S(Y)$$

Also,

$$
\begin{aligned}
T \circ S(rX) &= T(S(rX)) && \text{(definition of } T \circ S) \\
&= T(rS(X)) && (S \text{ is linear}) \\
&= r(T(S(X))) && (T \text{ is linear}) \\
&= r(T \circ S(X)) && \text{(definition of } T \circ S)
\end{aligned}
$$

Thus $T \circ S(rX) = r(T \circ S(X))$. Q.E.D.

Example 6.25. If $S: \mathbf{R}_4 \to \mathbf{R}_3$ and $T: \mathbf{R}_3 \to \mathbf{R}_2$ are linear transformations defined by

$$S\begin{bmatrix} x \\ y \\ z \\ w \end{bmatrix} = \begin{bmatrix} x + 2y \\ x - z \\ w + 2z \end{bmatrix} \text{ and } T\begin{bmatrix} x \\ y \\ z \end{bmatrix} = \begin{bmatrix} 2x + y \\ 3y + 4z \end{bmatrix}$$

then $T \circ S: \mathbf{R}_4 \to \mathbf{R}_2$. Let C, C', and C'' be the natural bases for $\mathbf{R}_4, \mathbf{R}_3$, and $\mathbf{R}_2$, respectively.

(a) Find $T \circ S \begin{bmatrix} x \\ y \\ z \\ w \end{bmatrix}$.

(b) Find $[S]_{CC'}$, $[T]_{C'C''}$, and $[T \circ S]_{CC''}$.

(c) Use $[T \circ S]_{CC''}$ to calculate $T \circ S \begin{bmatrix} x \\ y \\ z \\ w \end{bmatrix}$.

(d) Calculate the matrix product $[T]_{C'C''}[S]_{CC'}$.

Solution

(a) $T \circ S(X) = T(S(X)) = T \begin{bmatrix} x + 2y \\ x - z \\ w + 2z \end{bmatrix} = \begin{bmatrix} 2(x + 2y) + (x - z) \\ 3(x - z) + 4(w + 2z) \end{bmatrix}$

$$= \begin{bmatrix} 3x + 4y - z \\ 3x + 5z + 4w \end{bmatrix}.$$

(b) $[S]_{CC'} = [[S(E_1)]_{C'} | [S(E_2)]_{C'} | [S(E_3)]_{C'} | [S(E_4)]_{C'}] = \begin{bmatrix} 1 & 2 & 0 & 0 \\ 1 & 0 & -1 & 0 \\ 0 & 0 & 2 & 1 \end{bmatrix}$.

$[T]_{C'C''} = [[T(E_1)]_{C''} | [T(E_2)]_{C''} | [T(E_3)]_{C''}] = \begin{bmatrix} 2 & 1 & 0 \\ 0 & 3 & 4 \end{bmatrix}$.

$[T \circ S]_{CC''} = [[T \circ S(E_1)]_{C''} | [T \circ S(E_2)]_{C''} | [T \circ S(E_3)]_{C''} | [T \circ S(E_4)]_{C''}]$

$$= \begin{bmatrix} 3 & 4 & -1 & 0 \\ 3 & 0 & 5 & 4 \end{bmatrix}.$$

(c) $\begin{bmatrix} 3 & 4 & -1 & 0 \\ 3 & 0 & 5 & 4 \end{bmatrix} \begin{bmatrix} x \\ y \\ z \\ w \end{bmatrix} = \begin{bmatrix} 3x + 4y - z \\ 3x + 5z + 4w \end{bmatrix} = [T \circ S(X)]_{C''} = T \circ S(X).$

(d) $[T]_{C'C''}[S]_{CC'} = \begin{bmatrix} 2 & 1 & 0 \\ 0 & 3 & 4 \end{bmatrix} \begin{bmatrix} 1 & 2 & 0 & 0 \\ 1 & 0 & -1 & 0 \\ 0 & 0 & 2 & 1 \end{bmatrix} = \begin{bmatrix} 3 & 4 & -1 & 0 \\ 3 & 0 & 5 & 4 \end{bmatrix}$

$\qquad = [T \circ S]_{CC''}.$

The conclusion of Example 6.25(d) is true in general as the next theorem indicates.

Theorem 6.13. Let $S: \mathbf{V} \to \mathbf{W}$ and $T: \mathbf{W} \to \mathbf{U}$ be linear transformations. Let $C_1 = \{X_1, X_2, \ldots, X_n\}$, $C_2 = \{Y_1, Y_2, \ldots, Y_m\}$, and $C_3 = \{Z_1, Z_2, \ldots, Z_r\}$ be bases for $\mathbf{V}$, $\mathbf{W}$, and $\mathbf{U}$, respectively. If $[T]_{C_2C_3} = A = [a_{ij}]_{(r,m)}$ and $[S]_{C_1C_2} = B = [b_{ij}]_{(m,n)}$, then the matrix of $T \circ S: \mathbf{V} \to \mathbf{U}$ is $[T \circ S]_{C_1C_3} = AB$.

Proof: (Optional) We must show that the jth column of AB is the coordinate vector of $T \circ S(X_j)$ relative to the basis C_3 for $j = 1, 2, \ldots, n$. Since $[b_{ij}]_{(m,n)}$ is the matrix of S relative to the bases C_1 and C_2, we have $[S(X_j)]_{C_2} = $ the jth column of $[b_{ij}]$. Therefore,

$$S(X_j) = X_j' = b_{1j}Y_1 + b_{2j}Y_2 + \cdots + b_{mj}Y_m$$

and

$$T \circ S(X_j) = T(S(X_j)) = T(X_j') = T(b_{1j}Y_1 + b_{2j}Y_2 + \cdots + b_{mj}Y_m)$$
$$= b_{1j}T(Y_1) + b_{2j}T(Y_2) + \cdots + b_{mj}T(Y_m)$$

The matrix of T relative to the bases C_2 and C_3 is $[T]_{C_2C_3} = A$. Therefore, $[T(Y_k)]_{C_3} = A^{(k)}$ and $T(Y_k) = Y_k' = a_{1k}Z_1 + a_{2k}Z_2 + \cdots + a_{rk}Z_r$ for $k = 1, 2, \ldots, m$. So

$$T \circ S(X_j) = b_{1j}Y_1' + b_{2j}Y_2' + \cdots + b_{mj}Y_m'$$
$$= b_{1j}(a_{11}Z_1 + a_{21}Z_2 + \cdots + a_{r1}Z_r)$$
$$+ b_{2j}(a_{12}Z_1 + a_{22}Z_2 + \cdots + a_{r2}Z_r)$$
$$+ \cdots + b_{mj}(a_{1m}Z_1 + a_{2m}Z_2 + \cdots + a_{rm}Z_r)$$

After writing this last expression as a linear combination of $Z_1, Z_2, \ldots, Z_r$, we see that the coefficient of Z_1 is $b_{1j}a_{11} + b_{2j}a_{12} + \cdots + b_{mj}a_{1m} = A_1^T \cdot B^{(j)}$, and in general the coefficient of Z_i is $A_i^T \cdot B^{(j)}$ for $i = 1, 2, \ldots, r$. Thus the coordinate vector of $T \circ S(X_j)$ relative to the basis C_3 is

$$\begin{bmatrix} A_1^T \cdot B^{(j)} \\ A_2^T \cdot B^{(j)} \\ \cdot \\ \cdot \\ \cdot \\ A_r^T \cdot B^{(j)} \end{bmatrix}$$

for $j = 1, 2, \ldots, n$. But this is the jth column of AB. Hence $[T \circ S]_{C_1C_3} = AB$.

Q.E.D.

Exercise 6.6

1. If $T: \mathbf{R}_4 \longrightarrow \mathbf{R}_3$ and $S: \mathbf{R}_4 \longrightarrow \mathbf{R}_3$ are linear transformations, if $X = \begin{bmatrix} x \\ y \\ z \\ w \end{bmatrix}$, and

   if T and S are defined by

$$T(X) = \begin{bmatrix} 2x + y \\ 3x + w \\ w + x + y + z \end{bmatrix} \quad \text{and} \quad S(X) = \begin{bmatrix} w - x \\ x + y \\ y + z \end{bmatrix}$$

   (a) Find $T + S(X)$.
   (b) Find $(6T)(X)$.
   (c) Find $A = [T]_{CC'}$ and $[S]_{CC'} = B$ if C and C' are the natural bases for $\mathbf{R}_4$ and $\mathbf{R}_3$, respectively,
   (d) Find $(A + B)X$ and $(6A)X$. Compare the results with parts (a) and (b).
   (e) Is $S \circ T$ or $T \circ S$ defined? Is AB or BA defined?

2. If $T: \mathbf{P}_3 \longrightarrow \mathbf{P}_2$ and $S: \mathbf{P}_3 \longrightarrow \mathbf{P}_2$ are linear transformations defined by $T(f: a + bx + cx^2) = g_f: (2a + b + c) + (a - c)x$ and $S(f: a + bx + cx^2) = h_f: (a - 2b + 4c) + (2b + a)x$, and if C_1 and C_2 are the natural bases for $\mathbf{P}_3$ and $\mathbf{P}_2$ defined in Example 6.24, find
   (a) $[T]_{C_1 C_2}$.
   (b) $[S]_{C_1 C_2}$.
   (c) $[T + S]_{C_1 C_2}$ and show that $[T + S]_{C_1 C_2} = [T]_{C_1 C_2} + [S]_{C_1 C_2}$.
   (d) $[rT]_{C_1 C_2}$.
   (e) Use $[T + S]_{C_1 C_2}$ to calculate $(T + S)(f: a + bx + cx^2)$.

3. Same as Problem 2 but with $T(f: a + bx + cx^2) = g_f: a + bx$ and $S(f: a + bx + cx^2) = h_f: c + bx$.

4. If $S: \mathbf{R}_4 \longrightarrow \mathbf{R}_3$ and $T: \mathbf{R}_3 \longrightarrow \mathbf{R}_2$ are linear transformations, if $X = \begin{bmatrix} x \\ y \\ z \\ w \end{bmatrix}$,

$$S(X) = \begin{bmatrix} 2x + y + w \\ 3x - z + w \\ 2y + 4z \end{bmatrix}, \text{ and } T\begin{bmatrix} x \\ y \\ z \end{bmatrix} = \begin{bmatrix} x + y + z \\ x + 2y \end{bmatrix}, \text{ and if } C, C', \text{ and } C'' \text{ are}$$

   the natural bases for $\mathbf{R}_4$, $\mathbf{R}_3$, and $\mathbf{R}_2$, respectively,
   (a) Find $T \circ S(X)$.
   (b) Find $B = [S]_{CC'}$, $A = [T]_{C'C''}$, and $[T \circ S]_{CC''}$.
   (c) Show that $AB = [T \circ S]_{CC''}$.

5. Repeat Problem 4 but with $S(X) = \begin{bmatrix} 2x - 3z + 2y \\ x + y + z + w \\ x + 2y + 2w \end{bmatrix}$ and $T\begin{bmatrix} x \\ y \\ z \end{bmatrix} =$

   $\begin{bmatrix} x - y + 2z \\ 2x + 3y + z \end{bmatrix}$.

6. Repeat Problem 4 but with bases C, C', and C'' replaced by

$$C_1 = \left\{ X_1 = \begin{bmatrix} 1 \\ 0 \\ 0 \\ 0 \end{bmatrix},\ X_2 = \begin{bmatrix} 1 \\ 1 \\ 0 \\ 0 \end{bmatrix},\ X_3 = \begin{bmatrix} 1 \\ 1 \\ 1 \\ 0 \end{bmatrix},\ X_4 = \begin{bmatrix} 1 \\ 1 \\ 1 \\ 1 \end{bmatrix} \right\}$$

$$C_2 = \left\{ Y_1 = \begin{bmatrix} 1 \\ 0 \\ 0 \end{bmatrix},\ Y_2 = \begin{bmatrix} 1 \\ 1 \\ 0 \end{bmatrix},\ Y_3 = \begin{bmatrix} 1 \\ 1 \\ 1 \end{bmatrix} \right\}$$

$$C_3 = \left\{ Z_1 = \begin{bmatrix} 1 \\ 0 \end{bmatrix},\ Z_2 = \begin{bmatrix} 1 \\ 1 \end{bmatrix} \right\}$$

7. Repeat Problem 5 but with bases C, C', and C'' replaced by the bases C_1, C_2, and C_3 of Problem 6.

8. If $T: V \longrightarrow W$ and $S: V \longrightarrow W$ are linear transformations,
   (a) Prove that $T + S: V \longrightarrow W$ is a linear transformation.
   (b) Prove that $rT: V \longrightarrow W$ is a linear transformation.

9. If $T: V \longrightarrow W$ is a linear transformation and if $C_1 = \{X_1, X_2, \ldots, X_n\}$ and $C_2 = \{Y_1, Y_2, \ldots, Y_m\}$ are bases for V and W, respectively, and if $A = [T]_{C_1 C_2}$, verify that $rA = [rT]_{C_1 C_2}$.

10. It is clear geometrically that a rotation $R_\alpha: \mathbf{R}_2 \longrightarrow \mathbf{R}_2$ followed by a rotation $R_\theta: \mathbf{R}_2 \longrightarrow \mathbf{R}_2$ is a rotation $R_{\theta + \alpha}: \mathbf{R}_2 \longrightarrow \mathbf{R}_2$; that is, $R_\theta \circ R_\alpha = R_{\theta + \alpha}$.
    (a) If C is the natural basis for $\mathbf{R}_2$, find $[R_\theta]_{CC}$, $[R_\alpha]_{CC}$, and $[R_{\theta + \alpha}]_{CC}$.
    (b) Assuming that $R_\theta \circ R_\alpha = R_{\theta + \alpha}$ and using Theorem 6.13, verify the addition formulas for the sine and cosine functions ($\cos(\theta + \alpha) = \cos\theta\cos\alpha - \sin\theta\sin\alpha$ and $\sin(\theta + \alpha) = \sin\theta\sin\alpha + \sin\alpha\cos\theta$).

11. Let V and W be vector spaces and let $L = L(V, W)$ be the collection of all linear transformations $T: V \longrightarrow W$. *Prove:* $(L, +, sm)$ is a vector space if addition and scalar multiplication are defined as in Definition 6.8.

12. Let V be a vector space and let $L = L(V, V)$ be the collection of all linear transformations $T: V \longrightarrow V$. *Prove:* $(L, +, sm, \circ)$ is an algebra if addition and scalar multiplication are defined as in Definition 6.8 and the product is defined as in Definition 6.9.

13. Let $\mathbf{R}_{n \times n}$ be the collection of all square matrices of order n, and let L be the collection of all linear transformations $T: V \longrightarrow V$. *Prove:* If $B = \{X_1, X_2, \ldots, X_n\}$ is a basis for V, then the mapping $\alpha: L \longrightarrow \mathbf{R}_{n \times n}$ defined by $\alpha(T) = [T]_{BB}$ is an **algebra isomorphism**; that is, α is 1–1 onto $\mathbf{R}_{n \times n}$ and
    (i) $\alpha(T + S) = \alpha(T) + \alpha(S)$.
    (ii) $\alpha(rT) = r(\alpha(T))$.
    (iii) $\alpha(T \circ S) = (\alpha(T))(\alpha(S))$.

   In words, the algebra of square matrices of order n and the algebra of linear transformations (linear operators) on a vector space V of dimension n are essentially the same (algebraically the same) except for notation.

6.7 INVERTIBLE TRANSFORMATIONS

In this section we restrict our attention to linear transformations from a (finite-dimensional) vector space **V** into itself. Such linear transformations are called *linear operators*. The matrix counterpart of a linear operator $T: \mathbf{V} \longrightarrow \mathbf{V}$ is a square matrix. Corresponding to the identity matrix for matrix multiplication is an identity transformation for linear operators. Just as some matrices have inverses, we shall see that certain linear operators have inverses.

A simple but important transformation on a vector space **V** is the transformation $I: \mathbf{V} \longrightarrow \mathbf{V}$ defined by $I(X) = X$ for all X in **V**. In Problem 1, Exercise 6.7, the reader is asked to prove that this is a *linear* transformation. I is an identity for the collection $\mathbf{L(V)} = \mathbf{L}$ of all linear operators on **V** in the sense that $T \circ I = I \circ T = T$ for all $T: \mathbf{V} \longrightarrow \mathbf{V}$ in **L** (Problem 2, Exercise 6.7). Since we have an identity for **L**, it makes sense to ask the question, for a given $T: \mathbf{V} \longrightarrow \mathbf{V}$ in **L**, is there a linear transformation $S: \mathbf{V} \longrightarrow \mathbf{V}$ in **L** such that $T \circ S = S \circ T = I$? Such a transformation, if there is one, is called an inverse of T. We make the following formal definitions.

Definition 6.10. Let **V** be an n-dimensional vector space. Let $\mathbf{L(V)} = \mathbf{L}$ denote the collection of all linear operators $T: \mathbf{V} \longrightarrow \mathbf{V}$. The linear operator $I: \mathbf{V} \longrightarrow \mathbf{V}$ defined by $I(X) = X$ for all X in **V** is called the **identity transformation** on **V**. A linear operator $T: \mathbf{V} \longrightarrow \mathbf{V}$ is said to be **invertible** or **nonsingular** if and only if there is a transformation $S: \mathbf{V} \to \mathbf{V}$ such that $T \circ S = S \circ T = I$. S is called an **inverse** for T.

The inverse of a nonsingular linear operator T is unique (Problem 4, Exercise 6.7) and is denoted by T^{-1}.

Theorem 6.14. If $T: \mathbf{V} \longrightarrow \mathbf{V}$ is a nonsingular linear transformation, then T^{-1} is a *linear* transformation.

Proof: We must show that T^{-1} satisfies properties (i) and (ii) of Definition 6.2. Let $X \in \mathbf{V}$ and $Y \in \mathbf{V}$. Since $T^{-1}: \mathbf{V} \longrightarrow \mathbf{V}$ has domain **V**, there are vectors X' and Y' such that $X' = T^{-1}(X)$ and $Y' = T^{-1}(Y)$. Applying T to X' and Y', we get

$$T(X') = T(T^{-1}(X)) = T \circ T^{-1}(X) = I(X) = X$$

and

$$T(Y') = T(T^{-1}(Y)) = T \circ T^{-1}(Y) = I(Y) = Y$$

so

$$
\begin{aligned}
T^{-1}(X + Y) &= T^{-1}(T(X') + T(Y')) \\
&= T^{-1}(T(X' + Y')) \quad (T \text{ is } linear) \\
&= T^{-1} \circ T(X' + Y') \quad (\text{definition of } T^{-1} \circ T) \\
&= I(X' + Y') \quad (T^{-1} \circ T = I) \\
&= X' + Y' \quad (\text{definition of } I) \\
&= T^{-1}(X) + T^{-1}(Y)
\end{aligned}
$$

.

Therefore, $T^{-1}(X + Y) = T^{-1}(X) + T^{-1}(Y)$.

Now suppose $X \in \mathbf{V}$ and r is a scalar. As above, let $X' = T^{-1}(X)$. Then $T(X') = X$ and

$$T^{-1}(rX) = T^{-1}(rT(X')) \qquad \text{(substitution)}$$
$$= T^{-1}(T(rX')) \qquad \text{(T is linear)}$$
$$= T^{-1} \circ T(rX') \qquad \text{(definition of $T^{-1} \circ T$)}$$
$$= I(rX') = rX' = rT^{-1}(X)$$

Hence $T^{-1}(rX) = rT^{-1}(X)$, and T^{-1} is a linear transformation. Q.E.D.

We have already seen that there is a close relationship between a linear transformation $T: \mathbf{V} \longrightarrow \mathbf{W}$ and the matrix $[T]_{B_1 B_2} = A$. The matrix A depends on both a basis B_1 for $\mathbf{V}$ and a basis B_2 for $\mathbf{W}$. In working with a linear operator $T: \mathbf{V} \longrightarrow \mathbf{V}$, it is often desirable to use the same ordered basis for the codomain as that chosen for the domain. In this case $B_1 = B_2$ and we write $[T]_{B_1}$ in place of $[T]_{B_1 B_1}$. If $A = [T]_{B_1}$, we say that A is the *matrix of T relative to the basis B_1* or A is the *matrix representation of T relative to the basis B_1*.

The reader knows that not all (nonzero) square matrices have inverses. Because of the close connection between a transformation and its matrix, one might suspect that not all linear operators have inverses. In fact, a linear operator will have an inverse if and only if the matrix of the linear operator relative to any given basis has an inverse. We illustrate this fact with the following example.

Example 6.26.   Determine if the following linear transformations have inverses:

(a) $T: \mathbf{R}_3 \longrightarrow \mathbf{R}_3$ defined by $T \begin{bmatrix} x \\ y \\ z \end{bmatrix} = \begin{bmatrix} 2x \\ x + 2y \\ x + 3z \end{bmatrix}$.

(b) $T: \mathbf{R}_3 \longrightarrow \mathbf{R}_3$ defined by $T \begin{bmatrix} x \\ y \\ z \end{bmatrix} = \begin{bmatrix} 2x + 2y \\ x + y \\ x + y + z \end{bmatrix}$.

Solution: (a) From the definition of T,

$$T \begin{bmatrix} 1 \\ 0 \\ 0 \end{bmatrix} = \begin{bmatrix} 2 \\ 1 \\ 1 \end{bmatrix}, \quad T \begin{bmatrix} 0 \\ 1 \\ 0 \end{bmatrix} = \begin{bmatrix} 0 \\ 2 \\ 0 \end{bmatrix}, \quad T \begin{bmatrix} 0 \\ 0 \\ 1 \end{bmatrix} = \begin{bmatrix} 0 \\ 0 \\ 3 \end{bmatrix}$$

Therefore, the matrix of T relative to the standard basis for $\mathbf{R}_3$ is $A = \begin{bmatrix} 2 & 0 & 0 \\ 1 & 2 & 0 \\ 1 & 0 & 3 \end{bmatrix}$. Next we determine the inverse of A, if it has one.

$$\begin{bmatrix} 2 & 0 & 0 & \vdots & 1 & 0 & 0 \\ 1 & 2 & 0 & \vdots & 0 & 1 & 0 \\ 1 & 0 & 3 & \vdots & 0 & 0 & 1 \end{bmatrix} \xrightarrow[\substack{-R_1+R_2 \\ -R_1+R_3}]{(-1/2)R_1} \begin{bmatrix} 1 & 0 & 0 & \vdots & \frac{1}{2} & 0 & 0 \\ 0 & 2 & 0 & \vdots & -\frac{1}{2} & 1 & 0 \\ 0 & 0 & 3 & \vdots & -\frac{1}{2} & 0 & 1 \end{bmatrix}$$

$$\xrightarrow[\substack{(1/3)R_3}]{(1/2)R_2} \begin{bmatrix} 1 & 0 & 0 & \vdots & \frac{1}{2} & 0 & 0 \\ 0 & 1 & 0 & \vdots & -\frac{1}{4} & \frac{1}{2} & 0 \\ 0 & 0 & 1 & \vdots & -\frac{1}{6} & 0 & \frac{1}{3} \end{bmatrix}$$

So

$$A^{-1} = \begin{bmatrix} \frac{1}{2} & 0 & 0 \\ -\frac{1}{4} & \frac{1}{2} & 0 \\ -\frac{1}{6} & 0 & \frac{1}{3} \end{bmatrix}$$

Let $S: \mathbf{R}_3 \longrightarrow \mathbf{R}_3$ be the linear operator whose matrix relative to the standard (natural) basis B for $\mathbf{R}_3$ is A^{-1}. Then

$$S\begin{bmatrix} x \\ y \\ z \end{bmatrix} = A^{-1}\begin{bmatrix} x \\ y \\ z \end{bmatrix} = \begin{bmatrix} \frac{1}{2}x \\ -\frac{1}{4}x + \frac{1}{2}y \\ -\frac{1}{6}x + \frac{1}{3}z \end{bmatrix}$$

By Theorem 6.13,

$$[S \circ T]_B = [S]_B[T]_B = A^{-1}A = I_3$$

Therefore, $S \circ T(X) = I_3 X = X = I(X)$. Also, $[T \circ S]_B = [T]_B[S]_B = AA^{-1} = I_3$. So $(T \circ S)(X) = I_3 X = X = I(X)$. Hence S is the inverse of T. We can also check directly that $T \circ S = S \circ T = I$.

$$(T \circ S)(X) = T(S(X)) = T\begin{bmatrix} \frac{1}{2}x \\ -\frac{1}{4}x + \frac{1}{2}y \\ -\frac{1}{6}x + \frac{1}{3}z \end{bmatrix} = \begin{bmatrix} 2(\frac{1}{2}x) \\ \frac{1}{2}x + 2(-\frac{1}{4}x + \frac{1}{2}y) \\ \frac{1}{2}x + 3(-\frac{1}{6}x + \frac{1}{3}z) \end{bmatrix} = \begin{bmatrix} x \\ y \\ z \end{bmatrix}$$

$$(S \circ T)(X) = S(T(X)) = S\begin{bmatrix} 2x \\ x + 2y \\ x + 3z \end{bmatrix} = \begin{bmatrix} \frac{1}{2}(2x) \\ -\frac{1}{4}(2x) + \frac{1}{2}(x + 2y) \\ -\frac{1}{6}(2x) + \frac{1}{3}(x + 3z) \end{bmatrix} = \begin{bmatrix} x \\ y \\ z \end{bmatrix}$$

(b) From the definition of T, we get

$$T\begin{bmatrix} 1 \\ 0 \\ 0 \end{bmatrix} = \begin{bmatrix} 2 \\ 1 \\ 1 \end{bmatrix}, \quad T\begin{bmatrix} 0 \\ 1 \\ 0 \end{bmatrix} = \begin{bmatrix} 2 \\ 1 \\ 1 \end{bmatrix}, \quad T\begin{bmatrix} 0 \\ 0 \\ 1 \end{bmatrix} = \begin{bmatrix} 0 \\ 0 \\ 1 \end{bmatrix}$$

Thus the matrix of T relative to the standard basis B for $\mathbf{R}_3$ is $A = \begin{bmatrix} 2 & 2 & 0 \\ 1 & 1 & 0 \\ 1 & 1 & 1 \end{bmatrix}$. To determine A^{-1}, if it exists, we reduce $[A \mid I_3]$.

$$\begin{bmatrix} 2 & 2 & 0 & \vdots & 1 & 0 & 0 \\ 1 & 1 & 0 & \vdots & 0 & 1 & 0 \\ 1 & 1 & 1 & \vdots & 0 & 0 & 1 \end{bmatrix} \xrightarrow[-R_1 + R_2]{(1/2)R_1} \begin{bmatrix} 1 & 1 & 0 & \vdots & \frac{1}{2} & 0 & 0 \\ 0 & 0 & 0 & \vdots & -\frac{1}{2} & 1 & 0 \\ 1 & 1 & 1 & \vdots & 0 & 0 & 1 \end{bmatrix}$$

We need not go further in the reduction process because the row of zeros in the 3×3 matrix on the left of the dotted line tells us that A has no inverse. Now if T were nonsingular, we would have $[T \circ T^{-1}]_B = [T]_B [T^{-1}]_B = A[T^{-1}]_B$. Also, $[T \circ T^{-1}]_B = [I]_B = I_3$. So $A[T^{-1}]_B = I_3$. But this is impossible, because A has no inverse. Hence T is not invertible.

The proof of the next theorem follows the solution to Example 6.26.

Theorem 6.15. Let $T: V \longrightarrow V$ be a linear operator on a vector space V with a basis $B = \{X_1, X_2, \ldots, X_n\}$. T is nonsingular if and only if the matrix $[T]_B$ is nonsingular. Furthermore, $[T^{-1}]_B = [T]_B^{-1}$ whenever T^{-1} exists.

Proof: Let $[T]_B = A$, and suppose that A is nonsingular. Let $S: V \longrightarrow V$ be the linear transformation such that $[S]_B = A^{-1}$. Then

$$[T \circ S]_B = [T]_B [S]_B = AA^{-1} = I_n$$

and

$$[S \circ T]_B = [S]_B [T]_B = A^{-1}A = I_n$$

Therefore, the matrix product

$$I_n[X]_B = [X]_B = [S \circ T]_B [X]_B = [T \circ S]_B [X]_B$$

Hence

$$[S \circ T(X)]_B = [T \circ S(X)]_B = [X]_B$$

and

$$S \circ T(X) = T \circ S(X) = X$$

Thus $S = T^{-1}$ and T is nonsingular.

Conversely, suppose that T is nonsingular. Then $T \circ T^{-1} = I$ and $[T \circ T^{-1}]_B = [I]_B = I_n$ (see Problem 5, Exercise 6.7). Also, $[T \circ T^{-1}]_B = [T]_B [T^{-1}]_B$. Therefore, $[T]_B [T^{-1}]_B = I_n$. But this implies that $[T]_B^{-1}$ exists, and $[T]_B^{-1} = [T^{-1}]_B$.   Q.E.D.

Example 6.27. Show that the linear operator $T: R_3 \longrightarrow R_3$ defined by

$$T \begin{bmatrix} x \\ y \\ z \end{bmatrix} = \begin{bmatrix} 2x \\ x + 2y \\ x + 3z \end{bmatrix}$$ [Example 6.26(a)] is 1–1 and onto R_3. Use this informa-

tion to define T^{-1}.

Solution: To determine if T is 1–1 and onto R_3, we solve the vector

equation $T(X') = \begin{bmatrix} x \\ y \\ z \end{bmatrix} = X$ for X'. If there is a solution $X' = \begin{bmatrix} x' \\ y' \\ z' \end{bmatrix}$, then T

is *onto* R_3. If there is a *unique* solution then T is 1–1. The vector equation

$$T(X') = \begin{bmatrix} 2x' \\ x' + 2y' \\ x' + 3z' \end{bmatrix} = \begin{bmatrix} x \\ y \\ z \end{bmatrix}$$

is equivalent to

$$2x' = x$$
$$x' + 2y' = y$$
$$x' + 3z' = z$$

The augmented matrix $\begin{bmatrix} 2 & 0 & 0 & | & x \\ 1 & 2 & 0 & | & y \\ 1 & 0 & 3 & | & z \end{bmatrix}$ of this system reduces to

$$\begin{bmatrix} 1 & 0 & 0 & | & \frac{1}{2}x \\ 0 & 1 & 0 & | & -\frac{1}{4}x + \frac{1}{2}y \\ 0 & 0 & 1 & | & -\frac{1}{6}x + \frac{1}{3}z \end{bmatrix}$$

Therefore, the *unique solution* is

$$\begin{bmatrix} x' \\ y' \\ z' \end{bmatrix} = \begin{bmatrix} \frac{1}{2}x \\ -\frac{1}{4}x + \frac{1}{2}y \\ -\frac{1}{6}x + \frac{1}{3}z \end{bmatrix}$$

So T is 1–1 and onto $\mathbf{R_3}$. Now define $S: \mathbf{R_3} \longrightarrow \mathbf{R_3}$ by

$$S\begin{bmatrix} x \\ y \\ z \end{bmatrix} = X' = \begin{bmatrix} \frac{1}{2}x \\ -\frac{1}{4}x + \frac{1}{2}y \\ -\frac{1}{6}x + \frac{1}{3}z \end{bmatrix}$$

Then

$$T \circ S \begin{bmatrix} x \\ y \\ z \end{bmatrix} = T\begin{bmatrix} \frac{1}{2}x \\ -\frac{1}{4}x + \frac{1}{2}y \\ -\frac{1}{6}x + \frac{1}{3}z \end{bmatrix} = \begin{bmatrix} 2(\frac{1}{2}x) \\ (\frac{1}{2}x) + 2(-\frac{1}{4}x + \frac{1}{2}y) \\ (\frac{1}{2}x) + 3(-\frac{1}{6}x + \frac{1}{3}z) \end{bmatrix} = \begin{bmatrix} x \\ y \\ z \end{bmatrix}$$

Also

$$S \circ T \begin{bmatrix} x \\ y \\ z \end{bmatrix} = S\begin{bmatrix} 2x \\ x + 2y \\ x + 3z \end{bmatrix} = \begin{bmatrix} \frac{1}{2}(2x) \\ -\frac{1}{4}(2x) + \frac{1}{2}(x + 2y) \\ -\frac{1}{6}(2x) + \frac{1}{3}(x + 3z) \end{bmatrix} = \begin{bmatrix} x \\ y \\ z \end{bmatrix}$$

Therefore, $S = T^{-1}$.

The next theorem is a generalization of Example 6.27.

Theorem 6.16. Let $\mathbf{V}$ be a finite-dimensional vector space of dimension n. A linear operator $T: \mathbf{V} \longrightarrow \mathbf{V}$ is nonsingular if and only if T is 1–1 and onto $\mathbf{V}$.

Proof: (Optional) Suppose that T is 1–1 onto **V**. We define $S: \mathbf{V} \longrightarrow \mathbf{V}$ as follows. For each vector X in **V**, $S(X) = X'$, where X' is the unique vector in **V** such that $T(X') = X$. Note that X' exists because T is onto **V**, and it is unique because T is 1–1. We shall show that $S = T^{-1}$. For any vector X in **V**,

$$(T \circ S)(X) = T(S(X)) = T(X') = X = I(X)$$

So $(T \circ S)(X) = I(X)$ and $T \circ S = I$. Now consider $(S \circ T)(X) = S(T(X))$. If we let $Y = T(X)$, then by the definition of S, $S(Y)$ is the *unique* vector Y' such that $T(Y') = Y$. But $T(X)$ is Y. Therefore, $Y' = X$ and $S(T(X)) = S(T(Y')) = S(Y) = Y' = X$. Thus $(S \circ T)(X) = X = I(X)$ and $S \circ T = I$. Hence $S = T^{-1}$.

Conversely, suppose that T is nonsingular. Then T^{-1} exists, and $T \circ T^{-1} = T^{-1} \circ T = I$. We first show that T must be onto **V**. For any vector Y in **V**,

$$Y = I(Y) = T \circ T^{-1}(Y) = T(T^{-1}(Y))$$

If we set $X = T^{-1}(Y)$, then X is a vector in **V** such that $T(X) = Y$. Thus T is onto **V**.

Now suppose that X_1 and X_2 are vectors in **V** such that $T(X_1) = T(X_2)$. Since T^{-1} is a *transformation* with domain **V**, we have $T^{-1}(T(X_1)) = T^{-1}(T(X_2))$. But $T^{-1}(T(X_1)) = T^{-1} \circ T(X_1) = X_1$, and $T^{-1}(T(X_2)) = T^{-1} \circ T(X_2) = X_2$. Therefore, $X_1 = X_2$ and T is 1–1.   Q.E.D.

Corollary 6.2.   Let **V** be a finite-dimensional vector space of dimension n and let $T: \mathbf{V} \longrightarrow \mathbf{V}$ be a linear operator on **V**.
 (i) T is nonsingular if and only if T is onto **V**.
 (ii) T is nonsingular if and only if the rank of T is n.
(iii) T is nonsingular if and only if T is 1–1.

Conclusions (i) and (iii) follow from Theorem 6.16 and Problems 16 and 17 of Exercise 6.4. Conclusion (ii) follows from conclusion (i) and the definition of the rank of a linear transformation.

Exercise 6.7

1. If **V** is a vector space and if $I: \mathbf{V} \longrightarrow \mathbf{V}$ is defined by $I(X) = X$ for each X in **V**, prove that I is a *linear* transformation.
2. Prove that $T \circ I = I \circ T = T$ for all T in $\mathbf{L}(\mathbf{V})$.
3. Prove that if $I': \mathbf{V} \longrightarrow \mathbf{V}$ is a linear transformation having the property that $I' \circ T = T \circ I' = T$ for all T in $\mathbf{L}(\mathbf{V})$, then $I' = I$.
4. If T, S_1, and S_2 are in $\mathbf{L}(\mathbf{V})$, and if $T \circ S_1 = S_1 \circ T = I$ and $T \circ S_2 = S_2 \circ T = I$, prove that $S_2 = S_1$.
5. *Prove:* If $I: \mathbf{V} \longrightarrow \mathbf{V}$ is the identity operator on an n-dimensional vector space **V**, then relative to any basis $B = \{X_1, X_2, \ldots, X_n\}$ for **V**, $[I]_B$ is the identity matrix I_n.

In Problems 6–9, follow the solution to Example 6.26 to determine if the given linear operators have inverses. If the linear operator T has an inverse $S = T^{-1}$, find $S(X)$.

6. $T: \mathbf{R}_3 \longrightarrow \mathbf{R}_3,\ X = \begin{bmatrix} x \\ y \\ z \end{bmatrix},\ T(X) = \begin{bmatrix} x + y \\ 2x + z \\ 3x + y \end{bmatrix}.$

7. $T: \mathbf{R}_4 \longrightarrow \mathbf{R}_4,\ X = \begin{bmatrix} x \\ y \\ z \\ w \end{bmatrix},\ T(X) = \begin{bmatrix} x + y \\ 2y + z \\ z + w \\ 2y + z \end{bmatrix}.$

8. $T: \mathbf{P}_2 \longrightarrow \mathbf{P}_2,\ X = f: a + bx,\ T(X) = g_f: (2a + b) + (2b - a)x.$

9. $T: \mathbf{P}_3 \longrightarrow \mathbf{P}_3,\ X = f: a + bx + cx^2,\ T(X) = g_f: (a + b) + (b + a)x + 2cx^2.$

In Problems 10–13, as in Example 6.27, determine if the given linear operator T is 1–1 and onto the codomain. Use the information obtained to find T^{-1} if it exists.

10. $T: \mathbf{R}_3 \longrightarrow \mathbf{R}_3$ as defined in Problem 6.

11. $T: \mathbf{R}_4 \longrightarrow \mathbf{R}_4$ as in Problem 7.

12. $T: \mathbf{P}_2 \longrightarrow \mathbf{P}_2$ as in Problem 8.

13. $T: \mathbf{P}_3 \longrightarrow \mathbf{P}_3$ as in Problem 9.

14. If $B = \{X_1, X_2, \ldots, X_n\}$ is a basis for $\mathbf{V}$, and if $T: \mathbf{V} \longrightarrow \mathbf{V}$ is a nonsingular linear operator on $\mathbf{V}$, prove that $B' = \{T(X_1), T(X_2), \ldots, T(X_n)\}$ is a basis for $\mathbf{V}$.

15. If $T: \mathbf{V} \longrightarrow \mathbf{V}$ is a linear operator on $\mathbf{V}$, and if $B' = \{T(X_1), T(X_2), \ldots, T(X_n)\}$ is a basis for $\mathbf{V}$, prove that T is 1–1 and that $B = \{X_1, X_2, \ldots, X_n\}$ is a basis for $\mathbf{V}$.

Properties (i) and (iii) of Corollary 6.2 are not valid if $\mathbf{V}$ is an infinite-dimensional vector space, as the next problem indicates.

16. Let $\mathbf{V} = \mathbf{P}$ be the space of all polynomial functions. Let $D: \mathbf{P} \longrightarrow \mathbf{P}$ be the linear operator defined by

$$D(f: a_0 + a_1x + a_2x^2 + \cdots + a_nx^n) = f': a_1 + 2a_2x + \cdots + na_nx^{n-1}$$

and let $S: \mathbf{P} \longrightarrow \mathbf{P}$ be the linear operator defined by

$$S(f: a_0 + a_1x + a_2x^2 + \cdots + a_nx^n) = F: a_0x + \frac{a_1x^2}{2} + \frac{a_2x^3}{3} + \cdots + \frac{a_nx^{n+1}}{n + 1}$$

(a) Prove that D is onto $\mathbf{P}$ but not 1–1.
(b) Prove that S is 1–1 but not onto $\mathbf{P}$.
(c) Prove that $D \circ S: \mathbf{P} \longrightarrow \mathbf{P}$ is equal to $I: \mathbf{P} \longrightarrow \mathbf{P}$.
(d) Prove that $S \circ D: \mathbf{P} \longrightarrow \mathbf{P}$ is not equal to $I: \mathbf{P} \longrightarrow \mathbf{P}$.

17. *Prove:* If S and T are linear operators on an n-dimensional vector space $\mathbf{V}$, and if S is a left inverse for T in the sense that $S \circ T = I$, then $T \circ S = I$.

18. *Prove:* If S and T are linear operators on an n-dimensional vector space $\mathbf{V}$, and if S is a right inverse of T in the sense that $T \circ S = I$, then $S \circ T = I$.

Note: Problems 17 and 18 give the linear operator counterpart of Theorem 3.20 for $n \times n$ matrices.

†6.8 THE RANK OF A MATRIX

In this section we define the row rank and the column rank of an arbitrary $m \times n$ matrix A. We shall see that these numbers are equal, and thus we can refer to this common number as the rank of A. A major result of this section is that the rank of a linear transformation and the rank of its matrix (relative to a pair of bases) are the same. The reader may recall that the term "row rank of a reduced row echelon matrix" has already been used in Chapter 3. The definition that we give here is more general and includes that definition as a special case.

Definition 6.11. Let A be an $m \times n$ matrix, let A_i be the ith row of A, and let $A^{(j)}$ be the jth column of A. The **row space of** A is the space $\langle A_1, A_2, \ldots, A_m \rangle$ of all linear combinations of the $1 \times n$ row vectors $A_1, A_2, \ldots, A_m$. The **row rank of** A is the dimension of the row space of A. The **column space of** A is the space $\langle A^{(1)}, A^{(2)}, \ldots, A^{(n)} \rangle$ of all linear combinations of the $m \times 1$ column vectors $A^{(1)}, A^{(2)}, \ldots, A^{(n)}$. The **column rank of** A is the dimension of the column space of A.

The reader should note that the row rank of A does not exceed m and the column rank of A does not exceed n. In Definition 3.10, we defined the row rank of a reduced row echelon matrix E to be the number r of nonzero rows of E that is also the number of leading 1's of E. As row vectors, the rows of E that contain leading 1's are linearly independent (Exercise 6.8, Problem 15). Thus the rows having leading 1's form a basis for the row space of E, and the dimension of the row space of E is r. We now see that the definition of row rank given in Definition 6.11 includes Definition 3.10 as a special case.

We wish to show that the row rank of A equals the row rank of the reduced row echelon matrix of A obtained by applying a sequence of elementary row operations to A. The next theorem leads to this result.

Theorem 6.17. If F is an $m \times m$ elementary matrix and A is an $m \times n$ matrix, the row space of FA is equal to the row space of A.

Proof: Since there are three types of elementary matrices, we break the proof down into three cases.

---

†This section is optional, for the results are not used in other sections of the text.

Case 1: $F = F_{st}$.

The row space of $A = \langle A_1, \ldots, A_s, \ldots, A_t, \ldots, A_m \rangle$

$$= \langle A_1, \ldots, A_t, \ldots, A_s, \ldots, A_m \rangle$$

$$= \text{row space of } FA$$

Case 2: $F = F_s(k),\ k \neq 0$.

The row space of $A = \langle A_1, \ldots, A_s, \ldots, A_m \rangle = \mathbf{C}$, and

the row space of $F_s(k)A = \langle A_1, \ldots, kA_s, \ldots, A_m \rangle = \mathbf{D}$

It is clear that $\mathbf{D} \subseteq \mathbf{C}$. If

$$X = x_1 A_1 + \cdots + x_s A_s + \cdots + x_m A_m \in \mathbf{C}$$

then $x_s A_s = (x_s/k)(kA_s)$, and

$$X = x_1 A_1 + \cdots + \frac{x_s}{k}(kA_s) + \cdots + x_m A_m \in \mathbf{D}$$

Therefore, $\mathbf{C} \subseteq \mathbf{D}$ and $\mathbf{C} = \mathbf{D}$.

Case 3: $F = F_{st}(k)$.

The row space of $A = \langle A_1, \ldots, A_s, \ldots, A_t, \ldots, A_m \rangle = \mathbf{C}$, and

the row space of $F_{st}(k)A = \langle A_1, \ldots, A_s, \ldots, kA_s + A_t, \ldots, A_m \rangle = \mathbf{H}$

If

$$X = x_1 A_1 + \cdots + x_s A_s + \cdots + x_t(kA_s + A_t) + \cdots + x_m A_m \in \mathbf{H}$$

then

$$X = x_1 A_1 + \cdots + (x_s + kx_t)A_s + \cdots + x_t A_t + \cdots + x_m A_m \in \mathbf{C}$$

and $\mathbf{H} \subseteq \mathbf{C}$. Also, if

$$X = y_1 A_1 + \cdots + y_s A_s + \cdots + y_t A_t + \cdots + y_m A_m \in \mathbf{C}$$

then

$$X = y_1 A_1 + \cdots + (-ky_t + y_s)A_s + \cdots + y_t(kA_s + A_t)$$
$$+ \cdots + y_m A_m \in \mathbf{H}$$

and $\mathbf{C} \subseteq \mathbf{H}$. Thus $\mathbf{C} = \mathbf{H}$.

In each of the three cases we have the row space of A is equal to the row space of FA. Q.E.D.

Theorem 6.17 can be extended by mathematical induction as follows.

Theorem 6.18. If $F_1, F_2, \ldots, F_k$ are elementary matrices of order m and if A is an $m \times n$ matrix, then the row space of A is equal to the row space of $(F_k \ldots F_2 F_1)A$.

The proof of Theorem 6.18 is left as Problem 16, Exercise 6.8. We know that any matrix can be transformed to a reduced row echelon form by multiplying on the left by appropriate elementary matrices. Consequently, we have the following corollary to Theorem 6.18.

Corollary 6.3. Let A be an $m \times n$ matrix. Let $F_q, \ldots, F_1$ be elementary matrices such that $(F_q \ldots F_1)A = E$, and E is a reduced row echelon matrix. Then the row space of E is equal to the row space of A and, consequently, the row rank of E is equal to the row rank of A.

Example 6.28.   Find a basis for the space

$$W = \langle (1, 0, 3, 1), (1, -1, 7, -1), (2, 1, 2, 4), (5, 1, 11, 7) \rangle.$$

Solution: We form the matrix

$$A = \begin{bmatrix} 1 & 0 & 3 & 1 \\ 1 & -1 & 7 & -1 \\ 2 & 1 & 2 & 4 \\ 5 & 1 & 11 & 7 \end{bmatrix}$$

and then transform A to reduced row echelon form.

$$\begin{bmatrix} 1 & 0 & 3 & 1 \\ 1 & -1 & 7 & -1 \\ 2 & 1 & 2 & 4 \\ 5 & 1 & 11 & 7 \end{bmatrix} \xrightarrow[\substack{-R_1+R_2 \\ -2R_1+R_3 \\ -5R_1+R_4}]{} \begin{bmatrix} 1 & 0 & 3 & 1 \\ 0 & -1 & 4 & -2 \\ 0 & 1 & -4 & 2 \\ 0 & 1 & -4 & 2 \end{bmatrix} \xrightarrow[\substack{R_2+R_3 \\ R_2+R_4 \\ -R_2}]{} \begin{bmatrix} 1 & 0 & 3 & 1 \\ 0 & 1 & -4 & 2 \\ 0 & 0 & 0 & 0 \\ 0 & 0 & 0 & 0 \end{bmatrix} = E$$

Since the row space of A is the row space of E, and the rows of E that have leading 1's are linearly independent, we have

$$\langle (1, 0, 3, 1), (1, -1, 7, -1), (2, 1, 2, 4), (5, 1, 11, 7) \rangle$$
$$= \langle (1, 0, 3, 1), (0, 1, -4, 2) \rangle$$

and $B = \{(1, 0, 3, 1), (0, 1, -4, 2)\}$ is a basis for the given space **W**.

Theorem 6.19.   Let **V** and **W** be vector spaces such that $\dim \mathbf{V} = n$ and $\dim \mathbf{W} = m$. Let B_1 and B_2 be bases for **V** and **W**, respectively. If $T: \mathbf{V} \rightarrow \mathbf{W}$ is a linear transformation and if $A = [T]_{B_1 B_2}$, then the rank of T is equal to the row rank of A.

Proof: By Theorem 6.8, the rank of $T = \dim \mathbf{R}_T = \dim \mathbf{V} - \dim \mathbf{N}_T$. Consider

$$\mathbf{N}_T = \{X \in \mathbf{V} \mid T(X) = 0\} \quad \text{and} \quad \mathbf{H} = \{Y \mid Y \in \mathbf{R}_n \text{ and } AY = 0\}$$

We claim that the spaces $\mathbf{N}_T$ and $\mathbf{H}$ are isomorphic. For any X in $\mathbf{N}_T$, define $\alpha(X) = [X]_{B_1}$. Since $X \in \mathbf{N}_T$, we have $T(X) = \mathbf{0}$. So

$$A[X]_{B_1} = [T(X)]_{B_2} = \mathbf{0}_m = 0$$

Thus $\alpha(X) = [X]_{B_1}$ is in $\mathbf{H}$. The relation $\alpha: \mathbf{N_T} \longrightarrow \mathbf{H}$ is both a mapping and 1–1 because $X = Y$ if and only if $[X]_{B_1} = [Y]_{B_1}$. Next we show that α is onto

$\mathbf{H}$. Let $Z' = \begin{bmatrix} z_1 \\ z_2 \\ \cdot \\ \cdot \\ \cdot \\ z_n \end{bmatrix}$ and suppose that Z' is in $\mathbf{H}$. If $B_1 = \{X_1, X_2, \ldots, X_n\}$,

then $Z = z_1 X_1 + z_2 X_2 + \cdots + z_n X_n \in \mathbf{V}$. $[Z]_{B_1} = Z'$ and $[T(Z)]_{B_2} = A[Z]_{B_1} = AZ' = \mathbf{0}$. Therefore, $T(Z) = \mathbf{0}$ and $Z \in \mathbf{N_T}$. We have $\alpha(Z) = [Z]_{B_1} = Z'$, and thus α is onto $\mathbf{H}$. Next we show that α preserves addition and scalar multiplication. For any X and Y in $\mathbf{N_T}$ and any scalar r, we have

$$\alpha(X + Y) = [X + Y]_{B_1} = [X]_{B_1} + [Y]_{B_1} = \alpha(X) + \alpha(Y)$$

and

$$\alpha(rX) = [rX]_{B_1} = r[X]_{B_1} = r\alpha(X)$$

Therefore, α is an isomorphism, and consequently dim $\mathbf{N_T}$ = dim $\mathbf{H}$.

$\mathbf{H}$ is the solution set of the matrix equation $AY = \mathbf{0}$, which is equivalent to a system of m linear equations in n unknowns having coefficient matrix A. According to Theorem 5.13, the dimension of the solution space of this system is $n - r$, where r is the row rank of the reduced row echelon matrix of A. By Corollary 6.3, r is the row rank of A. Thus

dim $\mathbf{N_T}$ = dim $\mathbf{H}$ = $n - r$ = $n -$ (row rank of A)

Also,   dim $\mathbf{N_T}$ = dim $\mathbf{V}$ − (rank of T)

Hence the rank of T is equal to the row rank of A.   Q.E.D.

We are now in a position to show that the row rank and the column rank of a matrix are the same.

Theorem 6.20.   For any $m \times n$ matrix A, the row rank of A is equal to the column rank of A.

Proof: For any $m \times n$ matrix A, let $T: \mathbf{R_n} \longrightarrow \mathbf{R_m}$ be the linear transformation defined by $T(X) = AX$. If B and B' are the natural bases for $\mathbf{R_n}$ and $\mathbf{R_m}$, respectively, then $[T]_{BB'} = A = [T(E_1)|T(E_2)|\ldots|T(E_n)]$. Therefore, the column space of A is $\langle T(E_1), T(E_2), \ldots, T(E_n)\rangle$. But this is also the range of T. So the column rank of A is equal to the rank of T. By Theorem 6.19, the rank of T is equal to the row rank of A. Hence the row rank of A is equal to the column rank of A.   Q.E.D.

Definition 6.12.   If A is an $m \times n$ matrix, then the **rank of** A is the common value of the row rank of A and the column rank of A. We denote the rank of A by rank (A).

The next corollary is an immediate consequence of Theorems 6.19 and 6.20.

Corollary 6.4. If $T: \mathbf{V} \to \mathbf{W}$ is a linear transformation, if B_1 and B_2 are ordered bases for the finite-dimensional vector spaces $\mathbf{V}$ and $\mathbf{W}$, respectively, and if $A = [T]_{B_1 B_2}$, then rank $(A) =$ rank (T).

The result of Corollary 6.4 is remarkable in that a given linear transformation $T: \mathbf{V} \to \mathbf{W}$ (with $\mathbf{V}$ and $\mathbf{W}$ finite dimensional) has many matrix representations, each depending on the ordered bases chosen for $\mathbf{V}$ and $\mathbf{W}$. No matter what bases B_1 and B_2 are chosen, the resulting matrix $A = [T]_{B_1 B_2}$ has rank equal to the rank of T. Conversely, if we start with a given $m \times n$ matrix A, there are many different linear transformations T having A as a matrix representation. If B_1 and B_2 are ordered bases for vector spaces $\mathbf{V}$ and $\mathbf{W}$ of dimensions n and m, respectively, then a linear transformation $T: \mathbf{V} \to \mathbf{W}$ having the matrix representation A can be defined as follows. For any $X \in \mathbf{V}$, $T(X)$ is the vector in $\mathbf{W}$ such that $[T(X)]_{B_2} = A[X]_{B_1}$. According to Corollary 6.4, the rank of T must equal rank (A).

We close this section by giving a consistency condition for a system of linear equations in terms of the rank of the coefficient matrix of the system. Suppose that we wish to determine whether or not the system (6.1) is consistent.

$$a_{11}x_1 + a_{12}x_2 + \cdots + a_{1n}x_n = b_1$$
$$a_{21}x_1 + a_{22}x_2 + \cdots + a_{2n}x_n = b_2 \qquad (6.1)$$
$$\cdots$$
$$a_{m1}x_1 + a_{m2}x_2 + \cdots + a_{mn}x_n = b_m$$

If A is the coefficient matrix and $[A \mid B]$ is the augmented matrix of the system, we can write the equivalent vector equation

$$A^{(1)}x_1 + A^{(2)}x_2 + \cdots + A^{(n)}x_n = B$$

From this vector equation we see that X is a solution to (6.1) if and only if B is a linear combination of the columns of A. But this is true if and only if

$$\dim \langle A^{(1)}, A^{(2)}, \ldots, A^{(n)} \rangle = \dim \langle A^{(1)}, A^{(2)}, \ldots, A^{(n)}, B \rangle$$

that is, if and only if rank$(A) =$ rank$([A \mid B])$. We have proved the following result.

Theorem 6.21. The system (6.1) of m linear equations in n unknowns having coefficient matrix A and augmented matrix $[A \mid B]$ is consistent if and only if rank$(A) =$ rank$([A \mid B])$.

Exercise 6.8

In Problems 1–4, use the procedure of Example 6.28 to find a basis for the given space.

  1. $\mathbf{W} = \langle (1, 3, 2), (0, 1, 4), (-2, -3, -8) \rangle$.

  2. $\mathbf{W} = \langle (1, 2, 0, 1), (2, 3, 0, 1), (7, -2, 2, 5), (4, -1, 1, 2) \rangle$.

3. $W = \langle(1, 2), (3, 4), (2, 3)\rangle$.

4. $W = \langle(1, 2, 0), (2, 1, 1), (3, 0, 2), (3, 3, 1)\rangle$.

In Problems 5 and 6, find the rank of the given matrix.

5. $A = \begin{bmatrix} 1 & 2 & 3 & 4 & -2 \\ 2 & 0 & 1 & 2 & 1 \\ 4 & 1 & 0 & 2 & 1 \\ 5 & 2 & 10 & 14 & -5 \end{bmatrix}$.

6. $A = \begin{bmatrix} 3 & 5 & 2 & 0 & 4 \\ 1 & 2 & 0 & -1 & 0 \\ 2 & 3 & 2 & 1 & 4 \\ 2 & 1 & 2 & 0 & 2 \end{bmatrix}$.

7. Let $T: \mathbf{R}_3 \longrightarrow \mathbf{R}_2$ be the linear transformation defined by $T\begin{bmatrix} x \\ y \\ z \end{bmatrix} = \begin{bmatrix} 2x + y \\ x + 3z \end{bmatrix}$.

Let $B_1 = \left\{ X_1 = \begin{bmatrix} 1 \\ 0 \\ 0 \end{bmatrix}, X_2 = \begin{bmatrix} 1 \\ 1 \\ 0 \end{bmatrix}, X_3 = \begin{bmatrix} 1 \\ 1 \\ 1 \end{bmatrix} \right\}$, and $B_2 = \left\{ Y_1 = \begin{bmatrix} 1 \\ 0 \end{bmatrix}, \right.$

$Y_2 = \begin{bmatrix} 1 \\ 1 \end{bmatrix} \Big\}$, and let B and B' be the natural (ordered) bases for $\mathbf{R}_3$ and $\mathbf{R}_2$, respectively. Let $A = [T]_{B_1 B_2}$ and let $C = [T]_{BB'}$.
(a) Explain why rank (T), rank (A), and rank (C) must all be equal.
(b) Find rank (T) by finding a basis for the range of T.
(c) Find rank (A) by reducing A to reduced row echelon form.
(d) Find rank (C) by the reduction method.

8. If $T: \mathbf{R}_3 \longrightarrow \mathbf{R}_2$ is the linear transformation defined by $T\begin{bmatrix} x \\ y \\ z \end{bmatrix} = \begin{bmatrix} x + y + z \\ 2x + y + z \end{bmatrix}$,

and if $B_1, B_2, B,$ and B' are the bases given in Problem 7,
(a) Find rank (T) by finding a basis for the range of T.
(b) If $A = [T]_{B_1 B_2}$ and $C = [T]_{BB'}$, use the reduction method to find rank (A) and rank (C). Check your results with the conclusion of Corollary 6.4.

9. If $A = \begin{bmatrix} 1 & 2 & 1 \\ 3 & 2 & 7 \\ 2 & 4 & 2 \end{bmatrix}$ and if B_1 and B are the ordered bases for $\mathbf{R}_3$ given in Prob-

lem 7,
(a) Find rank (A).

(b) If $T: \mathbf{R}_3 \longrightarrow \mathbf{R}_3$ is a linear transformation such that $[T]_{B_1} = A$, find $T\begin{bmatrix} x \\ y \\ z \end{bmatrix}$.

(c) If $S: \mathbf{R}_3 \longrightarrow \mathbf{R}_3$ is a linear transformation such that $[S]_B = A$, find $S\begin{bmatrix} x \\ y \\ z \end{bmatrix}$.

(d) Explain why rank $(T) = $ rank (S). Check by finding rank (T) and rank (S).

10. If A is an $n \times n$ matrix, state a necessary and sufficient condition in terms of rank (A) that A be nonsingular.

11. If A is an $m \times n$ matrix, show that rank (A) does not exceed the minimum of the numbers m and n.

In Problems 12–14, determine whether or not the given system is consistent by finding the rank of the coefficient matrix and the rank of the augmented matrix and then applying Theorem 6.21.

12. $x + 2y + 4z = 2$
    $2x + 3y + 7z = 3$
    $3x - y + 5z = 1$

13. $x_1 + 3x_2 + 5x_3 + 10x_4 = 2$
    $-x_1 - 2x_3 - 4x_4 = 4$
    $2x_1 + 4x_2 + 8x_3 + 16x_4 = 0$
    $x_2 + x_3 + 2x_4 = 2$

14. $2x_1 + 3x_2 + x_3 = 3$
    $x_1 + 2x_2 + x_3 = 1$
    $-x_1 + 4x_2 = -2$

15. If E is an $m \times n$ reduced row echelon matrix having r nonzero rows (and consequently r leading 1's), prove that these r row vectors are linearly independent and thus form a basis for the row space of E.

16. Use mathematical induction and Theorem 6.17 to prove Theorem 6.18.

17. If A is an $m \times p$ matrix and B is a $p \times n$ matrix, prove
    (a) Rank $(AB) \leq$ rank (A).
    (b) Rank $(AB) \leq$ rank (B).
    (c) If A is nonsingular, then rank $(AB) =$ rank (B).
    (d) If B is nonsingular, then rank $(AB) =$ rank (A).
    Hint: Define linear transformations $T: \mathbf{R}_p \longrightarrow \mathbf{R}_m$ and $S: \mathbf{R}_n \longrightarrow \mathbf{R}_p$ by $T(X) = AX$ and $S(X) = BX$, and consider $T \circ S$.

6.9 CHANGE OF BASIS

We have seen that the matrix of a linear operator $T: \mathbf{V} \longrightarrow \mathbf{V}$ depends on the basis that one chooses for $\mathbf{V}$. For example, if $T: \mathbf{R}_3 \longrightarrow \mathbf{R}_3$ is defined by

$$T \begin{bmatrix} x \\ y \\ z \end{bmatrix} = \begin{bmatrix} x + 3y \\ y - z \\ x + z \end{bmatrix},$$ the matrix of T relative to the natural basis is

$\begin{bmatrix} 1 & 3 & 0 \\ 0 & 1 & -1 \\ 1 & 0 & 1 \end{bmatrix}$. However, the matrix of T relative to the basis $B = \left\{ Y_1 = \begin{bmatrix} 1 \\ 0 \\ 0 \end{bmatrix}, \right.$

$Y_2 = \begin{bmatrix} 1 \\ 1 \\ 0 \end{bmatrix}, Y_3 = \left. \begin{bmatrix} 1 \\ 1 \\ 1 \end{bmatrix} \right\}$ is $\begin{bmatrix} 1 & 3 & 4 \\ -1 & 0 & -2 \\ 1 & 1 & 2 \end{bmatrix}$. This is so because

$$T(Y_1) = \begin{bmatrix} 1 \\ 0 \\ 1 \end{bmatrix} = 1Y_1 - 1Y_2 + 1Y_3$$

$$T(Y_2) = \begin{bmatrix} 4 \\ 1 \\ 1 \end{bmatrix} = 3Y_1 + 0Y_2 + 1Y_3$$

$$T(Y_3) = \begin{bmatrix} 4 \\ 0 \\ 2 \end{bmatrix} = 4Y_1 - 2Y_2 + 2Y_3$$

Many of the applications of linear algebra depend on the fact that one can (sometimes) select a basis such that the matrix of the transformation will be of a certain form.

If a given linear operator $T: \mathbf{V} \longrightarrow \mathbf{V}$ has matrix A relative to some given basis and matrix B relative to another basis, the question arises, how are A and B related? The answer can easily be seen if, instead of concentrating our attention on the matrices A and B, we concentrate on the linear operator T. Before proceeding we shall make some remarks about notation. If $\{X_1, X_2, \ldots, X_n\}$ and $\{Y_1, Y_2, \ldots, Y_n\}$ are ordered bases for $\mathbf{V}$, we shall abbreviate them as $\{X_i\}$ and $\{Y_i\}$. The notation

$$T: \mathbf{V} \underset{\{X_i\} \quad C \quad \{Y_i\}}{\longrightarrow} \mathbf{V} \quad \text{or} \quad \mathbf{V} \underset{\{X_i\} \quad C \quad \{Y_i\}}{\overset{T}{\longrightarrow}} \mathbf{V}$$

will mean that we are considering a linear operator T on $\mathbf{V}$, and C is the matrix of the operator relative to the ordered pair of bases $\{X_i\}$ and $\{Y_i\}$.

Next we point out a notational problem that arises with function composition. If we compose two linear operators $T: \mathbf{V} \longrightarrow \mathbf{W}$ and $S: \mathbf{W} \longrightarrow \mathbf{U}$, we apply T first and then S; that is, $X \overset{T}{\longrightarrow} T(X) \overset{S}{\longrightarrow} S(T(X))$. However, when we write the symbol for the composition, we write $S \circ T$ and not $T \circ S$. We do this because the "X" is written on the right, and $S \circ T(X) = S(T(X))$, which reads, from *right to left*, apply T first and then apply S.

Let's look at the same notational problem in terms of a diagram.

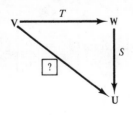

The diagram suggests that one can transform $\mathbf{V}$ into $\mathbf{U}$ by first transforming $\mathbf{V}$ into $\mathbf{W}$ via T, and then transforming $\mathbf{W}$ into $\mathbf{U}$ via S. The diagram also suggests that the mapping $\boxed{?}$ that sends $\mathbf{V}$ directly into $\mathbf{U}$ should be named $T \circ S$. In fact, the mapping $\boxed{?}$ that sends $\mathbf{V}$ directly into $\mathbf{U}$ is $S \circ T: \mathbf{V} \longrightarrow \mathbf{U}$, because $S \circ T(X) = S(T(X))$. Had we chosen the notation $(X)T$ instead of $T(X)$ (as some authors do), then the above composition would be written $(X)T \circ S = ((X)T)S$, which is still read "apply T first and then S."† Now let's consider a diagram as before, but where we keep track of both the linear transformations involved and the corresponding matrices. If $D = [T]_{\{X_i\}\{Y_i\}}$ and $C = [S]_{\{Y_i\}\{Z_i\}}$, then $CD = [S \circ T]_{\{X_i\}\{Z_i\}}$.

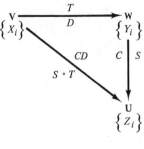

We are now ready to tackle the question, if $T: \mathbf{V} \longrightarrow \mathbf{V}$ is a linear operator on $\mathbf{V}$, if A is the matrix of T relative to some basis $\{X_i\}$, and if B is the matrix of T relative to another basis $\{Y_i\}$, how is the matrix B related to the matrix A? Consider the identity transformation on $\mathbf{V}$. If $\underset{\{Y_i\}\ \ P\ \ \{X_i\}}{I: \mathbf{V} \longrightarrow \mathbf{V}}$ and $\underset{\{X_i\}\ \ Q\ \ \{Y_i\}}{I: \mathbf{V} \longrightarrow \mathbf{V}}$, then the matrix

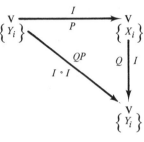

of $I \circ I$ relative to the bases $\{Y_i\}$, $\{Y_i\}$ is QP. But $I \circ I = I$ and the matrix of the identity operator on $\mathbf{V}$ relative to the basis $\{Y_i\}$ used twice is I_n (Problem 5, Exercise 6.7). Therefore, $QP = I_n$ and $Q = P^{-1}$.

Now consider the diagram in Figure 6.3. The linear transformations are indicated on the outside of the diagram, and the corresponding matrices are indicated on the inside of the diagram. Going clockwise around the outside of the diagram, starting at the lower left corner and stopping at the lower right corner, we have $I \circ (T \circ I)$. We get the same result by going across the bottom from left to right; that is, $I \circ (T \circ I) = T$. Now repeating the procedure, but

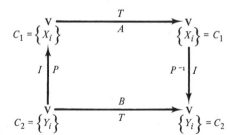

Figure 6.3

†The problem of choosing between the notation $T(X)$ and $(X)T$ for a linear transformation, when translated to matrices, becomes the problem of choosing between multiplication on the right of A by a *column vector* X (AX) or multiplying on the left by a *row vector* $(X^T A^T)$. There are advantages and disadvantages to both notations. The literature is not uniform on this point, and the reader is cautioned to determine which notation is being used when reading other texts in linear algebra.

going around the inside of the diagram, we obtain the corresponding matrix equation $P^{-1}(AP) = B$. We see that $B = P^{-1}AP$. The answer then to the question that was posed is, if the matrix of $T: \mathbf{V} \rightarrow \mathbf{V}$ relative to a basis $C_1 = \{X_i\}$ is A, and the matrix of T relative to a basis $C_2 = \{Y_i\}$ is B, then A and B are related by the equation $B = P^{-1}AP$, where $P = [I]_{C_2C_1}$ and $I: \mathbf{V} \rightarrow \mathbf{V}$ is the identity transformation.

If X is an arbitrary vector in $\mathbf{V}$, let us examine the effect of $P^{-1}AP$ on the coordinates of X. It is a good idea to look at Figure 6.3 to keep everything straight.

$$
\begin{aligned}
(P^{-1}AP)[X]_{C_2} &= P^{-1}A(P[X]_{C_2}) \\
&= (P^{-1}A)[X]_{C_1}
\end{aligned}
\left.\vphantom{\begin{aligned}a\\b\end{aligned}}\right\}
\quad
\begin{array}{l}
P \text{ transforms coordinates from } C_2 \\
\text{to } C_1 \text{ because } P = [I]_{C_2C_1}
\end{array}
$$

$$
\begin{aligned}
&= P^{-1}(A[X]_{C_1}) \\
&= P^{-1}[T(X)]_{C_1}
\end{aligned}
\left.\vphantom{\begin{aligned}a\\b\end{aligned}}\right\}
\quad
\begin{array}{l}
A \text{ does the work of } T \text{ in } C_1 \text{ coor-} \\
\text{dinates because } A = [T]_{C_1}
\end{array}
$$

$$
= [T(X)]_{C_2}
\left.\vphantom{\begin{aligned}a\\b\end{aligned}}\right\}
\quad
\begin{array}{l}
P^{-1} \text{ transforms coordinates from} \\
C_1 \text{ to } C_2 \text{ because } P^{-1} = [I]_{C_1C_2}
\end{array}
$$

$$
= B[X]_{C_2}
\left.\vphantom{\begin{aligned}a\\b\end{aligned}}\right\}
\quad
\begin{array}{l}
B \text{ does the work of } T \text{ in } C_2 \text{ coor-} \\
\text{dinates because } B = [T]_{C_2}
\end{array}
$$

We see that $P^{-1}AP$ performs the same work as B in the following manner: *P transforms the coordinates from C_2 to C_1; A does the work of T in C_1 coordinates; and P^{-1} transforms coordinates from C_1 to C_2. The net effect of $P^{-1}AP$ is to do the work of T in C_2 coordinates.*

Example 6.29.   Let $T: \mathbf{R}_3 \rightarrow \mathbf{R}_3$ be   the   linear   operator   defined   by $T\begin{bmatrix} x \\ y \\ z \end{bmatrix} = \begin{bmatrix} x + 3y \\ y - z \\ x + z \end{bmatrix}$. Let $C_1 = \{E_1, E_2, E_3\}$ (the natural basis) and C_2

$= \left\{ Y_1 = \begin{bmatrix} 1 \\ 0 \\ 0 \end{bmatrix}, Y_2 = \begin{bmatrix} 1 \\ 1 \\ 0 \end{bmatrix}, Y_3 = \begin{bmatrix} 1 \\ 1 \\ 1 \end{bmatrix} \right\}$. Then

$$
A = [T]_{C_1} = \begin{bmatrix} 1 & 3 & 0 \\ 0 & 1 & -1 \\ 1 & 0 & 1 \end{bmatrix} \quad \text{and} \quad B = [T]_{C_2} = \begin{bmatrix} 1 & 3 & 4 \\ -1 & 0 & -2 \\ 1 & 1 & 2 \end{bmatrix}
$$

(The reader should check this.)

(a) Find P and P^{-1} and show that $B = P^{-1}AP$.

(b) Show the effect of $P^{-1}AP$ on $[X]_{C_2}$ if $X = \begin{bmatrix} x \\ y \\ z \end{bmatrix}$.

Solution: (a) Since P is the matrix of the identity transformation relative to the ordered pair of bases C_2 and C_1, we have

$$P = [I]_{C_2C_1} = [[Y_1]_{C_1}|[Y_2]_{C_1}|[Y_3]_{C_1}] = [Y_1|Y_2|Y_3] = \begin{bmatrix} 1 & 1 & 1 \\ 0 & 1 & 1 \\ 0 & 0 & 1 \end{bmatrix}$$

$$P^{-1} = [I]_{C_1C_2} = [[E_1]_{C_2}|[E_2]_{C_2}|[E_3]_{C_2}] = \begin{bmatrix} 1 & -1 & 0 \\ 0 & 1 & -1 \\ 0 & 0 & 1 \end{bmatrix}$$

So

$$P^{-1}AP = P^{-1} \begin{bmatrix} 1 & 3 & 0 \\ 0 & 1 & -1 \\ 1 & 0 & 1 \end{bmatrix} \begin{bmatrix} 1 & 1 & 1 \\ 0 & 1 & 1 \\ 0 & 0 & 1 \end{bmatrix} = P^{-1} \begin{bmatrix} 1 & 4 & 4 \\ 0 & 1 & 0 \\ 1 & 1 & 2 \end{bmatrix}$$

$$= \begin{bmatrix} 1 & -1 & 0 \\ 0 & 1 & -1 \\ 0 & 0 & 1 \end{bmatrix} \begin{bmatrix} 1 & 4 & 4 \\ 0 & 1 & 0 \\ 1 & 1 & 2 \end{bmatrix}$$

$$= \begin{bmatrix} 1 & 3 & 4 \\ -1 & 0 & -2 \\ 1 & 1 & 2 \end{bmatrix} = B$$

(b) If $X = \begin{bmatrix} x \\ y \\ z \end{bmatrix}$, then $[X]_{C_2} = \begin{bmatrix} x - y \\ y - z \\ z \end{bmatrix}$, and

$$P^{-1}AP[X]_{C_2} = P^{-1}A \begin{bmatrix} 1 & 1 & 1 \\ 0 & 1 & 1 \\ 0 & 0 & 1 \end{bmatrix} \begin{bmatrix} x - y \\ y - z \\ z \end{bmatrix} = P^{-1}A \begin{bmatrix} x \\ y \\ z \end{bmatrix}$$

$$= P^{-1} \begin{bmatrix} 1 & 3 & 0 \\ 0 & 1 & -1 \\ 1 & 0 & 1 \end{bmatrix} \begin{bmatrix} x \\ y \\ z \end{bmatrix} = P^{-1} \begin{bmatrix} x + 3y \\ y - z \\ x + z \end{bmatrix}$$

$$= \begin{bmatrix} 1 & -1 & 0 \\ 0 & 1 & -1 \\ 0 & 0 & 1 \end{bmatrix} \begin{bmatrix} x + 3y \\ y - z \\ x + z \end{bmatrix} = \begin{bmatrix} x + 2y + z \\ -x + y - 2z \\ x + z \end{bmatrix}$$

$$= [T(X)]_{C_2}$$

Therefore,

$$T(X) = (x + 2y + z) \begin{bmatrix} 1 \\ 0 \\ 0 \end{bmatrix} + (-x + y - 2z) \begin{bmatrix} 1 \\ 1 \\ 0 \end{bmatrix} + (x + z) \begin{bmatrix} 1 \\ 1 \\ 1 \end{bmatrix}$$

$$= \begin{bmatrix} x + 3y \\ y - z \\ x + z \end{bmatrix},$$

which agrees with the given definition of T.

Definition 6.13.   Let $C_1 = \{X_1, X_2, \ldots, X_n\}$ and $C_2 = \{Y_1, Y_2, \ldots, Y_n\}$ be ordered bases for a vector space $\mathbf{V}$, and let $I: \mathbf{V} \to \mathbf{V}$ be the identity transformation. The matrix

$$Q = [I]_{C_1 C_2} = [[X_1]_{C_2} | [X_2]_{C_2} | \ldots | [X_n]_{C_2}]$$

is called the **matrix for the change of basis** from C_1 to C_2 (because $Q[X]_{C_1} = [X]_{C_2}$).

We summarize our results in the next theorem.

Theorem 6.22.   Let $T: \mathbf{V} \to \mathbf{V}$ be a linear operator on $\mathbf{V}$. Let A be the matrix of T relative to an ordered basis $C_1 = \{X_1, X_2, \ldots, X_n\}$ for $\mathbf{V}$. If $C_2 = \{Y_1, Y_2, \ldots, Y_n\}$ is any ordered basis for $\mathbf{V}$, then

(a) The matrix $P = [[Y_1]_{C_1} | [Y_2]_{C_1} | \ldots | [Y_n]_{C_1}]$ for the change of basis from C_2 to C_1 is invertible.

(b) $P^{-1} = [[X_1]_{C_2} | [X_2]_{C_2} | \ldots | [X_n]_{C_2}]$ is the matrix for the change of basis from C_1 to C_2.

(c) The matrix of T relative to $C_2 = \{Y_i\}$ is $B = P^{-1}AP$.

Example 6.30.   Given the "old" basis $C_1 = \{E_1, E_2, E_3\}$ for $\mathbf{R}_3$ and the

"new" basis $C_2 = \left\{ Y_1 = \begin{bmatrix} 1 \\ 2 \\ 0 \end{bmatrix}, \; Y_2 = \begin{bmatrix} 0 \\ 2 \\ 1 \end{bmatrix}, \; Y_3 = \begin{bmatrix} 1 \\ 0 \\ 0 \end{bmatrix} \right\}$,

(a) Find the matrix P for the change of basis from the new basis C_2 to the old basis C_1.

(b) Find the matrix P^{-1} for the change of basis from the old basis C_1 to the new basis C_2.

(c) Use P^{-1} to write $X = \begin{bmatrix} x \\ y \\ z \end{bmatrix}$ as a linear combination of the vectors in C_2.

Solution

(a) $P = [I]_{C_2 C_1} = [[Y_1]_{C_1} | [Y_2]_{C_1} | [Y_3]_{C_1}] = [Y_1 | Y_2 | Y_3] = \begin{bmatrix} 1 & 0 & 1 \\ 2 & 2 & 0 \\ 0 & 1 & 0 \end{bmatrix}$

(b) We find P^{-1} by inverting P.

$$\begin{bmatrix} 1 & 0 & 1 & | & 1 & 0 & 0 \\ 2 & 2 & 0 & | & 0 & 1 & 0 \\ 0 & 1 & 0 & | & 0 & 0 & 1 \end{bmatrix} \xrightarrow{-2R_1 + R_2} \begin{bmatrix} 1 & 0 & 1 & | & 1 & 0 & 0 \\ 0 & 2 & -2 & | & -2 & 1 & 0 \\ 0 & 1 & 0 & | & 0 & 0 & 1 \end{bmatrix}$$

$$\xrightarrow{R_{23}} \begin{bmatrix} 1 & 0 & 1 & | & 1 & 0 & 0 \\ 0 & 1 & 0 & | & 0 & 0 & 1 \\ 0 & 2 & -2 & | & -2 & 1 & 0 \end{bmatrix}$$

$$\xrightarrow{-2R_2+R_3} \begin{bmatrix} 1 & 0 & 1 & | & 1 & 0 & 0 \\ 0 & 1 & 0 & | & 0 & 0 & 1 \\ 0 & 0 & -2 & | & -2 & 1 & -2 \end{bmatrix}$$

$$\xrightarrow[-R_3+R_1]{-(1/2)R_3} \begin{bmatrix} 1 & 0 & 0 & | & 0 & \frac{1}{2} & -1 \\ 0 & 1 & 0 & | & 0 & 0 & 1 \\ 0 & 0 & 1 & | & 1 & -\frac{1}{2} & 1 \end{bmatrix}$$

Therefore, $P^{-1} = \begin{bmatrix} 0 & \frac{1}{2} & -1 \\ 0 & 0 & 1 \\ 1 & -\frac{1}{2} & 1 \end{bmatrix}$.

(c) $P^{-1}[X]_{C_1} = [X]_{C_2}$. Therefore,

$$\begin{bmatrix} 0 & \frac{1}{2} & -1 \\ 0 & 0 & 1 \\ 1 & -\frac{1}{2} & 1 \end{bmatrix} \begin{bmatrix} x \\ y \\ z \end{bmatrix} = \begin{bmatrix} \frac{1}{2}y - z \\ z \\ x - \frac{1}{2}y + z \end{bmatrix}$$

and

$$\begin{bmatrix} x \\ y \\ z \end{bmatrix} = (\tfrac{1}{2}y - z)\begin{bmatrix} 1 \\ 2 \\ 0 \end{bmatrix} + z\begin{bmatrix} 0 \\ 2 \\ 1 \end{bmatrix} + (x - \tfrac{1}{2}y + z)\begin{bmatrix} 1 \\ 0 \\ 0 \end{bmatrix}$$

Definition 6.14. A square matrix B of order n is **similar** to a square matrix A of order n if and only if there is an invertible matrix P such that $B = P^{-1}AP$. We write $B \underset{s}{\sim} A$.

We restate conclusion (c) of Theorem 6.22 as the following corollary.

Corollary 6.5. If $T: V \rightarrow V$ is a linear operator on V and if A and B are matrix representations of T but with respect to (possibly) different bases, then B is similar to A.

The converse of Corollary 6.5 is also true, and we state it as the final result of this section.

Theorem 6.23. If A and B are $n \times n$ matrices such that B is similar to A, then for any n-dimensional vector space V there is a linear operator $T: V \rightarrow V$ such that A and B are both matrix representations of T.

We illustrate a method of proof for Theorem 6.23 in the next example. A proof imitating the example is called for in Problem 17, Exercise 6.9.

Example 6.31. Let $A = \begin{bmatrix} 1 & 2 & 0 \\ 2 & 4 & -1 \\ 0 & -1 & 1 \end{bmatrix}$ and $P = \begin{bmatrix} 1 & 2 & 1 \\ 1 & 2 & 0 \\ 1 & 3 & 3 \end{bmatrix}$. Then

$P^{-1} = \begin{bmatrix} 6 & -3 & -2 \\ -3 & 2 & 1 \\ 1 & -1 & 0 \end{bmatrix}$ and $B = \begin{bmatrix} 3 & 7 & 3 \\ 1 & 1 & -2 \\ -2 & -3 & 2 \end{bmatrix} = P^{-1}AP$ is similar to A.

Show that there is a linear operator T on $\mathbf{V} = \mathbf{R}_3$ and bases C_1 and C_2 for $\mathbf{R}_3$ such that $[T]_{C_1} = A$ and $[T]_{C_2} = B$.

Solution: The first basis C_1 can be chosen arbitrarily and we choose

$$C_1 = \left\{ X_1 = \begin{bmatrix} 1 \\ 0 \\ 0 \end{bmatrix}, \ X_2 = \begin{bmatrix} 1 \\ 1 \\ 0 \end{bmatrix}, \ X_3 = \begin{bmatrix} 1 \\ 1 \\ 1 \end{bmatrix} \right\}$$

The idea behind the solution is to reconstruct the situation as depicted in Figure 6.3, where the known parts of the diagram are A, B, P, P^{-1}, and $C_1 = \{X_i\}$.

We wish to define a linear operator $T: \mathbf{R}_3 \rightarrow \mathbf{R}_3$ in such a way that the matrix of T relative to the basis C_1 is A. Thus for each X in $\mathbf{R}_3$ we must have $A[X]_{C_1} = [T(X)]_{C_1}$. It is clear then that we should *define* T by $T(X) = Y$, where $[Y]_{C_1} = A[X]_{C_1}$. We now check to see if $A = [T]_{C_1}$. The jth column of A is

$$A^{(j)} = AE_j = A[X_j]_{C_1} = [T(X_j)]_{C_1}$$

for $j = 1, 2, 3$. Thus

$$A = [[T(X_1)]_{C_1} | [T(X_2)_{C_1}] | [T(X_3)]_{C_1}] = [T]_{C_1}$$

Now we want a basis $C_2 = \{Y_1, Y_2, Y_3\}$ for $\mathbf{R}_3$ such that

$$P = [I]_{C_2 C_1} = [[Y_1]_{C_1} | [Y_2]_{C_1} | [Y_3]_{C_1}] = \begin{bmatrix} 1 & 2 & 1 \\ 1 & 2 & 0 \\ 1 & 3 & 3 \end{bmatrix}$$

Therefore, we *define* Y_1, Y_2, and Y_3 to be the vectors whose coordinates relative to C_1 are the respective columns of P; that is, $Y_1 = 1X_1 + 1X_2 + 1X_3$, $Y_2 = 2X_1 + 2X_2 + 3X_3$, and $Y_3 = 1X_1 + 0X_2 + 3X_3$. Since P is nonsingular, these vectors are linearly independent (Problem 16, Exercise 6.9) and

$$C_2 = \left\{ Y_1 = \begin{bmatrix} 3 \\ 2 \\ 1 \end{bmatrix}, \ Y_2 = \begin{bmatrix} 7 \\ 5 \\ 3 \end{bmatrix}, \ Y_3 = \begin{bmatrix} 4 \\ 3 \\ 3 \end{bmatrix} \right\}$$

is a basis for $\mathbf{R}_3$. Since $P = [I]_{C_2 C_1}$, we must have $P^{-1} = [I]_{C_1 C_2}$ and $P^{-1} = [[X_1]_{C_2} | [X_2]_{C_2} | [X_3]_{C_2}]$. We claim that $B = [T]_{C_2}$. The jth column of B for $j = 1, 2, 3$ is

$$B^{(j)} = BE_j = B[Y_j]_{C_2} = (P^{-1}AP)[Y_j]_{C_2} = P^{-1}A[Y_j]_{C_1}$$
$$= P^{-1}[T(Y_j)]_{C_1} = [T(Y_j)]_{C_2}$$

Hence $B = [T]_{C_2}$.

This *concludes our solution*, but we now find an explicit formula for T and use it to show that $[T]_{c_1} = A$ and $[T]_{c_2} = B$.

If $X = \begin{bmatrix} x \\ y \\ z \end{bmatrix}$,

then

$$[X]_{c_1} = \begin{bmatrix} x - y \\ y - z \\ z \end{bmatrix}$$

and

$$[T(X)]_{c_1} = \begin{bmatrix} 1 & 2 & 0 \\ 2 & 4 & -1 \\ 0 & -1 & 1 \end{bmatrix} \begin{bmatrix} x - y \\ y - z \\ z \end{bmatrix} = \begin{bmatrix} x + y - 2z \\ 2x + 2y - 5z \\ -y + 2z \end{bmatrix}$$

Thus

$$T(X) = (x + y - 2z)\begin{bmatrix} 1 \\ 0 \\ 0 \end{bmatrix} + (2x + 2y - 5z)\begin{bmatrix} 1 \\ 1 \\ 0 \end{bmatrix} + (-y + 2z)\begin{bmatrix} 1 \\ 1 \\ 1 \end{bmatrix}$$

$$= \begin{bmatrix} 3x + 2y - 5z \\ 2x + y - 3z \\ -y + 2z \end{bmatrix}$$

We now find $[T]_{c_1}$ and $[T]_{c_2}$.

$$T(X_1) = \begin{bmatrix} 3 \\ 2 \\ 0 \end{bmatrix} = 1X_1 + 2X_2 + 0X_3$$

$$T(X_2) = \begin{bmatrix} 5 \\ 3 \\ -1 \end{bmatrix} = 2X_1 + 4X_2 - 1X_3$$

$$T(X_3) = \begin{bmatrix} 0 \\ 0 \\ 1 \end{bmatrix} = 0X_1 - 1X_2 + 1X_3$$

Therefore,

$$[T]_{c_1} = \begin{bmatrix} 1 & 2 & 0 \\ 2 & 4 & -1 \\ 0 & -1 & 1 \end{bmatrix} = A$$

Also,

$$T(Y_1) = \begin{bmatrix} 8 \\ 5 \\ 0 \end{bmatrix} = 3Y_1 + 1Y_2 - 2Y_3$$

$$T(Y_2) = \begin{bmatrix} 16 \\ 10 \\ 1 \end{bmatrix} = 7Y_1 + 1Y_2 - 3Y_3$$

$$T(Y_3) = \begin{bmatrix} 3 \\ 2 \\ 3 \end{bmatrix} = 3Y_1 - 2Y_2 + 2Y_3$$

Hence $[T]_{C_2} = \begin{bmatrix} 3 & 7 & 3 \\ 1 & 1 & -2 \\ -2 & -3 & 2 \end{bmatrix} = B.$

We give a brief summary of the results of this section. Suppose that A and B are matrices of a linear operator $T: \mathbf{V} \longrightarrow \mathbf{V}$ relative to bases C_1 and C_2, respectively. If P is the matrix for the change of basis from C_2 to C_1 $(P = [I]_{C_2 C_1})$, then P^{-1} is the matrix for the change of basis from C_1 to C_2 $(P^{-1} = [I]_{C_1 C_2})$. Also,

$$P^{-1}AP[X]_{C_2} = P^{-1}A[X]_{C_1} = P^{-1}[T(X)]_{C_1} = [T(X)]_{C_2} = B[X]_{C_2}$$

and

$$P^{-1}AP = B$$

Conversely, if A and B are $n \times n$ matrices such that $P^{-1}AP = B$, then for any n-dimensional space $\mathbf{V}$ there is a linear operator $T: \mathbf{V} \longrightarrow \mathbf{V}$ such that A and B are both matrix representations of T.

Exercise 6.9

In Problems 1–5, you are given a vector space $\mathbf{V}$, a pair of ordered bases C_1 and C_2 for $\mathbf{V}$, and a vector X in $\mathbf{V}$.
(a) Find the matrix $P = [I]_{C_2 C_1}$ for the change of basis from C_2 to C_1.
(b) Find the matrix $Q = P^{-1} = [I]_{C_1 C_2}$ for the change of basis from C_1 to C_2.
(c) Use Q to write X as a linear combination of the vectors in C_2.

1. $\mathbf{V} = \mathbf{R}_2$. C_1 is the natural basis, $C_2 = \left\{ Y_1 = \begin{bmatrix} 1 \\ 2 \end{bmatrix}, Y_2 = \begin{bmatrix} 1 \\ 1 \end{bmatrix} \right\}$, and $X = \begin{bmatrix} x \\ y \end{bmatrix}$.

2. $V = R_3$. C_1 is the natural basis, $C_2 = \left\{ Y_1 = \begin{bmatrix} 1 \\ 1 \\ 0 \end{bmatrix}, Y_2 = \begin{bmatrix} 1 \\ 2 \\ 0 \end{bmatrix}, Y_3 = \begin{bmatrix} 4 \\ 1 \\ 1 \end{bmatrix} \right\}$,

and $X = \begin{bmatrix} x \\ y \\ z \end{bmatrix}$.

3. $V = R_2$. $C_1 = \left\{ X_1 = \begin{bmatrix} 1 \\ 4 \end{bmatrix}, X_2 = \begin{bmatrix} 0 \\ 2 \end{bmatrix} \right\}$, $C_2 = \left\{ Y_1 = \begin{bmatrix} 1 \\ 1 \end{bmatrix}, Y_2 = \begin{bmatrix} 2 \\ 0 \end{bmatrix} \right\}$,

and $X = \begin{bmatrix} x \\ y \end{bmatrix}$.

4. $V = R_3$. C_1 is the natural basis, $C_2 = \left\{ Y_1 = \begin{bmatrix} 3 \\ 1 \\ 1 \end{bmatrix}, Y_2 = \begin{bmatrix} 3 \\ 2 \\ 6 \end{bmatrix}, Y_3 = \begin{bmatrix} 4 \\ 1 \\ 0 \end{bmatrix} \right\}$,

and $X = \begin{bmatrix} x \\ y \\ z \end{bmatrix}$.

5. $V = P_3$. $C_1 = \{f_1: 1, f_2: x, f_3: x^2\}$,
$C_2 = \{g_1: 1 + x, g_2: 1 + 2x, g_3: 4 + x + x^2\}$, and $X = f: a + bx + cx^2$.

6. If $T\begin{bmatrix} x \\ y \end{bmatrix} = \begin{bmatrix} x + y \\ x - y \end{bmatrix}$, $C_1 = \left\{ X_1 = \begin{bmatrix} 1 \\ 1 \end{bmatrix}, X_2 = \begin{bmatrix} 0 \\ 1 \end{bmatrix} \right\}$,

and $C_2 = \left\{ Y_1 = \begin{bmatrix} 2 \\ 0 \end{bmatrix}, Y_2 = \begin{bmatrix} 1 \\ 2 \end{bmatrix} \right\}$,

(a) Find the matrix A of T relative to the basis C_1.
(b) Find the matrix B of T relative to the basis C_2.
(c) Find the matrix P for the change of basis from C_2 to C_1.
(d) Find the matrix P^{-1} for the change of basis from C_1 to C_2.
(e) Show that $P^{-1}AP = B$.
(f) Use P^{-1} to write $2X_1 + 3X_2$ as a linear combination of $\{Y_i\}$.
(g) Use P to write $3Y_1 + 4Y_2$ as a linear combination of $\{X_i\}$.

7. Given $T: R_4 \longrightarrow R_4$ defined by $T\begin{bmatrix} x \\ y \\ z \\ w \end{bmatrix} = \begin{bmatrix} x - y \\ x + 2z \\ z + w \\ y + z \end{bmatrix}$, C_1 is the standard basis,

and $C_2 = \left\{ Y_1 = \begin{bmatrix} 2 \\ 0 \\ 0 \\ 0 \end{bmatrix}, Y_2 = \begin{bmatrix} 1 \\ 1 \\ 0 \\ 0 \end{bmatrix}, Y_3 = \begin{bmatrix} 0 \\ 0 \\ 2 \\ 0 \end{bmatrix}, Y_4 = \begin{bmatrix} 0 \\ 0 \\ 1 \\ 1 \end{bmatrix} \right\}$.

(a)–(e) Same as Problem 6.

(f) Use P^{-1} to write $\begin{bmatrix} 3 \\ 1 \\ 2 \\ 1 \end{bmatrix}$ as a linear combination of the vectors in C_2.

(g) Use P to write $2Y_1 + 3Y_2 - 2Y_3 + 1Y_4$ as a linear combination of the vectors in C_1.

8. Prove that the relation $\underset{S}{\sim}$ defined in Definition 6.14 satisfies the properties:
   (a) For any $n \times n$ matrix A, $A \underset{S}{\sim} A$.
   (b) For any $n \times n$ matrices A and B, if $A \underset{S}{\sim} B$, then $B \underset{S}{\sim} A$.
   (c) For any $n \times n$ matrices A, B, and C, if $A \underset{S}{\sim} B$ and $B \underset{S}{\sim} C$, then $A \underset{S}{\sim} C$.

A relation satisfying the three properties (a)–(c) of Problem 8 is called an **equivalence relation** (on the set $\mathbf{R_{n \times n}}$).

9. If $P = \begin{bmatrix} 2 & 3 & 0 \\ 0 & 1 & 1 \\ 1 & 2 & 1 \end{bmatrix}$ is the matrix for the change of basis from $\{Y_i\}$ to $\{X_i\}$,
   (a) Write each of the vectors in $\{Y_i\}$ as a linear combination of vectors in $\{X_i\}$.
   (b) Write each of the vectors in $\{X_i\}$ as a linear combination of vectors in $\{Y_i\}$.

10. The transformation $T: \mathbf{R}_2 \longrightarrow \mathbf{R}_2$ defined by

$$T\begin{bmatrix} x \\ y \end{bmatrix} = \begin{bmatrix} \cos\theta & -\sin\theta \\ \sin\theta & \cos\theta \end{bmatrix}\begin{bmatrix} x \\ y \end{bmatrix} = \begin{bmatrix} x' \\ y' \end{bmatrix}$$

is a rotation of the plane through an angle θ. If $\theta = 45°$,
   (a) Find the matrix for the change of basis from the standard basis to the basis

$$C_2 = \left\{ Y_1 = T\begin{bmatrix} 1 \\ 0 \end{bmatrix}, \ Y_2 = T\begin{bmatrix} 0 \\ 1 \end{bmatrix} \right\}. \ (C_2 \text{ is a basis because } T \text{ is one to one.})$$

   (b) Write $\begin{bmatrix} x \\ y \end{bmatrix}$ as a linear combination of the "new" basis vectors.

11. If $P = \begin{bmatrix} 1 & 2 \\ 3 & 4 \end{bmatrix}$ is the matrix for a change of basis from a basis $C_2 = \{Y_i\}$ to a basis $C_1 = \{X_i\}$, and the matrix of $T: \mathbf{R}_2 \longrightarrow \mathbf{R}_2$ relative to C_1 is $A = \begin{bmatrix} 1 & 0 \\ 3 & 1 \end{bmatrix}$, find the matrix of T relative to the basis C_2.

12. Let $V = P_3$, let $C_1 = \{f_1: 1, f_2: x, f_3: x^2\}$, and let

$$C_2 = \left\{ g_1: \frac{(x-1)(x-2)}{2}, g_2: \frac{x(x-2)}{-1}, g_3: \frac{x(x-1)}{2} \right\}$$

   (a) If $f: a + bx + cx^2$, show that $f(x) = f(0)g_1(x) + f(1)g_2(x) + f(2)g_3(x)$, and thus $f = f(0)g_1 + f(1)g_2 + f(2)g_3$. (This proves that C_2 is a basis for P_3, since P_3 has dimension 3.)
   (b) Find the matrix for the change of basis from C_2 to C_1 and the matrix for the change of basis from C_1 to C_2.

The differentiation operator $D: \mathbf{P}_3 \longrightarrow \mathbf{P}_3$ defined by $D(f: a + bx + cx^2) = f': b + 2cx$ is a linear operator on $\mathbf{P}_3$.

(c) Find the matrix of D relative to C_1.

(d) Find the matrix of D relative to C_2.

13. If $A = \begin{bmatrix} 1 & 2 & 0 \\ 0 & 4 & -1 \\ 2 & 1 & 3 \end{bmatrix}$, $P = \begin{bmatrix} 1 & 1 & 4 \\ 1 & 2 & 1 \\ 0 & 0 & 1 \end{bmatrix}$, $B = P^{-1}AP$, and C_1 is the natural basis

for $\mathbf{R}_3$, find a basis $C_2 = \{Y_i\}$ for $\mathbf{R}_3$ and a linear transformation $T: \mathbf{R}_3 \longrightarrow \mathbf{R}_3$ such that $A = [T]_{C_1}$ and $B = [T]_{C_2}$. Find $T(Y_i)$ for $i = 1, 2, 3$ by calculating $B[Y_i]_{C_2} = T[Y_i]_{C_2}$, and then find $T(Y_i)$ by calculating $A[Y_i]_{C_1}$. Follow the solution of Example 6.31.

14. Repeat Problem 13 but with C_1 the "natural" basis for $\mathbf{P}_3$ as in Problem 5, and $T: \mathbf{P}_3 \longrightarrow \mathbf{P}_3$.

15. If A is an $n \times n$ matrix, if $\mathbf{V}$ is an n-dimensional vector space with ordered basis $C_1 = \{X_1, X_2, \ldots, X_n\}$, and $T: \mathbf{V} \longrightarrow \mathbf{V}$ is defined by $T(X) = Y$ if and only if $A[X]_{C_1} = [Y]_{C_1}$, prove that T is a linear operator.

16. If $\mathbf{V}$ is a vector space with ordered basis $C_1 = \{X_1, \ldots, X_n\}$, if P is an $n \times n$ nonsingular matrix, and if $Y_1, Y_2, \ldots, Y_n$ are vectors in $\mathbf{V}$ defined by $[Y_j]_{C_1} = P^{(j)}$ (the jth column of P), prove that $\{Y_1, Y_2, \ldots, Y_n\}$ is linearly independent and thus is a basis for $\mathbf{V}$.

17. Using Example 6.31 as a guide, prove Theorem 6.23.

6.10 CALCULATIONS WITH SIMILAR MATRICES

According to Definition 6.14, a square matrix B is similar to a square matrix A if there is an invertible matrix P such that $B = P^{-1}AP$. We have seen that A and B are similar if and only if they are matrix representations of the same linear operator. The similarity relationship can be used to simplify certain calculations, as we now proceed to illustrate. The connection between similar matrices and the linear operator that they represent will be further explored in Chapter 7.

Example 6.32. If $A = \begin{bmatrix} 1 & 0 \\ 0 & 2 \end{bmatrix}$, $P = \begin{bmatrix} 1 & 4 \\ 1 & 5 \end{bmatrix}$, $P^{-1} = \begin{bmatrix} 5 & -4 \\ -1 & 1 \end{bmatrix}$, and $B = P^{-1}AP$, find B^5.

Solution

$$P^{-1}AP = \begin{bmatrix} 5 & -4 \\ -1 & 1 \end{bmatrix}\begin{bmatrix} 1 & 0 \\ 0 & 2 \end{bmatrix}\begin{bmatrix} 1 & 4 \\ 1 & 5 \end{bmatrix} = \begin{bmatrix} -3 & -20 \\ 1 & 6 \end{bmatrix}$$

We will calculate B^5 by first calculating A^5. This calculation is easy because A is a **diagonal matrix**; that is, all the entries off the main diagonal of A are zero.

$$A^2 = \begin{bmatrix} 1 & 0 \\ 0 & 2 \end{bmatrix}\begin{bmatrix} 1 & 0 \\ 0 & 2 \end{bmatrix} = \begin{bmatrix} 1^2 & 0 \\ 0 & 2^2 \end{bmatrix}, \qquad A^3 = A^2A = \begin{bmatrix} 1^3 & 0 \\ 0 & 2^3 \end{bmatrix},$$

$$A^4 = \begin{bmatrix} 1^4 & 0 \\ 0 & 2^4 \end{bmatrix}, \qquad A^5 = \begin{bmatrix} 1^5 & 0 \\ 0 & 2^5 \end{bmatrix} = \begin{bmatrix} 1 & 0 \\ 0 & 32 \end{bmatrix}$$

Now

$$B^2 = (P^{-1}AP)(P^{-1}AP) = P^{-1}A(PP^{-1})AP = P^{-1}A^2P$$

$$B^3 = (P^{-1}A^2P)(P^{-1}AP) = P^{-1}A^3P, \qquad B^4 = P^{-1}A^4P$$

$$B^5 = P^{-1}A^5P = \begin{bmatrix} 5 & -4 \\ -1 & 1 \end{bmatrix}\begin{bmatrix} 1 & 0 \\ 0 & 32 \end{bmatrix}\begin{bmatrix} 1 & 4 \\ 1 & 5 \end{bmatrix} = \begin{bmatrix} 5 & -128 \\ -1 & 32 \end{bmatrix}\begin{bmatrix} 1 & 4 \\ 1 & 5 \end{bmatrix}$$

$$= \begin{bmatrix} -123 & -620 \\ 31 & 156 \end{bmatrix}$$

In general, one can show that if $B = P^{-1}AP$, then $B^k = P^{-1}A^kP$ or, equivalently, if $B = QAQ^{-1}$, then $B^k = QA^kQ^{-1}$ (Exercise 6.10, Problem 1).

Application: The following application is adapted from Ronald A. Knight, "Artificial Inbreeding," *Proceedings Summer Conference for College Teachers on Applied Mathematics* (University of Missouri-Rolla), pp. 188–192.

Suppose it is desired that a certain type of plant be grown and that all the plants are to be as much alike as possible. We wish to grow plants then that are similar homozygotes, that is, plants that are all of type AA or all are of type aa. We assume that the plants are subjected to continual self-fertilization. Let a_0 be the fraction of the number of original plants that are of type AA, let b_0 be the fraction of the original plants that are of type Aa, and let c_0 be the fraction of the original plants that are of type aa. Then $a_0 + b_0 + c_0 = 1$. Plants of type AA produce only type AA plants. Similarly, plants of type aa produce only type aa plants. Plants of type Aa produce plants one quarter of which are type AA, one quarter are type aa, and one half are of type Aa, as the diagram indicates.

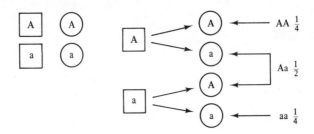

Let a_n, b_n, and c_n represent the proportion of plants produced in the nth generation of types AA, Aa, and aa, respectively. We wish to determine $X_n = \begin{bmatrix} a_n \\ b_n \\ c_n \end{bmatrix}$. The equations in the system of linear equations

$$a_1 = 1a_0 + \tfrac{1}{4}b_0 + 0c_0$$
$$b_1 = 0a_0 + \tfrac{1}{2}b_0 + 0c_0$$
$$c_1 = 0a_0 + \tfrac{1}{4}b_0 + 1c_0$$

represent the proportion of plants of types AA, Aa, and aa, respectively, that are produced in the first generation. This system of linear equations can be written in matrix form as

$$X_1 = \begin{bmatrix} a_1 \\ b_1 \\ c_1 \end{bmatrix} = \begin{bmatrix} 1 & \tfrac{1}{4} & 0 \\ 0 & \tfrac{1}{2} & 0 \\ 0 & \tfrac{1}{4} & 1 \end{bmatrix} \begin{bmatrix} a_0 \\ b_0 \\ c_0 \end{bmatrix} = BX_0$$

Similarly, $X_2 = BX_1 = B(BX_0) = B^2X_0$, $X_3 = BX_2 = B(B^2X_0) = B^3X_0$ and, in general, $X_n = B^nX_0$. We see then that we can calculate X_n by calculating B^n. If we set

$$D = \begin{bmatrix} 1 & 0 & 0 \\ 0 & \tfrac{1}{2} & 0 \\ 0 & 0 & 1 \end{bmatrix} \quad \text{and} \quad P = \begin{bmatrix} 1 & 1 & 0 \\ 0 & -2 & 0 \\ 0 & 1 & 1 \end{bmatrix}$$

direct calculations show that $P^{-1} = \begin{bmatrix} 1 & \tfrac{1}{2} & 0 \\ 0 & -\tfrac{1}{2} & 0 \\ 0 & \tfrac{1}{2} & 1 \end{bmatrix}$ and $B = PDP^{-1}$. (In Chapter 7 we shall see how the diagonal matrix D was determined.) Then

$$B^n = PD^nP^{-1} = \begin{bmatrix} 1 & 1 & 0 \\ 0 & -2 & 0 \\ 0 & 1 & 1 \end{bmatrix} \begin{bmatrix} 1 & 0 & 0 \\ 0 & \tfrac{1}{2^n} & 0 \\ 0 & 0 & 1 \end{bmatrix} \begin{bmatrix} 1 & \tfrac{1}{2} & 0 \\ 0 & -\tfrac{1}{2} & 0 \\ 0 & \tfrac{1}{2} & 1 \end{bmatrix}$$

$$= \begin{bmatrix} 1 & \tfrac{1}{2^n} & 0 \\ 0 & -\tfrac{1}{2^{n-1}} & 0 \\ 0 & \tfrac{1}{2^n} & 1 \end{bmatrix} \begin{bmatrix} 1 & \tfrac{1}{2} & 0 \\ 0 & -\tfrac{1}{2} & 0 \\ 0 & \tfrac{1}{2} & 1 \end{bmatrix} = \begin{bmatrix} 1 & \tfrac{1}{2} - \tfrac{1}{2^{n+1}} & 0 \\ 0 & \tfrac{1}{2^n} & 0 \\ 0 & \tfrac{1}{2} - \tfrac{1}{2^{n+1}} & 1 \end{bmatrix}$$

and

$$\begin{bmatrix} 1 & \frac{1}{2}-\frac{1}{2^{n+1}} & 0 \\ 0 & \frac{1}{2^n} & 0 \\ 0 & \frac{1}{2}-\frac{1}{2^{n+1}} & 1 \end{bmatrix}\begin{bmatrix} a_0 \\ b_0 \\ c_0 \end{bmatrix} = \begin{bmatrix} a_n \\ b_n \\ c_n \end{bmatrix},$$

Thus

$$a_n = a_0 + \left(\frac{1}{2} - \frac{1}{2^{n+1}}\right)b_0$$

$$b_n = \frac{1}{2^n}b_0$$

$$c_n = \left(\frac{1}{2} - \frac{1}{2^{n+1}}\right)b_0 + c_0$$

The following table gives values of a_n, b_n, and c_n rounded off to five places for $n = 0, 1, 5,$ and 10.

n	0	1	5	10
a_n	a_0	$a_0 + 0.25b_0$	$a_0 + 0.48438b_0$	$a_0 + 0.49951b_0$
b_n	b_0	$0.25b_0$	$0.03125b_0$	$0.00098b_0$
c_n	c_0	$c_0 + 0.25b_0$	$c_0 + 0.48438b_0$	$c_0 + 0.49951b_0$

We see that as n gets large

$$a_n \longrightarrow a_0 + \tfrac{1}{2}b_0$$

$$b_n \longrightarrow 0$$

$$c_n \longrightarrow \tfrac{1}{2}b_0 + c_0$$

The above solution assumes that each plant fertilizes itself.

Now suppose that a plant of type AA is selected, and that this plant is used to fertilize all the plants as well as succeeding generations of plants. Since plants of type AA crossed with plants of type AA produce all type AA plants, plants of type AA crossed with plants of type Aa produce plants one half of which are type AA and one half of which are type Aa, and plants of type AA crossed with plants of type aa produce all type Aa plants, we obtain the following linear system of equations.

$$a_1 = a_0 + \tfrac{1}{2}b_0 + 0c_0$$

$$b_1 = 0a_0 + \tfrac{1}{2}b_0 + 1c_0 \qquad (*)$$

$$c_1 = 0$$

If C is the coefficient matrix, this system can be written in the matrix form $X_1 = CX_0$. Also, $X_2 = CX_1 = C^2X_0$, and in general $X_n = C^nX_0$. If

$$Q = \begin{bmatrix} 1 & -1 & 1 \\ 0 & 1 & -2 \\ 0 & 0 & 1 \end{bmatrix}, \quad \text{and} \quad D = \begin{bmatrix} 1 & 0 & 0 \\ 0 & \frac{1}{2} & 0 \\ 0 & 0 & 0 \end{bmatrix}$$

then $C = QDQ^{-1}$ and

$$C^n = QD^nQ^{-1} = \begin{bmatrix} 1 & 1 - \dfrac{1}{2^n} & 1 - \dfrac{1}{2^{n-1}} \\ 0 & \dfrac{1}{2^n} & \dfrac{1}{2^{n-1}} \\ 0 & 0 & 0 \end{bmatrix}$$

The reader is asked to verify these results in Problem 6, Exercise 6.10. From the equation $X_n = C^nX_0$, we obtain

$$a_n = a_0 + \left(1 - \frac{1}{2^n}\right)b_0 + \left(1 - \frac{1}{2^{n-1}}\right)c_0$$

$$b_n = \frac{1}{2^n}b_0 + \frac{1}{2^{n-1}}c_0$$

$$c_n = 0$$

As n get large,

$$a_n \longrightarrow a_0 + b_0 + c_0$$

$$b_n \longrightarrow 0$$

$$c_n = 0$$

Eventually, almost all the plants in the nth generation will be of type AA.

Exercise 6.10

1. If $B = QAQ^{-1}$, use mathematical induction to prove that $B^k = QA^kQ^{-1}$ for all positive integers k.

2. If $Q = \begin{bmatrix} 1 & 2 \\ 2 & 3 \end{bmatrix}$ and $D = \begin{bmatrix} 2 & 0 \\ 0 & -1 \end{bmatrix}$,
   (a) Find Q^{-1}.
   (b) Find $B = QDQ^{-1}$.
   (c) Find B^5 by calculating $BBBBB$.
   (d) Find B^5 by calculating QD^5Q^{-1}.

3. Let $D = \begin{bmatrix} 1 & 0 & 0 \\ 0 & \frac{1}{2} & 0 \\ 0 & 0 & 2 \end{bmatrix}$ and $P = \begin{bmatrix} 1 & 0 & 1 \\ 0 & 1 & 1 \\ 1 & 0 & 2 \end{bmatrix}$.
   (a) Find P^{-1}.
   (b) Find $B = P^{-1}DP$.
   (c) Find B^5 by calculating $P^{-1}D^5P$.

4. Use Theorem 4.5 to prove that similar matrices have the same determinant.

5. Suppose that $D = [d_{ij}]$ is an $n \times n$ diagonal matrix; that is, $d_{ij} = 0$ if $i \neq j$.
   (a) Show that D^2 is a diagonal matrix with diagonal entries $f_{ii} = d_{ii}^2$ for $i = 1, 2, \ldots, n$. Hint: Let $D^2 = [f_{ij}]$, where $f_{ij} = D_i \cdot (D^{(j)})^T$.
   (b) Use mathematical induction to prove that D^k is a diagonal matrix with diagonal entries d_{ii}^k for $i = 1, 2, \ldots, n$.

6. Under the assumption that a plant of type AA is used to fertilize all existing plants as well as succeeding generations of plants, we obtained the linear system marked (*). The coefficient matrix of this system is $C = \begin{bmatrix} 1 & \frac{1}{2} & 0 \\ 0 & \frac{1}{2} & 1 \\ 0 & 0 & 0 \end{bmatrix}$. If $Q = \begin{bmatrix} 1 & -1 & 1 \\ 0 & 1 & -2 \\ 0 & 0 & 1 \end{bmatrix}$ and $D = \begin{bmatrix} 1 & 0 & 0 \\ 0 & \frac{1}{2} & 0 \\ 0 & 0 & 0 \end{bmatrix}$,

   (a) Find Q^{-1}.
   (b) Show that $C = QDQ^{-1}$.
   (c) Verify that

$$C^n = \begin{bmatrix} 1 & 1 - \dfrac{1}{2^n} & 1 - \dfrac{1}{2^{n-1}} \\ 0 & \dfrac{1}{2^n} & \dfrac{1}{2^{n-1}} \\ 0 & 0 & 0 \end{bmatrix}$$

7. In the plant fertilization problem, suppose that all plants and all succeeding generations of plants are fertilized by a plant of type aa. Determine the vector
$$X_n = \begin{bmatrix} a_n \\ b_n \\ c_n \end{bmatrix}.$$

8. If D is a diagonal matrix with diagonal entries $d_{11}, d_{22}, \ldots, d_{nn}$, and if A is similar to D, what is the determinant of A?

9. If A is similar to B, prove that A^T is similar to B^T.

The **trace** of a square matrix is defined to be the sum of its diagonal elements; that is, if $A = [a_{ij}]$ is an $n \times n$ matrix, then the trace of A is $a_{11} + a_{22} + \cdots + a_{nn}$. We denote the trace of A by Tr (A).

10. (a) Prove that, for $n \times n$ matrices A and B, Tr $(AB) =$ Tr (BA).
    (b) Use the result of part (a) to prove that, if A and B are similar, then Tr (A) = Tr (B).

7 Eigenvalues, Eigenvectors, and Diagonal Matrices

7.1 EIGENVALUES AND EIGENVECTORS OF A LINEAR OPERATOR

In the application to genetics in Section 6.10, the problem of finding the distribution of plants in the nth generation was reduced to calculating the nth power of a matrix B. We were able to calculate B^n because we were given a diagonal matrix D that was similar to B. Hopefully, the reader is curious as to how D was found. The main task of the first two sections of this chapter is to answer the question, given an $n \times n$ matrix A, is there a diagonal matrix D such that $D = P^{-1}AP$? If we determine that such a matrix exists, how do we find it? The answer is obtained by studying linear operators. We know that, under the condition stated, D must be similar to A. But matrices A and D are similar if and only if they are matrix representations of the same linear operator. Thus we attack the problem in the following way. We find a linear operator $T: \mathbf{V} \longrightarrow \mathbf{V}$ and an ordered basis C_1 for $\mathbf{V}$ such that A is the matrix of T relative to C_1. Then we look for another ordered basis C_2 for $\mathbf{V}$ such that the matrix of T relative to C_2 is a diagonal matrix D. If we are successful in finding C_2, then the matrix $D = [T]_{C_2}$ has the property $D = P^{-1}AP$, where P^{-1} is the matrix for the change of basis from C_1 to C_2. Thus the problem of finding a diagonal matrix D has been transferred over to the problem of finding an appropriate basis C_2 for a vector space.

To motivate the definition of eigenvector, we make the following observation. If $C_2 = \{Y_1, Y_2, \ldots, Y_n\}$ is an ordered basis for $\mathbf{V}$, and if $T: \mathbf{V} \longrightarrow \mathbf{V}$ is a linear operator on $\mathbf{V}$ such that

$$[T]_{C_2} = D = [[T(Y_1)]_{C_2} | \ldots | [T(Y_n)]_{C_2}]$$

is a diagonal matrix then

$$[T(Y_t)]_{c_s} = \begin{bmatrix} 0 \\ \cdot \\ \cdot \\ \cdot \\ d_{tt} \\ \cdot \\ \cdot \\ \cdot \\ 0 \end{bmatrix} \quad \text{and} \quad T(Y_t) = d_{tt}Y_t$$

We see that the linear operator T maps each vector Y_t into a scalar multiple of itself.

Definition 7.1. Let $T: \mathbf{V} \longrightarrow \mathbf{V}$ be a linear operator on $\mathbf{V}$. A vector $X \neq \mathbf{0}$ in $\mathbf{V}$ is an **eigenvector (characteristic vector) of** T if and only if there is a scalar λ such that $T(X) = \lambda X$. The scalar λ is called an **eigenvalue (characteristic value) of** T. We say that λ is an eigenvalue *corresponding* to the eigenvector X. We also say that the eigenvector X *belongs* to the eigenvalue λ. The terms **proper vector** and **latent vector** are also used as synonyms for eigenvector. Similarly, **proper value** and **latent value** are used as synonyms for eigenvalue.

Example 7.1. Find the eigenvectors and eigenvalues of the linear operator $T: \mathbf{R}_2 \longrightarrow \mathbf{R}_2$ defined by $T\begin{bmatrix} x \\ y \end{bmatrix} = \begin{bmatrix} 2x \\ x + 3y \end{bmatrix}$.

Solution and discussion: If $\begin{bmatrix} x \\ y \end{bmatrix}$ is an eigenvector of T, then $\begin{bmatrix} 2x \\ x + 3y \end{bmatrix} = \lambda \begin{bmatrix} x \\ y \end{bmatrix}$ for some scalar λ, and not both of x and y are zero. So $2x = \lambda x$ and $x + 3y = \lambda y$. If $x \neq 0$, then $\lambda = 2$ and $x + 3y = 2y$. Therefore, $x = -y$. Thus any vector of the form $\begin{bmatrix} x \\ -x \end{bmatrix}$ such that $x \neq 0$ is an eigenvector belonging to the eigenvalue 2. As a check, we use the definition of T and find

$$T\begin{bmatrix} x \\ -x \end{bmatrix} = \begin{bmatrix} 2x \\ -2x \end{bmatrix} = 2\begin{bmatrix} x \\ -x \end{bmatrix}$$

If $x = 0$, then $3y = \lambda y$ and $\lambda = 3$. Thus vectors of the form $\begin{bmatrix} 0 \\ y \end{bmatrix}$ and $y \neq 0$ are eigenvectors belonging to the eigenvalue 3. Checking, we see that

$$T\begin{bmatrix} 0 \\ y \end{bmatrix} = \begin{bmatrix} 0 \\ 3y \end{bmatrix} = 3\begin{bmatrix} 0 \\ y \end{bmatrix}$$

The set of eigenvectors

$$C_2 = \left\{ Y_1 = \begin{bmatrix} 1 \\ -1 \end{bmatrix}, Y_2 = \begin{bmatrix} 0 \\ 1 \end{bmatrix} \right\}$$

is a basis for $\mathbf{R}_2$. Furthermore, since

$$T\begin{bmatrix} 1 \\ -1 \end{bmatrix} = 2\begin{bmatrix} 1 \\ -1 \end{bmatrix} + 0\begin{bmatrix} 0 \\ 1 \end{bmatrix} \quad \text{and} \quad T\begin{bmatrix} 0 \\ 1 \end{bmatrix} = 0\begin{bmatrix} 1 \\ -1 \end{bmatrix} + 3\begin{bmatrix} 0 \\ 1 \end{bmatrix}$$

it follows that the matrix of T relative to the ordered basis C_2 is the diagonal matrix $\begin{bmatrix} 2 & 0 \\ 0 & 3 \end{bmatrix}$.

We can describe the situation geometrically as follows. The vectors Y_1 and Y_2 determine a coordinate system (see Figure 7.1). Relative to this coordinate system, T can be described in a rather simple way. Vectors along Y_1 are stretched by a factor of 2, and vectors along Y_2 are stretched by a factor of 3. If X is an arbitrary vector in $\mathbf{R}_2$, and the coordinate vector of X relative to C_2 is $\begin{bmatrix} a \\ b \end{bmatrix}$, then the coordinate vector of the image $T(X)$ is

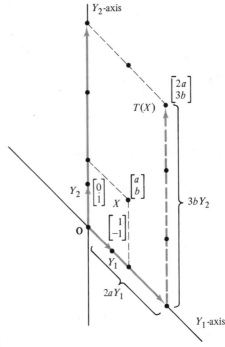

Figure 7.1

$$\begin{bmatrix} 2 & 0 \\ 0 & 3 \end{bmatrix} \begin{bmatrix} a \\ b \end{bmatrix} = \begin{bmatrix} 2a \\ 3b \end{bmatrix}.$$ Therefore, the effect of T is to stretch X by a factor of 2 along Y_1 and a factor of 3 along Y_2.

Theorem 7.1. Let $\mathbf{V}$ be an n-dimensional vector space, let $T: \mathbf{V} \to \mathbf{V}$ be a linear operator on $\mathbf{V}$, and let $C_1 = \{X_1, X_2, \ldots, X_n\}$ be an ordered basis for $\mathbf{V}$. Then the matrix of T relative to C_1 is a diagonal matrix if and only if each vector in the basis C_1 is an eigenvector of T. Furthermore, if $[T]_{C_1}$ is a diagonal matrix and the (i, i) entry of $[T]_{C_1}$ is d_{ii}, then d_{ii} is the eigenvalue corresponding to the eigenvector X_i.

Proof: Suppose that T is represented by a diagonal matrix

$$D = \begin{bmatrix} d_{11} & 0 & \cdots & 0 & 0 \\ 0 & & d_{22} & & 0 \\ \cdot & & \cdot & & \cdot \\ \cdot & & \cdot & & \cdot \\ \cdot & & \cdot & & \cdot \\ 0 & \cdots & 0 & \cdots & d_{nn} \end{bmatrix}$$

which we denote by $\mathrm{diag}(d_{11}, d_{22}, \ldots, d_{nn})$. For each vector $X_i, i = 1, 2, \ldots, n$, we have $D[X_i]_{C_1} = [T(X_i)]_{C_1}$. But $[X_i]_{C_1} = E_i$ (the $n \times 1$ column vector with ith entry 1 and all other entries 0); therefore, $D[X_i]_{C_1}$ is $D^{(i)} = d_{ii}E_i$ and $T(X_i) = d_{ii}X_i$. Thus X_i is an eigenvector of T for $i = 1, 2, \ldots, n$, and d_{ii} is an eigenvalue corresponding to the eigenvector X_i.

Conversely, suppose that C_1 is a basis of eigenvectors. Let λ_i be the eigenvalue corresponding to the eigenvector X_i. Then $T(X_i) = \lambda_i X_i$ and $[T(X_i)]_{C_1} = \lambda_i E_i$. Hence the matrix of T relative to C_1 is

$$[T]_{C_1} = [\lambda_1 E_1 \,|\, \lambda_2 E_2 \,|\, \cdots \,|\, \lambda_n E_n] = \mathrm{diag}(\lambda_1, \lambda_2, \ldots, \lambda_n) \quad \text{Q.E.D.}$$

In the next example we illustrate a technique for finding the eigenvalues and eigenvectors of a linear operator.

Example 7.2. Let $C_1 = \left\{ E_1 = \begin{bmatrix} 1 \\ 0 \end{bmatrix}, E_2 = \begin{bmatrix} 0 \\ 1 \end{bmatrix} \right\}$ be the natural basis for $\mathbf{R}_2$, and let $T: \mathbf{R}_2 \to \mathbf{R}_2$ be the linear operator defined by $T\begin{bmatrix} x \\ y \end{bmatrix} = \begin{bmatrix} x + 2y \\ -x + 4y \end{bmatrix}$. Use $[T]_{C_1}$ to find the eigenvectors and corresponding eigenvalues of T. Determine if there is a basis C_2 of eigenvectors for $\mathbf{R}_2$ and, if so, find $[T]_{C_2}$.

Solution: The matrix of T relative to C_1 is

$$[T]_{C_1} = [[T(E_1)]_{C_1} \,|\, [T(E_2)]_{C_1}] = \begin{bmatrix} 1 & 2 \\ -1 & 4 \end{bmatrix} = A$$

We shall use the symbol $\Longleftrightarrow$ to denote "equivalent" or "if and only if."
$X = \begin{bmatrix} x \\ y \end{bmatrix}$ is an eigenvector of $T \Longleftrightarrow T(X) = \lambda X$ for some

$$\lambda \Longleftrightarrow A[X]_{C_1} = [\lambda X]_{C_1} \Longleftrightarrow A\begin{bmatrix} x \\ y \end{bmatrix} = \begin{bmatrix} \lambda x \\ \lambda y \end{bmatrix} = \lambda \begin{bmatrix} x \\ y \end{bmatrix}$$

$$\Longleftrightarrow A\begin{bmatrix} x \\ y \end{bmatrix} - \lambda \begin{bmatrix} x \\ y \end{bmatrix} = \begin{bmatrix} 0 \\ 0 \end{bmatrix} \Longleftrightarrow (A - \lambda I_2)\begin{bmatrix} x \\ y \end{bmatrix} = \begin{bmatrix} 0 \\ 0 \end{bmatrix}$$

Now $A - \lambda I_2$ is the matrix of $T - \lambda I$ relative to the basis C_1. Since $X \neq \mathbf{0}$
and $(T - \lambda I)(X) = \mathbf{0}$, we see that $T - \lambda I$ is not invertible. But then $A - \lambda I_2$
is not invertible. This is true if and only if $\det(A - \lambda I_2) = 0$. Thus

$$\begin{vmatrix} 1 - \lambda & 2 \\ -1 & 4 - \lambda \end{vmatrix} = 0 \Longleftrightarrow (1 - \lambda)(4 - \lambda) + 2 = 0 \Longleftrightarrow \lambda^2 - 5\lambda + 6 = 0$$

$$\Longleftrightarrow (\lambda - 3)(\lambda - 2) = 0$$

Hence λ is an eigenvalue of T if and only if $\lambda = 2$ or $\lambda = 3$. We proceed to
find the eigenvectors belonging to 2. $X = \begin{bmatrix} x \\ y \end{bmatrix}$ is an eigenvector belonging to

$$2 \Longleftrightarrow A\begin{bmatrix} x \\ y \end{bmatrix} = 2\begin{bmatrix} x \\ y \end{bmatrix} \Longleftrightarrow (A - 2I_2)\begin{bmatrix} x \\ y \end{bmatrix} = \begin{bmatrix} 0 \\ 0 \end{bmatrix}$$

$$\Longleftrightarrow \begin{bmatrix} 1 - 2 & 2 \\ -1 & 4 - 2 \end{bmatrix}\begin{bmatrix} x \\ y \end{bmatrix} = \begin{bmatrix} 0 \\ 0 \end{bmatrix}$$

$$\Longleftrightarrow -x + 2y = 0 \Longleftrightarrow x = 2y$$

Therefore, the eigenvectors of T belonging to 2 are of the form $\begin{bmatrix} 2c \\ c \end{bmatrix}$, $c \neq 0$.
As a check, we find

$$T\begin{bmatrix} 2c \\ c \end{bmatrix} = \begin{bmatrix} 4c \\ 2c \end{bmatrix} = 2\begin{bmatrix} 2c \\ c \end{bmatrix}$$

Next we find the eigenvectors belonging to 3. $X = \begin{bmatrix} x \\ y \end{bmatrix}$ is an eigenvector belonging to

$$3 \Longleftrightarrow A\begin{bmatrix} x \\ y \end{bmatrix} = 3\begin{bmatrix} x \\ y \end{bmatrix} \Longleftrightarrow (A - 3I_2)\begin{bmatrix} x \\ y \end{bmatrix} = \begin{bmatrix} 0 \\ 0 \end{bmatrix}$$

$$\Longleftrightarrow \begin{bmatrix} 1 - 3 & 2 \\ -1 & 4 - 3 \end{bmatrix}\begin{bmatrix} x \\ y \end{bmatrix} = \begin{bmatrix} 0 \\ 0 \end{bmatrix} \Longleftrightarrow \begin{array}{l} -2x + 2y = 0 \\ -x + y = 0 \end{array}$$

$$\Longleftrightarrow x = y$$

So the eigenvectors belonging to 3 are of the form $\begin{bmatrix} c \\ c \end{bmatrix}$. Checking, we find that

$$T\begin{bmatrix} c \\ c \end{bmatrix} = \begin{bmatrix} 3c \\ 3c \end{bmatrix} = 3\begin{bmatrix} c \\ c \end{bmatrix}$$

The set of eigenvectors $C_2 = \left\{ X_1 = \begin{bmatrix} 2 \\ 1 \end{bmatrix}, X_2 = \begin{bmatrix} 1 \\ 1 \end{bmatrix} \right\}$ is a basis for $\mathbf{R}_2$.

Relative to this basis, the matrix of T is $[T]_{C_2} = \begin{bmatrix} 2 & 0 \\ 0 & 3 \end{bmatrix}$. For an arbitrary vec-

tor X with coordinate vector $[X]_{C_2} = \begin{bmatrix} a \\ b \end{bmatrix}$ $(X = aX_1 + bX_2)$, the coordinate

vector of the image $T(X)$ relative to C_2 is $\begin{bmatrix} 2 & 0 \\ 0 & 3 \end{bmatrix}\begin{bmatrix} a \\ b \end{bmatrix} = \begin{bmatrix} 2a \\ 3b \end{bmatrix}$.

In Definition 7.1 we stated that an eigenvector was a nonzero vector. A reason for excluding zero is that, no matter how a linear operator T is defined, $T(\mathbf{0}) = \mathbf{0} = \lambda\mathbf{0}$ *for any* scalar λ. If we allowed $\mathbf{0}$ to be an eigenvector, it would belong to every scalar λ, and no useful information would be obtained. However, the zero scalar is a permissible eigenvalue, and it does yield information about T. For example, if $T: \mathbf{R}_2 \longrightarrow \mathbf{R}_2$ has the matrix representation $\begin{bmatrix} 4 & 0 \\ 0 & 0 \end{bmatrix}$ relative to some basis $C_2 = \{Y_1, Y_2\}$, then Y_2 is an eigenvector belonging to 0. All vectors along the axis determined by $\mathbf{0}$ and Y_2 are mapped into $\mathbf{0}$, because $T(rY_2) = rT(Y_2) = r\mathbf{0} = \mathbf{0}$. Next we state and prove a theorem which generalizes Example 7.2.

Theorem 7.2. Let $T: \mathbf{V} \longrightarrow \mathbf{V}$ be a linear operator on $\mathbf{V}$, let $C_1 = \{X_1, X_2, \dots, X_n\}$ be an ordered basis for $\mathbf{V}$, and let $A = [T]_{C_1}$ be the matrix of T relative to C_1. Then

(i) A nonzero vector X in $\mathbf{V}$ is an eigenvector of T with corresponding eigenvalue λ_0 if and only if $A[X]_{C_1} = \lambda_0[X]_{C_1}$. This is true if and only if $(A - \lambda_0 I_n)[X]_{C_1} = \mathbf{0}$.

(ii) λ_0 is a (real) eigenvalue of T if and only if λ_0 is a (real) root of the equation $\det(A - \lambda I_n) = 0$.

Proof: (i) Let X be a nonzero vector in V. X is an eigenvector of T corresponding to

$$\lambda_0 \Longleftrightarrow T(X) = \lambda_0 X \Longleftrightarrow A[X]_{C_1} = [\lambda_0 X]_{C_1} \Longleftrightarrow A[X]_{C_1} = \lambda_0[X]_{C_1}$$

This is the first conclusion. Also,

$$A[X]_{C_1} = \lambda_0[X]_{C_1} \Longleftrightarrow A[X]_{C_1} = \lambda_0 I_n[X]_{C_1} \Longleftrightarrow (A - \lambda_0 I_n)[X]_{C_1} = \mathbf{0}$$

(ii) Suppose that λ_0 is an eigenvalue of T. This is true if and only if there is a nonzero vector X in $\mathbf{V}$ such that $T(X) = \lambda_0 X$. By part (i), $T(X) = \lambda_0 X$ if and only if $(A - \lambda_0 I_n)[X]_{C_1} = 0$. Since $[X]_{C_1} \neq 0$, the matrix $A - \lambda_0 I_n$ is singular (Exercise 3.8, Problem 13, or Exercise 7.1, Problem 12). By Theorem 4.4, $A - \lambda_0 I_n$ is singular if and only if $\det(A - \lambda_0 I_n) = 0$. Hence the scalar λ_0 is an eigenvalue of T if and only if λ_0 is a real root of the equation $\det(A - \lambda I_n) = 0$.   Q.E.D.

Example 7.3. If $T: \mathbf{R}_3 \longrightarrow \mathbf{R}_3$ is the linear operator defined by

$$T\begin{bmatrix} x \\ y \\ z \end{bmatrix} = \begin{bmatrix} 3x + 2y + 4z \\ 2x + 2z \\ 4x + 2y + 3z \end{bmatrix}, \text{ find the eigenvectors and corresponding eigen-}$$

values of T. Determine if there is a basis C_2 of eigenvectors of T, and if so find a diagonal matrix representation of T.

Solution: We let C_1 be the natural basis for $\mathbf{R}_3$ and use Theorem 7.2.

$$[T]_{C_1} = A = \begin{bmatrix} 3 & 2 & 4 \\ 2 & 0 & 2 \\ 4 & 2 & 3 \end{bmatrix}$$

$$\det(A - \lambda I_3) = \begin{vmatrix} 3 - \lambda & 2 & 4 \\ 2 & -\lambda & 2 \\ 4 & 2 & 3 - \lambda \end{vmatrix} = -\lambda^3 + 6\lambda^2 + 15\lambda + 8$$

$$= 0 \Longleftrightarrow \lambda^3 - 6\lambda^2 - 15\lambda - 8 = 0$$

We test to see if there are integer roots. (Integer roots may or may not exist.) If there is an integer root, it must be a divisor of -8. (The following is a more general test for possible rational roots: if $a_1, a_2, \ldots, a_n$ are integers and $a_n \neq 0$, and if p/q is a rational root of the equation $a_0 + a_1 x + \cdots + a_n x^n = 0$, then p is a divisor of a_0 and q is a divisor of a_n.) We see that $(-1)^3 - 6(-1)^2 - 15(-1) - 8 = 0$. Therefore, -1 is a root and $\lambda - (-1) = \lambda + 1$ is a factor of $\lambda^3 - 6\lambda^2 - 15\lambda - 8 = 0$. By long division we find that the other factor is $\lambda^2 - 7\lambda - 8 = (\lambda + 1)(\lambda - 8)$. The eigenvalues are seen to be -1 and 8.

We find the eigenvectors belonging to -1. Since C_1 is the natural basis, we have $[X]_{C_1} = X$. By Theorem 7.2(i), X belongs to $-1 \Longleftrightarrow (A - (-1)I_3)X = 0$. Solving for X by the Gaussian reduction method, we have

$$A + I_3 = \begin{bmatrix} 4 & 2 & 4 \\ 2 & 1 & 2 \\ 4 & 2 & 4 \end{bmatrix} \xrightarrow[\substack{-R_1 + R_3 \\ (1/2)R_1 \\ -R_1 + R_2}]{} \begin{bmatrix} 2 & 1 & 2 \\ 0 & 0 & 0 \\ 0 & 0 & 0 \end{bmatrix}$$

The general solution may be written as

$$\begin{bmatrix} x \\ y \\ z \end{bmatrix} = y \begin{bmatrix} -\frac{1}{2} \\ 1 \\ 0 \end{bmatrix} + z \begin{bmatrix} -1 \\ 0 \\ 1 \end{bmatrix}$$

Thus the collection of eigenvectors belonging to -1 is the space

$$\left\langle \begin{bmatrix} -\frac{1}{2} \\ 1 \\ 0 \end{bmatrix}, \begin{bmatrix} -1 \\ 0 \\ 1 \end{bmatrix} \right\rangle = \left\langle \begin{bmatrix} -1 \\ 2 \\ 0 \end{bmatrix}, \begin{bmatrix} -1 \\ 0 \\ 1 \end{bmatrix} \right\rangle$$

with $\mathbf{0}$ excluded.

Next we find the eigenvectors belonging to 8. Solving the equation $(A - 8I_3)X = \mathbf{0}$ by Gaussian reduction, we get

$$\begin{bmatrix} -5 & 2 & 4 \\ 2 & -8 & 2 \\ 4 & 2 & -5 \end{bmatrix} \xrightarrow[\substack{R_3+R_1 \\ 2R_1+R_2 \\ 4R_1+R_3}]{} \begin{bmatrix} -1 & 4 & -1 \\ 0 & 0 & 0 \\ 0 & 18 & -9 \end{bmatrix} \xrightarrow[\substack{(1/18)R_3 \\ R_{23} \\ -1R_1 \\ 4R_2+R_1}]{} \begin{bmatrix} 1 & 0 & -1 \\ 0 & 1 & -\frac{1}{2} \\ 0 & 0 & 0 \end{bmatrix}$$

Writing the general solution as

$$\begin{bmatrix} x \\ y \\ z \end{bmatrix} = z \begin{bmatrix} 1 \\ \frac{1}{2} \\ 1 \end{bmatrix}$$

we conclude that the set of eigenvectors belonging to 8 is the space

$$\left\langle \begin{bmatrix} 1 \\ \frac{1}{2} \\ 1 \end{bmatrix} \right\rangle = \left\langle \begin{bmatrix} 2 \\ 1 \\ 2 \end{bmatrix} \right\rangle$$

with zero excluded. The set of eigenvectors

$$C_2 = \left\{ X_1 = \begin{bmatrix} -1 \\ 2 \\ 0 \end{bmatrix}, X_2 = \begin{bmatrix} -1 \\ 0 \\ 1 \end{bmatrix}, X_3 = \begin{bmatrix} 2 \\ 1 \\ 2 \end{bmatrix} \right\}$$

is a basis for $\mathbf{R}_3$. Relative to this ordered basis the matrix of T is the diagonal matrix

$$[T]_{C_2} = \begin{bmatrix} -1 & 0 & 0 \\ 0 & -1 & 0 \\ 0 & 0 & 8 \end{bmatrix}$$

Geometrically, the situation is this. The vectors in C_2 determine a coordinate system in 3-space. Relative to this coordinate system, T can be described in the following way. Any vector along the X_1 axis (the line determined by $\mathbf{0}$

and X_1) is "stretched" by a factor of -1; that is, it is mapped onto its additive inverse. Similarly, for vectors along the X_2 axis. Vectors along the X_3 axis are stretched by a factor of 8. If $X = \begin{bmatrix} x \\ y \\ z \end{bmatrix}$ is a vector in 3-space with coordinate

vector $\begin{bmatrix} a \\ b \\ c \end{bmatrix}$ relative to C_2, then the image has coordinate vector

$$\begin{bmatrix} -1 & 0 & 0 \\ 0 & -1 & 0 \\ 0 & 0 & 8 \end{bmatrix} \begin{bmatrix} a \\ b \\ c \end{bmatrix} = \begin{bmatrix} -a \\ -b \\ 8c \end{bmatrix}$$

Example 7.4. Determine whether or not there is a basis C_2 for $\mathbf{R}_2$ such that the matrix of T relative to C_2 is a diagonal matrix, if $T: \mathbf{R}_2 \longrightarrow \mathbf{R}_2$ is defined by

$$T\begin{bmatrix} x \\ y \end{bmatrix} = \begin{bmatrix} x - 2y \\ 4x + 3y \end{bmatrix}.$$

Solution: The matrix of T relative to the standard basis is $A = \begin{bmatrix} 1 & -2 \\ 4 & 3 \end{bmatrix}$. If λ is an eigenvalue, then

$$\det(A - \lambda I_2) = \begin{vmatrix} 1 - \lambda & -2 \\ 4 & 3 - \lambda \end{vmatrix} = \lambda^2 - 4\lambda + 11 = 0$$

The discriminant $(b^2 - 4ac)$ is $16 - 44 < 0$. Therefore, $|A - \lambda I_2| = 0$ has no real solutions. Hence there are no eigenvectors for T. Applying Theorem 7.1, we see that there is no diagonal matrix that represents T.

Exercise 7.1

In Problems 1–9, use the method of Examples 7.2–7.4 to
(a) Find the real eigenvalues and eigenvectors of the given linear operators.
(b) Determine whether or not there is a basis C_2 of eigenvectors for the domain of T.
(c) Find a diagonal matrix that represents T if there is one.

1. $T: \mathbf{R}_2 \longrightarrow \mathbf{R}_2, \; T\begin{bmatrix} x \\ y \end{bmatrix} = \begin{bmatrix} 2x + y \\ 2x + 3y \end{bmatrix}.$

2. $T: \mathbf{R}_3 \longrightarrow \mathbf{R}_3, \; T\begin{bmatrix} x \\ y \\ z \end{bmatrix} = \begin{bmatrix} x \\ x + y \\ x + y + z \end{bmatrix}.$

3. $T: \mathbf{P}_2 \longrightarrow \mathbf{P}_2$, $T(f: a + bx) = g_f: (-2a + 2b) + (-2a + 3b)x$.

4. $T: \mathbf{R}_3 \longrightarrow \mathbf{R}_3$, $T\begin{bmatrix} x \\ y \\ z \end{bmatrix} = \begin{bmatrix} 2x \\ 2y \\ 3z \end{bmatrix}$.

5. $T: \mathbf{P}_2 \longrightarrow \mathbf{P}_2$, $T(f: a + bx) = f': b$.

6. $T: \mathbf{R}_3 \longrightarrow \mathbf{R}_3$, $T\begin{bmatrix} x \\ y \\ z \end{bmatrix} = \begin{bmatrix} 3x - \frac{1}{2}y + \frac{1}{2}z \\ 4x - z \\ 4x - 2y + z \end{bmatrix}$.

7. $T: \mathbf{P}_3 \longrightarrow \mathbf{P}_3$, $T(f: a + bx + cx^2) = f': b + 2cx$.

8. $T: \mathbf{R}_4 \longrightarrow \mathbf{R}_4$, $T\begin{bmatrix} x \\ y \\ z \\ w \end{bmatrix} = \begin{bmatrix} 3x \\ 2y \\ x + 2z \\ 2w \end{bmatrix}$.

9. $T: \mathbf{R}_3 \longrightarrow \mathbf{R}_3$, $T\begin{bmatrix} x \\ y \\ z \end{bmatrix} = \begin{bmatrix} x + z \\ 0 \\ -x + y + \frac{7}{2}z \end{bmatrix}$.

10. If $T_\theta: \mathbf{R}_2 \longrightarrow \mathbf{R}_2$ is defined by $T_\theta(X) = AX$ and $A = \begin{bmatrix} \cos\theta & -\sin\theta \\ \sin\theta & \cos\theta \end{bmatrix}$, then T_θ is a rotation counterclockwise through θ.
   (a) For what values of θ are there real eigenvalues?
   (b) Find the eigenvectors belonging to any eigenvalues found in part (a).
   (c) Is there a diagonal matrix representation for T_θ for some θ?
   (d) Interpret the results geometrically.

11. Let $T: \mathbf{R}_2 \longrightarrow \mathbf{R}_2$ be the reflection of the plane $\mathbf{R}_2$ through the line $y = x$.
   (a) Find the matrix representation for T relative to the natural basis.
   (b) Find the eigenvalues and the eigenvectors belonging to the eigenvalues.
   (c) Find a basis C_2 for $\mathbf{R}_2$ such that $[T]_{C_2}$ is a diagonal matrix.
   (d) Interpret the transformation geometrically relative to the coordinate system determined by C_2.

12. Use Theorems 6.7 and 6.15 to prove that an $n \times n$ matrix A is nonsingular if and only if the only solution to the matrix equation $AX = 0$ is $X = 0$.

7.2 EIGENVALUES AND EIGENVECTORS OF A SQUARE MATRIX

According to Theorem 7.2(ii), if $T: \mathbf{V} \longrightarrow \mathbf{V}$ is a linear operator on an n-dimensional vector space $\mathbf{V}$, and if A is the matrix of T relative to some ordered bases for $\mathbf{V}$, then the (real) eigenvalues of T are the real roots of the equation $\det(A - \lambda I_n) = 0$. This equation is a polynomial equation of the form

$$(-1)^n \lambda^n + b_1 \lambda^{n-1} + b_2 \lambda^{n-2} + \cdots + b_{n-1} \lambda + b_n = 0$$

where the b's are real numbers. The reader is probably aware of the fact that every polynomial equation of degree n with real coefficients has n or less complex roots (exactly n roots if we agree to count roots that are repeated k times as k roots). The nonreal complex roots of $\det(A - \lambda I_n) = 0$ are also called eigenvalues, but we shall consider only *real eigenvalues*. An eigenvalue of T is a scalar, and in our treatment of vector spaces a scalar is a real number. One can study vector spaces where the scalars are complex numbers; in fact, in more advanced treatments of vector spaces the only restriction on the collection of scalars is that they be elements of a mathematical system called a field.

If the matrix $A = [T]_{c_1}$ satisfies the hypothesis of Theorem 7.2, then $T(X) = \lambda(X)$ if and only if $A[X]_{c_1} = \lambda[X]_{c_1}$. For this reason we are led to define the eigenvalues and eigenvectors of a matrix.

Definition 7.2. If A is an $n \times n$ matrix a nonzero $n \times 1$ column vector X is called an **eigenvector (characteristic vector)** of A if and only if $AX = \lambda X$ for some scalar λ. λ is called an **eigenvalue (characteristic value)** of A that corresponds to the vector X, and we say that X belongs to the eigenvalue λ.

The proof of the next theorem, which is essentially a rephrasing of Theorem 7.2, is left as an exercise for the reader (Problem 10, Exercise 7.2).

Theorem 7.3. Let A be an $n \times n$ matrix and let X be a nonzero $n \times 1$ column vector.
(i) X is an eigenvector of A belonging to λ_0 if and only if $(A - \lambda_0 I_n)X = \mathbf{0}$.
(ii) A scalar λ_0 is an eigenvalue of A if and only if λ_0 is a (real) root of the polynomial equation $\det(A - \lambda_0 I_n) = 0$.

Definition 7.3. Let A be an $n \times n$ matrix. The nth-degree polynomial $\det(A - \lambda I_n)$ is called the **characteristic polynomial** of A, and $\det(A - \lambda I_n) = 0$ is called the **characteristic equation** of A. The real roots of the characteristic equation of A are the (real) eigenvalues or characteristic values of A.

Note: Some authors define the characteristic equation of A to be the equation $\det(\lambda I_n - A) = 0$. Since $\det(A - \lambda I_n) = \pm\det(\lambda I_n - A)$, it is clear that under either definition of characteristic equation we obtain the same characteristic values.

Example 7.5. Find the characteristic values and characteristic vectors of the matrix $A = \begin{bmatrix} -1 & 2 & 0 \\ -4 & 5 & 0 \\ -4 & 2 & 3 \end{bmatrix}$. Determine if A is similar to a diagonal matrix D; if it is, find P and P^{-1} such that $D = P^{-1}AP$.

Solution: According to Theorem 7.3(ii), the characteristic values of A are the real roots of $\det(A - \lambda I_3) = 0$.

$$|A - \lambda I_3| = \begin{vmatrix} -1-\lambda & 2 & 0 \\ -4 & 5-\lambda & 0 \\ -4 & 2 & 3-\lambda \end{vmatrix}$$

$$= (\lambda^2 - 4\lambda + 3)(3 - \lambda) = 0 \Longleftrightarrow \lambda = 3 \quad \text{or} \quad \lambda = 1$$

Therefore, the eigenvalues of A are 3 and 1. To determine the eigenvectors belonging to 3, we find the nonzero solutions to $(A - 3I_3)X = \mathbf{0}$.

$$A - 3I_3 = \begin{bmatrix} -4 & 2 & 0 \\ -4 & 2 & 0 \\ -4 & 2 & 0 \end{bmatrix} \longrightarrow \begin{bmatrix} 1 & -\frac{1}{2} & 0 \\ 0 & 0 & 0 \\ 0 & 0 & 0 \end{bmatrix}$$

Using the method of Example 5.26, we see that the eigenvectors are the nonzero solutions in the space

$$\left\langle \begin{bmatrix} \frac{1}{2} \\ 1 \\ 0 \end{bmatrix}, \begin{bmatrix} 0 \\ 0 \\ 1 \end{bmatrix} \right\rangle = \left\langle \begin{bmatrix} 1 \\ 2 \\ 0 \end{bmatrix}, \begin{bmatrix} 0 \\ 0 \\ 1 \end{bmatrix} \right\rangle$$

Next we determine the eigenvectors belonging to 1.

$$A - I_3 = \begin{bmatrix} -2 & 2 & 0 \\ -4 & 4 & 0 \\ -4 & 2 & 2 \end{bmatrix} \xrightarrow[\substack{4R_1+R_2 \\ 4R_1+R_3}]{-(1/2)R_1} \begin{bmatrix} 1 & -1 & 0 \\ 0 & 0 & 0 \\ 0 & -2 & 2 \end{bmatrix} \xrightarrow[\substack{R_2 + R_1}]{\substack{R_{23} \\ -(1/2)R_2}} \begin{bmatrix} 1 & 0 & -1 \\ 0 & 1 & -1 \\ 0 & 0 & 0 \end{bmatrix}$$

Therefore, the eigenvectors are the nonzero solutions in the space $\left\langle \begin{bmatrix} 1 \\ 1 \\ 1 \end{bmatrix} \right\rangle$.

$$C_2 = \left\{ Y_1 = \begin{bmatrix} 1 \\ 2 \\ 0 \end{bmatrix}, Y_2 = \begin{bmatrix} 0 \\ 0 \\ 1 \end{bmatrix}, Y_3 = \begin{bmatrix} 1 \\ 1 \\ 1 \end{bmatrix} \right\}$$

is a basis of eigenvectors, and thus A is similar to a diagonal matrix. Let C_1 be the natural basis for $\mathbf{R}_3$. The matrix for the change of basis from C_2 to C_1 is

$$P = [I]_{c_2 c_1} = [[I(Y_1)]_{c_1} \,|\, [I(Y_2)]_{c_1} \,|\, [I(Y_3)]_{c_1}] = [Y_1 \,|\, Y_2 \,|\, Y_3] = \begin{bmatrix} 1 & 0 & 1 \\ 2 & 0 & 1 \\ 0 & 1 & 1 \end{bmatrix}$$

Also, $E_1 = -Y_1 - 2Y_2 + 2Y_3$, $E_2 = Y_1 + Y_2 - Y_3$, and $E_3 = Y_2$. Thus the matrix for the change of basis from C_1 to C_2 is

$$P^{-1} = [I]_{c_1 c_2} = \begin{bmatrix} -1 & 1 & 0 \\ -2 & 1 & 1 \\ 2 & -1 & 0 \end{bmatrix}$$

Hence

$$D = P^{-1}AP = \begin{bmatrix} -1 & 1 & 0 \\ -2 & 1 & 1 \\ 2 & -1 & 0 \end{bmatrix} \begin{bmatrix} -1 & 2 & 0 \\ -4 & 5 & 0 \\ -4 & 2 & 3 \end{bmatrix} \begin{bmatrix} 1 & 0 & 1 \\ 2 & 0 & 1 \\ 0 & 1 & 1 \end{bmatrix}$$

$$= \begin{bmatrix} -3 & 3 & 0 \\ -6 & 3 & 3 \\ 2 & -1 & 0 \end{bmatrix} \begin{bmatrix} 1 & 0 & 1 \\ 2 & 0 & 1 \\ 0 & 1 & 1 \end{bmatrix} = \begin{bmatrix} 3 & 0 & 0 \\ 0 & 3 & 0 \\ 0 & 0 & 1 \end{bmatrix}$$

We note that if $T: \mathbf{R}_3 \longrightarrow \mathbf{R}_3$ is the linear operator on $\mathbf{R}_3$ defined by $T(X) = AX$, then the matrix of T relative to the natural basis is A. The matrix of T relative to the basis C_2 of eigenvectors is the diagonal matrix D.

The next theorem enables us to define the characteristic polynomial of a linear transformation.

Theorem 7.4. Let $\mathbf{V}$ be an n-dimensional vector space, and let $T: \mathbf{V} \longrightarrow \mathbf{V}$ be a linear operator defined on $\mathbf{V}$. If A and B are any two matrix representations of T, then A and B have the same characteristic polynomial.

Proof: Suppose that A and B are $n \times n$ matrices and that C_1 and C_2 are ordered bases for $\mathbf{V}$ such that $A = [T]_{C_1}$ and $B = [T]_{C_2}$. By Theorem 6.22, $B = P^{-1}AP$, where P is the matrix for the change of basis from C_2 to C_1. Then

$$\det(B - \lambda I_n) = \det(P^{-1}AP - \lambda I_n) = \det(P^{-1}AP - \lambda P^{-1}I_nP)$$
$$= \det(P^{-1}(AP - \lambda I_nP)) = \det(P^{-1}(A - \lambda I_n)P)$$
$$= \det P^{-1} \det(A - \lambda I_n) \det P = \det P^{-1} \det P \det(A - \lambda I_n)$$
$$= \det(P^{-1}P) \det(A - \lambda I_n) = \det I_n \det(A - \lambda I_n)$$
$$= \det(A - \lambda I_n) \qquad\qquad \text{Q.E.D.}$$

Definition 7.4. The **characteristic polynomial** of a linear operator $T: \mathbf{V} \longrightarrow \mathbf{V}$ on an n-dimensional vector space $\mathbf{V}$ is the characteristic polynomial of any one of the matrix representations of T.

If A and B are similar matrices, the proof of Theorem 7.4 shows that they have the same characteristic polynomial. The converse is *not* true, as the next example illustrates.

Example 7.6. Show that $A = \begin{bmatrix} 0 & 0 \\ 0 & 0 \end{bmatrix}$ and $B = \begin{bmatrix} 0 & 1 \\ 0 & 0 \end{bmatrix}$ have the same characteristic polynomial, but that A is not similar to B.

Solution

$$\det(A - \lambda I_2) = \begin{vmatrix} -\lambda & 0 \\ 0 & -\lambda \end{vmatrix} = \lambda^2 = \begin{vmatrix} -\lambda & 1 \\ 0 & -\lambda \end{vmatrix} = \det(B - \lambda I_2)$$

So A and B have the same characteristic polynomial. For any nonsingular 2×2 matrix P, $P^{-1}AP = 0 \neq B$. Therefore, B is not similar to A.

As we have seen, a linear operator may not have a matrix representation that is a diagonal matrix. (Equivalently, an $n \times n$ matrix may not be similar to a diagonal matrix.) However, if T is a linear operator on a vector space of dimension n, and if T has n distinct eigenvalues, we can show that T has a diagonal matrix representation. The next lemma paves the way for this result.

Lemma 7.1. If $\lambda_1, \lambda_2, \ldots, \lambda_k$ are distinct eigenvalues of a linear operator $T: \mathbf{V} \longrightarrow \mathbf{V}$, and if $X_1, X_2, \ldots, X_k$ are eigenvectors such that X_i belongs to λ_i for $i = 1, 2, \ldots, k$, then $C = \{X_1, X_2, \ldots, X_k\}$ is a linearly independent set.

Proof: We use an indirect proof to obtain the conclusion. If the nonzero eigenvectors are linearly dependent, we can apply Theorem 5.5 to obtain a first vector (scanning the vectors from left to right) X_t, such that X_t is a linear combination of the preceding vectors. Then

$$X_t = a_1X_1 + a_2X_2 + \cdots + a_{t-1}X_{t-1}$$

Multiplying each side of this equation by λ_t, we obtain

$$\lambda_t X_t = \lambda_t a_1 X_1 + \lambda_t a_2 X_2 + \cdots + \lambda_t a_{t-1} X_{t-1}$$

Also,

$$T(X_t) = T(a_1X_1 + a_2X_2 + \cdots + a_{t-1}X_{t-1})$$
$$= a_1T(X_1) + a_2T(X_2) + \cdots + a_{t-1}T(X_{t-1})$$

and $T(X_i) = \lambda_i X_i$ for $i = 1, 2, \ldots, t$ implies that

$$\lambda_t X_t = a_1\lambda_1 X_1 + a_2\lambda_2 X_2 + \cdots + a_{t-1}\lambda_{t-1}X_{t-1}$$

Therefore,

$$\lambda_t a_1 X_1 + \lambda_t a_2 X_2 + \cdots + \lambda_t a_{t-1}X_{t-1} = a_1\lambda_1 X_1 + a_2\lambda_2 X_2$$
$$+ \cdots + a_{t-1}\lambda_{t-1}X_{t-1}$$

But the vectors $X_1, X_2, \ldots, X_{t-1}$ are linearly independent because X_t is the first vector in the list that is a linear combination of the preceding vectors. Thus,

$$\lambda_t a_i = a_i\lambda_i \quad \text{and} \quad (\lambda_t - \lambda_i)a_i = 0, \quad \text{for } i = 1, 2, \ldots, t-1$$

Since $\lambda_t - \lambda_i \neq 0$, we have $a_i = 0$ for $i = 1, 2, \ldots, t-1$. But then $X_t = \mathbf{0}$, which is a contradiction since X_t is an eigenvector. Hence $C = \{X_1, X_2, \ldots, X_t\}$ is a linearly independent set. Q.E.D.

Theorem 7.5. If $T: \mathbf{V} \longrightarrow \mathbf{V}$ is a linear operator on an n-dimensional vector space $\mathbf{V}$, and if T has n distinct eigenvalues $\lambda_1, \lambda_2, \ldots, \lambda_n$, then T has a diagonal matrix representation. In fact, if $C = \{X_1, X_2, \ldots, X_n\}$ is a set of vectors such that $T(X_i) = \lambda_i X_i$ for $i = 1, 2, \ldots, n$, then C is a basis for V and $[T]_C = \operatorname{diag}(\lambda_1, \lambda_2, \ldots, \lambda_n)$.

Proof: Let $\lambda_1, \lambda_2, \ldots, \lambda_n$ be n distinct eigenvalues of T. Then there are nonzero eigenvectors $X_1, X_2, \ldots, X_n$ in V such that $T(X_i) = \lambda_i X_i$ for $i = 1, 2, \ldots, n$. By Lemma 7.1, the set $C = \{X_1, X_2, \ldots, X_n\}$ is linearly independent. Since V has dimension n, C is a basis for V. The matrix $[T]_C$ has jth column $[T(X_j)]_C = \lambda_j E_j$. Hence $[T]_C = \operatorname{diag}(\lambda_1, \lambda_2, \ldots, \lambda_n)$.   Q.E.D.

The following corollary is a restatement of Theorem 7.5 in terms of similar matrices.

Corollary 7.1.   If A is an $n \times n$ matrix and if the characteristic equation $\det(A - \lambda I_n) = 0$ has n distinct real roots, then A is similar to a diagonal matrix D, where the diagonal entries of D are the roots of the characteristic equation.

Definition 7.5.   The set of all distinct characteristic values of a linear operator $T: V \longrightarrow V$ is called the **spectrum** of T.

If a linear operator T has a matrix representation that is a diagonal matrix, the next theorem states that not only are the diagonal entries of this matrix characteristic values of T, but the diagonal entries form the complete spectrum of T.

Theorem 7.6.   If a linear operator $T: V \longrightarrow V$ has a matrix representation $D = \operatorname{diag}(\lambda_1, \lambda_2, \ldots, \lambda_n)$ relative to some basis C, then the elements $\lambda_1, \lambda_2, \ldots, \lambda_n$ (which are not necessarily all different) represent the complete set of characteristic values of T.

Proof: Let $C = \{X_1, X_2, \ldots, X_n\}$ be an ordered basis for V such that $[T]_C = \operatorname{diag}(\lambda_1, \lambda_2, \ldots, \lambda_n)$. Then the jth column of $[T]_C$ is $[T(X_j)]_C = \lambda_j E_j$, and $T(X_j) = \lambda_j X_j$. Thus every entry on the diagonal of $[T]_C$ is an eigenvalue. Now suppose that λ is a characteristic value of T different from all the diagonal entries. Then there is a nonzero vector X such that $T(X) = \lambda X$. Since C is a basis for V, there are scalars $a_1, a_2, \ldots, a_n$ such that $X = a_1 X_1 + a_2 X_2 + \cdots + a_n X_n$. So

$$\lambda X = \lambda a_1 X_1 + \lambda a_2 X_2 + \cdots + \lambda a_n X_n$$

Also

$$\lambda X = T(X) = a_1 T(X_1) + a_2 T(X_2) + \cdots + a_n T(X_n)$$

$$= a_1 \lambda_1 X_1 + a_2 \lambda_2 X_2 + \cdots + a_n \lambda_n X_n$$

The vector λX is a unique linear combination of the basis vectors. Thus $\lambda a_i = \lambda_i a_i$ and $(\lambda - \lambda_i) a_i = 0$ for $i = 1, 2, \ldots, n$. Since we are assuming that $\lambda \neq \lambda_i$, we must have $a_i = 0$ for $i = 1, 2, \ldots, n$. But then $X = \mathbf{0}$, which is a contradiction. Hence there are no characteristic values of T other than those on the diagonal of $[T]_C$. Q.E.D.

Applications: Model for Automobile Demand†

The following model for automobile demand considers only the factors of distribution by age and the existing stock. Suppose that all automobiles are 1, 2, or 3 years old. Let $x_{i,t}$ be the number of cars of age i in the year t for

$i = 1, 2,$ or 3. The vector $X_t = \begin{bmatrix} x_{1,t} \\ x_{2,t} \\ x_{3,t} \end{bmatrix}$ gives the age distribution of the cars in

the year t. We let X_0 be the initial distribution according to age. Whenever a car is 3 years old, it is replaced by a new car. Also, let us assume that 10% of the cars must be replaced when they are 2 years old. Then for $t \geq 1$, we have

$$x_{1,t} = 0.1x_{2,t-1} + x_{3,t-1}$$

$$x_{2,t} = x_{1,t-1}$$

$$x_{3,t} = 0.9x_{2,t-1}$$

In matrix form, $X_t = AX_{t-1}$, where $A = \begin{bmatrix} 0 & 0.1 & 1 \\ 1 & 0 & 0 \\ 0 & 0.9 & 0 \end{bmatrix}$ is the coefficient matrix

of the system. The number of cars in circulation does not change because we continue to replace those that are removed. One may wish to know if there are certain age distributions that perpetuate themselves. If so, then $X_{t+1} = X_t = AX_t$. In terms of eigenvalues, we are asking if the matrix A has the number 1 as an eigenvalue. The characteristic equation of A is

$$\det(A - \lambda I_3) = \begin{vmatrix} -\lambda & 0.1 & 1 \\ 1 & -\lambda & 0 \\ 0 & 0.9 & -\lambda \end{vmatrix} = 0 \Longleftrightarrow -\lambda^3 + 0.1\lambda + 0.9 = 0$$

We see that $\lambda = 1$ is a solution. We now find a vector X that belongs to the eigenvalue 1. To do this, we transform the coefficient matrix of the system $(A - I_3)X = 0$ to reduced row echelon form.

$$\begin{vmatrix} -1 & 0.1 & 1 \\ 1 & -1 & 0 \\ 0 & 0.9 & -1 \end{vmatrix} \xrightarrow[-1R_1]{R_1+R_2} \begin{vmatrix} 1 & -0.1 & -1 \\ 0 & -0.9 & 1 \\ 0 & 0.9 & -1 \end{vmatrix} \xrightarrow[(1/10)R_2+R_1]{\substack{R_2+R_3 \\ (-10/9)R_2}} \begin{vmatrix} 1 & 0 & -\frac{10}{9} \\ 0 & 1 & -\frac{10}{9} \\ 0 & 0 & 0 \end{vmatrix}$$

†From S. R. Searle and W. H. Hausman, *Matrix Algebra for Business and Economics* (New York: Wiley–Interscience/A Division of John Wiley & Sons, Inc., 1970), pp. 297–299. Reprinted by permission of John Wiley & Sons, Inc.

Therefore,

$$x_{1,t} = \tfrac{10}{9} x_{3,t-1}$$

$$x_{2,t} = \tfrac{10}{9} x_{3,t-1}$$

$$x_{3,t} = x_{3,t-1}$$

Since the x's are all integers, we see that $x_{3,t-1}$ must be a multiple of 9. For example, if the initial distribution by age is 100 one-year-old cars, 100 two-year-old cars, and 90 three-year-old cars, then the distribution perpetuates itself.

The model above does not allow for growth in the number of automobiles. Let's alter the model to allow for growth, and let's assume that the rate of growth of the number of cars is proportional to the existing number of cars. Then the production needed in the year t to account for growth is $k(x_{1,t-1} + x_{2,t-1} + x_{3,t-1})$, where k is the *growth proportionality factor*. The total production of new cars in the year t will be the above number plus $0.1x_{2,t-1} + x_{3,t-1}$, which is

$$kx_{1,t-1} + (0.1 + k)x_{2,t-1} + (k + 1)x_{3,t-1}$$

Our age distribution vector X_t is now of the form $X_t = BX_{t-1}$, where

$$B = \begin{bmatrix} k & 0.1+k & 1+k \\ 1 & 0 & 0 \\ 0 & 0.9 & 0 \end{bmatrix}.$$ Since the number of cars increases each year, we

will not have $X_t = X_{t-1}$; that is, the distribution will not perpetuate itself. However, we can ask, is there a vector X_t such that $X_{t+1} = \lambda X_t$ for some scalar $\lambda \neq 1$? To answer this question, we consider

$$\det(B - \lambda I) = (k - \lambda)(-\lambda)(-\lambda) + 0.9(k + 1) - (0.1 + k)(-\lambda)$$

The characteristic equation $\det(B - \lambda I) = 0$ is equivalent to the equation $\lambda^3 - k\lambda^2 - (k + 0.1)\lambda - 0.9(k + 1) = 0$. Substituting $k + 1$ for λ, we get

$$(k + 1)^3 - k(k + 1)^2 - (0.1 + k)(1 + k) - 0.9(1 + k) = 0$$

Therefore, $k + 1$ is a characteristic value. To find a characteristic vector belonging to $k + 1$, we transform the coefficient matrix of $(B - (k + 1)I_3)X = \mathbf{0}$ to reduced row echelon form.

$$\begin{bmatrix} -1 & k+0.1 & k+1 \\ 1 & -k-1 & 0 \\ 0 & 0.9 & -k-1 \end{bmatrix} \xrightarrow[R_2+R_3]{R_1+R_2} \begin{bmatrix} -1 & k+0.1 & k+1 \\ 0 & -0.9 & 1+k \\ 0 & 0 & 0 \end{bmatrix}$$

$$\xrightarrow[(-10/9)R_2]{\substack{-1R_1 \\ R_2+R_1}} \begin{bmatrix} 1 & -1-k & 0 \\ 0 & 1 & -\tfrac{10}{9}(1+k) \\ 0 & 0 & 0 \end{bmatrix} \xrightarrow{(1+k)R_2+R_1} \begin{bmatrix} 1 & 0 & -\tfrac{10}{9}(1+k)^2 \\ 0 & 1 & -\tfrac{10}{9}(1+k) \\ 0 & 0 & 0 \end{bmatrix}$$

Therefore, if

$$x_1 = \tfrac{10}{9}(1 + k)^2 x_3$$

$$x_2 = \tfrac{10}{9}(1 + k)x_3$$

$$x_3 = x_3$$

then $X = \begin{bmatrix} x_1 \\ x_2 \\ x_3 \end{bmatrix}$ is an eigenvector belonging to $k + 1$. If, for example, we let

$k = 0.1$, $x_3 = 900$, and $X = X_0$, then the initial distribution is $X_0 = \begin{bmatrix} 1210 \\ 1100 \\ 900 \end{bmatrix}$.

The matrix B, which depends on k, is $B = \begin{bmatrix} 0.1 & 0.2 & 1.1 \\ 1 & 0 & 0 \\ 0 & 0.9 & 0 \end{bmatrix}$ when $k = 0.1$.

Computing $X_1 = BX_0$, we find that

$$X_1 = \begin{bmatrix} 0.1 & 0.2 & 1.1 \\ 1 & 0 & 0 \\ 0 & 0.9 & 0 \end{bmatrix} \begin{bmatrix} 1210 \\ 1100 \\ 900 \end{bmatrix} = \begin{bmatrix} 1331 \\ 1210 \\ 990 \end{bmatrix} = 1.1 \begin{bmatrix} 1210 \\ 1100 \\ 900 \end{bmatrix}$$

which agrees with the fact that X_0 is an eigenvector belonging to $k + 1 = 1.1$. Also,

$$X_2 = BX_1 = B(BX_0) = B(1.1X_0) = 1.1BX_0 = 1.1X_1$$

In general, $X_n = 1.1X_{n-1}$ (see Problem 11, Exercise 7.2). We see that the number of cars in any year will be 10% more than in the previous year.

Model for the Working Condition of a Machine (Example of a Markov Chain)†

Before studying this example the reader may wish to review Example 3.22. Suppose that a certain machine is always in one of three states: (1) broken beyond repair (B), (2) in need of adjustment (N), or (3) working properly (W). Let a_{ij} denote the probability that a machine which is in state j will be in state i one time period later. The matrix

$$A = [a_{ij}] = \begin{array}{ccc} \text{B} & \text{N} & \text{W} \\ \begin{bmatrix} 1 & \tfrac{1}{4} & \tfrac{1}{18} \\ 0 & \tfrac{1}{2} & \tfrac{8}{18} \\ 0 & \tfrac{1}{4} & \tfrac{9}{18} \end{bmatrix} & \begin{array}{c} \text{B} \\ \text{N} \\ \text{W} \end{array} \end{array}$$

†Searle and Hausman, *Matrix Algebra for Business and Economics*, pp. 305–307.

gives us the various transition probabilities that a machine being in state j will be in state i one time period later. For example, $a_{11} = 1$ means that it is certain that a machine which is broken beyond repair will be in the same condition one time period later. $a_{23} = \frac{8}{18}$ means that the probability that a machine working properly will be in need of adjustment one time period later is $\frac{8}{18}$. We assume that the probability of being in any one of the states at the end of a time period depends only on the state that the process (in this case a machine) was in at the beginning of the time period (i.e., the end of the previous time period).

Let $X_t = \begin{bmatrix} x_{1,t} \\ x_{2,t} \\ x_{3,t} \end{bmatrix}$, where $x_{i,t}$ is the probability that the machine is in state

i at the beginning of the tth time period (which equals the end of the $t - 1$ time period for $t \geq 1$). Then $AX_{t-1} = X_t$ for $t = 1, 2, 3, \ldots$. The reader is asked to show (Problem 12, Exercise 7.2) that $X_n = A^n X_0$ for all positive integers n. As an example, if the machine is working properly at the beginning of the initial time period, then

$$X_0 = \begin{bmatrix} 0 \\ 0 \\ 1 \end{bmatrix} \quad \text{and} \quad \begin{bmatrix} 1 & \frac{1}{4} & \frac{1}{18} \\ 0 & \frac{1}{2} & \frac{8}{18} \\ 0 & \frac{1}{4} & \frac{9}{18} \end{bmatrix} \begin{bmatrix} 0 \\ 0 \\ 1 \end{bmatrix} = \begin{bmatrix} \frac{1}{18} \\ \frac{8}{18} \\ \frac{9}{18} \end{bmatrix} = X_1$$

At the end of one time period, $\frac{1}{18}$ of the machines are broken beyond repair, $\frac{4}{9}$ of the machines need adjustment, and $\frac{1}{2}$ of the machines are working properly.

To calculate the probability vector $X_n = A^n X_0$, which gives the probabilities of the machine being in state 1, 2, or 3 at the end of n time periods, we need to be able to calculate A^n. This can be done if A is **diagonalizable**, that is, if A is similar to a diagonal matrix (see Section 6.10). The reader is asked in Problem 13, Exercise 7.2, to find a diagonal matrix similar to A, and then use it to calculate A^n. The behavior of the machine as given by the matrix A, and the sequence of vectors $X_0, X_1, \ldots, X_n, \ldots$ is called a *Markov chain*.

Exercise 7.2

1. Let $B = \begin{bmatrix} 1 & \frac{1}{4} & 0 \\ 0 & \frac{1}{2} & 0 \\ 0 & \frac{1}{4} & 1 \end{bmatrix}$

(a) Find the real eigenvalues and eigenvectors of B.

(b) Find nonsingular matrices P and P^{-1} and a diagonal matrix D such that $D = P^{-1}BP$.

2. Repeat Problem 1 for the matrix $C = \begin{bmatrix} 1 & \frac{1}{2} & 0 \\ 0 & \frac{1}{2} & 1 \\ 0 & 0 & 0 \end{bmatrix}$.

Remark: The matrices B and C are the matrices that appear in the plant reproduction problem in Section 6.10.

3. If $A = \begin{bmatrix} 1 & 0 & 0 \\ 0 & -2 & 0 \\ 0 & 0 & 0 \end{bmatrix}$,

    (a) What are the eigenvalues of A?
    (b) What are the eigenvectors of A?
    (c) Find the characteristic polynomial of A.

4. Let $A = \begin{bmatrix} 1 & 0 \\ 0 & 1 \end{bmatrix}$ and $B = \begin{bmatrix} 1 & 1 \\ 0 & 1 \end{bmatrix}$.

    (a) Show that A and B have the same characteristic polynomial.
    (b) If V is any vector space with ordered basis $C_1 = \{X_1, X_2\}$, and $T: V \longrightarrow V$ is a linear operator on V such that $A = [T]_{C_1}$ and $X = a_1 X_1 + a_2 X_2$, find $T(X)$.
    (c) Is there a basis C_2 for V such that $[T]_{C_2} = B$ with T as in part (b)? Verify your answer.
    (d) Find the eigenvectors of B and show that there is no basis of eigenvectors of B for $\mathbf{R}_2$.

5. If A is a 4×4 matrix, use the theorems of Sections 7.1 and 7.2 to decide whether A is (i) definitely similar to a diagonal matrix, (ii) possibly similar to a diagonal matrix, or (iii) definitely not similar to a diagonal matrix, if the characteristic equation of A is
    (a) $\lambda(\lambda - 1)(\lambda + 3)(\lambda + 5) = 0$.
    (b) $(\lambda - 2)^2(\lambda + 1)(\lambda - 1) = 0$.
    (c) $\lambda^4 + 1 = 0$.
    (d) $(\lambda^2 - 1)(\lambda^2 - 4) = 0$.

6. (a) Show that the matrix $A = \begin{bmatrix} 2 & 8 \\ -1 & 3 \end{bmatrix}$ is not similar to a diagonal matrix.

    (b) Show that the matrix $A = \begin{bmatrix} 1 & 0 \\ 3 & 1 \end{bmatrix}$ is not similar to a diagonal matrix.

7. What are the eigenvalues of the $n \times n$ zero matrix? What are the eigenvectors of the zero matrix? Is the zero matrix diagonalizable?

8. If $A = \begin{bmatrix} 0 & 0 & 1 \\ 0 & 1 & 0 \\ 1 & 0 & 0 \end{bmatrix}$, (a) find the eigenvalues and eigenvectors of A; (b) find a matrix P so that $P^{-1}AP$ is a diagonal matrix.

9. Repeat Problem 8 if $A = \begin{bmatrix} 2 & -3 & 5 \\ 0 & -1 & 5 \\ 0 & 0 & 4 \end{bmatrix}$.

10. Prove Theorem 7.3.

11. In the model for automobile demand, with a growth proportionality factor of k, we found a vector $X = X_0$ such that $BX_0 = (k + 1)X_0 = X_1$. (X_0 is an eigenvector of B belonging to $k + 1$). Also, $BX_{n-1} = X_n$. Use mathematical induction to prove that $X_n = (1 + k)X_{n-1}$.

12. In the model for the working condition of a machine, we have $AX_{t-1} = X_t$ for any positive integer t. Use mathematical induction to verify that $A^n X_0 = X_n$ for all positive integers n.

13. With A as in Problem 12 (A is called the transition probability matrix),
   (a) Show that $1, \frac{1}{6},$ and $\frac{5}{6}$ are eigenvalues for A, and thus that A is diagonalizable.
   (b) Find the eigenvectors for A and from these select the basis

$$C_2 = \left\{ Y_1 = \begin{bmatrix} 1 \\ 0 \\ 0 \end{bmatrix}, Y_2 = \begin{bmatrix} 1 \\ -4 \\ 3 \end{bmatrix}, Y_3 = \begin{bmatrix} -7 \\ 4 \\ 3 \end{bmatrix} \right\}.$$

   (c) Find the matrix P for the change of basis from C_2 to the natural basis C_1 for $\mathbf{R}_3$, find P^{-1}, and show that $D = P^{-1}AP$ is the diagonal matrix, $\mathrm{diag}(1, \frac{1}{6}, \frac{5}{6})$.
   (d) Find $A^n = P^{-1}D^nP$, and show that

$$A^n = \begin{bmatrix} 1 & 1 - (\frac{1}{8})(\frac{1}{6})^n - (\frac{7}{8})(\frac{5}{6})^n & 1 + (\frac{1}{8})(\frac{1}{6})^n - (\frac{7}{8})(\frac{5}{6})^n \\ 0 & (\frac{1}{6})^n + (\frac{1}{2})(\frac{5}{6})^n & (-\frac{2}{3})(\frac{1}{6})^n + (\frac{2}{3})(\frac{5}{6})^n \\ 0 & (-\frac{3}{8})(\frac{1}{6})^n + (\frac{3}{8})(\frac{5}{6})^n & (\frac{1}{2})(\frac{1}{6})^n + (\frac{1}{2})(\frac{5}{6})^n \end{bmatrix}$$

   (e) As the number of time periods $n \longrightarrow +\infty$, what is the limiting value of A^n. Interpret your result.

14. In the model for the working condition of a machine, suppose that the matrix of transition probabilities is

$$A = \begin{array}{c} \\ \\ B \\ N \\ W \end{array} \begin{array}{c} \text{B} \quad \text{N} \quad \text{W} \\ \begin{bmatrix} 1 & \frac{1}{2} & 0 \\ 0 & \frac{1}{4} & \frac{1}{2} \\ 0 & \frac{1}{4} & \frac{1}{2} \end{bmatrix} \end{array} \begin{array}{c} \text{B} \\ \text{N} \\ \text{W} \end{array}$$

Determine matrices P and P^{-1} such that $P^{-1}AP = D$ is a diagonal matrix. Calculate $A^n = PD^nP^{-1}$.

15. Verify that A and A^T have the same characteristic polynomial if A is an $n \times n$ matrix.

16. Let A be an $n \times n$ matrix, and suppose that λ is an eigenvalue of A. Prove that the set of all eigenvectors belonging to λ together with $\mathbf{0}$ is a subspace of $\mathbf{R}_n$.

17. *Prove:* If λ is a characteristic value of a nonsingular matrix A, then λ^{-1} is a characteristic value of A^{-1}.

7.3 QUADRATIC FORMS

The functions that we have studied to this point have been linear functions from one vector space to another. If $T: \mathbf{R}_n \longrightarrow \mathbf{R}_1$ (we identify $\mathbf{R}_1$ with $\mathbf{R}$ and write "a" in place of "$[a]$") is a linear function, and $T(E_i) = a_i$ for $i = 1, 2,$

$\ldots, n$ are the images of the vectors in the natural basis for $\mathbf{R_n}$, then, for an arbitrary vector $X = x_1E_1 + x_2E_2 + \cdots + x_nE_n$, we have $T(X) = a_1x_1 + a_2x_2 + \cdots + a_nx_n$. Thus $T(X)$ is a **sum of linear terms**. We now wish to consider an arbitrary function $Q: \mathbf{R_n} \rightarrow \mathbf{R}$ such that $Q(X)$ is a **sum of quadratic terms**, that is, a sum of terms of the form $a_{ij}x_ix_j$. A function of this type is called a *quadratic form*.

Quadratic forms have applications in several areas, including statistics, mechanics, and operations research. Our reasons for studying quadratic forms at this particular point are the following. We shall see that associated with a quadratic form $Q(X)$ is a matrix that depends on the ordered basis chosen for $\mathbf{R_n}$. Relative to a properly chosen basis, the matrix of the quadratic form is a *diagonal matrix*. When the coordinates of X are expressed in terms of this new basis, we shall see that a relatively simple expression for the quadratic form results.

Example 7.7 Examples of Quadratic Forms. (a) The expression $3x^2 + 4xy + 6y^2$ defines a quadratic form, which we can think of as a function of the *two variables* x and y. We may also think of the quadratic form as a function $Q: \mathbf{R_2} \rightarrow \mathbf{R}$ defined by $Q(X) = 3x^2 + 4xy + 6y^2$, where $X = \begin{bmatrix} x \\ y \end{bmatrix}$. Viewed in this manner, the quadratic form Q is a function of the *vector* $X = \begin{bmatrix} x \\ y \end{bmatrix}$.

(b) The expression $2x^2 + 3y^2 + z^2 - 2xy + 3yz$ defines a quadratic form. In function notation, the quadratic form is the function $Q: \mathbf{R_3} \rightarrow \mathbf{R}$, where

$$Q(X) = 2x^2 + 3y^2 + z^2 - 2xy + 3yz \quad \text{and} \quad X = \begin{bmatrix} x \\ y \\ z \end{bmatrix}$$

(c) The expression $2x_1^2 + 3x_4^2 + 2x_1x_2 + 4x_3x_4$ can be used to define a quadratic form $Q: \mathbf{R_4} \rightarrow \mathbf{R}$ such that for

$$X = \begin{bmatrix} x_1 \\ x_2 \\ x_3 \\ x_4 \end{bmatrix}, \qquad Q(X) = 2x_1^2 + 3x_4^2 + 2x_1x_2 + 4x_3x_4$$

Note that in each of the above examples, each term is of degree 2. A term such as $2x_1x_2$ is called a *cross term*; while a term such as $2x_1^2$ is called a *square term*.

Example 7.8 Examples of Nonquadratic Forms. (a) $2x^2 + 3y^2 + 6$ does not define a quadratic form because the term 6 has degree 0.

(b) The expression $3xy + 4xz + 7xyz$ does not define a quadratic form because the term $7xyz$ has degree 3.

(c) $3x + 4yz + z^2$ defines neither a linear form (linear functional) or a quadratic form.

Each of the quadratic forms in Example 7.7 can be represented by a matrix product, as we now illustrate.

Example 7.9   Matrix of a Quadratic Form (Relative to the Natural Basis). (a) The quadratic form $3x^2 + 4xy + 6y^2$ can be expressed as the matrix product $[x \ \ y] \begin{bmatrix} 3 & 2 \\ 2 & 6 \end{bmatrix} \begin{bmatrix} x \\ y \end{bmatrix}$. This matrix product is the 1×1 matrix $[3x^2 + 4xy + 6y^2]$. Although there is a difference between a 1×1 matrix and its entry, we do not distinguish them here.

(b) The quadratic form $2x^2 + 3y^2 + z^2 - 2xy + 3yz$ can be expressed as the matrix product $[x \ \ y \ \ z] \begin{bmatrix} 2 & -1 & 0 \\ -1 & 3 & \frac{3}{2} \\ 0 & \frac{3}{2} & 1 \end{bmatrix} \begin{bmatrix} x \\ y \\ z \end{bmatrix}$.

(c) The quadratic form $2x_1^2 + 3x_4^2 + 2x_1x_2 + 4x_3x_4$ can be expressed as the matrix product $X^T A X$, where $X^T = [x_1 \ x_2 \ x_3 \ x_4]$ and

$$A = [a_{ij}] = \begin{bmatrix} 2 & 1 & 0 & 0 \\ 1 & 0 & 0 & 0 \\ 0 & 0 & 0 & 2 \\ 0 & 0 & 2 & 3 \end{bmatrix}$$

The reader should convince himself that the single entry in the product $X^T A X$ is the given quadratic form. In each of parts (a)–(c) we have written the given quadratic form as a matrix expression $X^T A X$, where the matrix A has the property that the (i, j) entry is equal to the (j, i) entry. Geometrically, A is symmetric with respect to the main diagonal. Algebraically, $A = A^T$. We repeat the definition given in Chapter 3.

Definition 7.6.   An $n \times n$ matrix A having the property that $A^T = A$ is called a **symmetric matrix**.

Definition 7.7.   A function $Q: \mathbf{R}_n \rightarrow \mathbf{R}$ is a (real) **quadratic form** if and only if there exist real constants a_{ij}, for $i \leq j$ and $i, j = 1, 2, \ldots, n$, such that for each $n \times 1$ column vector X with coordinates x_i, $i = 1, 2, \ldots, n$ (relative to the natural basis)

$$\begin{aligned} Q(X) = \ &a_{11}x_1^2 + 2a_{12}x_1x_2 + 2a_{13}x_1x_3 & + \cdots + 2a_{1n}x_1x_n \\ &+ a_{22}x_2^2 + 2a_{23}x_2x_3 & + \cdots + 2a_{2n}x_2x_n \\ &+ a_{33}x_3^2 + 2a_{34}x_3x_4 + \cdots + 2a_{3n}x_3x_n \\ &\qquad\qquad \vdots \\ &\qquad\qquad\qquad\qquad\qquad + a_{nn}x_n^2 \end{aligned}$$

$Q(X)$ is a sum of terms each of degree 2. The terms $a_{ij}x_i x_j$ for $i \neq j$ are called **cross terms**, and the terms $a_{ii}x_i^2$ are called **square terms**. The expression for $Q(X)$ may be written in matrix form as

$$Q(X) = [x_1 \quad x_2 \dots x_n] \begin{bmatrix} a_{11} & a_{12} & \cdots & a_{1n} \\ a_{12} & a_{22} & \cdots & a_{2n} \\ \cdot & \cdot & & \cdot \\ \cdot & \cdot & & \cdot \\ \cdot & \cdot & & \cdot \\ a_{1n} & a_{2n} & & a_{nn} \end{bmatrix} \begin{bmatrix} x_1 \\ x_2 \\ \cdot \\ \cdot \\ \cdot \\ x_n \end{bmatrix}$$

or

$$Q(X) = X^T A X$$

The **symmetric matrix** A is called the **matrix of the quadratic form relative to the standard basis for $\mathbf{R}_n$**. We shall refer to $Q(X)$ as the quadratic form, although technically $Q(X)$ is the value of the quadratic form at X.

Consider $Q: \mathbf{R}_2 \longrightarrow \mathbf{R}$ defined by $Q(X) = 3x^2 + 4xy + 6y^2$. This quadratic form can be expressed as $Q(X) = [x \quad y] \begin{bmatrix} 3 & 2 \\ 2 & 6 \end{bmatrix} \begin{bmatrix} x \\ y \end{bmatrix}$, where x and y are the coordinates of X relative to the standard basis $C_1 = \left\{ E_1 = \begin{bmatrix} 1 \\ 0 \end{bmatrix}, E_2 = \begin{bmatrix} 0 \\ 1 \end{bmatrix} \right\}$. Suppose we take a different basis, say $C_2 = \left\{ Y_1 = \begin{bmatrix} 1 \\ 1 \end{bmatrix}, Y_2 = \begin{bmatrix} -1 \\ 1 \end{bmatrix} \right\}$. The quadratic form $Q(X)$ can also be expressed as a matrix product relative to this basis, as we now illustrate.

Let P be the matrix for the change of basis from C_2 to C_1. Then

$$P = [I]_{C_2 C_1} = [[Y_1]_{C_1} \,|\, [Y_2]_{C_1}] = [Y_1 \,|\, Y_2] = \begin{bmatrix} 1 & -1 \\ 1 & 1 \end{bmatrix}$$

If X has coordinates $\begin{bmatrix} u \\ v \end{bmatrix}$ relative to C_2, then $P \begin{bmatrix} u \\ v \end{bmatrix} = \begin{bmatrix} x \\ y \end{bmatrix}$ and

$$[x \quad y] = \begin{bmatrix} x \\ y \end{bmatrix}^T = \left(P \begin{bmatrix} u \\ v \end{bmatrix} \right)^T = [u \quad v]P^T$$

Also, if $A = \begin{bmatrix} 3 & 2 \\ 2 & 6 \end{bmatrix}$ is the matrix of Q relative to the standard basis, then

$$Q(X) = [x \quad y]A \begin{bmatrix} x \\ y \end{bmatrix} = [u \quad v]P^T A \begin{bmatrix} x \\ y \end{bmatrix} = [u \quad v]P^T A P \begin{bmatrix} u \\ v \end{bmatrix}$$

Computing, we find that $P^T A P = \begin{bmatrix} 13 & 3 \\ 3 & 5 \end{bmatrix}$, which is also a symmetric matrix.

Thus

$$Q(X) = [u \ \ v] \begin{bmatrix} 13 & 3 \\ 3 & 5 \end{bmatrix} \begin{bmatrix} u \\ v \end{bmatrix} = 13u^2 + 6uv + 5v^2$$

We say that the matrix $B = \begin{bmatrix} 13 & 3 \\ 3 & 5 \end{bmatrix}$ represents the quadratic form relative

to the basis C_2. If we wish to calculate the value of $Q(X)$, say at $X = \begin{bmatrix} 1 \\ 2 \end{bmatrix}$, by

using the matrix B, we must first find the coordinates of X relative to C_2.

We find that $\begin{bmatrix} 1 \\ 2 \end{bmatrix} = \frac{3}{2}Y_1 + \frac{1}{2}Y_2$ and $\begin{bmatrix} 1 \\ 2 \end{bmatrix}_{C_2} = \begin{bmatrix} \frac{3}{2} \\ \frac{1}{2} \end{bmatrix}$. Hence

$$Q \begin{bmatrix} 1 \\ 2 \end{bmatrix} = [\frac{3}{2} \ \ \frac{1}{2}] \begin{bmatrix} 13 & 3 \\ 3 & 5 \end{bmatrix} \begin{bmatrix} \frac{3}{2} \\ \frac{1}{2} \end{bmatrix} = 35$$

Using the matrix of Q relative to the standard basis, we have

$$Q \begin{bmatrix} 1 \\ 2 \end{bmatrix} = [1 \ \ 2] \begin{bmatrix} 3 & 2 \\ 2 & 6 \end{bmatrix} \begin{bmatrix} 1 \\ 2 \end{bmatrix} = 35$$

which agrees with our previous answer.

Definition 7.8. Let $Q: \mathbf{R}_n \longrightarrow \mathbf{R}$ be a quadratic form, and suppose that $C_2 = \{Y_1, Y_2, \ldots, Y_n\}$ is a basis for $\mathbf{R}_n$. The **symmetric matrix B is the matrix of the quadratic form relative to the basis C_2** if and only if $Q(X) = [X]_{C_2}^T B[X]_{C_2}$, where $[X]_{C_2}$ is the coordinate vector of X relative to the basis C_2.

The next theorem generalizes the example preceding Definition 7.8.

Theorem 7.7. Let $Q: \mathbf{R}_n \longrightarrow \mathbf{R}$ be a quadratic form, and let A be the matrix of the form $Q(X)$ relative to the standard basis C_1. If $C_2 = \{Y_1, Y_2, \ldots, Y_n\}$ is an ordered basis for $\mathbf{R}_n$, and if P is the matrix for the change of basis from C_2 to C_1, then $B = P^T A P$ is the matrix of Q relative to the basis C_2.

Proof: Since A is the matrix of Q relative to the standard basis C_1, we have $Q(X) = X^T A X$, and A is a symmetric matrix. Let P be the matrix for the change of basis from C_2 to C_1, and let $Y = [X]_{C_2}$. Then $PY = P[X]_{C_2} = [X]_{C_1} = X$. Since the transpose of a matrix product is the product of the transposes in reverse order (Theorem 3.10), we have $X^T = (PY)^T = Y^T P^T$. So

$$Q(X) = X^T A X = (Y^T P^T)AX = Y^T(P^T A)X = Y^T(P^T A)PY = Y^T(P^T A P)Y$$

The proof will be complete if we can verify that $P^T A P$ is a symmetric matrix. We consider the transpose of $P^T A P$.

$$(P^T A P)^T = P^T(P^T A)^T = P^T A^T(P^T)^T = P^T A^T P$$

But A is symmetric. Therefore, $A^T = A$ and $(P^T A P)^T = P^T A P$. Hence $B = P^T A P$ is symmetric and $Q(X) = Y^T B Y = [X]_{C_2}^T B[X]_{C_2}$.   Q.E.D.

Suppose that we have a quadratic form $Q(X)$, and that somehow we are able to find a basis C_2 such that the matrix of the quadratic form relative to this basis is a diagonal matrix. If we express $Q(X)$ in coordinates relative to C_2 ($[X]_{C_2} = U$), then all the cross terms (terms of the form $b_{ij}u_iu_j$, $i \neq j$) have zero coefficients. This means that when $Q(X)$ is expressed in terms of the C_2 coordinates it is a sum of square terms (terms of the form $b_{ii}u_i^2$), and $Q(X)$ has a relatively simple form. Before proceeding to show how such a basis can be found, we define an orthonormal basis. Recall that two vectors in $\mathbf{R}_n$ are orthogonal or perpendicular if their dot product is zero (Section 2.2). Also recall that the length of a vector in $\mathbf{R}_n$ is the square root of the dot product of the vector with itself ($X = \sqrt{X \cdot X}$).

Definition 7.9. A basis $\{X_1, X_2, \ldots, X_n\}$ for $\mathbf{R}_n$ is an **orthonormal basis** if and only if (i) the vectors in the basis are **mutually orthogonal**, that is, $X_i \cdot X_j = 0$ if $i \neq j$, and (ii) each vector in the basis has length 1, that is, $X_i \cdot X_i = 1$.

Example 7.10   Some Orthonormal Bases.   (a) The standard or natural basis for $\mathbf{R}_n$ is an orthonormal basis.

(b) The basis

$$\left\{ X_1 = \begin{bmatrix} \dfrac{1}{\sqrt{2}} \\ \dfrac{1}{\sqrt{2}} \end{bmatrix}, X_2 = \begin{bmatrix} -\dfrac{1}{\sqrt{2}} \\ \dfrac{1}{\sqrt{2}} \end{bmatrix} \right\}$$

is a orthonormal basis for $\mathbf{R}_2$ because $X_1 \cdot X_1 = \frac{1}{2} + \frac{1}{2} = X_2 \cdot X_2 = 1$ and $X_1 \cdot X_2 = -\frac{1}{2} + \frac{1}{2} = 0$.

(c) The basis

$$C_2 = \left\{ Y_1 = \tfrac{1}{3}\begin{bmatrix} 1 \\ -2 \\ 2 \end{bmatrix}, Y_2 = \tfrac{1}{3}\begin{bmatrix} 2 \\ -1 \\ -2 \end{bmatrix}, Y_3 = \tfrac{1}{3}\begin{bmatrix} 2 \\ 2 \\ 1 \end{bmatrix} \right\}$$

is an orthonormal basis for $\mathbf{R}_3$.

Verification

$$Y_1 \cdot Y_2 = \tfrac{1}{9}(2 + 2 - 4) = 0, \qquad Y_1 \cdot Y_3 = \tfrac{1}{9}(2 - 4 + 2) = 0$$

$$Y_2 \cdot Y_3 = \tfrac{1}{9}(4 - 2 - 2) = 0, \qquad Y_1 \cdot Y_1 = \tfrac{1}{9}(1 + 4 + 4) = 1$$

$$Y_2 \cdot Y_2 = \tfrac{1}{9}(4 + 1 + 4) = 1, \qquad Y_3 \cdot Y_3 = \tfrac{1}{9}(4 + 4 + 1) = 1$$

Example 7.11.   Show that the matrix P for the change of basis from the orthonormal basis C_2 in Example 7.10(c) to the natural basis C_1 has the property that $P^TP = I_3$; that is, $P^T = P^{-1}$.

Solution: Since $[Y_i]_{C_1} = Y_i$, the matrix for the change of basis from C_2 to C_1 is

$$P = [I]_{C_2C_1} = [Y_1 \,|\, Y_2 \,|\, Y_3] = \tfrac{1}{3}\begin{bmatrix} 1 & 2 & 2 \\ -2 & -1 & 2 \\ 2 & -2 & 1 \end{bmatrix}$$

We have factored out $\tfrac{1}{3}$ in order to simplify the entries of P and also to simplify the calculations in finding P^TP. Thus

$$P^TP = \tfrac{1}{9}\begin{bmatrix} 1 & -2 & 2 \\ 2 & -1 & -2 \\ 2 & 2 & 1 \end{bmatrix}\begin{bmatrix} 1 & 2 & 2 \\ -2 & -1 & 2 \\ 2 & -2 & 1 \end{bmatrix} = \tfrac{1}{9}\begin{bmatrix} 9 & 0 & 0 \\ 0 & 9 & 0 \\ 0 & 0 & 9 \end{bmatrix} = I_3$$

Definition 7.10.　An $n \times n$ matrix A is an **orthogonal matrix** if and only if $A^TA = I_n$.

Example 7.11 is a special case of the following theorem, which provides a reason for the name "orthogonal matrix."

Theorem 7.8.　Let $C_2 = \{Y_1, Y_2, \ldots, Y_n\}$ be an ordered basis for $\mathbf{R}_n$, and let $P = [p_{ij}]$ be the matrix for the change of basis from C_2 to the natural basis C_1. Then C_2 is an orthonormal basis if and only if P is an orthogonal matrix.

Proof:　Since P is the matrix for the change of basis from C_2 to the natural basis C_1, we have $P^{(j)} = [Y_j]_{C_1} = Y_j$; that is, the jth column of P is the vector Y_j. So for any i and j from 1 to n, we have

$$Y_i = p_{1i}E_1 + p_{2i}E_2 + \cdots + p_{ni}E_n$$

and

$$Y_j = p_{1j}E_1 + p_{2j}E_2 + \cdots + p_{nj}E_n$$

Let $P^T = [q_{ij}]$. If $P^TP = [c_{ij}]$, then

$$c_{ij} = q_{i1}p_{1j} + q_{i2}p_{2j} + \cdots + q_{in}p_{nj}$$
$$= p_{1i}p_{1j} + p_{2i}p_{2j} + \cdots + p_{ni}p_{nj} = Y_i \cdot Y_j$$

Thus $P^TP = [c_{ij}]$ and $c_{ij} = Y_i \cdot Y_j$. C_2 is an orthonormal basis if and only if $Y_i \cdot Y_i = 1$ and $Y_i \cdot Y_j = 0$ for $i \neq j$. But this is true if and only if $c_{ii} = 1$ and $c_{ij} = 0$ for $i \neq j$. Hence C_2 is an orthonormal basis if and only if $P^TP = I_n$.
　　　　　　　　　　　　　　　　　　　　　　　　　　　　　　Q.E.D.

Exercise 7.3

In Problems 1 and 2, decide which of the given expressions can be used to define a quadratic form. If the expression does define a quadratic form $Q: \mathbf{R}_n \to \mathbf{R}$ for some n, write $Q(X)$ in the form described in Definition 7.7, and then write $Q(X) = X^TAX$ for an appropriate symmetric matrix A and column vector X.

1. (a) $2x^2 + 6xy + 5y^2$.
   (b) $3x^2 + 4xy + 3$.
   (c) $2xy + 4yz + 3xz$.
   (d) $2x^2 + 3xy + 4xz + 6yz + 3z^2$.
   (e) $3x^2 + 3xy + 5yx + z^2$.
   (f) $2x^2 - 4yz + 2xyz - 3z^2$.

2. (a) $6x^2 + 3x + 5y^2$.
     (d) $x_1^2 + 3x_1x_2 - 5x_2x_4 + 3x_3x_4 + x_3^2 - x_4^2$.
   (b) $3^4xy - 2x^2 + 6y^2$.
     (e) $2x_1^2 + 5^4x_1x_3 - 14x_2x_3 + x_2^2$.
   (c) $y^2 + 2xy + 3yz + 4z^2$.

3. If $Q(X) = 3x^2 + 4xy + 7y^2 + 2yw + 3w^2$ and $X^T = [x \ y \ z \ w]$, express $Q(X)$ in the form X^TAX. Check your answer by performing the matrix multiplication.

In Problems 4 and 5, you are given the matrix of a quadratic form relative to the natural basis. Write the quadratic form as a sum of terms of degree 2.

4. $A = \begin{bmatrix} 0 & 2 & 1 \\ 2 & 1 & -3 \\ 1 & -3 & 5 \end{bmatrix}$.
       5. $A = \begin{bmatrix} 2 & 0 & 0 & 0 \\ 0 & 3 & 0 & 0 \\ 0 & 0 & 1 & 0 \\ 0 & 0 & 0 & 4 \end{bmatrix}$.

6. If $Q: \mathbf{R}_2 \rightarrow \mathbf{R}$ is defined by $Q(X) = X^TAX$, $X = \begin{bmatrix} x \\ y \end{bmatrix}$, and $A = \begin{bmatrix} 3 & 2 \\ 2 & 1 \end{bmatrix}$, find a symmetric matrix B such that $Q(X) = [u \ v]B\begin{bmatrix} u \\ v \end{bmatrix}$, where $\begin{bmatrix} u \\ v \end{bmatrix} = \begin{bmatrix} x \\ y \end{bmatrix}_{C_2}$ and $C_2 = \left\{ Y_1 = \begin{bmatrix} 1 \\ 1 \end{bmatrix}, Y_2 = \begin{bmatrix} 2 \\ 0 \end{bmatrix} \right\}$.

7. If $Q: \mathbf{R}_3 \rightarrow \mathbf{R}$ is defined by X^TAX, $X = \begin{bmatrix} x \\ y \\ z \end{bmatrix}$, and $A = \begin{bmatrix} 1 & -1 & 0 \\ -1 & 3 & 2 \\ 0 & 2 & -1 \end{bmatrix}$, find a symmetric matrix B such that $Q(X) = [u \ v \ w]B\begin{bmatrix} u \\ v \\ w \end{bmatrix}$, where

$\begin{bmatrix} u \\ v \\ w \end{bmatrix} = \begin{bmatrix} x \\ y \\ z \end{bmatrix}_{C_2}$ and, $C_2 = \left\{ Y_1 = \begin{bmatrix} 1 \\ 1 \\ 0 \end{bmatrix}, Y_2 = \begin{bmatrix} 1 \\ 1 \\ 1 \end{bmatrix}, Y_3 = \begin{bmatrix} 0 \\ 1 \\ 1 \end{bmatrix} \right\}$

Use B to compute $Q\begin{bmatrix} 1 \\ 2 \\ 3 \end{bmatrix}$ and then use A to compute $Q\begin{bmatrix} 1 \\ 2 \\ 3 \end{bmatrix}$.

8. If $Q(X) = [x \ y]\begin{bmatrix} 1 & -3 \\ -3 & 2 \end{bmatrix}\begin{bmatrix} x \\ y \end{bmatrix}$ and $C_2 = \left\{ Y_1 = \begin{bmatrix} 2 \\ 0 \end{bmatrix}, Y_2 = \begin{bmatrix} 3 \\ 3 \end{bmatrix} \right\}$, find a symmetric matrix B such that $Q(X) = [u \ v]B\begin{bmatrix} u \\ v \end{bmatrix}$, where $\begin{bmatrix} u \\ v \end{bmatrix} = \begin{bmatrix} x \\ y \end{bmatrix}_{C_2}$.
Use B to calculate $Q\begin{bmatrix} 3 \\ 4 \end{bmatrix}$.

9. If

$$Q(X) = [x \ y \ z]\begin{bmatrix} 1 & 3 & -1 \\ 3 & 0 & 2 \\ -1 & 2 & 1 \end{bmatrix}\begin{bmatrix} x \\ y \\ z \end{bmatrix},$$

$$C_2 = \left\{ Y_1 = \begin{bmatrix} 1 \\ 1 \\ 0 \end{bmatrix}, Y_2 = \begin{bmatrix} 0 \\ 0 \\ 2 \end{bmatrix}, Y_3 = \begin{bmatrix} 2 \\ 1 \\ 0 \end{bmatrix} \right\},$$

and

$$\begin{bmatrix} u \\ v \\ w \end{bmatrix} = \begin{bmatrix} x \\ y \\ z \end{bmatrix}_{C_2}$$

find a symmetric matrix B such that $Q(X) = [u \ \ v \ \ w]B \begin{bmatrix} u \\ v \\ w \end{bmatrix}$. Use B to cal-

culate $Q \begin{bmatrix} 1 \\ 1 \\ 2 \end{bmatrix}$ and then use A to perform the calculation.

10. Verify that $C_2 = \left\{ Y_1 = (1/\sqrt{5}) \begin{bmatrix} 1 \\ -2 \end{bmatrix}, \ Y_2 = (1/\sqrt{5}) \begin{bmatrix} 2 \\ 1 \end{bmatrix} \right\}$ is an orthonormal

basis. Find the matrix P for the change of basis from C_2 to the natural basis. Verify that $P^T P = I_2$.

11. Verify that

$$C_2 = \left\{ Y_1 = \frac{1}{\sqrt{3}} \begin{bmatrix} 1 \\ 1 \\ 1 \end{bmatrix}, \ Y_2 = \frac{1}{\sqrt{6}} \begin{bmatrix} -2 \\ 1 \\ 1 \end{bmatrix}, \ Y_3 = \frac{1}{\sqrt{2}} \begin{bmatrix} 0 \\ -1 \\ 1 \end{bmatrix} \right\}$$

is an orthonormal basis. Find the matrix P for the change of basis from C_2 to the natural basis for $\mathbf{R}_3$. Verify that P is an orthogonal matrix.

Let A and B be $n \times n$ matrices. A is **congruent** to B if and only if there exists a non-singular matrix P such that $B = P^T AP$.

12. For $n \times n$ matrices A, B, and C verify that
   (a) A is congruent to A.
   (b) If A is congruent to B, then B is congruent to A.
   (c) If A is congruent to B and B is congruent to C, then A is congruent to C.

An $n \times n$ matrix D is said to be **skew-symmetric** if and only if $D = -D^T$.

13. For any $n \times n$ matrix A, let $C = \frac{1}{2}(A + A^T)$ and let $D = \frac{1}{2}(A - A^T)$.
   (a) Show that $A = C + D$.
   (b) Prove that C is symmetric and D is skew-symmetric.

7.4 DIAGONALIZATION OF QUADRATIC FORMS; APPLICATIONS TO CENTRAL CONICS AND QUADRIC SURFACES

In analytic geometry it is shown that the general equation of a central conic (ellipse or hyperbola with center at the origin of a rectangular coordinate system) is of the form $ax^2 + bxy + cy^2 = d$. Furthermore, if the axes are rotated through an appropriate angle, the equation of the conic with respect to the coordinate system determined by the new axes is "free" of the cross term. We wish to accomplish this same result by using our knowledge of quadratic forms.

The left side of the equation of the central conic defines a quadratic form $Q(X)$. Thus the problem of eliminating the cross term can be transferred to the problem of finding a diagonal matrix for Q. If $Q(X) = X^T A X$, the matrix of Q relative to an ordered basis C_2 is B, where $B = P^T A P$ and P is the matrix for the change of basis from C_2 to the natural basis. We wish to find C_2 so that the matrix B is a diagonal matrix. From our knowledge of linear operators, we know that if C_2 is a basis of eigenvectors of A then the matrix $D = P^{-1} A P$ is a diagonal matrix. Our problem would be solved if P were an orthogonal matrix, for then $P^T = P^{-1}$ and $B = D$. According to Theorem 7.8, P is an orthogonal matrix if and only if C_2 is an orthonormal basis. Thus what we need to solve the problem is an orthonormal basis of eigenvectors. It turns out that, for a symmetric matrix A, if each vector in a basis of eigenvectors of A belongs to a different eigenvalue, then the eigenvectors in the basis are mutually orthogonal. It is a simple matter to then normalize the vectors and produce an orthonormal basis. We illustrate the procedure in the next example.

Example 7.12. Given the central conic with equation $5x^2 + 4xy + 2y^2 = 4$. Find an orthonormal basis C_2 such that the matrix of the quadratic form $5x^2 + 4xy + 2y^2$ relative to C_2 is a diagonal matrix. Sketch the graph of the conic relative to the new axes determined by the basis C_2.

Solution: The matrix of the quadratic form relative to the standard basis is $A = \begin{bmatrix} 5 & 2 \\ 2 & 2 \end{bmatrix}$. We determine the eigenvalues of A. The characteristic equation of A is

$$\det(A - \lambda I_2) = 0 \Longleftrightarrow \begin{vmatrix} 5 - \lambda & 2 \\ 2 & 2 - \lambda \end{vmatrix} = 0$$

$$\Longleftrightarrow 10 - 7\lambda + \lambda^2 - 4 = 0$$

$$\Longleftrightarrow (\lambda - 6)(\lambda - 1) = 0$$

Therefore, the eigenvalues of A are 6 and 1.

The eigenvectors belonging to 6 are the nonzero solutions to the matrix equation $(A - 6I_2)X = 0$.

$$A - 6I_2 = \begin{bmatrix} -1 & 2 \\ 2 & -4 \end{bmatrix} \xrightarrow[-1R_1]{2R_1 + R_2} \begin{bmatrix} 1 & -2 \\ 0 & 0 \end{bmatrix}$$

Therefore, the eigenvectors belonging to 6 are the nonzero vectors in the space $\left\langle \begin{bmatrix} 2 \\ 1 \end{bmatrix} \right\rangle$. Also

$$A - I_2 = \begin{bmatrix} 4 & 2 \\ 2 & 1 \end{bmatrix} \xrightarrow{-2R_2+R_1} \begin{bmatrix} 0 & 0 \\ 2 & 1 \end{bmatrix} \xrightarrow[(1/2)R_1]{R_{12}} \begin{bmatrix} 1 & \frac{1}{2} \\ 0 & 0 \end{bmatrix}$$

Thus the eigenvectors belonging to 1 are the nonzero vectors in the space $\left\langle \begin{bmatrix} -\frac{1}{2} \\ 1 \end{bmatrix} \right\rangle = \left\langle \begin{bmatrix} -1 \\ 2 \end{bmatrix} \right\rangle$. The vectors $\begin{bmatrix} 2 \\ 1 \end{bmatrix}$ and $\begin{bmatrix} -1 \\ 2 \end{bmatrix}$ are orthogonal eigenvectors. Normalizing by multiplying each vector by the reciprocal of its length, we obtain

$$C_2 = \left\{ Y_1 = \frac{1}{\sqrt{5}} \begin{bmatrix} 2 \\ 1 \end{bmatrix}, Y_2 = \frac{1}{\sqrt{5}} \begin{bmatrix} -1 \\ 2 \end{bmatrix} \right\}$$

which is an orthonormal basis of eigenvectors. The matrix P for the change of basis from C_2 to C_1 (the natural basis) is $P = (1/\sqrt{5}) \begin{bmatrix} 2 & -1 \\ 1 & 2 \end{bmatrix}$. If $T : \mathbf{R}_2 \rightarrow \mathbf{R}_2$ is the linear operator defined by $T(X) = AX$, then A is the matrix of T relative to the standard basis C_1. By Theorem 7.1, $P^{-1}AP = [T]_{C_2}$ is the diagonal matrix $D = \begin{bmatrix} 6 & 0 \\ 0 & 1 \end{bmatrix}$ and by Theorem 7.8 $P^{-1} = P^T$. Hence the matrix of the quadratic form $Q(X)$ relative to the basis C_2 is the diagonal matrix D. If $X = \begin{bmatrix} x \\ y \end{bmatrix}$, has coordinates $\begin{bmatrix} u \\ v \end{bmatrix}$ relative to the basis C_2, then $Q(X) = [u \ \ v] D \begin{bmatrix} u \\ v \end{bmatrix} = 6u^2 + v^2$. Therefore, the given central conic has the equation

$$6u^2 + v^2 = 4 \quad \text{or} \quad \frac{u^2}{2/3} + \frac{v^2}{4} = 1$$

relative to the coordinate system determined by C_2. We recognize this equation to be the equation of an ellipse with major axis of length 4 along the Y_2 axis, and minor axis of length $2\sqrt{2/3}$ along the Y_1 axis ($X = uY_1 + vY_2$). A sketch of the conic is given in Figure 7.2.

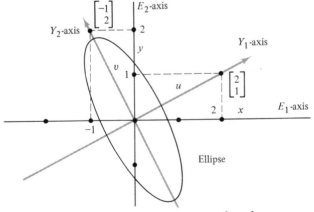

Figure 7.2 $5x^2 + 4xy + 2y^2 = 4 \rightleftharpoons \dfrac{u^2}{2/3} + \dfrac{v^2}{4} = 1$

In the preceding example the eigenvectors of the symmetric matrix A that belonged to different eigenvalues (6 and 1) were orthogonal. As we mentioned earlier, this is true in general, but to prove it we need the following lemma.

Lemma 7.2. If X and Y are column vectors in $\mathbf{R}_n$ and A is an $n \times n$ matrix, then $AX \cdot Y = X \cdot A^T Y$. (The dot denotes the dot product of column vectors, and juxtaposition AX denotes the matrix product.)

Proof: First note that for any two column vectors X and Z in $\mathbf{R}_n$ the dot product $X \cdot Z$ is equal to the scalar k if and only if the matrix product $X^T Z$ is equal to the 1×1 matrix $[k]$. Now suppose $AX \cdot Y = k$. Then $[k] = (AX)^T Y = (X^T A^T)Y = X^T(A^T Y)$. Therefore, $X \cdot A^T Y = k$. We have proved that $AX \cdot Y = X \cdot A^T Y$. Q.E.D.

Theorem 7.9. If A is a symmetric matrix and if X and Y are eigenvectors of A that belong to distinct characteristic values, then X and Y are orthogonal.

Proof: Suppose that X and Y are eigenvectors of the symmetric matrix A belonging to distinct eigenvalues λ_1 and λ_2, respectively. Then $A^T = A$, $AX = \lambda_1 X$, and $AY = \lambda_2 Y$. So

$$\lambda_1(X \cdot Y) = (\lambda_1 X) \cdot Y = AX \cdot Y = X \cdot A^T Y = X \cdot AY = X \cdot \lambda_2 Y = \lambda_2(X \cdot Y)$$

Therefore, $\lambda_1(X \cdot Y) = \lambda_2(X \cdot Y)$. Since $\lambda_1 \neq \lambda_2$, we have $X \cdot Y = 0$. Hence X and Y are orthogonal. Q.E.D.

We now have a process for "diagonalizing" a quadratic form $Q(X) = X^T A X$, where the $n \times n$ symmetric matrix A is the matrix of the quadratic form relative to the standard basis, and A has n *distinct* eigenvalues. We summarize the process and the reasoning behind the process.

Summary: Let $Q(X) = X^T A X$ be a quadratic form, where A is the $n \times n$ symmetric matrix of Q relative to the standard basis C_1.

Step 1: Find the n distinct eigenvalues of A; find the corresponding eigenvectors and normalize them. The normalized vectors are still eigenvectors (Problem 16, Exercise 7.2), and by Theorem 7.5 they form a basis C_2 for $\mathbf{R}_n$. According to Theorem 7.9, C_2 is a basis of orthogonal vectors, and thus C_2 is an orthonormal basis.

Step 2: Now find the matrix P for the change of basis from the basis C_2 to the standard basis C_1. If $C_2 = \{Y_1, Y_2, \ldots, Y_n\}$, then the jth column of P is the vector Y_j. By Theorem 7.7, $P^T A P = B$ is the matrix of the quadratic form $Q(X)$ relative to the basis C_2. If we think of A as the matrix (relative to C_1) of a linear operator $T: \mathbf{R}_n \rightarrow \mathbf{R}_n$ defined by $T(X) = AX$, then $P^{-1} A P = D$

is the matrix of T relative to the basis C_2, and D is a diagonal matrix. Since C_2 is an orthonormal basis, Theorem 7.8 tells us that $P^T = P^{-1}$, thus $D = B$, and the matrix of $Q(X)$ relative to C_2 is a diagonal matrix having the n distinct eigenvalues on the diagonal. If $U = [X]_{C_2}$, then

$$Q(X) = U^T B U = \sum_{i=1}^{n} \lambda_i u_i^2$$

which is a sum of "square" terms where the λ_i are the eigenvalues of A. Furthermore, $U = P^T [X]_{C_1} = P^T X = [X]_{C_2}$, because $P^T = P^{-1}$ is the matrix for the change of basis from C_1 to C_2.

We shall illustrate the process of diagonalizing a quadratic form in three variables in the next two examples, but first we give a definition.

Definition 7.11. Let A be an $n \times n$ matrix and let λ_0 be a (real) eigenvalue of A. The set of all solutions to the matrix equation $(A - \lambda_0 I_n)X = 0$ is called the **eigenspace of A associated with λ_0.**

The eigenspace of A associated with λ_0 is a subspace of $\mathbf{R_n}$ (Problem 16, Exercise 7.2). The nonzero members of this space are the eigenvectors of A belonging to λ_0.

Example 7.13. Identify the graph of the equation $2x^2 + y^2 - 4xy - 4yz = 4$ by using the procedure outlined in the above summary to diagonalize the quadratic form appearing on the left side of the equation.

Solution: The matrix of the quadratic form relative to the standard basis C_1 is $A = \begin{bmatrix} 2 & -2 & 0 \\ -2 & 1 & -2 \\ 0 & -2 & 0 \end{bmatrix}$. The characteristic equation of A is $\lambda^3 - 3\lambda^2 - 6\lambda + 8 = 0$. The roots of this equation are the distinct eigenvalues $1, -2$, and 4. The eigenspaces associated with $1, -2$, and 4, respectively, are

$$\left\langle \begin{bmatrix} -2 \\ -1 \\ 2 \end{bmatrix} \right\rangle, \left\langle \begin{bmatrix} 1 \\ 2 \\ 2 \end{bmatrix} \right\rangle, \text{ and } \left\langle \begin{bmatrix} 2 \\ -2 \\ 1 \end{bmatrix} \right\rangle.$$

Normalizing a vector from each of these spaces, we obtain an orthonormal basis of eigenvectors

$$C_2 = \left\{ \tfrac{1}{3} \begin{bmatrix} -2 \\ -1 \\ 2 \end{bmatrix}, \tfrac{1}{3} \begin{bmatrix} 1 \\ 2 \\ 2 \end{bmatrix}, \tfrac{1}{3} \begin{bmatrix} 2 \\ -2 \\ 1 \end{bmatrix} \right\}$$

The matrix for the change of basis from C_2 to C_1 is

$$P = \tfrac{1}{3} \begin{bmatrix} -2 & 1 & 2 \\ -1 & 2 & -2 \\ 2 & 2 & 1 \end{bmatrix}$$

and

$$D = P^T A P = \frac{1}{9} \begin{bmatrix} -2 & -1 & 2 \\ 1 & 2 & 2 \\ 2 & -2 & 1 \end{bmatrix} \begin{bmatrix} 2 & -2 & 0 \\ -2 & 1 & -2 \\ 0 & -2 & 0 \end{bmatrix} \begin{bmatrix} -2 & 1 & 2 \\ -1 & 2 & -2 \\ 2 & 2 & 1 \end{bmatrix}$$

$$= \begin{bmatrix} 1 & 0 & 0 \\ 0 & -2 & 0 \\ 0 & 0 & 4 \end{bmatrix}$$

If the coordinate vector of $X = \begin{bmatrix} x \\ y \\ z \end{bmatrix}$ is $\begin{bmatrix} u \\ v \\ w \end{bmatrix} = U$ relative to the basis C_2,
then $Q(X) = u^2 - 2v^2 + 4w^2$, where $PU = X$. Relative to the coordinate system determined by the vectors in C_2, the given equation is transformed to

$$u^2 - 2v^2 + 4w^2 = 4 \quad \text{or} \quad \frac{u^2}{4} - \frac{v^2}{2} + w^2 = 1$$

The cross section of the graph in the uv plane obtained by setting $w = 0$ is the hyperbola $(u^2/4) - (v^2/2) = 1$, the cross section of the graph in the uw plane obtained by setting $v = 0$ is the ellipse $(u^2/4) + (w^2/1) = 1$, and the cross section of the graph in the vw plane is the hyperbola $-(v^2/2) + (w^2/1) = 1$. The graph is called a *hyperboloid of one sheet*. The axes determined by the vectors in C_2 are called the **principal axes** of the hyperboloid.

In 3-space a surface with equation $ax^2 + by^2 + cz^2 + dxy + exz + fyz = g$ is called a **quadric surface**. In Figure 7.3, various quadric surfaces are sketched where the principal axes are the axes determined by the natural basis for $\mathbf{R_3}$.

In Examples 7.12 and 7.13, the eigenvalues were all different, and there were as many eigenvalues as there were variables in the quadratic form. It can be shown that the characteristic roots (values) of the characteristic equation of a real symmetric matrix are always *real* numbers, but there may be repeated roots, as the next example shows.

Example 7.14. The equation of a quadric surface is $Q(X) = 3x^2 + 6y^2 + 3z^2 - 4xy - 8xz - 4yz = 14$. Diagonalize the quadratic form $Q(X)$ and identify the quadric surface.

Solution: The matrix of the quadratic form relative to the standard basis is $A = \begin{bmatrix} 3 & -2 & -4 \\ -2 & 6 & -2 \\ -4 & -2 & 3 \end{bmatrix}$. The characteristic polynomial is

$$\det(A - \lambda I_3) = -\lambda^3 + 12\lambda^2 - 21\lambda - 98 = -(\lambda + 2)(\lambda - 7)^2$$

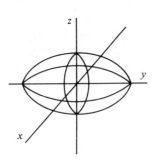

Ellipsoid

$$\frac{x^2}{a^2} + \frac{y^2}{b^2} + \frac{z^2}{c^2} = 1$$

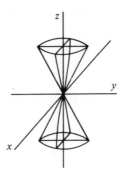

Elliptic cone

$$\frac{x^2}{a^2} + \frac{y^2}{b^2} = z^2$$

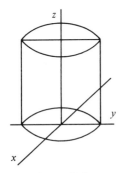

Elliptic cylinder

$$\frac{x^2}{a^2} + \frac{y^2}{b^2} = 1$$

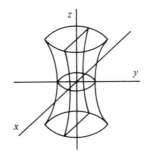

Hyperboloid of one sheet

$$\frac{x^2}{a^2} + \frac{y^2}{b^2} - \frac{z^2}{c^2} = 1$$

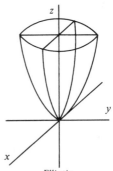

Elliptic paraboloid

$$\frac{x^2}{a^2} + \frac{y^2}{b^2} = cz$$

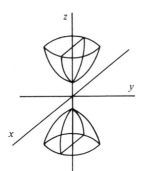

Hyperboloid of two sheets

$$\frac{z^2}{c^2} - \frac{x^2}{a^2} - \frac{y^2}{b^2} = 1$$

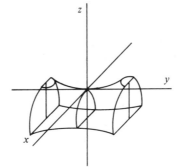

Hyperbolic paraboloid

$$\frac{-x^2}{a^2} + \frac{y^2}{b^2} = cz$$

Figure 7.3

The characteristic values are -2 and 7. Transforming $A + 2I_3$ to reduced row echelon form, one finds that the eigenspace of vectors associated with -2 is $\left\langle \begin{bmatrix} 2 \\ 1 \\ 2 \end{bmatrix} \right\rangle$. Transforming $A - 7I$ to reduced row echelon form, we find that the eigenspace associated with 7 is the two-dimensional space $\left\langle \begin{bmatrix} -1 \\ 2 \\ 0 \end{bmatrix}, \begin{bmatrix} -1 \\ 0 \\ 1 \end{bmatrix} \right\rangle$.

All vectors in this space are of the form $a \begin{bmatrix} -1 \\ 2 \\ 0 \end{bmatrix} + b \begin{bmatrix} -1 \\ 0 \\ 1 \end{bmatrix} = \begin{bmatrix} -a - b \\ 2a \\ b \end{bmatrix}$.

We look for one that is perpendicular to $\begin{bmatrix} -1 \\ 0 \\ 1 \end{bmatrix}$. We must have $\begin{bmatrix} -1 \\ 0 \\ 1 \end{bmatrix} \cdot \begin{bmatrix} -a - b \\ 2a \\ b \end{bmatrix} = 0$, which is equivalent to $a + 2b = 0$. Taking $a = -2$ and $b = 1$, we get the vector $\begin{bmatrix} 1 \\ -4 \\ 1 \end{bmatrix}$. We have found two eigenvectors belonging to 7 that are orthogonal. By Theorem 7.9, each of these vectors is orthogonal to any eigenvector belonging to -2. Thus $Y_1' = \begin{bmatrix} 2 \\ 1 \\ 2 \end{bmatrix}$, $Y_2' = \begin{bmatrix} -1 \\ 0 \\ 1 \end{bmatrix}$, and $Y_3' = \begin{bmatrix} 1 \\ -4 \\ 1 \end{bmatrix}$ are mutually orthogonal. Normalizing these vectors, we obtain the orthonormal basis $C_2 = \{Y_1 = \frac{1}{3}Y_1', Y_2 = (1/\sqrt{2})Y_2', Y_3 = (1/3\sqrt{2})Y_3'\}$. The matrix for the change of basis from C_2 to the standard basis C_1 is $P = \frac{1}{3\sqrt{2}} \begin{bmatrix} 2\sqrt{2} & -3 & 1 \\ \sqrt{2} & 0 & -4 \\ 2\sqrt{2} & 3 & 1 \end{bmatrix}$, and $P^T A P = \begin{bmatrix} -2 & 0 & 0 \\ 0 & 7 & 0 \\ 0 & 0 & 7 \end{bmatrix} = D$. If a vector $X = \begin{bmatrix} x \\ y \\ z \end{bmatrix}$ has coordinates $\begin{bmatrix} u \\ v \\ w \end{bmatrix}$ relative to the basis C_2, then $Q(X) = [u \; v \; w] D \begin{bmatrix} u \\ v \\ w \end{bmatrix} = -2u^2 + 7v^2 + 7w^2$. We see that, relative to the coordinate system determined by the vectors in C_2, the quadric surface has the equation

$$-2u^2 + 7v^2 + 7w^2 = 14 \quad \text{or} \quad \frac{-u^2}{7} + \frac{v^2}{2} + \frac{w^2}{2} = 1$$

It is now clear that the quadric surface is again a hyperboloid of one sheet. The principal axes of this quadric surface are the three mutually perpendicular lines determined by the vectors in C_2 and by the origin.

Exercise 7.4

1. In Example 7.12,
   (a) Show that $P^T A P = \text{diag}\,(6, 1)$ by carrying out the matrix multiplication.
   (b) Let θ be the angle between the vectors E_1 and $Y_1 = \begin{bmatrix} 2/\sqrt{5} \\ 1/\sqrt{5} \end{bmatrix}$. If R_θ is the rotation counterclockwise through θ, show that $R_\theta(X) = PX$.

2. In Example 7.13,
   (a) Verify that the characteristic polynomial of A is $-(\lambda - 1)(\lambda + 2)(\lambda - 4)$.
   (b) Verify that the eigenspaces associated with 1, -2, and 4 are the given spaces.

3. In Example 7.14,
   (a) Verify that the characteristic polynomial of A is $-(\lambda + 2)(\lambda - 7)^2$.
   (b) Verify that the eigenspaces of A associated with -2 and 7 are the given eigenspaces.
   (c) Verify that $P^T A P = \text{diag}\,(-2, 7, 7)$.

In Problems 4–9, find an orthogonal matrix P such that $P^T A P$ is a diagonal matrix.

4. $A = \begin{bmatrix} 2 & 2 \\ 2 & 5 \end{bmatrix}$.   5. $A = \begin{bmatrix} 0 & 1 \\ 1 & 0 \end{bmatrix}$.   6. $A = \begin{bmatrix} 3 & -1 \\ -1 & 3 \end{bmatrix}$.

7. $A = \begin{bmatrix} 10 & 12 \\ 12 & 17 \end{bmatrix}$.   8. $A = \begin{bmatrix} 5 & 6 \\ 6 & 10 \end{bmatrix}$.   9. $A = \begin{bmatrix} 5 & 2 \\ 2 & 8 \end{bmatrix}$.

In Problems 10–15, each of the given equations is the equation of a central conic section. Diagonalize the quadratic form defined by the left side of the equation and identify the central conic. Identify the orthonormal vectors that determine the principal axes and sketch the graph of the central conic relative to the principal axes. The matrix of the quadratic form (relative to the standard basis) in each of Problems 10–15 is the matrix in each of Problems 4–9, respectively.

10. $2x^2 + 4xy + 5y^2 = 6$.   11. $2xy = 1$.

12. $3x^2 - 2xy + 3y^2 = 32$.   13. $10x^2 + 24xy + 17y^2 = 26$.

14. $5x^2 + 12xy + 10y^2 = 28$.   15. $5x^2 + 4xy + 8y^2 = 36$.

In Problems 16–18, find an orthogonal matrix P such that $P^T A P$ is a diagonal matrix.

16. $A = \begin{bmatrix} 1 & -2 & 0 \\ -2 & 2 & -2 \\ 0 & -2 & 3 \end{bmatrix}$.

17. $A = \begin{bmatrix} 5 & -2 & -2 \\ -2 & 6 & 0 \\ -2 & 0 & 4 \end{bmatrix}$.

18. $A = \begin{bmatrix} 2 & 2 & -2 \\ 2 & 5 & -4 \\ -2 & -4 & 5 \end{bmatrix}$.

In Problems 19–21, each of the given equations is the equation of a quadric surface. Diagonalize the quadratic form defined by the left side of the equation, and use Figure 7.3 to identify the quadric surface. Write the new coordinates $\begin{bmatrix} u \\ v \\ w \end{bmatrix}$ in terms of the old coordinates $\begin{bmatrix} x \\ y \\ z \end{bmatrix}$. The quadratic forms correspond to the matrices in Problems 16–18.

19. $x^2 + 2y^2 + 3z^2 - 4xy - 4yz = 10$.

20. $5x^2 + 6y^2 + 4z^2 - 4xy - 4xz = 40$.

21. $2x^2 + 5y^2 + 5z^2 + 4xy - 4xz - 8yz = 10$.

22. Transform the quadratic form $xy - xz + yz$ to the form $au^2 + bv^2 + cw^2$ by using the diagonalization process.

23. Let A be an $n \times n$ matrix, let λ_1 and λ_2 be eigenvalues of A, and let E_{λ_i} denote the eigenspace of A associated with λ_i. Prove: $E_{\lambda_1} + E_{\lambda_2} = E_{\lambda_1} \oplus E_{\lambda_2}$ if $\lambda_1 \neq \lambda_2$.

†7.5   THE SPECTRAL THEOREM AND THE HAMILTON–CAYLEY THEOREM

We close this chapter by discussing some results that we feel the reader should be aware of, but whose proofs are left for more advanced texts. We have seen that the task of diagonalizing a quadratic form $Q(X)$ with matrix A relative to the standard basis can be accomplished by finding an orthonormal basis of eigenvectors for A. In the case that the $n \times n$ matrix A has n distinct eigenvalues, the task is fairly easy, because any n vectors belonging to the n different eigenvalues will be a basis of mutually orthogonal vectors. It is a simple matter to normalize these vectors to obtain the desired orthonormal basis. As was mentioned earlier, the characteristic values (roots) of a real symmetric matrix A are all real numbers; however, the nth degree characteristic equation

†This section is optional, for the results are not used in other sections of the text.

of A need not have n *distinct* roots. The next theorem states that an $n \times n$ (real) symmetric matrix can be diagonalized regardless of whether or not there are n different eigenvalues.

Theorem 7.10. **Principal Axes Theorem or Spectral Theorem.** If A is an $n \times n$ symmetric matrix, then there exists an orthogonal matrix P such that $P^T A P = D$ is a diagonal matrix and the diagonal entries of D are the characteristic values (the spectrum of A) of A. P is the matrix for the change of basis from an orthonormal basis of eigenvectors of A to the standard basis.

It is always possible then (theoretically at least) to diagonalize a symmetric matrix. Example 7.14 indicates how one can proceed to diagonalize an $n \times n$ symmetric matrix that has less than n characteristic values. In that example the characteristic polynomial was $-(\lambda + 2)(\lambda - 7)^2$. The number 7 is a repeated root of multiplicity 2 (the power of $\lambda - 7$). The eigenspace associated with 7 had dimension 2, and we were able to find an orthonormal basis for the space. In general, if the characteristic polynomial of a symmetric matrix has a factor $(\lambda - \lambda_0)^k$, then λ_0 is a root of multiplicity k. The eigenspace associated with λ_0 will have dimension k. A general method for transforming a basis for a subspace of $\mathbf{R}_n$ to an orthonormal basis is discussed in Chapter 8. This method is called the Gram–Schmidt process. Using the Gram–Schmidt process, one can find an orthonormal basis for each of the eigenspaces, put them together, and thus find an orthonormal basis of eigenvectors of A.

The reader might ask the question, how can we determine whether or not a nonsymmetric matrix is diagonalizable? We have seen examples of matrices that were not similar to a diagonal matrix and thus not diagonalizable. But in general how does one tell? Is a (real) matrix that has a characteristic equation with some nonreal roots ever diagonalizable? The answer can be obtained by studying the minimum polynomial of a matrix, which we define shortly. To lead up to this definition, we give a brief discussion of polynomial functions.

Consider a polynomial $f(x)$ over $\mathbf{R}$, that is, a polynomial $f(x) = a_0 + a_1 x + \cdots + a_k x^k$, where the coefficients $a_0, a_1, \ldots, a_k$ are real numbers. If A is an $n \times n$ square matrix, *we define* $f(A)$ *to be the expression* $f(A) = a_0 I + a_1 A + \cdots + a_k A^k$. This expression is obtained by replacing x by A and a_0 by $a_0 I$, where I is the $n \times n$ identity matrix. Since each term in the sum defining $f(A)$ is an $n \times n$ square matrix, we see that $f(A)$ is an $n \times n$ square matrix. One can show that for any two polynomials $f(x)$ and $g(x)$ over $\mathbf{R}$ and for any scalar r, (i) $(f + g)(A) = f(A) + g(A)$, (ii) $(fg)(A) = f(A)g(A) = g(A)f(A)$, and (iii) $rf(A) = r(f(A))$.

Example 7.15. If $f(x) = 2 + 3x + 4x^2$, and $g(x) = 3 + 5x$, and A is an $n \times n$ matrix, show that $f(A)g(A) = g(A)f(A) = (fg)(A)$.

Solution: We use the fact that the collection of $n \times n$ matrices is an algebra.

$$f(A)g(A) = (2I + 3A + 4A^2)(3I + 5A)$$
$$= (2I + 3A + 4A^2)(3I) + (2I + 3A + 4A^2)(5A)$$
$$= (6I^2 + 9AI + 12A^2I) + (10IA + 15A^2 + 20A^3)$$
$$= 6I + 19A + 27A^2 + 20A^3$$

Also,

$$g(A)f(A) = (3I + 5A)(2I + 3A + 4A^2)$$
$$= (3I)(2I + 3A + 4A^2) + (5A)(2I + 3A + 4A^2)$$
$$= (6I^2 + 9IA + 12IA^2) + (10AI + 15A^2 + 20A^3)$$
$$= 6I + 19A + 27A^2 + 20A^3$$

So $f(A)g(A) = g(A)f(A) = (fg)(A)$. (The reader should check the last equality.)

Although matrix multiplication is not commutative, property (ii) states that $f(A)g(A) = g(A)f(A)$ for any two polynomials $f(x)$ and $g(x)$ over **R**. The basic reason that (ii) holds is because *powers of A do commute*; that is, $A^m A^p = A^{m+p} = A^p A^m$ for all nonnegative integers m and p.

Definition 7.12. An $n \times n$ matrix A is a **zero of the polynomial** $f(x)$ if and only if $f(A) = \mathbf{0}$ (the zero $n \times n$ matrix). We say that A satisfies the equation $f(x) = 0$ if it is true that $f(A) = \mathbf{0}$.

An important theorem in the theory of eigenvalues and eigenvectors is the following theorem.

Theorem 7.11 (Hamilton–Cayley Theorem). An $n \times n$ square matrix A is a zero of its characteristic polynomial. In symbols, if $f(\lambda) = \det(A - \lambda I_n)$, then $f(A) = \mathbf{0}$.

Example 7.16. Show that the matrix $A = \begin{bmatrix} 5 & 2 \\ 2 & 2 \end{bmatrix}$ satisfies its characteristic polynomial.

Solution: The matrix A is the matrix of Example 7.12. Its characteristic polynomial is $\lambda^2 - 7\lambda + 6 = f(\lambda)$.

$$f(A) = \begin{bmatrix} 5 & 2 \\ 2 & 2 \end{bmatrix}\begin{bmatrix} 5 & 2 \\ 2 & 2 \end{bmatrix} - 7\begin{bmatrix} 5 & 2 \\ 2 & 2 \end{bmatrix} + 6\begin{bmatrix} 1 & 0 \\ 0 & 1 \end{bmatrix}$$
$$= \begin{bmatrix} 29 & 14 \\ 14 & 8 \end{bmatrix} + \begin{bmatrix} -35 & -14 \\ -14 & -14 \end{bmatrix} + \begin{bmatrix} 6 & 0 \\ 0 & 6 \end{bmatrix} = \begin{bmatrix} 0 & 0 \\ 0 & 0 \end{bmatrix}$$

Applying the Hamilton–Cayley theorem, we are assured that for any $n \times n$ matrix A there is always a polynomial of positive degree that is satisfied by A, namely, the characteristic polynomial. There may be polynomials $f(x)$

of degree smaller than the degree of the characteristic polynomial that are satisfied by A. We single out the polynomial of least degree that has a leading coefficient of 1. By *leading coefficient* we mean the nonzero coefficient of the highest power of the polynomial.

Definition 7.13. The **minimum polynomial** of a square matrix A is the monic polynomial (the leading coefficient is 1) $m(x)$ of smallest positive degree such that $m(A) = \mathbf{0}$.

By choosing $m(x)$ to be a monic polynomial, we force the minimum polynomial to be unique.

Theorem 7.12. If $m(x)$ is the minimum polynomial of an $n \times n$ matrix A and if $f(x)$ is a polynomial of positive degree such that $f(A) = \mathbf{0}$, then $m(x)$ is a divisor of $f(x)$; that is, $f(x) = q(x)m(x)$ for some polynomial $q(x)$. In particular, the minimum polynomial is a divisor of the characteristic polynomial.

Proof: Let $f(x)$ be a polynomial of positive degree such that $f(A) = \mathbf{0}$, and let $m(x)$ be the minimum polynomial of A. By the division algorithm, we can divide $f(x)$ by $m(x)$ (long division) and obtain a quotient $q(x)$ and a remainder $r(x)$ such that $f(x) = q(x)m(x) + r(x)$, and the degree of $r(x)$ is less than the degree of $m(x)$ or $r(x) = 0$. Then $f(A) = q(A)m(A) + r(A)$, and $r(A) = f(A) - q(A)m(A) = \mathbf{0} - q(A)\mathbf{0} = \mathbf{0}$. So $r(A) = \mathbf{0}$. If $r(x) \neq 0$, then $r(x)$ is a polynomial of positive degree that is satisfied by A and its degree is less than the degree of $m(x)$. This contradicts the definition of minimum polynomial. Hence $r(x) = 0$ and $f(x) = q(x)m(x)$. This establishes the theorem.   Q.E.D.

Example 7.17. Find the minimum polynomial of the matrix
$$A = \begin{bmatrix} 3 & -2 & -4 \\ -2 & 6 & -2 \\ -4 & -2 & 3 \end{bmatrix}.$$

Solution: The characteristic polynomial of A is the polynomial $C(x) = -x^3 + 12x^2 - 21x - 98 = -(x + 2)(x - 7)^2$ (Example 7.14). By Theorem 7.12, $m(x)$ is a divisor of $C(x)$. The only monic polynomials of positive degree that divide $C(x)$ are the polynomials $x + 2$, $x - 7$, $(x + 2)(x - 7)$, and $-C(x)$. $A + 2I \neq \mathbf{0}$, $A - 7I \neq \mathbf{0}$, and
$$(A + 2I)(A - 7I) = \begin{bmatrix} 5 & -2 & -4 \\ -2 & 8 & -2 \\ -4 & -2 & 5 \end{bmatrix} \begin{bmatrix} -4 & -2 & -4 \\ -2 & -1 & -2 \\ -4 & -2 & -4 \end{bmatrix} = \begin{bmatrix} 0 & 0 & 0 \\ 0 & 0 & 0 \\ 0 & 0 & 0 \end{bmatrix}$$

Thus the minimum polynomial of A is
$$m(x) = (x + 2)(x - 7) = x^2 - 5x - 14.$$

Every polynomial over **R** is divisible by a nonzero real number k or has k as a factor. For example, $f(x) = k(1/k)f(x)$. We say that a polynomial over **R** is *irreducible* if its only factors (over **R**) are the nonzero real numbers. As an illustration, $x^2 + 1$ is irreducible over **R**, but

$$x^2 - 2 = (x - \sqrt{2})(x + \sqrt{2})$$

is not irreducible over **R**.

Theorem 7.13.   If A is an $n \times n$ matrix, the minimum polynomial of A and the characteristic polynomial of A have the same irreducible factors.

Theorem 7.14.   A (real) $n \times n$ matrix A is diagonalizable if and only if the minimum polynomial of A is a product of distinct linear polynomials (polynomials of degree 1) with real coefficients.

Example 7.18.   If the characteristic polynomial of A is

$$(-\lambda + 2)(\lambda^2 - \lambda + 3) = C(\lambda)$$

find the minimum polynomial of A and determine if A is diagonalizable.

Solution: The polynomial $\lambda^2 - \lambda + 3$ has discriminant $b^2 - 4ac = -11$. Therefore, the roots of $\lambda^2 - \lambda + 3 = 0$ are not real numbers, and $\lambda^2 - \lambda + 3$ is irreducible. Since the minimum polynomial and the characteristic polynomial of A have the same irreducible factors (Theorem 7.13), it follows that $m(\lambda) = (\lambda - 2)(\lambda^2 - \lambda + 3)$. By Theorem 7.14, A is not diagonalizable.

If an $n \times n$ matrix A is not diagonalizable, that is, if A is not similar to a diagonal matrix, we might ask the question, under the relation of similarity, how close can we come to diagonalizing A? The attempt to answer this question leads to a study of canonical forms. The reader who is interested in pursuing this topic or in finding proofs for the theorems of this section may consult the following references: Daniel T. Finkbeiner, II, *Introduction to Matrices and Linear Transformations*, 2nd ed. (San Francisco: W. H. Freeman and Company, 1966), or Kenneth Hoffman and Ray Kunze, *Linear Algebra*, 2nd ed. (Englewood Cliffs, N. J.: Prentice-Hall, Inc., 1971).

We close this chapter with an application that shows how the method of diagonalizing a matrix can be applied to solve a system of linear differential equations.

Application to Differential Equations

(For those who have studied calculus.) We wish to apply our knowledge of similar matrices to the problem of solving a **system of linear differential equations**

$$f_1' = \frac{df_1}{dx} = a_{11}f_1 + a_{12}f_2 + \cdots + a_{1n}f_n$$

$$f_2' = \frac{df_2}{dx} = a_{21}f_1 + a_{22}f_2 + \cdots + a_{2n}f_n$$

$$\vdots \qquad\qquad \vdots \tag{7.1}$$

$$f_n' = \frac{df_n}{dx} = a_{n1}f_1 + a_{n2}f_2 + \cdots + a_{nn}f_n$$

A solution to the system (7.1) is an n-tuple of functions $(f_1, f_2, \ldots, f_n)$ such that

$$\frac{df_i(x)}{dx} = a_{i1}f_1(x) + a_{i2}f_2(x) + \cdots + a_{in}f_n(x), \qquad \text{for } i = 1, 2, \ldots, n$$

If we let

$$F = \begin{bmatrix} f_1 \\ f_2 \\ \cdot \\ \cdot \\ \cdot \\ f_n \end{bmatrix} \quad \text{and } F' = \frac{dF}{dx} = \begin{bmatrix} f_1' \\ f_2' \\ \cdot \\ \cdot \\ \cdot \\ f_n' \end{bmatrix}$$

then (7.1) may be written in the equivalent matrix form $F' = dF/dx = AF$, where $A = [a_{ij}]$ is the $n \times n$ coefficient matrix.

Before tackling the general solution of (7.1) we consider some examples. We know from calculus that the single differential equation $f' = bf$ has the solution $f: e^{bx}$. Furthermore, all solutions are of the form $g: ce^{bx}$, where c is any constant. Using this information, it is a simple matter to solve a system of linear differential equations such as

$$f_1' = b_1 f_1$$
$$f_2' = b_2 f_2$$
$$f_3' = b_3 f_3$$

or equivalently

$$\begin{bmatrix} f_1' \\ f_2' \\ f_3' \end{bmatrix} = \begin{bmatrix} b_1 & 0 & 0 \\ 0 & b_2 & 0 \\ 0 & 0 & b_3 \end{bmatrix} \begin{bmatrix} f_1 \\ f_2 \\ f_3 \end{bmatrix}$$

The solutions are $f_1: c_1 e^{b_1 x}, f_2: c_2 e^{b_2 x}$, and $f_3: c_3 e^{b_3 x}$. Notice that we are able to solve this system because each equation has only one unknown, or, in terms of matrices, because the coefficient matrix of the system is a *diagonal* matrix.

Now consider the system of linear differential equations

$$f_1' = -1f_1 + 2f_2$$
$$f_2' = -4f_1 + 5f_2$$
$$f_3' = -4f_1 + 2f_2 + 3f_3$$

which is equivalent to

$$\begin{bmatrix} f_1' \\ f_2' \\ f_3' \end{bmatrix} = \begin{bmatrix} -1 & 2 & 0 \\ -4 & 5 & 0 \\ -4 & 2 & 3 \end{bmatrix} \begin{bmatrix} f_1 \\ f_2 \\ f_3 \end{bmatrix}$$

The coefficient matrix A has eigenvalues 3 and 1 and

$$B = \begin{bmatrix} 3 & 0 & 0 \\ 0 & 3 & 0 \\ 0 & 0 & 1 \end{bmatrix} = P^{-1}AP, \quad \text{where } P = \begin{bmatrix} 1 & 0 & 1 \\ 2 & 0 & 1 \\ 0 & 1 & 1 \end{bmatrix}, P^{-1} = \begin{bmatrix} -1 & 1 & 0 \\ -2 & 1 & 1 \\ 2 & -1 & 0 \end{bmatrix}$$

(see Example 7.5). Let

$$G = P^{-1}F = \begin{bmatrix} -1f_1 + 1f_2 \\ -2f_1 + 1f_2 + 1f_3 \\ 2f_1 - 1f_2 \end{bmatrix} = \begin{bmatrix} g_1 \\ g_2 \\ g_3 \end{bmatrix}$$

Then $F = PG$, and from the equation $F' = AF$, we obtain $F' = A(PG)$. Multiplying each side of this matrix equation (on the left) by P^{-1} and reassociating, we obtain $P^{-1}F' = (P^{-1}AP)G = BG$. But

$$P^{-1}F' = \begin{bmatrix} -1f_1' + 1f_2' \\ -2f_1' + 1f_2' + 1f_3' \\ 2f_1' - 1f_2' \end{bmatrix} = \begin{bmatrix} g_1' \\ g_2' \\ g_3' \end{bmatrix} = \frac{dG}{dx} = G'$$

Thus

$$G' = \begin{bmatrix} g_1' \\ g_2' \\ g_3' \end{bmatrix} = BG = \begin{bmatrix} 3g_1 \\ 3g_2 \\ g_3 \end{bmatrix}$$

and $g_1 : c_1e^{3x}$, $g_2 : c_2e^{3x}$, and $g_3 : c_3e^x$. Since $F = PG$, we have

$$\begin{bmatrix} f_1 \\ f_2 \\ f_3 \end{bmatrix} = \begin{bmatrix} 1g_1 & + 1g_3 \\ 2g_1 & + 1g_3 \\ 1g_2 + 1g_3 \end{bmatrix}$$

and $f_1 : c_1e^{3x} + c_3e^x$, $f_2 : 2c_1e^{3x} + c_3e^x$, and $f_3 : c_2e^{3x} + c_3e^x$. As a check on our work, we find

$$f_1'(x) = 3c_1e^{3x} + c_3e^x = -1f_1(x) + 2f_2(x)$$
$$f_2'(x) = 6c_1e^{3x} + c_3e^x = -4f_1(x) + 5f_2(x)$$
$$f_3'(x) = 3c_2e^{3x} + c_3e^x = -4f_1(x) + 2f_2(x) + 3f_3(x)$$

The solution to (7.1) is easily obtained by following the solution to our last example (provided that A is diagonalizable). If A is diagonalizable, then there is an invertible matrix P such that $B = P^{-1}AP$ is a diagonal matrix with the eigenvalues of A as the diagonal entries. Let $G = P^{-1}F$. Then $F = PG$, and from the equation $F' = AF$ we obtain $F' = A(PG)$. Multiplying each side of this matrix equation (on the left) by P^{-1}, we get $P^{-1}F' = P^{-1}APG = BG$. The reader is asked to show (Problem 14, Exercise 7.5) that $P^{-1}F' = P^{-1}(dF/dx) = (dP^{-1}F/dx) = G'$. Thus we have

$$G' = BG = \begin{bmatrix} \lambda_1 & 0 & 0 & \cdots & 0 \\ 0 & \lambda_2 & 0 & \cdots & 0 \\ \cdot & \cdot & \cdot & & \cdot \\ \cdot & \cdot & \cdot & & \cdot \\ \cdot & \cdot & \cdot & & \cdot \\ 0 & 0 & 0 & \cdots & \lambda_n \end{bmatrix} \begin{bmatrix} g_1 \\ g_2 \\ \cdot \\ \cdot \\ \cdot \\ g_n \end{bmatrix}$$

The equivalent system of differential equations is

$$g_1' = \lambda_1 g_1$$
$$g_2' = \lambda_2 g_2$$
$$\cdot$$
$$\cdot$$
$$\cdot$$
$$g_n' = \lambda_n g_n$$

and its solutions are of the form

$$g_1 : c_1 e^{\lambda_1 x}$$
$$g_2 : c_2 e^{\lambda_2 x}$$
$$\cdot$$
$$\cdot$$
$$\cdot$$
$$g_n : c_n e^{\lambda_n x}$$

Now using the fact that $F = PG$, we get

$$f_1 : p_{11} c_1 e^{\lambda_1 x} + p_{12} c_2 e^{\lambda_2 x} + \ldots + p_{1n} c_n e^{\lambda_n x}$$
$$f_2 : p_{21} c_1 e^{\lambda_1 x} + p_{22} c_2 e^{\lambda_2 x} + \ldots + p_{2n} c_n e^{\lambda_n x}$$
$$\cdot$$
$$\cdot$$
$$\cdot$$
$$f_n : p_{n1} c_1 e^{\lambda_1 x} + p_{n2} c_2 e^{\lambda_2 x} + \ldots + p_{nn} c_n e^{\lambda_n x}$$

We apply our results to the system of linear differential equations

$$f_1' = 2f_1 - 2f_2$$
$$f_2' = -2f_1 + 1f_2 - 2f_3$$
$$f_3' = -2f_2$$

From Example 7.13, we see that the eigenvalues of the coefficient matrix A are 1, -2, and 4, and $B = \begin{bmatrix} 1 & 0 & 0 \\ 0 & -2 & 0 \\ 0 & 0 & 4 \end{bmatrix} = P^{-1}AP$, where $P = \frac{1}{3}\begin{bmatrix} -2 & 1 & 2 \\ -1 & 2 & -2 \\ 2 & 2 & 1 \end{bmatrix}$.

From our work above we see that $G' = BG$ and $g_1' = 1g_1$, $g_2' = -2g_2$, and $g_3' = 4g_3$. Thus $g_1 : c_1 e^x$, $g_2 : c_2 e^{-2x}$, and $g_3 : c_3 e^{4x}$. Since $F = PG$, the solutions to the given system are

$$f_1 : \tfrac{1}{3}(-2g_1 + 1g_2 + 2g_3)$$
$$f_2 : \tfrac{1}{3}(-1g_1 + 2g_2 - 2g_3)$$
$$f_3 : \tfrac{1}{3}(2g_1 + 2g_2 + 1g_3)$$

Hence

$$f_1 : \tfrac{1}{3}(-2c_1 e^x + c_2 e^{-2x} + 2c_3 e^{4x})$$
$$f_2 : \tfrac{1}{3}(-c_1 e^x + 2c_2 e^{-2x} - 2c_3 e^{4x})$$
$$f_3 : \tfrac{1}{3}(2c_1 e^x + 2c_2 e^{-2x} + c_3 e^{4x})$$

If we are given the **initial conditions** that $f_1(0) = 3$, $f_2(0) = 0$, and $f_3(0) = 6$, then

$$f_1(0) = \tfrac{1}{3}(-2c_1 + c_2 + 2c_3) = 3$$
$$f_2(0) = \tfrac{1}{3}(-c_1 + 2c_2 - 2c_3) = 0$$
$$f_3(0) = \tfrac{1}{3}(2c_1 + 2c_2 + c_3) = 6$$

This linear system has coefficient matrix P, and thus the unique solution is

$$P^{-1}\begin{bmatrix} 3 \\ 0 \\ 6 \end{bmatrix} = \tfrac{1}{3}\begin{bmatrix} -2 & -1 & 2 \\ 1 & 2 & 2 \\ 2 & -2 & 1 \end{bmatrix}\begin{bmatrix} 3 \\ 0 \\ 6 \end{bmatrix} = \tfrac{1}{3}\begin{bmatrix} 6 \\ 15 \\ 12 \end{bmatrix} = \begin{bmatrix} 2 \\ 5 \\ 4 \end{bmatrix} = \begin{bmatrix} c_1 \\ c_2 \\ c_3 \end{bmatrix}$$

We see that the initial conditions determine the unique solution (f_1, f_2, f_3), where

$$f_1 : \tfrac{1}{3}(-4e^x + 5e^{-2x} + 8e^{4x})$$
$$f_2 : \tfrac{1}{3}(-2e^x + 10e^{-2x} - 8e^{4x})$$
$$f_3 : \tfrac{1}{3}(4e^x + 10e^{-2x} + 4e^{4x})$$

Our final example indicates how an nth-order ($n = 3$ in our example) linear differential equation can be converted to an equivalent system of n first-order linear differential equations, and thus the above method can be applied to obtain the solutions to the nth-order linear differential equation.

Consider the third-order linear differential equation

$$f''' - 7f'' + 14f' - 8f = 0$$

Let $f_1 = f$, $f_2 = f'$, and $f_3 = f''$. Then

$$f_1' = \frac{df_1}{dx} = f_2$$

$$f_2' = \frac{df_2}{dx} = f_3$$

$$f_3' = \frac{df_3}{dx} = 8f_1 - 14f_2 + 7f_3$$

is a system of linear differential equations that is equivalent to the given third-order differential equation. The equivalent matrix form of this system is

$$F' = \begin{bmatrix} f_1' \\ f_2' \\ f_3' \end{bmatrix} = \begin{bmatrix} 0 & 1 & 0 \\ 0 & 0 & 1 \\ 8 & -14 & 7 \end{bmatrix} \begin{bmatrix} f_1 \\ f_2 \\ f_3 \end{bmatrix} = AF$$

The reader is asked to solve this system and thus the equivalent third-order differential equation in Problem 19, Exercise 7.5.

We make one final remark concerning the eigenvalues of a matrix. Although the eigenvalues are the real roots of the characteristic equation of the matrix, it is often difficult or inefficient to find them by solving this equation. In physical problems, other means are often used to find the eigenvalues and corresponding eigenvectors. The interested reader may consult Ben Noble, *Applied Linear Algebra* (Englewood Cliffs, N.J.: Prentice-Hall, Inc., 1969), pp. 299–301, to see how this is done.

Exercise 7.5

1. If $A = \begin{bmatrix} 3 & -1 \\ -1 & 3 \end{bmatrix}$,

    (a) Show that $f(A) = 0$ if $f(x) = x^3 - 8x^2 + 20x - 16$.
    (b) Find $C(x) = \det(A - xI)$.
    (c) Find the minimum polynomial $m(x)$ of A.
    (d) Show that $m(x)$ is a divisor of $f(x)$.

2. If $A = \begin{bmatrix} 5 & 2 \\ 2 & 8 \end{bmatrix}$,

    (a) Verify by direct calculation that A satisfies its characteristic polynomial.
    (b) Find the minimum polynomial of A.

3. If the characteristic polynomial of A is $C(x) = \det(A - xI)$
    $= -(x - 2)^2(x + 3)^2(x - 1)$,

    (a) What are the possible answers for the minimum polynomial $m(x)$?
    (b) What must $m(x)$ be if A is diagonalizable?

4. If the characteristic polynomial of A is $C(x) = -(x^2 + 1)(x - 3)$,

    (a) What is the minimum polynomial $m(x)$?
    (b) Is A diagonalizable?

5. *Prove:* λ_0 is an eigenvalue of an $n \times n$ matrix A if and only if $x - \lambda_0$ is a factor of the minimum polynomial $m(x)$ of A.

6. Consider the subspace of functions $W = \langle f_1 : \sin x, f_2 : \cos x \rangle$ and the linear operator $D: W \longrightarrow W$ defined by $D(a_1 f_1 + a_2 f_2) = a_1 f_2 - a_2 f_1$ (D is the derivative operator).
   (a) Find the matrix of D relative to the ordered basis $C_1 = \{f_1, f_2\}$ and call it A.
   (b) Find both the characteristic and minimum polynomials for A. Is A diagonalizable?

7. Consider the linear operator $D: P_3 \longrightarrow P_3$ defined by $D(f : a + bx + cx^2) = f' : b + 2cx$. If $C_2 = \{g_1 : x^2, g_2 : 2x, g_3 : 2\}$,
   (a) Find A, the matrix of D with respect to C_2.
   (b) Find both the characteristic and minimum polynomials for A. Is A diagonalizable?

8. Find the minimum polynomial of $A = \begin{bmatrix} 0 & 1 \\ 2 & -1 \end{bmatrix}$. Is A diagonalizable?

9. If $A = \begin{bmatrix} 0 & -4 & -4 \\ \frac{1}{2} & 3 & 1 \\ \frac{1}{2} & 1 & 3 \end{bmatrix}$, show that the minimum polynomial of A is $(x - 2)^2$, and thus conclude that A is not diagonalizable. Verify this conclusion by finding the eigenvectors of A and determining that there is no basis of eigenvectors of A for R_3.

10. If $A = \begin{bmatrix} 3 & -2 & -2 \\ -1 & 2 & 1 \\ 3 & -3 & -2 \end{bmatrix}$, find the minimum polynomial of A and determine if A is diagonalizable. Verify your answer by determining if there is a basis of eigenvectors of A for R_3.

11. Consider the linear operator $D: P_4 \longrightarrow P_4$ defined by $D(f : a + bx + cx^2 + dx^3) = f' : b + 2cx + 3dx^2$ (the derivative of f). If

$$C_2 = \{g_1 : x^3, g_2 : 3x^2, g_3 : 6x, g_4 : 6\}$$

   (a) Find the matrix of D relative to C_2 and call it A.
   (b) Find the characteristic and minimum polynomials of A and determine if A is diagonalizable.

A **linear operator** T **is** called **nilpotent** if $T^k = 0$ for some positive integer k. The index of a nilpotent operator T is q if $T^q = 0$, but T^{q-1} is not zero. A **nilpotent matrix** is defined in a similar way.

12. Show that D in Problem 11 is nilpotent and determine the index.

13. (a) Use the fact that $R_{n \times n}$ has dimension n^2 to verify that, for any $n \times n$ matrix $A \neq 0$, the matrices $A^0 = I, A, A^2, \ldots, A^{n^2}$ are linearly dependent.
   (b) Show that there is a least positive integer k such that $A^k = a_0 I + a_1 A + \cdots + a_{k-1} A^{k-1}$ is a linear combination of the preceding matrices.
   (c) Verify that $f(x) = x^k - a_{k-1} x^{k-1} - \cdots - a_1 x - a_0$ is the minimum polynomial of A.

14. Prove that

$$P^{-1}\frac{dF}{dx} = \frac{dP^{-1}F}{dx}, \qquad \text{where } P^{-1} = [q_{ij}]_{(n,n)} \text{ and } \quad F = \begin{bmatrix} f_1 \\ f_2 \\ \cdot \\ \cdot \\ \cdot \\ f_n \end{bmatrix}$$

15. (a) Solve the system of linear differential equations

$$f_1' = 2f_1 - 3f_2 + 5f_3$$
$$f_2' = -1f_2 + 5f_3$$
$$f_3' = 4f_3$$

(b) Find the unique solution that satisfies the initial conditions $f_1(0) = 9$, $f_2(0) = 5$, and $f_3(0) = 3$.

16. (a) Solve the system of linear differential equations

$$f_1' = 5f_1 + 6f_2$$
$$f_2' = 6f_1 + 10f_2$$

(b) Find the unique solution that satisfies the initial conditions $f_1(0) = 1$ and $f_2(0) = 8$.

17. (a) Solve the system of linear differential equations

$$f_1' = 5f_1 + 2f_2$$
$$f_2' = 2f_1 + 8f_2$$

(b) Find the unique solution that satisfies the initial conditions $f_1(0) = -1$ and $f_2(0) = 3$.

18. (a) Solve the system of linear differential equations

$$f_1' = 2f_1 + 2f_2 - 2f_3$$
$$f_2' = 2f_1 + 5f_2 - 4f_3$$
$$f_3' = -2f_1 - 4f_2 + 5f_3$$

(b) Find the unique solution that satisfies the initial conditions $f_1(0) = 6$, $f_2(0) = 9$, and $f_3(0) = 3$.

19. (a) Solve the third-order differential equation $f''' - 7f'' + 14f' - 8f = 0$ by solving the equivalent system of differential equations given immediately preceding Exercise 7.5.

(b) Solve the above equation given the initial conditions $f(0) = 3$, $f'(0) = 2$, and $f''(0) = 6$.

Euclidean Spaces

8.1 INNER PRODUCTS AND EUCLIDEAN SPACE

In Chapter 1 we introduced the notion of vector as a quantity that had length and direction. We were led by this geometrical notion of vector to define coordinate vectors in 1-, 2-, and 3-space and then to define coordinate vectors in n-space. Although our definition of a vector in $\mathbf{R_n}$ was independent of the notion of length or angle, we still noted that these concepts could be defined in $\mathbf{R_n}$. From that point on, however, our study of vectors has led us to such objects as matrices, polynomial functions, and linear transformations. In studying these objects, as well as in our study of abstract vector spaces, no mention was made of either length or angle. We have now come full circle. In this chapter we shall study the notion of an inner product on an abstract vector space. We shall see that an inner product is precisely the structure that is needed to enable us to define the concepts of length, distance, and angle between two vectors. Our effort will be more than just an exercise in studying abstract mathematics, although we feel this too is important. We shall see that there are applications of an inner product that are far removed from our usual geometric notion of vector.

For convenience, we now choose to write vectors in $\mathbf{R_n}$ as row vectors rather than column vectors. If $A = (a_1, a_2, \ldots, a_n)$ and $B = (b_1, b_2, \ldots, b_n)$ are vectors in $\mathbf{R_n}$, we have already defined the dot product $A \cdot B = a_1 b_1 + a_2 b_2 + \cdots + a_n b_n$. In Section 2.2 we listed the following five properties of the dot product.

For all A, B, and C in $\mathbf{R_n}$ and for all scalars r in $\mathbf{R}$,

1I. The dot product of vectors A and B in $\mathbf{R}_n$ is a uniquely determined scalar.
2I. $A \cdot B = B \cdot A$.
3I. $rA \cdot B = A \cdot rB = r(A \cdot B)$.
4I. $(A + B) \cdot C = A \cdot C + B \cdot C$.
5I. $A \cdot A > 0$ if A is not the zero vector. $A \cdot A = 0$ if $A = \mathbf{0}$.

In the presence of 2I, properties 3I and 4I are equivalent to the single statement $(rA + sB) \cdot C = r(A \cdot C) + s(B \cdot C)$ (Exercise 8.1, Problem 11). We define an inner product to be a function that satisfies properties equivalent to 1I–5I.

Definition 8.1. Let $\mathbf{V}$ be a vector space defined over the real numbers (finite or infinite-dimensional). An **inner product** on $\mathbf{V}$ is a function that maps each ordered pair of vectors X, Y in $\mathbf{V}$ into a real number, denoted by (X, Y), such that the following properties hold.

1P. $(X, Y) = (Y, X)$.
2P. $(aX + bY, Z) = a(X, Z) + b(Y, Z)$.
3P. $(X, X) > 0$ for $X \neq \mathbf{0}$ and $(\mathbf{0}, \mathbf{0}) = 0$.

A scalar-valued function of several variables is called a *form*. (We have already encountered quadratic forms.) Property 1P states that the form defined in Definition 8.1 is **symmetric**. Because of Property 2P and Problem 19, the form is **bilinear**, and property 3P states that the form is **positive definite**. The inner product is said to be a **positive definite symmetric bilinear form**.

We now give some examples of inner products. Of course, the dot product on $\mathbf{R}_n$ is one example. Our next example requires a knowledge of elementary calculus.

Example 8.1. Show that the function defined by $(f, g) = \int_0^1 f(x)g(x) \, dx$ is an inner product on the space $\mathbf{V}$ of all continuous functions defined on the closed interval $[0, 1]$ [see Example 5.1(i)].

Solution: Let f and g be in $\mathbf{V}$. The product of two continuous functions is continuous, and a function that is continuous on a closed interval is integrable. Thus (f, g) is a (uniquely determined) real number. Since $f(x)g(x) = g(x)f(x)$, it is clear that $(f, g) = (g, f)$ and property 1P holds. Let a and b be scalars. Then

$$(af + bg, h) = \int_0^1 (af + bg)(x)h(x) \, dx = \int_0^1 [af(x)h(x) + bg(x)h(x)] \, dx$$

$$= a \int_0^1 f(x)h(x) \, dx + b \int_0^1 g(x)h(x) \, dx$$

$$= a(f, h) + b(g, h)$$

and property 2P holds. If $f = 0$, then $(f, f) = \int_0^1 0 \, dx = 0$. Now suppose that $f \neq 0$. Then $f(x_0) \neq 0$ for some x_0 in $[0, 1]$. Since f is continuous on $[0, 1]$, there must be a small closed interval $[c, d]$ containing x_0 such that $f(x) \neq 0$ for all x in $[c, d]$. Thus, for each x in $[0, 1]$, $f^2(x) \geq 0$, and for each x in $[c, d]$, $f^2(x) > 0$. The function f^2 is continuous on $[c, d]$ and therefore assumes a minimum value m; that is, $f^2(x) \geq m$ for all x in $[c, d]$, and $f(x') = m$ for some x' in $[c, d]$. So

$$\int_0^1 f^2(x) \, dx = \int_0^c f^2(x) \, dx + \int_c^d f^2(x) \, dx + \int_d^1 f^2(x) \, dx$$

$$\geq 0 + \int_c^d f^2(x) \, dx + 0 \geq \int_c^d m \, dx = m(d - c)$$

Since both m and $d - c$ are positive, we have $m(d - c) > 0$, and consequently $\int_0^1 f^2(x) \, dx > 0$. Hence property 3P also holds and (f, g) is an inner product on **V**.

Example 8.2. Let **V** be a finite-dimensional vector space. Let $C_2 = \{X_1, X_2, \ldots, X_n\}$ be any ordered basis for **V**. If X and Y are arbitrary vectors in **V**, we define

$$(X, Y) = a_1 b_1 + a_2 b_2 + \cdots + a_n b_n$$

where $a_1, a_2, \ldots, a_n$ and $b_1, b_2, \ldots, b_n$ are the coordinates of X and Y, respectively, relative to the ordered basis C_2. In symbols, $(X, Y) = [X]_{C_2} \cdot [Y]_{C_2}$.

(a) Show that (X, Y) is an inner product for $n = 3$. (We have chosen the dimension of **V** to be 3 for purposes of illustration. The reader is asked to verify the result for arbitrary n in Exercise 8.1, Problem 18.)

(b) If $\mathbf{V} = \mathbf{R}_3$ and $C_2 = \{X_1 = (1, 0, 0), X_2 = (1, 1, 0), X_3 = (1, 1, 1)\}$, find (X, Y) if $X = (3, 4, 6)$ and $Y = (2, -1, 3)$.

(c) If C_2 is the natural basis for $\mathbf{R}_3$ and X and Y are the vectors in part (b), find (X, Y) and compare this result with the answer in part (b).

Solution: (a) For any vectors X and Y in **V**, the coordinate vectors relative to an *ordered* basis are uniquely determined; thus (X, Y) is a uniquely determined real number. Since multiplication of real numbers is commutative it is clear that $(X, Y) = (Y, X)$. We verify property 2P. Let $X = a_1 X_1 + a_2 X_2 + a_3 X_3$, $Y = b_1 X_1 + b_2 X_2 + b_3 X_3$, and let $Z = c_1 X_1 + c_2 X_2 + c_3 X_3$. For any scalars r and s,

$$rX + sY = ra_1 X_1 + ra_2 X_2 + ra_3 X_3 + sb_1 X_1 + sb_2 X_2 + sb_3 X_3$$

$$= (ra_1 + sb_1)X_1 + (ra_2 + sb_2)X_2 + (ra_3 + sb_3)X_3$$

Thus

$$(rX + sY, Z) = (ra_1 + sb_1)c_1 + (ra_2 + sb_2)c_2 + (ra_3 + sb_3)c_3$$

$$= ra_1 c_1 + ra_2 c_2 + ra_3 c_3 + sb_1 c_1 + sb_2 c_2 + sb_3 c_3$$

$$= r(a_1 c_1 + a_2 c_2 + a_3 c_3) + s(b_1 c_1 + b_2 c_2 + b_3 c_3)$$

$$= r(X, Z) + s(Y, Z)$$

Thus property 2P holds. If $X = a_1X_1 + a_2X_2 + a_3X_3$, then $(X, X) = a_1^2 + a_2^2 + a_3^2$. We see that $(X, X) \geq 0$ and $(X, X) = 0$ if and only if $X = \mathbf{0}$. Hence property 3P holds and (X, Y) is an inner product.

(b) First we must find the coordinates of X and Y relative to C_2. We have

$$X = (3, 4, 6) = -1X_1 + -2X_2 + 6X_3$$

and

$$Y = (2, -1, 3) = 3X_1 - 4X_2 + 3X_3.$$

Therefore,

$$(X, Y) = [X]_{c_2} \cdot [Y]_{c_2} = (-1, -2, 6) \cdot (3, -4, 3) = -3 + 8 + 18 = 23$$

(c) Since C_2 is the natural basis, we have

$$(X, Y) = [X]_{c_2} \cdot [Y]_{c_2} = X \cdot Y = 6 - 4 + 18 = 20$$

We see that the answers to parts (b) and (c) are different. *Thus the inner product defined in Example 7.2 depends on the basis chosen.*

Definition 8.2. A vector space **V** together with an inner product (,) defined on **V** is called a **Euclidean space.**

We proceed to show how a notion of length and distance can be defined in a Euclidean space.

Definition 8.3. Let **V**, (,) be a Euclidean space. The **norm** (or **length**) of a vector A in **V** is the nonnegative real number $\| A \| = \sqrt{(A, A)}$.

Theorem 8.1 (Schwarz's Inequality). Let **V**, (,) be a Euclidean space. Then for all A and B in **V**, $|(A, B)| \leq \| A \| \| B \|$.

Proof: The proof is a little tricky but is fairly easy to follow. Let r be any real number, and let A and B be arbitrary vectors in **V**. Consider the inner product of the vector $A - rB$ with itself. We have

$$
\begin{aligned}
(A - rB, A - rB) &= (1A + (-r)B, A - rB) \\
&= 1(A, A - rB) + (-r)(B, A - rB) \quad \text{[Property 2P]} \\
&= (A - rB, A) - r(A - rB, B) \quad \text{[Property 1P]} \\
&= (A, A) - r(B, A) - r(A, B) + r^2(B, B) \quad \text{[Property 2P]} \\
&= (A, A) - 2r(A, B) + r^2(B, B) \quad \text{[Property 1P]}
\end{aligned}
$$

Therefore,

$$(A - rB, A - rB) = (A, A) - 2r(A, B) + r^2(B, B)$$

Using Definition 8.3 and writing this equation in terms of norms, we have

$$\| A - rB \|^2 = \| A \|^2 - 2r(A, B) + r^2 \| B \|^2$$

Therefore

$$r^2 \| B \|^2 - 2r(A, B) + \| A \|^2 \geq 0, \qquad \text{for all } r$$

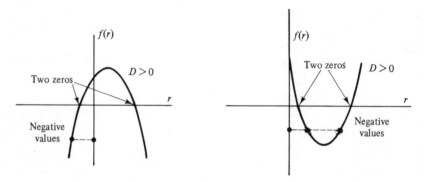

Figure 8.1

The left side of this inequality is a quadratic polynomial in r *which is nonnega-tive for all real values of* r and which has discriminant $D = 4(A, B)^2 - 4\|A\|^2\|B\|^2$ (the discriminant of $ax^2 + bx + c$ is $b^2 - 4ac$). We claim that $D \leq 0$. If the discriminant $D > 0$ (see Figure 8.1), then the quadratic polynomial function in r has two real zeros. This means that part of the graph of $f(r) = \|B\|^2 r^2 - 2(A, B)r + \|A\|^2$ is below the horizontal axis. But then $f(r) < 0$ for some r, and this contradicts the fact that $f(r) \geq 0$ for all r. Hence $D \leq 0$ and $(A, B)^2 \leq \|A\|^2\|B\|^2$. Consequently, $|(A, B)| \leq \|A\|\|B\|$.

Q.E.D.

The reader might wish to compare this "coordinate-free" proof of Schwarz's inequality with Problems 24 and 25 of Exercise 2.2. Schwarz's inequality has several important consequences. In the next theorem we derive several properties of the norm of a vector. The key result, which in words states that "the norm of a sum of two vectors is less than or equal to the sum of the norms of the vectors," depends heavily on the Schwarz inequality.

Theorem 8.2.   Let $V, (\quad, \quad)$ be a Euclidean space. The norm of a vector as defined in Definition 8.3 has the following properties:

1N. $\|A\| > 0$ if $A \neq 0$ and $\|0\| = 0$.
2N. $\|rA\| = |r|\|A\|$.
3N. $\|A + B\| \leq \|A\| + \|B\|$.
4N. $\|A\| = \|-A\|$.
5N. $|\|A\| - \|B\|| \leq \|A - B\|$.†

   Proof: 1N. This is essentially a restatement of property 3P of the inner product. If $A \neq 0$, then $(A, A) > 0$. Therefore, $\|A\| = \sqrt{(A, A)} > 0$. If $A = 0$, then $\|0\| = \sqrt{(0, 0)} = 0$.

---

†Properties 1N–3N are often taken as the defining properties of a norm.

2N.

$$\|rA\|^2 = (rA, rA) = r(A, rA) = r(rA, A) = r^2(A, A) = r^2\|A\|^2$$

Taking the nonnegative square root of each side, we get $\|rA\| = |r|\|A\|$.

3N.

$$\|A + B\|^2 = (A + B, A + B) = (A, A) + 2(A, B) + (B, B)$$
$$= \|A\|^2 + 2(A, B) + \|B\|^2 \le \|A\|^2 + 2\|A\|\|B\| + \|B\|^2$$

The last inequality follows from Schwarz's inequality. We have

$$\|A + B\|^2 \le \|A\|^2 + 2\|A\|\|B\| + \|B\|^2 = (\|A\| + \|B\|)^2$$

and thus $\|A + B\| \le \|A\| + \|B\|$.

4N.

$$\|-A\|^2 = (-A, -A) = -(A, -A) = -[-(A, A)] = (A, A) = \|A\|^2$$

Therefore, $\|-A\| = \|A\|$.

5N.

$$\|A\| = \|(A - B) + B\| \le \|A - B\| + \|B\| \quad \text{[from conclusion 3N]}$$

Therefore, $\|A\| - \|B\| \le \|A - B\|$. Also, $\|B\| = \|(B - A) + A\| \le \|B - A\| + \|A\|$. So

$$\|B\| - \|A\| \le \|B - A\| = \|A - B\| \quad \text{[from conclusion 4N]}$$

Hence $\big|\|A\| - \|B\|\big| \le \|A - B\|$. Q.E.D.

Definition 8.4. Let **V** be a vector space and let d be a function that assigns to each ordered pair of vectors in **V** a unique real number. d is a **distance function** on **V** if and only if d satisfies the following properties. For all A, B in **V**,

1D. $d(A, B) > 0$ if $A \ne B$ and $d(A, A) = 0$.
2D. $d(A, B) = d(B, A)$.
3D. $d(A, C) \le d(A, B) + d(B, C)$ (the triangle inequality).

Note: The symbol $d(A, B)$ should be read "the distance from A to B."

In a Euclidean space the inner product can be used to define a norm. The norm in turn can be used to define a distance function, as the next theorem shows.

Theorem 8.3. Let **V**, (,) be a Euclidean space. The function d, defined by $d(A, B) = \|A - B\|$ for all A and B in **V**, is a distance function.

Proof: We must verify that d satisfies properties 1D–3D of Definition 8.4. For any vectors A and B in **V**, the number $\|A - B\|$ is nonnegative, and

$\|A - B\| = 0$ if and only if $A = B$. So $d(A, B) > 0$ if $A \neq B$ and $d(A, A) = 0$. Thus property 1D holds. Also, $d(A, B) = \|A - B\| = \|B - A\| = d(B, A)$, and property 2D holds. Finally,

$$d(A, C) = \|A - C\|$$
$$= \|(A - B) + (B - C)\|$$
$$\leq \|A - B\| + \|B - C\| \quad \text{[Theorem 8.2 (3N)]}$$
$$= d(A, B) + d(B, C)$$

Q.E.D.

We close this section by giving a brief summary of what is happening in a Euclidean space and then illustrate the concepts in ordinary 1-, 2-, and 3-space. An inner product on a vector space **V** is a positive definite symmetric bilinear form denoted by $(\ ,\ )$. A vector space **V** that has an inner product defined on it is called a Euclidean space. In a Euclidean space **V**, $(\ ,\ )$, the inner product is used to define the norm or length of a vector by $\|A\| = \sqrt{(A, A)}$. The norm satisfies properties 1N–3N of Theorem 8.2; because of this, it can be used to define a distance function d, where $d(A, B) = \|A - B\|$ for all A and B in **V**.

If **V** is $\mathbf{R}_1$, $\mathbf{R}_2$, or $\mathbf{R}_3$, and if the inner product is the dot product, then the norm of A is the usual distance from the point A to the origin or the length of the vector A.

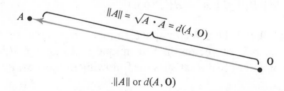

$$\|A\| = \sqrt{A \cdot A} = d(A, O)$$

$\|A\|$ or $d(A, O)$

If we view the vectors A and B as points, then $d(A, B)$ is the usual distance between the points A and B. Also, it is the length of the vector $A - B$ or $B - A$.

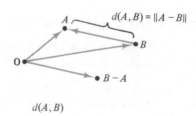

$$d(A, B) = \|A - B\|$$

$d(A, B)$

In one-, two-, or three-coordinate space the triangle inequality can be interpreted as: the sum of the lengths of two sides of a triangle is greater than or equal to the length of the third side. Of course, this is the reason for the name "triangle inequality."

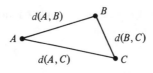

$$d(A, C) \leqslant d(A, B) + d(B, C)$$

Property 5N of Theorem 8.2 can be interpreted geometrically as: the difference between the lengths of two sides of a triangle does not exceed the length of the third side.

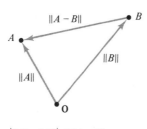

$$\left| \|A\| - \|B\| \right| \leqslant \|A - B\|$$

Exercise 8.1

In Problems 1–6, for the given vector space V and for the given ordered basis C_2, define an inner product by $(X, Y) = [X]_{C_2} \cdot [Y]_{C_2}$ as in Example 8.2. Find the required norms and distances relative to this inner product.

1. $V = R_2$, $C_2 = \{X_1 = (1, 0), X_2 = (1, 1)\}$. If $A = (a, b)$, $B = (c, d)$, $C = (3, 1)$, and $D = (2, 0)$,
   (a) Find $\|A\|$, $\|C\|$, $\|X_1\|$, and $\|X_2\|$.
   (b) Find $d(A, B)$, $d(C, D)$, and $d(X_1, X_2)$.
2. Repeat Problem 1 but let C_2 be the natural basis.
3. $V = R_3$, $C_2 = \{X_1 = (1, 1, 0), X_2 = (\frac{1}{2}, 0, 0), X_3 = (0, 0, 1)\}$. If $A = (a, b, c)$, $B = (e, f, g)$, $C = (1, 1, 3)$, and $D = (1, 1, 4)$
   (a) Find $\|A\|$, $\|C\|$, $\|X_1\|$, $\|X_2\|$, and $\|X_3\|$.
   (b) Find $d(A, B)$, $d(C, D)$, and $d(X_1, X_2)$.
4. Repeat Problem 3 but let C_2 be the natural basis for R_3.
5. $V = P_2$, $C_2 = \{f_1: 2x, f_2: x + 1\}$. If $f: a + bx$, $g: c + dx$, and $h: 2x + 3$,
   (a) Find $\|f\|$, $\|h\|$, $\|f_1\|$, and $\|f_2\|$.
   (b) Find $d(f, g)$, $d(h, f_1)$, and $d(f_1, f_2)$.

6. $V = R_4$, $C_2 = \{X_1 = (1, 0, 0, 0),\ X_2 = (1, 1, 0, 0),\ X_3 = (1, 1, 1, 0),\ X_4 = (1, 1, 1, 1)\}$. If $A = (1, 2, 0, 1)$ and $B = (-2, 0, 1, 2)$, find (a) (A, B); (b) $\|A\|$; (c) $d(A, B)$.

7. With the inner product defined as in Example 8.1 and V the vector space of all continuous functions defined on $[0, 1]$, find
   (a) $(f: x^2 + 1, g: 2x)$.
   (b) $\|h: x^2 - 1\|$.
   (c) $d(f_1: x + 1, f_2: 3x - 1)$.

8. On R_2 define $(A, B) = a_1 b_1 - a_2 b_1 - a_1 b_2 + 2a_2 b_2$, where $A = (a_1, a_2)$ and $B = (b_1, b_2)$.
   (a) Show that (A, B) is an inner product.
   (b) What is $\|A\|$ using this inner product?
   (c) Using the above inner product, verify that Schwarz's inequality holds for the vectors $C = (2, 3)$ and $D = (1, -2)$.

9. If $V = R_1$, identify the ordered 1-tuple $A = (a)$ with the real number a. The dot product $(a_1) \cdot (a_2) = a_1 a_2$ is real number multiplication.
   (a) Find $\|a_1\|$.
   (b) Find $d(a_1, a_2)$. Interpret properties 3N and 5N of Theorem 8.2 in terms of absolute value.

10. Let V be an n-dimensional vector space, and let $C_2 = \{X_1, \ldots, X_n\}$ be an ordered basis for V. If $(X, Y) = [X]_{C_2} \cdot [Y]_{C_2}$ is the inner product of Example 8.2, show that the norm of each X_i is 1.

11. Let V be a vector space, and let (X, Y) denote a unique real number for each X and Y in V. Prove that properties 1P and 2P of Definition 8.1 are equivalent to the properties (i) $(X, Y) = (Y, X)$, (ii) $(rX, Y) = (X, rY) = r(X, Y)$, and (iii) $(X + Y, Z) = (X, Z) + (Y, Z)$.

12. *Prove:* (The parallelogram theorem) If $V, (\ ,\ )$ is a Euclidean space, then $\|A + B\|^2 + \|A - B\|^2 = 2\|A\|^2 + 2\|B\|^2$. Draw a parallelogram with vertices 0, A, B, and $A + B$ and interpret the result geometrically.

13. Determine if $(f, g) = \int_0^1 x f(x) g(x)\, dx$ defines an inner product on the space of all polynomial functions defined on $[0, 1]$.

14. Which of the functions below defines an inner product on R_3? If $A = (a_1, a_2, a_3)$ and $B = (b_1, b_2, b_3)$,
   (a) $(A, B) = a_1^2 b_1^2 + a_2^2 b_2^2 + a_3^2 b_3^2$.
   (b) $(A, B) = a_1 b_1 + a_2 b_2 + a_3 b_3 + 2a_2 b_3 - 2a_3 b_2$.

15. *Prove:* For any inner product, $(A, 0) = (0, A) = 0$.

16. If $V = R_{3 \times 2}$, prove that $(A, B) = \mathrm{tr}\,(B^T A)$ is an inner product. (The trace of a square matrix is the sum of the diagonal elements.)

17. Let $V, (\ ,\ )$ be a Euclidean space. Use mathematical induction to prove that for all positive integers n, if $Y, X_1, X_2, \ldots, X_n$ are vectors in V and $a_1, a_2, \ldots, a_n$ are scalars, then

$$(a_1 X_1 + a_2 X_2 + \cdots + a_n X_n, Y) = a_1(X_1, Y) + a_2(X_2, Y) + \cdots + a_n(X_n, Y)$$

18. *Prove:* If **V** is an n-dimensional vector space, then the function (X, Y) defined in Example 8.2 is an inner product.

19. *Prove:* In any Euclidean space **V**, (,), $(X, bY + cZ) = b(X, Y) + c(X, Z)$ for all X, Y, and Z in **V** and for all scalars b and c.

20. Extend the result in Problem 19 to n vectors. Prove by mathematical induction that the extension is true (see Problem 17).

8.2 THE GRAM–SCHMIDT PROCESS

We have seen that in a Euclidean space we can define both the length of a vector and the distance between two vectors. In this section we show that we can also define the angle between two vectors. We focus our attention on vectors that are orthogonal or perpendicular, that is, vectors for which the angle between them is 90°. The major topic of this section is the Gram–Schmidt process, which provides a method for taking an arbitrary basis for a finite-dimensional Euclidean vector space and converting it into an orthonormal basis.

The reader may recall that in Section 2.2 we defined the angle θ between two nonzero *coordinate vectors* A and B by the formula

$$\cos \theta = \frac{A \cdot B}{\|A\| \|B\|}$$

This made sense because we were able to prove that

$$-1 \le \frac{A \cdot B}{\|A\| \|B\|} \le 1$$

From the Schwarz inequality, if **V**, (,) is any Euclidean space, we have $|(A, B)| \le \|A\| \|B\|$, and for nonzero vectors A and B this is equivalent to

$$-1 \le \frac{(A, B)}{\|A\| \|B\|} \le 1$$

With this information before us we are ready to define the angle between nonzero vectors in a Euclidean space.

Definition 8.5. Let **V**, (,) be a Euclidean vector space. Let A and B be any two nonzero vectors in **V**. The **measure of the angle between the vectors** A and B is θ, where $0 \le \theta \le \pi$ and $\cos \theta = (A, B)/\|A\| \|B\|$.

Since the range of the cosine function is the closed interval $[-1, 1]$ we see that $\cos \theta$ is always defined; furthermore θ is uniquely determined, because we have insisted that θ be in the interval $[0, \pi]$. Note that the angle between nonzero vectors A and B has measure $\pi/2$ (90°) if and only if $(A, B) = 0$.

Definition 8.6. Let **V**, (,) be a Euclidean space. Vectors A and B in **V** are **orthogonal** if and only if $(A, B) = 0$. An **orthogonal set of vectors** is a collection of vectors such that any two distinct vectors in the set are orthogonal. We say that the vectors in the set are **mutually orthogonal**. An **orthonormal set of vectors** is an orthogonal set of vectors such that each vector in the set has norm 1.

Note: Definition 8.6 allows for the possibility of an infinite set of orthogonal vectors. Also, any nonzero vector A in a Euclidean space can be **normalized** in the sense that $(1/\|A\|)A$ is a vector with norm 1.

In Section 2.3, for vectors in $\mathbf{R}_n$ we defined the vector projection of A along B ($B \neq 0$) to be the vector $\dfrac{A \cdot B}{B \cdot B} B$. We wish to use this terminology in a Euclidean space, so we make the following definition.

Definition 8.7. In a Euclidean space $\mathbf{V}$, $(\ ,\ )$, if A and B are vectors and $B \neq 0$, the **projection of A along B** is the vector $\dfrac{(A, B)}{(B, B)} B$.

We are now ready to consider the problem of converting a basis for a finite-dimensional Euclidean space into an orthonormal basis, that is, a basis of mutually orthogonal vectors each of which is a unit vector. It is important to note that the space $\mathbf{W}$ for which a basis is known may be a subspace of some larger space $\mathbf{V}$. What we want to accomplish is to come up with an orthonormal set of vectors that generates the *same space* $\mathbf{W}$. For example, in Section 7.3 we wanted to convert a basis for an eigenspace into an orthonormal basis for that eigenspace.

Example 8.3. Given a basis $\{X_1, X_2\}$ for a two-dimensional subspace $\mathbf{W}$ of a Euclidean space $\mathbf{V}$, $(\ ,\ )$, find an orthonormal basis for $\mathbf{W}$.

Solution: The solution given is a special case of the Gram–Schmidt process. To motivate the procedure, the reader is encouraged to think geometrically although the solution is algebraic. The situation is depicted in Figure 8.2. The objective is to trade the vectors X_1 and X_2 for the orthogonal vectors Y_1 and Y_2 and then normalize them.

Step 1: Let $Y_1 = X_1$. Y_1 is not zero because it is a member of a basis.

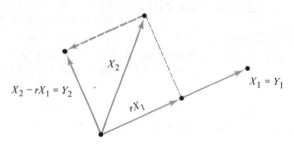

$$X_2 - rX_1 = Y_2$$

$$X_2$$

$$X_1 = Y_1$$

$$rX_1$$

Figure 8.2

Step 2: The vector projection of X_2 onto $Y_1 \neq 0$ is

$$rY_1 = \frac{(X_2, Y_1)}{(Y_1, Y_1)} Y_1$$

We set

$$Y_2 = X_2 - \frac{(X_2, Y_1)}{(Y_1, Y_1)} Y_1$$

which is X_2 minus the projection of X_2 along X_1. $Y_2 \neq 0$ because $Y_2 = 1X_2 - rX_1$ and X_1 and X_2 are linearly independent. We claim that $\langle X_1, X_2 \rangle = \langle Y_1, Y_2 \rangle$ and Y_1 and Y_2 are orthogonal. $Y_1 = X_1$ and $Y_2 \in \langle X_1, X_2 \rangle$. Therefore, all linear combinations of Y_1 and Y_2 are in $\langle X_1, X_2 \rangle$ and $\langle Y_1, Y_2 \rangle \subseteq \langle X_1, X_2 \rangle$. Also, $X_2 = Y_2 + rX_1 = Y_2 + rY_1$. So X_1 and X_2 are both in $\langle Y_1, Y_2 \rangle$, and consequently we have $\langle X_1, X_2 \rangle = \langle Y_1, Y_2 \rangle$. We show (without appealing to Figure 8.2) that Y_1 and Y_2 are orthogonal.

$$(Y_1, Y_2) = (X_1, X_2 - rX_1) = (X_1, X_2) - r(X_1, X_1)$$
$$= (X_1, X_2) - \frac{(X_2, X_1)}{(X_1, X_1)}(X_1, X_1) = 0$$

Therefore, Y_1 and Y_2 are orthogonal.

Step 3: Normalizing these vectors, we get an orthonormal basis $\{(1/\|Y_1\|)Y_1, (1/\|Y_2\|)Y_2\}$ for **W**.

Example 8.4. Given a basis $\{X_1, X_2, X_3\}$ for a subspace **W** of a Euclidean space **V**, (,), find an orthonormal basis for **W**.

Solution: As in Example 8.3, we encourage the reader to think geometrically. Keep in mind, however, that the solution is for an arbitrary Euclidean space and thus must be independent of the geometrical interpretation. Figure 8.3 is provided as a guide, and the reader should refer to it as he studies the solution.

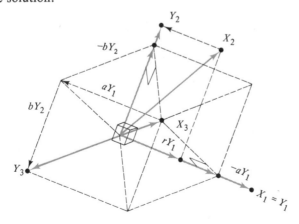

Figure 8.3

Step 1: Let $Y_1 = X_1$. Y_1 is not zero because it is a member of a basis.

Step 2: The vector projection of X_2 onto the nonzero vector Y_1 is

$$rY_1 = \frac{(X_2, Y_1)}{(Y_1, Y_1)}Y_1$$

We set

$$Y_2 = X_2 - \frac{(X_2, Y_1)}{(Y_1, Y_1)}Y_1$$

which is X_2 minus the projection of X_2 along $Y_1 = X_1$. The vector Y_2 is not zero since $Y_2 = 1X_2 - rX_1$ and the X's are linearly independent. As in Example 8.3, $\langle Y_1, Y_2 \rangle = \langle X_1, X_2 \rangle$ and Y_1 and Y_2 are orthogonal.

Step 3: We wish to find a vector Y_3 that is orthogonal to each of the vectors Y_1 and Y_2 and such that $\langle Y_1, Y_2, Y_3 \rangle = \langle X_1, X_2, X_3 \rangle$. Let $Y_3 = aY_1 + bY_2 + cX_3$. Our task is to determine a, b, and c so that the stated conditions on Y_3 hold. Thus we solve the equations $(Y_3, Y_1) = 0$ and $(Y_3, Y_2) = 0$.

$$\begin{aligned}
0 = (Y_3, Y_1) &= (aY_1 + bY_2 + cX_3, Y_1) \\
&= a(Y_1, Y_1) + b(Y_2, Y_1) + c(X_3, Y_1) \\
&= a(Y_1, Y_1) + 0 + c(X_3, Y_1)
\end{aligned}$$

Setting $c = 1$ and solving for a, we obtain

$$a = -\frac{(X_3, Y_1)}{(Y_1, Y_1)}$$

Next,

$$\begin{aligned}
0 = (Y_3, Y_2) &= (aY_1 + bY_2 + X_3, Y_2) \\
&= a(Y_1, Y_2) + b(Y_2, Y_2) + (X_3, Y_2) \\
&= 0 + b(Y_2, Y_2) + (X_3, Y_2)
\end{aligned}$$

Solving for b, we obtain

$$b = -\frac{(X_3, Y_2)}{(Y_2, Y_2)}.$$

Thus the vector

$$Y_3 = X_3 - \frac{(X_3, Y_1)}{(Y_1, Y_1)}Y_1 - \frac{(X_3, Y_2)}{(Y_2, Y_2)}Y_2$$

(see Figure 8.3) which is X_3 minus the projections of X_3 along Y_1 and Y_2 respectively, is a vector orthogonal to each of the vectors Y_1 and Y_2. Since $Y_3 \in \langle Y_1, Y_2, X_3 \rangle$ and $\langle Y_1, Y_2 \rangle = \langle X_1, X_2 \rangle$, it is clear that $Y_3 \in \langle X_1, X_2, X_3 \rangle$. Also, from the above equation for Y_3 we see that $X_3 \in \langle Y_1, Y_2, Y_3 \rangle$. We have determined that if

$$Y_1 = X_1, \quad Y_2 = X_2 - \frac{(X_2, Y_1)}{(Y_1, Y_1)}Y_1, \quad Y_3 = X_3 - \frac{(X_3, Y_1)}{(Y_1, Y_1)}Y_1 - \frac{(X_3, Y_2)}{(Y_2, Y_2)}Y_2$$

then $\mathbf{W} = \langle X_1, X_2, X_3 \rangle = \langle Y_1, Y_2, Y_3 \rangle$, and the vectors Y_1, Y_2, and Y_3 are mutually orthogonal.

Step 4: Normalizing these vectors, we get an orthonormal basis $\{(1/\| Y_1 \|)Y_1, (1/\| Y_2 \|)Y_2, (1/\| Y_3 \|)Y_3\}$ for $\mathbf{W}$.

The process described in Examples 8.3 and 8.4 can be generalized and is known as the *Gram–Schmidt orthogonalization process*.

Theorem 8.4. If $\mathbf{W} \neq \langle \mathbf{0} \rangle$ is a subspace of a finite-dimensional Euclidean space $\mathbf{V}$, $(\ ,\ )$, and if $C_1 = \{X_1, X_2, \ldots, X_n\}$ is a basis for $\mathbf{W}$, then the set of vectors $C_2 = \{Y_1, Y_2, \ldots, Y_n\}$ is an orthogonal basis for $\mathbf{W}$ if

$$Y_1 = X_1$$

$$Y_2 = X_2 - \frac{(X_2, Y_1)}{(Y_1, Y_1)} Y_1$$

$$Y_3 = X_3 - \frac{(X_3, Y_1)}{(Y_1, Y_1)} Y_1 - \frac{(X_3, Y_2)}{(Y_2, Y_2)} Y_2$$

$$\vdots$$

$$Y_n = X_n - \frac{(X_n, Y_1)}{(Y_1, Y_1)} Y_1 - \frac{(X_n, Y_2)}{(Y_2, Y_2)} Y_2 - \cdots - \frac{(X_n, Y_{n-1})}{(Y_{n-1}, Y_{n-1})} Y_{n-1}$$

A formal proof of this theorem can be given by using mathematical induction (Exercise 8.2, Problem 15). The case when $n = 1$ is a degenerate case, and the theorem is verified for $n = 2$ and $n = 3$ in Examples 8.3 and 8.4. Note that the vector Y_j for $j \geq 2$ is the vector X_j minus the projections of X_j along the vectors $Y_1, Y_2, \ldots$, and Y_{j-1}, respectively. Normalizing each of the Y's, we obtain an orthonormal basis. We state this result as a corollary.

Corollary 8.1. Every finite-dimensional Euclidean vector space $(\neq \langle \mathbf{0} \rangle)$ has an orthonormal basis.

If the vector space under discussion is, a subspace of $\mathbf{R}_n$, we shall assume unless explicitly stated otherwise that the inner product used is the dot product.

Example 8.5. Use the Gram–Schmidt orthogonalization process to convert $C_1 = \{X_1 = (1, 0, 0, 1), X_2 = (0, 0, 1, 1), X_3 = (0, 1, 1, 0), X_4 = (1, 0, 1, 0)\}$ into an orthonormal basis.

Solution: Let $Y_1 = X_1$. Then

$$Y_2 = X_2 - \frac{(X_2, Y_1)}{(Y_1, Y_1)} Y_1 = (0, 0, 1, 1) - \frac{1}{2}(1, 0, 0, 1)$$

$$= \frac{1}{2}(-1, 0, 2, 1)$$

$$Y_3 = X_3 - \frac{(X_3, Y_1)}{(Y_1, Y_1)}Y_1 - \frac{(X_3, Y_2)}{(Y_2, Y_2)}Y_2$$

$$= (0, 1, 1, 0) - 0(1, 0, 0, 1) - \frac{2}{3}\left(-\frac{1}{2}, 0, 1, \frac{1}{2}\right)$$

$$= \frac{1}{3}(1, 3, 1, -1)$$

$$Y_4 = X_4 - \frac{(X_4, Y_1)}{(Y_1, Y_1)}Y_1 - \frac{(X_4, Y_2)}{(Y_2, Y_2)}Y_2 - \frac{(X_4, Y_3)}{(Y_3, Y_3)}Y_3$$

$$= (1, 0, 1, 0) - \frac{1}{2}(1, 0, 0, 1) - \frac{1}{3}\left(-\frac{1}{2}, 0, 1, \frac{1}{2}\right) - \frac{1}{2}\left(\frac{1}{3}, 1, \frac{1}{3}, -\frac{1}{3}\right)$$

$$= \frac{1}{2}(1, -1, 1, -1)$$

The basis $\{Y_1, Y_2, Y_3, Y_4\}$ is orthogonal. Normalizing, we obtain an orthonormal basis

$$C_2 = \left\{\frac{1}{\sqrt{2}}(1, 0, 0, 1), \frac{1}{\sqrt{6}}(-1, 0, 2, 1), \frac{1}{\sqrt{12}}(1, 3, 1, -1),\right.$$

$$\left.\frac{1}{2}(1, -1, 1, -1)\right\}$$

Note: Nonzero vectors X and Y are orthogonal if and only if X and rY are orthogonal and $r \neq 0$. Also $\langle X, Y \rangle = \langle X, rY \rangle$. For these reasons we could have replaced Y_2 with the vector $2Y_2 = (-1, 0, 2, 1)$, which has integer coordinates. Proceeding in this manner, we could have found a set of mutually orthogonal vectors having integer coordinates. The final step would be to normalize these vectors.

The Gram–Schmidt process shows us how to take any basis for a finite-dimensional Euclidean vector space and convert it into an orthonormal basis. We have already seen that orthonormal bases play a key role in the diagonalization of quadratic forms. Another reason for working with an orthonormal basis is suggested by the corollary to the next theorem. As we shall see from the corollary, it is a rather simple matter to express the coordinates of a vector in terms of an orthonormal basis.

Theorem 8.5. If $C = \{Y_1, Y_2, \ldots, Y_n\}$ is an orthogonal basis for a Euclidean vector space **V**, (,), and if X is any vector in **V**, then

$$X = \frac{(X, Y_1)}{(Y_1, Y_1)}Y_1 + \frac{(X, Y_2)}{(Y_2, Y_2)}Y_2 + \cdots + \frac{(X, Y_n)}{(Y_n, Y_n)}Y_n$$

Proof: Let X be any vector in **V**. The fact that C is a basis for **V** tells us that $X = a_1Y_1 + a_2Y_2 + \cdots + a_nY_n$ for certain scalars $a_1, a_2, \ldots, a_n$. We will show that $a_i = (X, Y_i)/(Y_i, Y_i)$ for $i = 1, 2, \ldots, n$.

$$(X, Y_i) = (a_1Y_1 + a_2Y_2 + \cdots + a_nY_n, Y_i)$$

and so, from Problem 17, Exercise 8.1, we have

$$(X, Y_i) = a_1(Y_1, Y_i) + a_2(Y_2, Y_i) + \cdots + a_n(Y_n, Y_i)$$

Since C is an orthogonal basis, $(Y_j, Y_i) = 0$ if $j \neq i$. Thus we have (X, Y_i) $= a_i(Y_i, Y_i)$. The scalar $(Y_i, Y_i) \neq 0$ because $Y_i \neq \mathbf{0}$. Therefore

$$a_i = \frac{(X, Y_i)}{(Y_i, Y_i)}$$

Q.E.D.

If the basis C in Theorem 8.5 is an orthonormal basis, then $(Y_i, Y_i) = 1$ and we have the following corollary.

Corollary 8.2. If $C = \{Y_1, Y_2, \dots, Y_n\}$ is an orthonormal basis for a Euclidean vector space $\mathbf{V}, (\quad, \quad)$, then for any vector X in $\mathbf{V}$,

$$X = (X, Y_1)Y_1 + (X, Y_2)Y_2 + \cdots + (X, Y_n)Y_n.$$

Example 8.6. Express the vector $(5, 20, 10)$ as a linear combination of the vectors $Y_1 = (\frac{3}{5}, \frac{4}{5}, 0)$, $Y_2 = (\frac{4}{5}, -\frac{3}{5}, 0)$, and $Y_3 = (0, 0, 1)$.

Solution: It is easy to see that the vectors Y_1, Y_2, and Y_3 are orthogonal and each has length 1. The reader can show that any set of n orthogonal vectors is linearly independent (Exercise 8.2, Problem 14). Therefore, $C = \{Y_1, Y_2, Y_3\}$ is a basis for $\mathbf{R}_3$. If we let $X = (5, 20, 10)$, then (X, Y_1) $= 19, (X, Y_2) = -8,$ and $(X, Y_3) = 10$. According to Corollary 8.2, we have

$$X = (5, 20, 10) = 19(\tfrac{3}{5}, \tfrac{4}{5}, 0) - 8(\tfrac{4}{5}, -\tfrac{3}{5}, 0) + 10(0, 0, 1)$$

The reader may have noted that each of the terms $[(X, Y_i)/(Y_i, Y_i)]Y_i$ in Theorem 8.5 is the vector projection of X along Y_i. Thus the conclusion of this theorem may be stated as: X is the sum of its projections along the vectors $Y_1, Y_2, \dots, Y_n$.

Exercise 8.2

In Problems 1–4, use the Gram–Schmidt orthogonalization process to convert the given set of linearly independent vectors into an orthonormal set of vectors that generates the same space.

1. $C_1 = \{X_1 = (1, 0, 2), X_2 = (1, 0, 1)\}$.

2. $C_1 = \{X_1 = (1, 0, 0, 1), X_2 = (1, 1, 0, 0), X_3 = (0, 1, 0, 1)\}$.

3. $C_1 = \{X_1 = (1, 1, 1, 0), X_2 = (1, 1, 0, 1), X_3 = (1, 0, 1, 1), X_4 = (0, 1, 1, 1)\}$.

4. $C_1 = \{X_1 = (1, 0, 1, 0), X_2 = (0, 1, -1, 1), X_3 = (1, 1, 2, 1)\}$.

5. Extend $\{X_1 = (1, 2, 0), X_2 = (-2, 1, 1)\}$ to an orthogonal basis for $\mathbf{R}_3$ and then find the associated orthonormal basis.

6. In Example 7.14 we found the eigenspace associated with 7 to be the space $\langle X_1 = (-1, 2, 0), X_2 = (-1, 0, 1) \rangle = \mathbf{W}$. (Actually, the space was generated

by the transpose of X_1 and of X_2.) Use the Gram–Schmidt process to convert the given basis for **W** to an orthonormal basis for **W**.

7. Prove that in any Euclidean space **V**, (,), if A and B are orthogonal vectors, then $\| A - B \|^2 = \| A \|^2 + \| B \|^2$. Sketch a figure in $\mathbf{R}_2$ and explain why this might be called the Pythagorean theorem.

8. Let $C_2 = \{X_1, X_2, \ldots, X_n\}$ be a basis for a vector space **V**. Define (,) as in Example 8.2; that is, $(X, Y) = [X]_{C_2} \cdot [Y]_{C_2}$. Show that the basis C_2 used to define the inner product is an orthonormal basis relative to this inner product.

9. The function defined by $(f, g) = \int_{-\pi}^{\pi} f(x)g(x)\, dx$ is an inner product on the vector space of all continuous functions on $[-\pi, \pi]$ (the proof is similar to that in Example 8.1).
   (a) Show that $\cos x$ and $\sin x$ are orthogonal in this Euclidean space.
   (b) If f and g are functions in the above space, what is $d(f, g)$?

10. If P_3 is the space of all polynomial functions of degree ≤ 2, define an inner product by $(f, g) = \int_{-1}^{1} f(x)g(x)\, dx$. Use the Gram–Schmidt process to convert the basis $C_1 = \{f_1 : 1, f_2 : x, f_3 : x^2\}$ into an orthogonal basis. The orthogonal functions obtained are called *Legendre polynomials*.

11. If **V**, (,) is a Euclidean space and if $\| A \| = \| B \|$, prove that $A + B$ and $A - B$ are orthogonal. Interpret the result geometrically in 2-space.

12. Use Theorem 8.5 to write $(3, 2, 1)$ as a linear combination of the vectors in the orthogonal basis $C = \{Y_1 = (1, 2, 2),\ Y_2 = (2, -2, 1),\ Y_3 = (2, 1, -2)\}$.

13. Use Theorem 8.5 to write $(6, 24, 12, 8)$ as a linear combination of the vectors in the orthogonal basis $C = \{Y_1 = (1, 0, 0, 1),\ Y_2 = (-1, 0, 2, 1),\ Y_3 = (1, 3, 1, -1),\ Y_4 = (1, -1, 1, -1)\}$.

14. Prove that in a Euclidean space **V**, (,), if $C = \{X_1, X_2, \ldots, X_n\}$ is an orthogonal set, then C is a linearly independent set. *Hint:* Assume that $a_1 X_1 + a_2 X_2 + \cdots + a_n X_n = \mathbf{0}$ and expand $(\mathbf{0}, X_k) = (a_1 X_1 + a_2 X_2 + \cdots + a_n X_n, X_k)$ for $k = 1, 2, \ldots, n$.

15. Use mathematical induction to prove Theorem 8.4. *Hint:* For $n = 1$, choose any nonzero vector and normalize it. Let **W** be an n-dimensional subspace of **V**, and suppose that any basis of $n - 1$ vectors can be converted into an orthogonal basis by the formulas of Theorem 8.4. Let $\{X_1, X_2, \ldots, X_n\}$ be a basis for **W**, and define $Y_1, Y_2, \ldots, Y_n$ by the formulas of Theorem 8.4. By the induction hypothesis, $\langle X_1, X_2, \ldots, X_{n-1}\rangle = \langle Y_1, Y_2, \ldots, Y_{n-1}\rangle$, and the Y's are mutually orthogonal. As in Example 8.4, show that $Y_n \in \langle Y_1, Y_2, \ldots, Y_{n-1}, X_n\rangle = \langle X_1, X_2, \ldots, X_{n-1}, X_n\rangle$ and $X_n \in \langle Y_1, Y_2, \ldots, Y_n\rangle$, and also show that $(Y_n, Y_k) = 0$ for $k = 1, 2, \ldots, n - 1$.

†8.3 APPLICATIONS OF INNER PRODUCTS

At this point the reader may have asked the question, Why in Definition 8.1 did we state a general definition of inner product when primarily what we had in mind was the dot product of vectors in $\mathbf{R_n}$? One answer is that generaliza-

†This section is optional.

tion and abstraction are important parts of mathematics. The results that we obtain may or may not be useful in the real world. In the next two examples, however, we hope to convince the reader that there are useful inner products other than the dot product on $\mathbf{R}_n$, and also that there are useful inner products on vector spaces other than $\mathbf{R}_n$.

Before we look at the specific examples, we return to the definition of inner product (Definition 8.1) and how it led to the definition of norm and distance. A key result, the Schwarz inequality, led to the triangle inequality. In examining the proof of the Schwarz inequality, we note that it depends only on properties 1P and 2P of the inner product. The "near" inner products that we use in our examples will satisfy properties 1P and 2P of Definition 8.1, but not property 3P.

Example 8.7 Einstein's Theory of Special Relativity: A Prediction. One of the major differences between Newtonian mechanics and Einstein's special theory of relativity concerns the way in which time is viewed. We illustrate this difference with the following prediction from Einstein's theory of special relativity. Edward and Edwin are identical twins. On their twentieth birthdays astronaut Edward takes off in a very fast rocket ship (at nearly the speed of light) for the nearest star Alpha Centauri (about 4 light years away). Upon reaching Alpha Centauri, the rocket ship immediately begins its return trip to earth.

Meanwhile, followers of Newton (Newtonian mechanics) and followers of Einstein (special theory of relativity) have been debating the issue of the age of the twins upon Edward's return to earth. Newton's followers claim that time is absolute and that both twins will be age 30. Einstein's followers claim that time is relative, and their prediction is that Edwin, the earth twin, will be 30 years of age and Edward, the rocket twin, will be 26 years old!

We shall verify the prediction of the followers of Einstein by using the following "near" norm. On $\mathbf{R}_4$, *we define the following form*: for any $X = (x_1, x_2, x_3, x_4)$ and $Y = (y_1, y_2, y_3, y_4)$ in $\mathbf{R}_4$, $(X, Y) = -x_1y_1 - x_2y_2 - x_3y_3 + x_4y_4$. The reader is asked in Exercise 8.3, Problem 1 to verify that, for all X, Y, and $Z = (z_1, z_2, z_3, z_4)$ in $\mathbf{R}_4$ and for all a and b in $\mathbf{R}$,

1P. $(X, Y) = (Y, X)$.
2P. $(aX + bY, Z) = a(X, Z) + b(Y, Z)$.
3'P. $|(X, X)| \geq 0$ and $(0, 0) = 0$.

Property 3P of Definition 8.1 does not hold. First, it is clear that (X, X) may be negative, but in addition we may have $(X, X) = 0$ and $X \neq 0$. For example, if $X = (3, 4, 0, 5)$, then $(X, X) = -9 - 16 + 25 = 0$. The "near" norm resulting from this bilinear form is defined by $\| X \| = \sqrt{|(X, X)|}$. Following the proof of Theorem 8.2, we see that

1'N. $\|X\| \geq 0$ and $\|\mathbf{0}\| = 0$.
2N. $\|rX\| = |r| \|X\|$. However,
3N. $\|X + Y\| \leq \|X\| + \|Y\|$ *does not hold* as we shall see in our example.

Note that $X = (3, 4, 0, 5) \neq (0, 0, 0, 0)$, but $\|X\| = 0$. So $X \neq \mathbf{0}$ does not imply that $\|X\| > 0$. For this reason, $\|X\|$ is called a *pseudonorm*. Defining a function d by the formula $d(X, Y) = \|X - Y\|$ for all X and Y in $\mathbf{R}_4$, we find, as in the proof of Theorem 8.3, that

1'D. $d(X, Y) \geq 0$ if $X \neq Y$ and $d(X, X) = 0$.
2D. $d(X, Y) = d(Y, X)$. However, because we do not have 3N,
3D. $d(X, Z) \leq d(X, Y) + d(Y, Z)$ *does not hold*.

A function d satisfying these three properties is called a *pseudodistance function* because it allows for the possibility for X and Y to be different and yet have $d(X, Y) = 0$. For the function just defined, we have, for example, $X = (3, 4, 0, 5)$, $Y = (0, 0, 0, 0)$, and $d(X, Y) = 0$.

We are now ready to verify the prediction from the theory of special relativity that the twin Edward, who rocketed through space to Alpha Centauri and back, will be 4 years younger than his identical twin brother Edwin, who remained on earth. An event (something happening suddenly at a point) will be described by a vector $X = (x_1, x_2, x_3, x_4)$ in $\mathbf{R}_4$. The first three coordinates (x_1, x_2, x_3) locate the point X in space, and the fourth coordinate x_4 locates the point X in time. We shall assume that Alpha Centauri lies in the direction of the X_1-axis. As Edward rockets to Alpha Centauri, he advances toward Alpha Centauri and he also advances in time, whereas Edwin advances in time only. We can suppress the second and third coordinates since there is no motion in either the X_2 or the X_3 direction. If A and B are points in $\mathbf{R}_4$, then the physical interpretation of $\|B - A\| = d(A, B)$ is the actual time recorded by an observer in moving along the path from A to B. We assume that the earth twin Edwin is exactly 30 years old when his brother Edward returns. The origin $\mathbf{0} = (0, 0, 0, 0)$ describes the event "rocket takes off for Alpha Centauri," and the point $Q = (0, 0, 0, 10)$ describes the event "rocket returns to earth." This is so because the observer on earth advances through time only (see Figure 8.4). Since it takes exactly the same time to get to Alpha Centauri as it does to return, the event "arrives at Alpha Centauri" will have coordinates $P = (4, 0, 0, 5)$. Then

$$d(\mathbf{0}, P) = \|P\| = \sqrt{|-16 + 25|} = 3$$

is the actual time recorded by Edward as he rockets from the earth to Alpha Centauri, and

$$d(P, Q) = \|P - Q\| = \|(4, 0, 0, -5)\| = \sqrt{|-16 + 25|} = 3$$

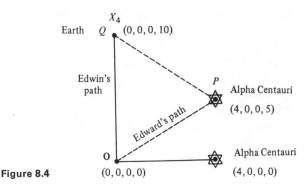

Figure 8.4

is the actual time recorded by Edward as he rockets from Alpha Centauri to the earth. Thus the total time (in years) recorded by Edward is $d(0, P) + d(P, Q) = 6$. His age upon returning to earth is therefore 26. The actual time recorded by Edwin as he remains on earth but advances in time is $d(0, Q) = 10$. Consequently, Edwin's age is 30 years when Edward steps from the rocket.

The reader might ask, Which of the two predictions, the prediction of the followers of Newton or the prediction of the followers of Einstein, is really correct? It is probably impossible to perform the experiment of actually sending a person to Alpha Centauri at the speed indicated. However, other experiments can and have been carried out to test whether time is absolute (independent of space as in Newtonian mechanics) or relative (as in the special theory of relativity). In every case, the prediction based on the theory of special relativity has proved correct. The interested reader who wishes to explore the theory of special relativity may consult Chapter 18 of John L. Synge and Byron A. Griffith, *Principles of Mechanics*, 3rd ed., (New York: McGraw-Hill Book Company, 1959). A book by Hermann Bondi, *Relativity and Common Sense, A New Approach to Einstein* (Garden City, N.Y.: Doubleday & Company, Inc., 1964) is an account of Einstein's theory of relativity intended to provide a survey for the young student or layman. An interesting account of "the clock or the twin effect" of the theory of relativity is given in Alfred Schild, "The Clock Paradox in Relativity Theory," *American Mathematical Monthly*, Volume 66, 1959, pp. 1–18.

Fourier Series

A complete understanding of Fourier series requires a background in advanced calculus, which we are *not assuming* that the reader has. We think that the ideas presented here are accessible to the reader who has had some experience with integration as taught in a beginning calculus course. Our intent is to indicate how the notion of inner product can be applied to Fourier series. By doing so we wish to give the reader additional evidence that there are inner products, other than the dot product, that are useful. We also hope to convince the reader that there are good reasons to think of a vector space in a more general sense than that encountered in coordinate spaces.

Roughly speaking, trigonometric functions are to Fourier series what polynomial functions are to power series. In studying power series, one attempts to approximate a given function by a polynomial function of degree n. If the polynomial approximation gets "better and better" as n gets large, we say that the infinite polynomial represents the function. For example, the reader may know that

$$e^x \simeq 1 + \frac{x}{1!} + \frac{x^2}{2!} + \cdots + \frac{x^n}{n!}$$

(read $\simeq$ as "is approximately"). Furthermore, the approximation gets better as n increases, and with the proper definition of infinite polynomial we have

$$e^x = 1 + \frac{x}{1!} + \frac{x^2}{2!} + \cdots + \frac{x^n}{n!} + \cdots$$

In Fourier analysis the basic building blocks are the functions with values 1, $\cos x$, $\sin x$, $\cos (2x)$, $\sin (2x)$, ..., $\cos (nx)$, $\sin (nx)$, ... We shall write **cos n** and **sin k** to mean the functions with values $\cos (nx)$ and $\sin (kx)$, respectively, in order to stress that it is the function rather than the value of the function that is under consideration. We shall show that, if f is (Riemann) integrable on $[-\pi, \pi]$, then $f \simeq a_0 \mathbf{1} + a_1 \cos + b_1 \sin + \cdots + a_n \cos \mathbf{n} + b_n \sin \mathbf{n}$ for appropriate constants $a_0, a_1, b_1, \ldots, a_n, b_n$. How close the linear combination on the right of the equation is to the function f on the left side of the equation will be measured in terms of the distance function obtained from the inner product $(f, g) = \int_{-\pi}^{\pi} f(x)g(x)\, dx$. The fact that for a large class of functions there are constant a's and b's such that $f = a_0\mathbf{1} + a_1 \cos + b_1 \sin + \cdots + a_n \cos \mathbf{n} + b_n \sin \mathbf{n} + \ldots$ is proved in advanced calculus texts. The question of whether or not a function could be written in this way arose in an attempt to solve a *wave equation* that models the motion of a vibrating string. In 1753, Daniel Bernoulli (1700–1782) gave a series solution involving sines and cosines. In studying another physical problem, the problem of the conduction of heat through metal bars, Joseph Fourier (1758–1830) in 1807 announced that an "arbitrary" function could be represented by what is now known as a Fourier series. Although Fourier claimed too much, his statement was correct for a large class of functions.

Let $\mathbf{W} = \mathbf{W}_{[a,b]}$ be the vector space of all continuous functions defined on an interval $[a, b]$, and let $\mathbf{V} = \mathbf{V}_{[a,b]}$ be the collection of all (Riemann) integrable functions defined on $[a, b]$. $\mathbf{W}$ and $\mathbf{V}$ are vector spaces [Examples 5.1(i) and (k)], and since every continuous function is integrable it follows that **W** is a subspace of **V**. A form (,) is defined on **W** as follows. For all functions f and g in **W**,

$$(f, g) = \int_a^b f(x)g(x)\, dx \tag{P}$$

We have already seen (Example 8.1) that (f, g) is an *inner product on* **W**: that is, we have properties 1P: $(f, g) = (g, f)$; 2P: $(af + bg, h) = a(f, h) + b(g, h)$, and 3P: $(f, f) > 0$ for $f \neq \mathbf{0}$ and $(\mathbf{0}, \mathbf{0}) = 0$. If the form (f, g) is defined on the superspace **V**, properties 1P and 2P hold, but property 3P does not. It is true that for all f in **V**, we have the property 3'P: $(f, f) \geq 0$ and $(\mathbf{0}, \mathbf{0}) = 0$. However, in **V** it is possible for $(f, f) = 0$ and $f \neq \mathbf{0}$. For example, if f^* is defined by $f^*(x) = 0$ if $x \neq (a + b)/2$, and $f^*((a + b)/2) = 2$, then $f^* \neq \mathbf{0}$ but

$$(f^*, f^*) = \int_a^b f^*(x)f^*(x)\, dx = 0$$

A form satisfying properties 1P, 2P, and 3'P is called a *nonnegative semidefinite bilinear* form. On **W**, the product defined by (P) determines the norm

$$\| f \| = \sqrt{(f, f)} \tag{N}$$

This norm has the properties 1N: $\| f \| > 0$ if $f \neq \mathbf{0}$ and $\| \mathbf{0} \| = 0$, 2N: $\| af \| = |a|\| f \|$, and 3N: $\| f + g \| \leq \| f \| + \| g \|$. The norm (N) determines a distance function

$$d(f, g) = \| f - g \| \tag{D}$$

which has the properties 1D: $d(f, g) > 0$ if $f \neq g$ and $d(f, f) = 0$, 2D: $d(f, g) = d(g, f)$, and 3D: $d(f, h) \leq d(f, g) + d(g, h)$. On the space **V**, we obtain the same results except that property 1N is replaced by property 1'N: $\| f \| \geq 0$ and $\| \mathbf{0} \| = 0$. Also, property 1D is replaced by property 1'D: $d(f, g) \geq 0$ and $d(\mathbf{0}, \mathbf{0}) = 0$. The reader is asked in Problem 6, Exercise 8.3, to verify that properties 1N and 1D do not hold in **V**.

Theorem 8.6. In the vector space $\mathbf{W} = \mathbf{W}_{[-\pi, \pi]}$ of all continuous functions defined on $[-\pi, \pi]$, the set

$$F' = \{1, \cos, \sin, \cos 2, \sin 2, \dots, \cos n, \sin n, \dots\}$$

is an orthogonal set of functions relative to the inner product defined in equation (P).

Solution: The solution will depend on the trigonometric identities

$$\sin \alpha \sin \beta = \tfrac{1}{2}[\cos (\alpha - \beta) - \cos (\alpha + \beta)]$$
$$\cos \alpha \cos \beta = \tfrac{1}{2}[\cos (\alpha + \beta) + \cos (\alpha - \beta)]$$

Also, we remark that an *odd function* f $(f(-x) = -f(x))$ that is integrable on $[-a, a]$ has the property that $\int_{-a}^a f(x)\, dx = 0$, and an *even function* f $(f(-x)$

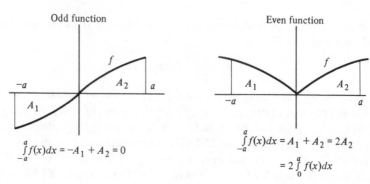

Odd function

Even function

$$\int_{-a}^{a} f(x)dx = -A_1 + A_2 = 0$$

$$\int_{-a}^{a} f(x)dx = A_1 + A_2 = 2A_2$$

$$= 2 \int_{0}^{a} f(x)dx$$

Figure 8.5

$= f(x)$) that is integrable on $[-a, a]$ has the property that $\int_{-a}^{a} f(x)\, dx = 2 \int_{0}^{a} f(x)\, dx$ (see Figure 8.5).

We must show that $(1, \cos n) = (1, \sin n) = (\cos n, \sin m) = 0$, and that $(\cos n, \cos k) = (\sin n, \sin k) = 0$ if $n \neq k$.

1. $(1, \cos n) = \displaystyle\int_{-\pi}^{\pi} \cos (nx)\, dx = 2 \int_{0}^{\pi} \cos (nx)\, dx = \dfrac{2}{n} \Big[\sin (nx) \Big]_{0}^{\pi} = 0.$

2. $(1, \sin n) = \displaystyle\int_{-\pi}^{\pi} \sin (nx)\, dx = 0$, since $\sin n$ is an odd function.

3. $(\cos n, \sin m) = \displaystyle\int_{-\pi}^{\pi} \cos (nx) \sin (mx)\, dx = 0$, since $(\cos n)(\sin m)$ is an odd function.

4. $(\cos n, \cos k) = \displaystyle\int_{-\pi}^{\pi} \cos (nx) \cos (kx)\, dx$

$\qquad\qquad = \displaystyle\int_{-\pi}^{\pi} \tfrac{1}{2}[\cos ((n + k)x) + \cos ((n - k)x)]\, dx$

$\qquad\qquad = 0$

by part (1), since $n - k \neq 0$.

5. $(\sin n, \sin k) = \displaystyle\int_{-\pi}^{\pi} \sin (nx) \sin (kx)\, dx$

$\qquad\qquad = \displaystyle\int_{-\pi}^{\pi} \tfrac{1}{2}[\cos ((n - k)x) - \cos ((n + k)x)]\, dx$

$\qquad\qquad = 0$

by part (1), since $n - k \neq 0$.

Thus, F' is an orthogonal set.   Q.E.D.

Note: An infinite set of vectors is linearly independent if and only if every nonempty finite subset is linearly independent. It follows from Problem 14, Exercise 8.2, that F' is linearly independent.

Converting the set F' to an orthonormal, set we have the following corollary.

Corollary 8.3. In the vector space **W** of all continuous functions on $[-\pi, \pi]$, the set

$$F = \left\{ \frac{1}{\sqrt{2\pi}}, \frac{\cos}{\sqrt{\pi}}, \frac{\sin}{\sqrt{\pi}}, \frac{\cos 2}{\sqrt{\pi}}, \frac{\sin 2}{\sqrt{\pi}}, \ldots, \frac{\cos n}{\sqrt{\pi}}, \frac{\sin n}{\sqrt{\pi}}, \ldots \right\}$$

is an orthonormal set relative to the inner product (P).

Proof: The notation $\frac{\cos}{k}$ means $(1/k)$ *cos*. We normalize each of the

vectors in F':

$$(1, 1) = \int_{-\pi}^{\pi} dx = 2\pi$$

Therefore, $\| 1 \| = \sqrt{2\pi}$.

$$(\cos k, \cos k) = \int_{-\pi}^{\pi} \cos (kx) \cos (kx) \, dx = \int_{-\pi}^{\pi} \tfrac{1}{2}[\cos (2kx) + \cos 0] \, dx$$

$$= \tfrac{1}{2} \int_{-\pi}^{\pi} [\cos (2kx) \, dx + \tfrac{1}{2} \int_{-\pi}^{\pi} dx = 0 + \pi$$

Therefore, $\| \cos k \| = \sqrt{\pi}$.

$$(\sin k, \sin k) = \int_{-\pi}^{\pi} \sin (kx) \sin (kx) \, dx = \int_{-\pi}^{\pi} \tfrac{1}{2}[\cos 0 - \cos (2kx)] \, dx$$

$$= \tfrac{1}{2} \int_{-\pi}^{\pi} dx - \tfrac{1}{2} \int_{-\pi}^{\pi} \cos (2kx) \, dx = \pi - 0 = \pi$$

Therefore, $\| \sin k \| = \sqrt{\pi}$. It now follows that F is an orthonormal set.

Q.E.D.

We intend to apply the next theorem to the set F above.

Theorem 8.7. Let $C = \{h_1, h_2, \ldots, h_n\}$ be an orthonormal set of functions in the space **V** of all (Riemann) integrable functions defined on a closed interval $[a, b]$. Let f be any function in **V**. Then the function g in $U = \langle h_1, h_2, \ldots, h_n \rangle$ that best approximates f, in the sense that for all h in **U**, $d(f, g)$ $\leq d(f, h)$, is the function $g = (f, h_1)h_1 + (f, h_2)h_2 + \cdots + (f, h_n)h_n$. Furthermore, the coefficients $(f, h_1), (f, h_2), \ldots, (f, h_n)$ are uniquely determined.

Proof: Let $g = a_1 h_1 + a_2 h_2 + \cdots + a_n h_n \in U$. We wish to determine the coefficients $a_1, a_2, \ldots, a_n$ such that $d(f, g)$ is minimal.

$$(d(f, g))^2 = \| f - g \|^2 = (f - g, f - g) = (f, f) - 2(f, g) + (g, g)$$

Also, $(f, f) = \| f \|^2$. This part is arbitrary but *fixed* once f is chosen.

$$(f, g) = (f, a_1 h_1 + a_2 h_2 + \cdots + a_n h_n) = a_1(f, h_1) + a_2(f, h_2)$$
$$+ \cdots + a_n(f, h_n)$$

The numbers $(f, h_1), (f, h_2), \ldots, (f, h_n)$ are also *fixed* once f is chosen.

$$(g, g) = (g, a_1 h_1 + a_2 h_2 + \cdots + a_n h_n)$$
$$= a_1(g, h_1) + a_2(g, h_2) + \cdots + a_n(g, h_n)$$

We calculate $a_k(g, h_k)$ for $k = 1, 2, \ldots, n$.

$$a_k(g, h_k) = a_k(a_1 h_1 + a_2 h_2 + \cdots + a_n h_n, h_k)$$
$$= a_k[a_1(h_1, h_k) + a_2(h_2, h_k) + \cdots + a_n(h_n, h_k)]$$
$$= a_k(0 + \cdots + 0 + a_k(h_k, h_k) + 0 + \cdots + 0)((h_i, h_k) = 0 \text{ if } i \neq k)$$
$$= a_k^2 \qquad ((h_k, h_k) = 1 \text{ since } h \text{ has length 1})$$

Thus $a_k(g, h_k) = a_k^2$ for $k = 1, 2, \ldots, n$. Therefore,

$$(d(f, g))^2 = (f, f) - 2(f, g) + (g, g)$$
$$= \| f \|^2 - 2[a_1(f, h_1) + a_2(f, h_2) + \cdots + a_n(f, h_n)]$$
$$\quad + a_1^2 + a_2^2 + \cdots + a_n^2$$
$$= ([a_1 - (f, h_1)]^2 + [a_2 - (f, h_2)]^2 + \cdots + [a_n - (f, h_n)]^2)$$
$$\quad + (\| f \|^2 - [(f, h_1)^2 + (f, h_2)^2 + \cdots + (f, h_n)^2])$$
$$= \sum_{i=1}^{n} [a_i - (f, h_i)]^2 + (\text{a constant})$$

From our equation for $(d(f, g))^2$, we see that the value of $d(f, g)$ depends on the value of $\sum_{i=1}^{n} [a_i - (f, h_i)]^2$. It is clear that the minimum value of this sum is zero, and that the sum is zero if and only if $(f, h_i) = a_i$ for $i = 1, 2, \ldots, n$. Hence $d(f, g)$ is minimal if and only if the coefficients are $a_1 = (f, h_1), a_2 = (f, h_2), \ldots, a_n = (f, h_n)$.   Q.E.D.

If g is the best approximation to f in the sense of Theorem 8.7, we shall write $f \simeq g$.

Example 8.8. Let F be the orthonormal set of functions in Corollary 8.3, and let **V** be the space of all integrable functions defined on $[-\pi, \pi]$. For any function f in **V**, show that the best approximation to f with respect to the first $2n + 1$ functions in the orthonormal set F is

$$f \simeq a_0 \mathbf{1} + a_1 \cos + b_1 \sin + \cdots + a_n \cos \mathbf{n} + b_n \sin \mathbf{n}$$

where

$$a_0 = \frac{1}{2\pi} \int_{-\pi}^{\pi} f(x)\, dx, \qquad a_k = \frac{1}{\pi} \int_{-\pi}^{\pi} f(x) \cos(kx)\, dx,$$

$$b_k = \frac{1}{\pi} \int_{-\pi}^{\pi} f(x) \sin(kx)\, dx$$

for $k = 1, 2, \ldots, n$.

Solution: By Theorem 8.7,

$$f \simeq \left(f, \frac{1}{\sqrt{2\pi}}\right)\frac{1}{\sqrt{2\pi}} + \left(f, \frac{\cos}{\sqrt{\pi}}\right)\frac{\cos}{\sqrt{\pi}} + \left(f, \frac{\sin}{\sqrt{\pi}}\right)\frac{\sin}{\sqrt{\pi}} + \cdots$$

$$+ \left(f, \frac{\cos n}{\sqrt{\pi}}\right)\frac{\cos n}{\sqrt{\pi}} + \left(f, \frac{\sin n}{\sqrt{\pi}}\right)\frac{\sin n}{\sqrt{\pi}}$$

Each term is a function of the form rh, where r is a scalar and $h = \dfrac{1}{\sqrt{2\pi}}$ or

$h = \dfrac{\cos k}{\sqrt{\pi}}$ or $h = \dfrac{\sin k}{\sqrt{\pi}}$ for $k = 1, 2, \ldots, n$. We transform each of these terms.

$$\left(f, \frac{1}{\sqrt{2\pi}}\right)\frac{1}{\sqrt{2\pi}} = \left(\frac{1}{\sqrt{2\pi}}\int_{-\pi}^{\pi} f(x)\,dx\right)\left(\frac{1}{\sqrt{2\pi}}\mathbf{1}\right) = \left(\frac{1}{2\pi}\int_{-\pi}^{\pi} f(x)\,dx\right)\mathbf{1}$$

Also

$$\left(f, \frac{\cos k}{\sqrt{\pi}}\right)\frac{\cos k}{\sqrt{\pi}} = \left(\frac{1}{\sqrt{\pi}}\int_{-\pi}^{\pi} f(x)\cos(kx)dx\right)\left(\frac{1}{\sqrt{\pi}}\cos \mathbf{k}\right)$$

$$= \left(\frac{1}{\pi}\int_{-\pi}^{\pi} f(x)\cos(kx)\,dx\right)\cos \mathbf{k}$$

and

$$\left(f, \frac{\sin k}{\sqrt{\pi}}\right)\frac{\sin k}{\sqrt{\pi}} = \left(\frac{1}{\sqrt{\pi}}\int_{-\pi}^{\pi} f(x)\sin(kx)\,dx\right)\left(\frac{1}{\sqrt{\pi}}\sin \mathbf{k}\right)$$

$$= \left(\frac{1}{\pi}\int_{-\pi}^{\pi} f(x)\sin(kx)\,dx\right)\sin \mathbf{k}$$

Hence

$$f \simeq \left(\frac{1}{2\pi}\int_{-\pi}^{\pi} f(x)\,dx\right)\mathbf{1} + \left(\frac{1}{\pi}\int_{-\pi}^{\pi} f(x)\cos x\,dx\right)\cos$$

$$+ \left(\frac{1}{\pi}\int_{-\pi}^{\pi} f(x)\sin x\,dx\right)\sin + \cdots + \left(\frac{1}{\pi}\int_{-\pi}^{\pi} f(x)\cos(nx)\,dx\right)\cos \mathbf{n}$$

$$+ \left(\frac{1}{\pi}\int_{-\pi}^{\pi} f(x)\sin(nx)\,dx\right)\sin \mathbf{n}$$

Definition 8.8. Let $C = \{h_1, h_2, \ldots, h_n, \ldots\}$ be a finite or an infinite set of orthonormal functions in the space $\mathbf{V}$ of all (Riemann) integrable functions defined on $[a, b]$, and let f be any function in $\mathbf{V}$. The **Fourier coefficients of** f **with respect to the orthonormal set** C are the numbers

$$(f, h_k) = \int_a^b f(x)h_k(x)\,dx, \qquad \text{for } k = 1, 2, \ldots, n, \ldots$$

The **generalized Fourier series of** f with respect to C is the series $\sum_{n=1}^{+\infty} (f, h_n)h_n$. We write $f \simeq \sum_{n=1}^{+\infty} (f, h_n)h_n$ regardless of whether or not the series converges.

The **Fourier series of** f is the series obtained when $C = F$, the orthonormal set in Corollary 8.3. Thus the Fourier series of f is the series

$$f \simeq \left(f, \frac{1}{\sqrt{2\pi}}\right)\frac{1}{\sqrt{2\pi}} + \left(f, \frac{\cos}{\sqrt{\pi}}\right)\frac{\cos}{\sqrt{\pi}} + \left(f, \frac{\sin}{\sqrt{\pi}}\right)\frac{\sin}{\sqrt{\pi}} + \left(f, \frac{\cos 2}{\sqrt{\pi}}\right)\frac{\cos 2}{\sqrt{\pi}}$$

$$+ \left(f, \frac{\sin 2}{\sqrt{\pi}}\right)\frac{\sin 2}{\sqrt{\pi}} + \cdots + \left(f, \frac{\cos n}{\sqrt{\pi}}\right)\frac{\cos n}{\sqrt{\pi}} + \left(f, \frac{\sin n}{\sqrt{\pi}}\right)\frac{\sin n}{\sqrt{\pi}} + \cdots$$

Using the result of Example 8.8, we see that the Fourier series of f may be written as

$$f \simeq a_0 \mathbf{1} + a_1 \cos + b_1 \sin + a_2 \cos 2 + b_2 \sin 2$$
$$+ \cdots + a_n \cos n + b_n \sin n + \cdots$$

where

$$a_0 = \frac{1}{2\pi}\int_{-\pi}^{\pi} f(x)\,dx, \qquad a_k = \frac{1}{\pi}\int_{-\pi}^{\pi} f(x)\cos(kx)\,dx$$

$$b_k = \frac{1}{\pi}\int_{-\pi}^{\pi} f(x)\sin(kx)\,dx, \qquad \text{for } k = 1, 2, \ldots$$

Example 8.8 shows that a function f that is integrable on $[-\pi, \pi]$ can be approximated by adding a finite number of terms of the Fourier series of F. It seems reasonable that for certain functions the approximation of f may get better as we increase n, and that the Fourier series of f may actually be f. This is the case for a large class of functions.

It can be shown that *if f is continuous on $[-\pi, \pi]$ except for a finite number of finite jump discontinuities* (points like b_2 and b_3 in Figure 8.6), *and if f is sectionally smooth* (f is differentiable except at points like b_1), *then the Fourier series of f converges to a function h on $[-\pi, \pi]$. Also, $h(-\pi) = h(\pi)$* $= [f(-\pi) + f(\pi)]/2$, *and $h(x) = [f(x_+) + f(x_-)]/2$ for all x in $(-\pi, \pi)$*. Here, $f(x_+)$ and $f(x_-)$ are the right and left hand limits of f at x. Thus $h(x) = f(x)$ if f **is continuous at** x **and** x **is in** $(-\pi, \pi)$ (see Figure 8.6). **The value of** h **at the endpoints** $-\pi$ **and** π **is the average value of** f **at** $-\pi$ **and** π.

Example 8.9. Find the Fourier series for the function f defined by $f(x) = -1$ if $x < 0$, and $f(x) = 1$ if $x \geq 0$. Sketch the graphs of the approximations of f given by (a) the first term, (b) the first two terms, and (c) the first three terms of the Fourier series. Find the points where f and its Fourier series disagree.

Solution: First we note that $f(-x) = -f(x)$, and f is an odd function. If h is the Fourier series of f then

$$h(x) = a_0 + a_1 \cos x + b_1 \sin x + a_2 \cos(2x) + b_2 \sin(2x) + \cdots$$

where

$$a_0 = \frac{1}{2\pi}\int_{-\pi}^{\pi} f(x)\,dx, \qquad a_k = \frac{1}{\pi}\int_{-\pi}^{\pi} f(x)\cos(kx)\,dx$$

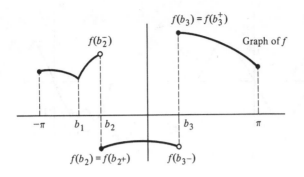

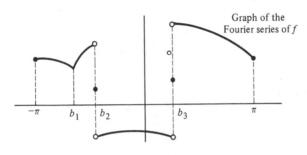

Figure 8.6

$$b_k = \frac{1}{\pi} \int_{-\pi}^{\pi} f(x) \sin(kx) \, dx$$

(Example 8.8). f and $f \cos \mathbf{k}$ are odd functions, and therefore each of the a's is 0. The function $f \sin \mathbf{k}$ is even and

$$b_k = \frac{2}{\pi} \int_{0}^{\pi} \sin(kx) \, dx = -\frac{2}{k\pi} \cos(kx) \Big]_{0}^{\pi} = \begin{cases} 0 & \text{if } k \text{ is even} \\ \dfrac{4}{k\pi} & \text{if } k \text{ is odd} \end{cases}$$

Therefore,

$$h(x) = b_1 \sin x + b_3 \sin(2x) + b_5 \sin(3x) + \cdots$$

$$= \frac{4}{\pi} \left(\sin x + \frac{\sin(3x)}{3} + \frac{\sin(5x)}{5} + \cdots \right)$$

$h(x) = f(x)$ for all $x \neq 0$, $-\pi$, and π.

$$h(0) = \frac{f(0_+) + f(0_-)}{2} = \frac{1 - 1}{2} = 0$$

and

$$h(\pi) = h(-\pi) = \frac{-1 + 1}{2} = 0$$

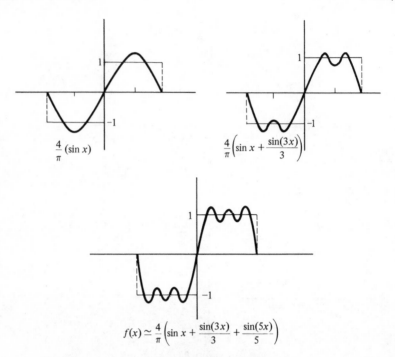

$$\frac{4}{\pi}(\sin x)$$

$$\frac{4}{\pi}\left(\sin x + \frac{\sin(3x)}{3}\right)$$

$$f(x) \simeq \frac{4}{\pi}\left(\sin x + \frac{\sin(3x)}{3} + \frac{\sin(5x)}{5}\right)$$

Note from the sketches that the approximation of f by the terms of its Fourier series is not good with one term, fair with two terms, and better with three terms. The graph of h, the Fourier series of f, is identical to the graph of f, except that $h(-\pi) = h(\pi) = h(0) = 0$, whereas $f(-\pi) = -1$ and $f(\pi) = f(0) = 1$.

We close this section by giving two references for the reader who may wish to explore further the study of Fourier series. The text by N. W. Gowar and J. E. Baker, *Fourier Series* (London: Chatto & Windus Ltd., with William Collins Sons and Co. Ltd., 1974) is a very readable text written for second and third year students. The reader with a stronger calculus background may consult John M. H. Olmsted, *Advanced Calculus* (Englewood Cliffs, N.J., Prentice-Hall, Inc., 1961).

Exercise 8.3

In Problems 1–4, let (X, Y) be the form defined in Example 8.7.

1. Verify properties 1P: $(X, Y) = (Y, X)$; 2P: $(aX + bY, Z) = a(X, Y) + b(X, Z)$; and 3'P: $|(X, X)| \geq 0$ and $(\mathbf{0}, \mathbf{0}) = 0$.

2. Verify that $\|X\| = \sqrt{|(X, X)|}$ satisfies 1'N and 2N.

3. Verify that $d(X, Y) = \|X - Y\|$ satisfies 1'D and 2D.

4. Verify that every point of the type $X = (x, 0, 0, x)$ is orthogonal to itself. If $X = (x, 0, 0, x)$, what is $d(0, X)$? If Edward had flown on the path $\{(x, 0, 0, x) \,|\, 0 \leq x \leq 4\}$, how fast would he be going? How old would he be when he returned to earth (keeping this constant speed)?

5. The brightest star in the heavens, Sirius, is 8.6 light-years away from earth. For ease of computation we round this off to 9 light-years. The time that elapses on earth while a rocket ship goes to Sirius and back is 22 years. If the pilot of the rocket ship is 20 years old when he leaves, how old will he be when he returns? Draw a sketch similar to Figure 8.4.

6. If **V** is the vector space of functions integrable on $[a, b]$, verify that the "norm" defined by equation (N) does not satisfy property 1N. Verify that the "distance" defined by equation (D) does not satisfy property 1D.

7. Find the Fourier series of the function f defined by $f(x) = x$, $x \in [-\pi, \pi]$. Sketch the graph of the approximations to f given by (a) the first term of the Fourier series of f; (b) the sum of the first three terms of the Fourier series of f.

8. Find the Fourier series of the function f defined by $f(x) = |x|$, $x \in [-\pi, \pi]$. Sketch the graph of the approximations to f given by (a) the first term of the Fourier series of f; (b) the sum of the first two terms of the Fourier series of f.

9. In the Fourier series for $f(x) = x$ obtained in Problem 7, substitute $x = \pi/2$ to show that

$$\frac{\pi}{4} = 1 - \frac{1}{3} + \frac{1}{5} - \frac{1}{7} + \cdots .$$

10. Find the Fourier series of the constant function $f(x) = 2$, $x \in [-\pi, \pi]$.

11. Find the Fourier series for the function f defined by $f(x) = 1$ if $-\pi \leq x < 0$, $f(x) = 0$ if $0 < x \leq \pi$ and $f(0) = \frac{1}{2}$. Sketch the graph of the approximations to f given by (a) the sum of the first two terms of the Fourier series of f; (b) the sum of the first four terms of the Fourier series of f.

Bibliography

The following textbooks and references are written at about the level of this text.

BRADLEY, GERALD L. *A Primer of Linear Algebra* (Englewood Cliffs, N.J.: Prentice-Hall, Inc., 1975).

CAMPBELL, HUGH G. *Linear Algebra with Applications Including Linear Programming* (Englewood Cliffs, N.J.: Prentice-Hall, Inc., 1971). A number of applications are mentioned along with an extensive list of references to texts in other fields that give applications of matrix algebra.

KRAUSE, EUGENE F. *Introduction to Linear Algebra* (New York: Holt, Rinehart and Winston, Inc., 1970).

MOORE, JOHN T. *Elementary Linear and Matrix Algebra: The Viewpoint of Geometry* (New York: McGraw-Hill Book Company, 1972).

Proceedings Summer Conference for College Teachers on Applied Mathematics (Rolla, Mo.: University of Missouri-Rolla, 1971; supported by the National Science Foundation, 1973), pp. 60–192. An extensive bibliography of sources on applications of linear algebra is given on pp. 193–198.

WILLIAMS, GARETH. *Computational Linear Algebra with Models.* (Boston: Allyn & Bacon, Inc., 1975).

The following texts are written at an intermediate or advanced level.

FINKBEINER, DANIEL T., II. *Introduction to Matrices and Linear Transformations*, 2nd ed. (San Francisco: W. H. Freeman and Company, 1966).

HOFFMAN, KENNETH, and RAY KUNZE. *Linear Algebra*, 2nd ed. (Englewood Cliffs, N.J.: Prentice-Hall, Inc., 1971).

JACOBSON, NATHAN. *Lectures in Abstract Algebra, Volume II* (New York: Van Nostrand Reinhold Company, 1953).

NERING, EVAR D. *Linear Algebra and Matrix Theory.* (New York: John Wiley & Sons, Inc., 1963).

NOBLE, BEN. *Applied Linear Algebra* (Englewood Cliffs, N.J.: Prentice-Hall, Inc., 1969).

Answers
to Odd-Numbered
Exercises

CHAPTER 1

Exercise 1.1

1. (a) The opposite sides of a parallelogram are equal in length and parallel. Since dir $\overrightarrow{AD}$ = dir $\overrightarrow{BC}$, we have $\overrightarrow{AD} = \overrightarrow{BC}$.
   (b) Similar to (a). (c) $\overrightarrow{AC} = \bar{u} + \bar{v}$. (d) $\overrightarrow{AC} = -\bar{x} + -\bar{w}$.
   (e) $\bar{x} + \bar{u} = \overrightarrow{BD}$, $(\bar{x} + \bar{u}) + \bar{v} = \overrightarrow{BC}$, and $((\bar{x} + \bar{u}) + \bar{v}) + \bar{w} = \overrightarrow{BB} = \bar{0}$.

3.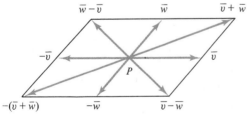

5. $\overrightarrow{BC} = \bar{v} - \bar{u}$, $\overrightarrow{CB} = \bar{u} - \bar{v}$, $\overrightarrow{CA} = -\bar{v}$.
   (b) $\overrightarrow{CD} = -\bar{u}$, $\overrightarrow{DA} = -\bar{v}$, $\overrightarrow{DB} = \bar{u} - \bar{v}$, and $\overrightarrow{AC} = \bar{u} + \bar{v}$.

9. (a) The sum of the lengths of two sides of a triangle does not exceed the length of the third side. (b) dir $\bar{a}$ = dir $\bar{b}$.

13. (b) (ground speed)^2 = $40^2 + 500^2 - 40,000 \cos 40° \approx 220,958.22$. Thus, the ground speed is approximately 470.06 km/h. The direction of travel is about 43°8′ west of north.

Exercise 1.2

1. $\bar{g} = \frac{1}{16}(5\bar{a} + 3\bar{b} + 8\bar{c})$.
3. $\frac{1}{20}(3\bar{a} + 2\bar{b} + 5\bar{c} + 4\bar{d} + 6\bar{e})$.
7. $\overrightarrow{AB} = \bar{b} - \bar{a}$, $\overrightarrow{BC} = \bar{c} - \bar{b}$, $\overrightarrow{CD} = \bar{d} - \bar{c}$, $\overrightarrow{DA} = \bar{a} - \bar{d}$.

Exercise 1.3

1. (b) The head of the position vector $\bar{p} - \bar{q}$ is the point $P - Q$.
   (c) $\bar{p} - \bar{q} = \overrightarrow{QP}$.
5. $(4, 3)$.
9. (a) $(2, 4, 4)$. (b) $(-3, 5, -1)$. (c) $(13, 0, 11)$. (d) $(4, 1, 8)$. (e) $(4, 1, 8)$.
   (f) $(15, 10, 5)$. (g) $(15, 10, 5)$. (h) $(6, 12, 12)$. (i) $(6, 12, 12)$.

CHAPTER 2

Exercise 2.1

1. (a) $\sqrt{2}$. (b) $\sqrt{19}$. (c) $\sqrt{21}$. (d) $\sqrt{55}$.
3. (a) $\cos \alpha = 2/\sqrt{14}$, $\cos \beta = 1/\sqrt{14}$, $\cos \gamma = 3/\sqrt{14}$.
   (b) $\cos \alpha = \frac{1}{2}$, $\cos \beta = \frac{3}{4}$, $\cos \gamma = \sqrt{3}/4$.
   (c) $\cos \alpha = -\frac{3}{5}$, $\cos \beta = \frac{4}{5}$.
9. (a) $\cos \theta = 0$. (b) $\cos \theta = (1 + \sqrt{2})/3$. (c) $\cos \theta = 0$. In (a) and (c), $\theta = 90°$.
   In (b), $\theta = \cos^{-1} (1 + \sqrt{2})/3 \approx 36.4°$.
11. $a_1b_1 + a_2b_2 + a_3b_3 = 0$.

Exercise 2.2

1. (a) 7. (b) -30. (c) $40(17) = 680$.
3. (a) One answer is $(-2, 1)$. (b) One answer is $(2, 0, 1)$.
5. $(A - B)\cdot(C - B) = 0$. Therefore, $\angle B$ is a right angle.
7. One answer is $A = (1, 2, 3)$ and $B = (-3, 0, 1)$.
9. $\sqrt{10} < \sqrt{18} + \sqrt{14}$.
11. No, $A\cdot B$ is a scalar r and $r\cdot C$ is undefined.
13. (a) $\frac{1}{3}A$. (b) $(1/\sqrt{29})A$.
15. (a) $\sqrt{5}$. (b) $\sqrt{78}$. (c) $\sqrt{26}$.
17. $175(217) = 37,975$.

31. $C = (233.75, 342.50, 445)$.

33. $M = (m_1, m_2, \ldots, m_n)$, $P = (p_1, p_2, \ldots, p_n)$,
$$M \cdot P = m_1 p_1 + m_2 p_2 + \cdots + m_n p_n.$$
$M \cdot P$ is the cost of the raw materials needed to produce one unit of the product.

Exercise 2.3

1. (a) $3x + 4y + 2z - 17 = 0$. (b) $2x - 2y + z = 0$.

3. (a) $3\bar{i} + 4\bar{j} - 6\bar{k}$. (b) $3\bar{i} + 2\bar{j} - 6\bar{k}$.

5. $\frac{27}{7}$.

7. $10/\sqrt{29}$.

15. 44.

17. $\frac{15}{2}(\cos 30° \, \bar{i} + \sin 30° \, \bar{j}) = \frac{15\sqrt{3}}{4}\bar{i} + \frac{15}{4}\bar{j}$, $W = 300$ ft-lb.

Exercise 2.4

3. (a) $A \times B = (-1, -5, -1)$. (b) $\|A \times B\| = \sqrt{27}$.

5. Area $= \sqrt{35}/2$.

7. $\sqrt{523}/2 \approx 11.4$.

13. The cross product is not associative.

15. $\Omega = \frac{20}{7}(6\bar{i} - 3\bar{j} + 2\bar{k})$ and $\bar{v} = \frac{20}{7}(-13\bar{i} - 16\bar{j} + 15\bar{k})$. $\|\bar{v}\| =$ linear speed $= 20/7\sqrt{650} \approx 72.8$.

Exercise 2.5

1. Vector equation: $(x, y, z) = (3, 4, 2) + t(1, 3, 1)$.
Parametric equations: $x = 3 + t$
$\qquad\qquad\qquad\quad y = 4 + 3t$ Two points on the line are $(4, 7, 3)$ and
$\qquad\qquad\qquad\quad z = 2 + t$ $\quad(2, 1, 1)$.

3. $\dfrac{x - 1}{1} = \dfrac{y - 2}{-3} = \dfrac{z - 3}{1}$.

5. A vector equation: $(x, y) = (2, 4) + t(4, -3)$
Parametric equations: $x = 2 + 4t$
$\qquad\qquad\qquad\quad y = 4 - 3t$

7. (a) $(x, y, z) = (-1, 2, 2) + s(1, 1, -1) + t(1, -2, 0)$ is a vector equation.
Parametric equations: $x = -1 + s + t$
$\qquad\qquad\qquad\quad y = 2 + s - 2t$
$\qquad\qquad\qquad\quad z = 2 - s$

(b) Vector equation: $(x, y, z) = (3, 1, 0) + s(-1, 0, 1) + t(-3, 1, -4)$.
Parametric equations: $x = 3 - s - 3t$
$$y = 1 + t$$
$$z = s - 4t$$

9. (a) Yes. (b) No.

11. (b) $x = 1 + t$
$y = 2 - 3t$ and $-2 \le t \le 3$.
$z = 3 + t$

CHAPTER 3

Exercise 3.1

1. (a) $\begin{bmatrix} 2 & 3 & 4 & | & 1 \\ 1 & 0 & -3 & | & 4 \\ 4 & 1 & -1 & | & 6 \end{bmatrix}$, $\begin{bmatrix} 2 & 3 & 4 \\ 1 & 0 & -3 \\ 4 & 1 & -1 \end{bmatrix}$.

(b) $\begin{bmatrix} 1 & 3 & 0 & | & 1 \\ -1 & 2 & -7 & | & 1 \\ 2 & 3 & -1 & | & -2 \end{bmatrix}$, $\begin{bmatrix} 1 & 3 & 0 \\ -1 & 2 & -7 \\ 2 & 3 & -1 \end{bmatrix}$.

3. (a) $x - y + 3z = 2$ (b) $x - 2y = 1$
$\quad\quad 2x - 2z = 1$ $\quad\quad 3x + 4y = 0$
$\quad 3x + y + 4z = 0$
$\quad\quad x - y + 4z = 5$

5. $S = \{(5 - k, -2 - 2k, k) | k \in \mathbf{R}\}$. A particular solution is $(5, -2, 0)$.

7. $(2, 0, -1)$ is the unique solution.

9. No solutions.

11. $S = \{(-2s - 4t - 4, -s - 2t + 2, s, t) | s, t \in \mathbf{R}\}$. A particular solution is $(-4, 2, 0, 0)$.

13. $(3, 1, 1)$ is the unique solution.

15. The complete solution is
$(x_1, x_2, x_3, x_4, x_5) = (\frac{1}{3}t, r + s - \frac{5}{3}t, r, s, t), r, s, t \in \mathbf{R}$.

17. $S = \{(-\frac{1}{3}t + \frac{2}{3}, -\frac{7}{6}t - \frac{1}{6}, -\frac{5}{6}t + \frac{1}{6}, t) | t \in \mathbf{R}\}$. A particular solution is $(\frac{2}{3}, -\frac{1}{6}, \frac{1}{6}, 0)$.

Exercise 3.2

1. (a) $-1, 8, 3, -2, 9$. (b) $m = 3, n = 6$. (c) $\begin{bmatrix} 7 \\ -2 \\ -7 \end{bmatrix}$.

(d) $[4 \quad 6 \quad 0 \quad -2 \quad -3 \quad 8]$.

3. (a) All. (b) 3. (c) 3.

7. (a) $x_1 + 2x_3 - x_5 - 4x_6 - 6x_7 = 3$
$$x_4 - 2x_5 + 3x_6 + x_7 = 2$$
$$x_8 = -1$$
(b) $m = 5, n = 8$. (c) 3. (d) The system has infinitely many solutions. There are five parameters.

(e) $(x_1, x_2, x_3, x_4, x_5, x_6, x_7, x_8)$
$$= (-2q + r + 4s + 6t + 3, p, q, 2r - 3s - t + 2, r, s, t, -1).$$

9. (a) $x_1 = 1, \quad x_2 = 3, \quad x_3 = 2$. (b) $m = n = 3$. (c) 3. (d) Unique solution. (e) $(1, 3, 2)$ is the unique solution. (f) Yes, the system has three equations and three unknowns.

11. A system of n linear equations in n unknowns has a unique solution if and only if the identity matrix I_n is row equivalent to the coefficient matrix of the system.

Exercise 3.3

1. $i_1 = \frac{77}{47} \approx 1.64$ Amperes, $i_2 = \frac{48}{47} \approx 1.02$ A, $i_3 = \frac{29}{47} \approx 0.617$ A.

3. $4\frac{1}{3}$ gallons of brand X, $\frac{4}{3}$ gallons of brand Y, and $1\frac{2}{3}$ gallons of brand Z.

5. If a, b, and c are the number of trucks of type A, B, and C, respectively, then $(a, b, c) \in \{(11, 10, 0), (12, 6, 2), (13, 2, 4)\}$. The most economical solution is $(13, 2, 4)$.

7. $(p_1, p_2, p_3, p_4) = k(62, 37, 26, 23)$.

9. $(-2x_3 - 4x_4 - 4, -x_3 - 2x_4 + 2, x_3, x_4) | x_3, x_4 \in \mathbf{R}\}$.

11. (a) $(13, -10)$. (b) $\quad x_1 + x_2 = 3$
$$x_1 + 1.000x_2 = 2.999$$
The system in (b) is inconsistent. (c) The system is illconditioned.

13. (a) $(-202, 200)$. (b) $(-152, 150)$. (c) $(-302, 300)$. (d) Yes.

Exercise 3.4

1. (a) $\{1, 4\}$. (b) $A^T = \begin{bmatrix} 1 & 2 \\ 3 & 4 \\ 8 & 5 \\ 7 & 6 \end{bmatrix}$.

3. (a) $A + B = \begin{bmatrix} -1 & 2 & 0 & 3 \\ 5 & 5 & 0 & 1 \\ 1 & 1 & 0 & -4 \end{bmatrix}$. (b) $A + B = \begin{bmatrix} -2 & 3 \\ -3 & 5 \end{bmatrix}$.

5. (a) $\begin{bmatrix} 0 & 0 \\ 0 & 0 \end{bmatrix}$. (b) $\begin{bmatrix} -9 & 3 \\ 0 & 3 \end{bmatrix}$. (c) $\begin{bmatrix} 4 & 8 \\ -12 & 16 \end{bmatrix}$. (d) $\begin{bmatrix} -5 & 11 \\ -12 & 19 \end{bmatrix}$.

7. $\begin{bmatrix} -1 & 3 & -4 \\ 3 & 2 & 5 \\ -4 & 5 & 2 \end{bmatrix}.$

9. (a) $[a_{ij}]_{(n,n)} = [a_{ji}]_{(n,n)}$. (b) Yes, $A = C$, $B = A^T$, $A \neq B$ if $a_{ij} \neq a_{ji}$ for some i, j. $B = \begin{bmatrix} 1 & 4 & 7 \\ 2 & 5 & 8 \\ 3 & 6 & 9 \end{bmatrix}$. $C = \begin{bmatrix} 1 & 2 & 3 \\ 4 & 5 & 6 \\ 7 & 8 & 9 \end{bmatrix}.$

Exercise 3.5

1. (a) $\begin{bmatrix} 1 & 3 & -1 & 5 \\ 1 & 0 & -4 & 0 \\ 1 & 7 & 1 & 2 \end{bmatrix} \begin{bmatrix} x_1 \\ x_2 \\ x_3 \\ x_4 \end{bmatrix} = \begin{bmatrix} 3 \\ 1 \\ 8 \end{bmatrix}$. (b) $\begin{bmatrix} 3 & 2 & -4 \\ 1 & 3 & -2 \end{bmatrix} \begin{bmatrix} x \\ y \\ z \end{bmatrix} = \begin{bmatrix} 9 \\ 4 \end{bmatrix}.$

3. (a)
$$2x_1 + 3x_2 + x_3 = 5$$
$$-x_1 + 3x_3 + 4x_4 = 8$$
$$2x_1 + x_2 + 5x_3 + 4x_4 = 3$$
(b)
$$2x + 3y = -1$$
$$4x + 7y = 6$$

5. $x_1 = -11z_1 + 8z_2$
$x_2 = -5z_1 - 10z_2$
$x_3 = -5z_1 + z_2$

7. (a) 4×2. (b) 1×5. (c) Not defined. (d) 4×4.

9. $AX = \begin{bmatrix} -8 \\ -17 \\ -4 \end{bmatrix}$, $AY = \begin{bmatrix} -4 \\ -11 \\ 11 \end{bmatrix}$, $A(2X + 3Y) = \begin{bmatrix} -28 \\ -67 \\ 25 \end{bmatrix}.$

$2(AX) + 3(AY) = \begin{bmatrix} -28 \\ -67 \\ 25 \end{bmatrix}.$

11. (a) $AB = AC = \begin{bmatrix} 8 & 9 \\ 24 & 27 \end{bmatrix}$. (b) $AB = AC$ and $A \neq 0$ does not imply $B = C$.

17. $A^3B \doteq \begin{bmatrix} 115.76 \\ 119.10 \\ 122.50 \end{bmatrix}$, $A^4B \doteq \begin{bmatrix} 121.55 \\ 126.25 \\ 131.08 \end{bmatrix}$. The entries are rounded off to give the prices to the nearest penny.

19. $\begin{bmatrix} 0.312 \\ 0.367 \\ 0.321 \end{bmatrix}.$

Exercise 3.6

1. (a) $a_{i1}b_{1j} + a_{i2}b_{2j} + a_{i3}b_{3j} + a_{i4}b_{4j}$. k is the index of summation.
   (b) 1st sum $= (a_{11} + a_{21} + a_{31}) + (a_{12} + a_{22} + a_{32})$.
   2nd sum $= (a_{11} + a_{12}) + (a_{21} + a_{22}) + (a_{31} + a_{32})$. The two sums are equal.
   (c) 1st sum $= a_{i1}(b_{1j} + c_{1j}) + a_{i2}(b_{2j} + c_{2j})$.
   2nd sum $= (a_{i1}b_{1j} + a_{i1}c_{1j}) + (a_{i2}b_{2j} + a_{i2}c_{2j})$.
   3rd sum $= (a_{i1}b_{1j} + a_{i2}b_{2j}) + (a_{i1}c_{1j} + a_{i2}c_{2j})$. The three sums are equal.

3. $10\begin{bmatrix} 8 & -11 & 19 & 9 \\ -7 & -5 & 7 & 10 \\ -1 & -9 & 7 & 8 \end{bmatrix}$.

Exercise 3.7

1. (a) $\begin{bmatrix} 2 \\ -14 \\ 21 \end{bmatrix}$. (b) $\begin{bmatrix} 1 \\ -7 \\ 13 \end{bmatrix}$. (c) $\begin{bmatrix} 1 \\ -7 \\ 15 \end{bmatrix}$. (d) $\begin{bmatrix} 0 \\ -1 \\ 2 \end{bmatrix}$.

3. (a) $\begin{bmatrix} 1 & 0 & 0 & 0 \\ 0 & 1 & 0 & 0 \\ 0 & \frac{1}{2} & 1 & 0 \\ 0 & 0 & 0 & 1 \end{bmatrix}$. (b) $\begin{bmatrix} 0 & 0 & 0 & 1 \\ 0 & 1 & 0 & 0 \\ 0 & 0 & 1 & 0 \\ 1 & 0 & 0 & 0 \end{bmatrix}$. (c) $\begin{bmatrix} 1 & 0 & 0 & 0 \\ 0 & -3 & 0 & 0 \\ 0 & 0 & 1 & 0 \\ 0 & 0 & 0 & 1 \end{bmatrix}$.

5. (a) $\begin{bmatrix} 1 & 2 & 2 \\ 3 & 0 & 4 \\ \frac{7}{2} & 1 & 5 \\ 1 & -1 & 5 \end{bmatrix}$. (b) $\begin{bmatrix} 1 & -1 & 5 \\ 3 & 0 & 4 \\ 2 & 1 & 3 \\ 1 & 2 & 2 \end{bmatrix}$. (c) $\begin{bmatrix} 1 & 2 & 2 \\ 3 & 0 & 4 \\ 8 & 4 & 12 \\ 1 & -1 & 5 \end{bmatrix}$.

7. (a) $\begin{bmatrix} 1 & 0 & 0 & 0 \\ 0 & 1 & 0 & 0 \\ 0 & -\frac{1}{2} & 1 & 0 \\ 0 & 0 & 0 & 1 \end{bmatrix}$. (b) $\begin{bmatrix} 0 & 0 & 0 & 1 \\ 0 & 1 & 0 & 0 \\ 0 & 0 & 1 & 0 \\ 1 & 0 & 0 & 0 \end{bmatrix}$. (c) $\begin{bmatrix} 1 & 0 & 0 & 0 \\ 0 & 1 & 0 & 0 \\ 0 & 0 & \frac{1}{4} & 0 \\ 0 & 0 & 0 & 1 \end{bmatrix}$.

9. (a) $\begin{bmatrix} 0 & 1 & 0 \\ 1 & 0 & 0 \\ 4 & 0 & -2 \end{bmatrix}$. (b) $\begin{bmatrix} 0 & 1 & 0 \\ 1 & 0 & 0 \\ 0 & 2 & -\frac{1}{2} \end{bmatrix}$.

Exercise 3.8

1. $A^{-1} = \begin{bmatrix} 2 & -1 & -7 \\ -1 & 1 & 3 \\ 0 & 0 & 1 \end{bmatrix} = F_{31}(-7)F_{32}(3)F_{21}(-1)F_{12}(-1)$.

$A = F_{12}(1)F_{21}(1)F_{32}(-3)F_{31}(7)$.

Note: There is more than one sequence of elementary matrices that will work in each of Problems 1–6.

3. $A^{-1} = \begin{bmatrix} -6 & 24 & -5 \\ 1 & -4 & 1 \\ 4 & -15 & 3 \end{bmatrix}$

$= F_{23}F_{21}(-1)F_{31}(-2)F_{32}(3)F_{23}(1)F_{12}F_{23}(-1)F_{21}(-3)$.

$A = F_{21}(3)F_{23}(1)F_{12}F_{23}(-1)F_{32}(-3)F_{31}(2)F_{21}(1)F_{23}$.

5. $A^{-1} = \begin{bmatrix} 2 & \frac{3}{2} & -\frac{1}{2} & -\frac{3}{2} \\ 0 & 1 & 0 & -1 \\ -1 & -\frac{3}{2} & \frac{1}{2} & \frac{3}{2} \\ 0 & 0 & 0 & 1 \end{bmatrix}$.

$A^{-1} = F_{31}(-1)F_3(\frac{1}{2})F_{42}(-1)F_{43}(3)F_{23}(-3)F_{13}(-2)$.

$A = F_{13}(2)F_{23}(3)F_{43}(-3)F_{42}(1)F_3(2)F_{31}(1)$.

7. $A^{-1} = \begin{bmatrix} 1 & 0 & 0 & 0 & 0 \\ 0 & 1 & 0 & 0 & 0 \\ -2 & -3 & 1 & 0 & 0 \\ 2 & 7 & -3 & 1 & 0 \\ 6 & -3 & 2 & -2 & 1 \end{bmatrix}$.

9. $A = F_{12}(4)F_{21}(-3)F_2(-1)F_{12} = F_{12}(3)F_{21}(1)F_{12}(2)$. There are other answers.

11. (b) $A^{-1} = (1/ad - bc)\begin{bmatrix} d & -b \\ -c & a \end{bmatrix}$.

CHAPTER 4

Exercise 4.1

1. $ad - bc$.

3. $-x^2 + 3x$.

5. (a) 18.

7. $- \lambda^3 - 7\lambda^2 + 7\lambda + 20)$.

19. (a) Minus. (b) Minus. (c) Plus. (d) Minus.

Exercise 4.2

3. (a) 24. (b) -31. (c) 32.

5. (a) -35. (b) 166. (c) -95.

7. 24.

9. (a) $A^{-1} = \frac{1}{10} \begin{bmatrix} 4 & -3 & 4 \\ 10 & -5 & 0 \\ -12 & 9 & -2 \end{bmatrix}$. det $A = 10$. det $A^{-1} = \frac{1}{10}$.

Exercise 4.3

1. $a_{11}C_{31} + a_{12}C_{32} + a_{13}C_{33} = 1(-4) + 3(-2) + 2(5) = 0$.

3. (a) $\tilde{A} = \begin{bmatrix} 2 & -3 & -4 \\ -2 & 7 & 8 \\ -2 & 9 & 12 \end{bmatrix}$. (b) adj $A = \begin{bmatrix} 2 & -2 & -2 \\ -3 & 7 & 9 \\ -4 & 8 & 12 \end{bmatrix}$.

(c) $A(\text{adj } A) = (4)I_3$. (d) det $A = 4$. (e) $A^{-1} = \frac{1}{4} \begin{bmatrix} 2 & -2 & -2 \\ -3 & 7 & 9 \\ -4 & 8 & 12 \end{bmatrix}$.

5. $|A| = -1$. Therefore, A^{-1} exists. adj $A = \begin{bmatrix} 1 & -1 \\ -2 & 1 \end{bmatrix}$, $A^{-1} = \begin{bmatrix} -1 & 1 \\ 2 & -1 \end{bmatrix}$.

7. $(x_1, x_2, x_3) = (\frac{39}{27}, \frac{21}{27}, -\frac{36}{27})$.

9. $(x, y) = ((ed - bf)/(ad - bc), (af - ec)/(ad - bc))$.

11. (a) 26. (b) 264.

CHAPTER 5

Exercise 5.1

1. Not a vector space. Properties 2A, 3A, 4A, 5A, and 8M fail.

3. Not a vector space. Properties 7M and 10M fail.

5. A vector space.

7. A vector space.

Exercise 5.2

1. (a) $(10, 3, 15) = 3(-2, 1, -3) + 4(4, 0, 6)$.
(b) Yes, $(-4, 5, -6) = 5(-2, 1, -3) + \frac{3}{2}(4, 0, 6)$.

3. (a) For example, $(4, 2, 3)$. (b) **W** is the plane determined by the three points $(0, 0, 0)$, $(3, 2, 0)$, and $(1, 0, 3)$. (c) $(3, 4, 1)$ is not in **W**.

5. $\left\{ \begin{bmatrix} a & 0 & b \\ 0 & c & 0 \end{bmatrix} \middle| a, b, c \in \mathbf{R} \right\}$.

7. (c) Not a subspace, not closed under addition. (f) Not a subspace, not closed under scalar multiplication.

13. (a) Spans R_3. (b) Does not span R_3. (c) Does not span R_3.

15. (a) A subspace of $R_{2\times3}$. (b) A subspace of $R_{2\times3}$.

19. For example, $(1, 0, 2) + (1, 0, 0) \notin U \cup W$.

Exercise 5.3

1. (a) Dependent, $-3X_1 - 1X_2 + 2X_3 = 0$.
   (b) Dependent, $-2X_1 + 0X_2 + X_3 = 0$.
   (c) Independent.

3. (a) Dependent, $-6f_1 + 2f_2 + 3f_3 = 0$. (b) Independent.
   (c) Dependent, $11f_1 - 7f_2 + 5f_3 = 0$. (d) Independent.

5. (a) Independent. (b) Independent. (c) Dependent, $\frac{13}{5}X_1 - X_2 = 0$.
   (d) Independent.

7. $(0, 18, 6) = -2(2, 3, 1) + 4(1, 6, 2)$.

9. $\{X_1 = (1, -2, -3), X_2 = (0, 1, 2)\}$.

11. $\{f_1, f_2\}$.

Exercise 5.4

1. (a) $(x, y) = \dfrac{x + 2y}{11}(3, 4) + \dfrac{4x - 3y}{11}(2, -1)$. Therefore, A generates R_2. If
   $a(3, 4) + b(2, -1) = (0, 0)$, then $\begin{array}{c} 3a + 2b = 0 \\ 4a - b = 0 \end{array}$ Since $\det \begin{bmatrix} 3 & 2 \\ 4 & -1 \end{bmatrix} \neq 0$,
   the unique solution is $a = b = 0$.
   (b) $(x, y) = (x - y)(1, 0) + y(1, 1)$, Also, $a(1, 0) + b(1, 1) = (0, 0)$ implies
   $a = b = 0$.

3. (a) $\begin{bmatrix} a & b \\ c & d \end{bmatrix} = aX_1 + bX_2 + cX_3 + dX_4$ and A generates $R_{2\times2}$. If $aX_1 + bX_2$
   $+ cX_3 + dX_4 = 0$, then $\begin{bmatrix} a & b \\ c & d \end{bmatrix} = \begin{bmatrix} 0 & 0 \\ 0 & 0 \end{bmatrix}$ and $a = b = c = d = 0$.
   (b) $\begin{bmatrix} a & b \\ c & d \end{bmatrix} = (a - b)X_1 + (b - c)X_2 + (c - d)X_3 + dX_4$ and B generates
   $R_{2\times2}$. If $aX_1 + bX_2 + cX_3 + dX_4 = 0$, then $\begin{array}{c} a + b + c + d = 0 \\ b + c + d = 0 \\ c + d = 0 \\ d = 0 \end{array}$ Hence
   $a = b = c = d = 0$.

5. (a) $Y_1 = 2X_1 + 0X_2 + 0X_3$, $Y_2 = 3X_1 + 1X_2 + 0X_3$.
   (b) $W = \langle Y_1, Y_2, X_3 \rangle$.

7. (a) $Y_1 = 0X_1 + 0X_2 + \frac{1}{4}X_3 + 0X_4$, $Y_2 = 1X_1 + 0X_2 + \frac{1}{4}X_3 + 0X_4$.
   (b) $W = \langle Y_1, Y_2, X_1, X_4 \rangle$.

9. Let E_{st} be the $m \times n$ matrix with a 1 in the (s, t) entry and zeroes elsewhere. The set of mn matrices of the form E_{st}, where $1 \leq s \leq m$ and $1 \leq t \leq n$, is a basis for $\mathbf{R}_{m \times n}$.

11.

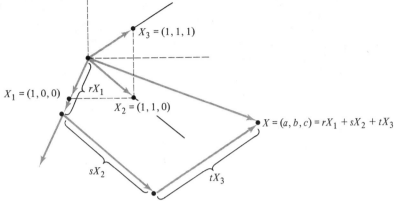

13. A basis is $\{1, i\}$. Therefore, the dimension is 2.

19. If $y = f(x)$ is a solution, then $y = ke^{x/3}$ for some constant k. The solution space is the one-dimensional space $\langle g \rangle$, where $g(x) = e^{x/3}$.

Exercise 5.5

1. (a) B is a basis. (b) B is not linearly independent. (c) B is not linearly independent. (d) B is not a spanning set for $\mathbf{R}_2$.

3. (a) B does not span $\mathbf{P}_4$. (b) B is a basis. (c) B is not linearly independent. (d) B is a basis.

5. (a) A has more than two vectors and $\mathbf{V}$ has dimension 2. Thus A is linearly dependent.
(b) A has less than three vectors and $\mathbf{V}$ has dimension 3. Thus A cannot span $\mathbf{V}$.
(c) A has more than four vectors and $\mathbf{V}$ has dimension 4. Thus A is linearly dependent.
(d) A has less than four vectors and $\mathbf{V}$ has dimension 4. Thus A cannot span $\mathbf{V}$.

7. For example, $B = \{X = (2, 1, 0, 0), \quad Y = (0, 1, 0, 1), \quad E_1 = (1, 0, 0, 0), \quad E_3 = (0, 0, 1, 0)\}$.

9. (a) $\{(0, 2, 1)\}$. (b) dim $\mathbf{U} + \mathbf{W} = 3$. (c) $\mathbf{U}$ is the plane determined by $(0, 0, 0)$, $(1, 2, 0)$, and $(0, 2, 1)$. $\mathbf{W}$ is the plane determined by $(0, 0, 0)$, $(0, 0, 2)$, and $(0, 1, 0)$. $\mathbf{U} \cap \mathbf{W}$ is the line determined by $(0, 0, 0)$ and $(0, 2, 1)$. $\mathbf{U} + \mathbf{W}$ is all of $\mathbf{R}_3$.

13. (a) S is the solution set to a system of one linear equation in three unknowns. Thus it is a subspace of $\mathbf{R}_3$.
(b) $\{(-2, 1, 0), (-2, 0, 1)\}$.

15. dim $\mathbf{W} = 3$.

Exercise 5.6

1. $B = \left\{ \begin{bmatrix} -3 \\ 1 \\ 0 \\ 0 \\ 0 \end{bmatrix}, \begin{bmatrix} -\frac{2}{3} \\ 0 \\ -\frac{1}{3} \\ 1 \\ 0 \end{bmatrix}, \begin{bmatrix} -1 \\ 0 \\ -\frac{1}{2} \\ 0 \\ 1 \end{bmatrix} \right\},$ 3.

3. $B = \left\{ \begin{bmatrix} -3 \\ 1 \\ 0 \\ 0 \end{bmatrix}, \begin{bmatrix} -2 \\ 0 \\ -4 \\ 1 \end{bmatrix} \right\},$ 2.

5. $B = \left\{ \begin{bmatrix} -2 \\ 1 \\ 0 \\ 0 \end{bmatrix}, \begin{bmatrix} -4 \\ 0 \\ -3 \\ 1 \end{bmatrix} \right\},$ 2.

7. $\mathbf{W} = \left\langle \begin{bmatrix} -3 \\ 1 \\ 0 \\ 0 \end{bmatrix}, \begin{bmatrix} -2 \\ 0 \\ 1 \\ 0 \end{bmatrix} \right\rangle,$ $C_0 = \begin{bmatrix} 7 \\ 0 \\ 0 \\ -2 \end{bmatrix},$ $S = \left\{ \begin{bmatrix} 7 \\ 0 \\ 0 \\ -2 \end{bmatrix} + Y \mid Y \in \mathbf{W} \right\}.$

9. $S = \left\{ \begin{bmatrix} 3 \\ -2 \\ 0 \end{bmatrix} + Y \mid Y \in \mathbf{W} \right\},$ $\mathbf{W} = \left\langle \begin{bmatrix} -3 \\ -1 \\ 1 \end{bmatrix} \right\rangle.$ S is a line through $\begin{bmatrix} 3 \\ -2 \\ 0 \end{bmatrix}$ and parallel to the line through $\begin{bmatrix} 0 \\ 0 \\ 0 \end{bmatrix}$ and $\begin{bmatrix} -3 \\ -1 \\ 1 \end{bmatrix}.$

11. (a) X, Y, and $\mathbf{0}$ are collinear.
(b) X and Y are nonzero, and X, Y, and $\mathbf{0}$ are noncollinear.
(c) X, Y, and Z are coplanar, but no two of the points are collinear with $\mathbf{0}$.
(d) X, Y, and Z are nonzero vectors that are not coplanar.

CHAPTER 6

Exercise 6.1

1. $\mathbf{Z}$ = domain = codomain, range $f = \{n \in \mathbf{Z} \mid n = 3k + 1$ for some $k \in \mathbf{Z}\}$. f is not onto $\mathbf{Z}$, f is 1–1.

3. $\mathbf{R}$ = domain = codomain, range $f = \{\sqrt{x^2 + 1} \mid x \in \mathbf{R}\}$. f is not onto $\mathbf{R}$, f is not 1–1.

5. $\mathbf{N}$ = domain = codomain, range $f = \mathbf{N}$, f is onto $\mathbf{N}$, f is not 1–1.

7. $\mathbf{R} =$ domain, $\mathbf{Z} =$ codomain, range $f = \mathbf{Z}$, f is onto $\mathbf{Z}$, f is not 1–1.

9. $\mathbf{R}_2 =$ domain $=$ codomain, range $T = \mathbf{R}_2$, T is onto $\mathbf{R}_2$, T is 1–1.

11. $\mathbf{R}_{2 \times 2} =$ domain, $\mathbf{R} =$ codomain, range $T = \mathbf{R}$. T is onto $\mathbf{R}$, T is not 1–1.

13. (a) $1, -2$. (b) $(3, 2), (-1, 9)$. (c) $\begin{bmatrix} 7 \\ 14 \end{bmatrix}, \begin{bmatrix} 3 \\ 6 \end{bmatrix}$.

15. $f(\frac{1}{2}) = 3$ and $f(\frac{2}{4}) = 6$. Since $\frac{1}{2} = \frac{2}{4}$, the relation is not single valued.

17. (9) $T = \{((x, y), (x + 2, y - 3)) \mid x, y \in \mathbf{R}\}$.

(10) $T = \left\{ \left(\begin{bmatrix} x \\ y \\ z \end{bmatrix}, \begin{bmatrix} x + 2z \\ 2x + 4z \end{bmatrix} \right) \Big| x, y, z \in \mathbf{R} \right\}$.

(11) $T = \left\{ \left(\begin{bmatrix} a & b \\ c & d \end{bmatrix}, ad - bc \right) a, b, c, d \in \mathbf{R} \right\}$.

19. $f(4) = 10$, $g(4) = 10$.

Exercise 6.2

1. (a) Linear. (b) Linear.

3. Linear.

5. Not linear, $T(2(1, 2)) = (4, 6) \neq 2T(1, 2) = (2, 6)$.

7. Linear.

9. Linear.

13. (a) $T(\pi/6 + \pi/6) = \sqrt{3} \neq T(\pi/6) + T(\pi/6) = 2/\sqrt{3}$.
(b) $T(10 + 10) = \log 20 = \log 10 + \log 2 = 1 + \log 2 \neq T(10) + T(10) = 2$.

15. $((x + 3y)/2, (y - x)/2)$.

Exercise 6.3

1. (a) $(3x + y - z, 4x - 2y + 3z)$. (b) $(11, -14)$.

3. (a) $(x + y - 2z, y - z, x + 2y - 3z, y - x)$. (b) $(2, -1, 1, -4)$.

5. (a) $T(f) = g_f: 4bx^2 + (3a + 2d)x + 3c$. (b) $T(f) = g_f: 4x + 9$.

7. (a) $(2a + 3b) + (3a - 4b)i$. (b) $3 + 30i$.

9. $T(x, y) = (x, x)$.

11. $T(f: ax^3 + bx^2 + cx + d) = (a, b, c, d)$.

13. $T(x, y, z, w) = \begin{bmatrix} x & y \\ z & w \end{bmatrix}$.

17. (a) $T(X) = \begin{bmatrix} 6 & 8 & 4 \\ 3 & 4 & 2 \\ 1 & 2 & 1 \end{bmatrix} \begin{bmatrix} x_1 \\ x_2 \\ x_3 \end{bmatrix} = \begin{bmatrix} y_1 \\ y_2 \\ y_3 \end{bmatrix}$. (b) $T \begin{bmatrix} 30 \\ 20 \\ 25 \end{bmatrix} = \begin{bmatrix} 440 \\ 220 \\ 95 \end{bmatrix}$.

(c) $\begin{bmatrix} 0 & 1 & 6 \\ 0.5 & 0 & 4 \end{bmatrix} \begin{bmatrix} y_1 \\ y_2 \\ y_3 \end{bmatrix} = Z = \begin{bmatrix} z_1 \\ z_2 \end{bmatrix}$. (d) $\begin{bmatrix} 790 \\ 600 \end{bmatrix} = \begin{bmatrix} 0 & 1 & 6 \\ 0.5 & 0 & 4 \end{bmatrix} \begin{bmatrix} 6 & 8 & 4 \\ 3 & 4 & 2 \\ 1 & 2 & 1 \end{bmatrix} \begin{bmatrix} 30 \\ 20 \\ 25 \end{bmatrix}$

Exercise 6.4

1. $\mathbf{R}_T = \langle (3, 4), (1, -2), (-1, 3) \rangle$, $B = \{(3, 4), (1, -2)\}$. $\mathbf{N}_T = \langle (-1, 13, 10) \rangle$. dim $\mathbf{R}_3 = 3 = $ dim $\mathbf{N}_T + $ dim $\mathbf{R}_T = 1 + 2$.

3. $\mathbf{R}_T = \langle (1, 0, 1, -1), (2, 1, 3, 0), (0, 0, 0, 0) \rangle$, $B = \{(1, 0, 1, -1), (2, 1, 3, 0)\}$. $\mathbf{N}_T = \langle (1, 1, 1) \rangle$, dim $\mathbf{R}_3 = 1 + 2 = $ dim $\mathbf{N}_T + $ dim $\mathbf{R}_T$.

5. $\mathbf{R}_T = \langle g_1: 3x, g_2: 4x^2, g_3: 3, g_4: 2x \rangle$, $B = \{g_1, g_2, g_3\}$. $\mathbf{N}_T = \langle h: -2x^3 + 3 \rangle$. dim $\mathbf{P}_4 = 1 + 3 = $ dim $\mathbf{N}_T + $ dim $\mathbf{R}_T$.

7. $\mathbf{R}_T = \langle 2 + 3i, 3 - 4i \rangle$, $B = \{2 + 3i, 3 - 4i\}$. $\mathbf{N}_T = \langle 0 \rangle$. There is no basis for $\mathbf{N}_T$. dim $\mathbf{V} = 2 = 0 + 2 = $ dim $\mathbf{N}_T + $ dim $\mathbf{R}_T$.

9. (a) $0 \le$ rank $T \le 3$. (b) $4 \le$ nullity of $T \le 7$.

11. (a) $\mathbf{N}_T = \langle (1, -1, 0) \rangle$. (b) T is not 1-1. (c) Rank $T = 2$.

13. $T(X) = (a_1, a_2, \ldots, a_n)$.

19. $\mathbf{N}_T = \langle h_1: 1, h_2: x \rangle$, $\{h_1, h_2\}$ is a basis for $\mathbf{N}_T$. $\mathbf{R}_T = \langle g_1: 1 \rangle$, $\{g_1\}$ is a basis for $\mathbf{R}_T$.

Exercise 6.5

1. $X_{B_1} = \begin{bmatrix} a \\ b \\ c \end{bmatrix}$, $X_{B_2} = \begin{bmatrix} c \\ b \\ a \end{bmatrix}$, $X_{B_3} = \begin{bmatrix} b \\ c \\ a \end{bmatrix}$, $X_{B_4} = \begin{bmatrix} a \\ c \\ b \end{bmatrix}$.

3. (a) $A = \begin{bmatrix} 3 & 0 & 1 \\ 2 & 3 & -1 \\ -3 & 4 & 2 \\ 4 & -1 & 0 \end{bmatrix}$. (b) $(3a + c)Y_1 + (2a + 3b - c)Y_2 + (-3a + 4b + 2c)Y_3 + (4a - b)Y_4$.

5. (a) $T(f_1) = h_1: x^2 + 3x + 1$. (b) $T(f_2) = h_2: 2x^2 + 1$. (c) $T(f) = h: (2a + b)/3)x^2 + 3bx + (a + 2b)/3$.

7. (a) $\begin{bmatrix} \frac{2}{3} & 0 & \frac{4}{3} \\ -\frac{1}{3} & 2 & \frac{10}{3} \end{bmatrix}$. (b) $T(Z_1) = \begin{bmatrix} 12 \\ \frac{16}{3} \end{bmatrix}$, $T(Z_2) = \begin{bmatrix} 10 \\ \frac{28}{3} \end{bmatrix}$.

9. (a) $\begin{bmatrix} 2 & 1 & 0 & 3 \\ \frac{3}{2} & 2 & \frac{3}{2} & 0 \\ 2 & 6 & 0 & 1 \\ 2 & -6 & 2 & 1 \end{bmatrix}$. (b) $T(Z_1) = \begin{bmatrix} 8 \\ 19 \\ 16 \\ 11 \end{bmatrix}$, $T(Z_2) = \begin{bmatrix} 8 \\ 15 \\ 18 \\ 8 \end{bmatrix}$.

11. (a) $A = \begin{bmatrix} -1 & 0 & -\frac{3}{2} \\ 1 & \frac{7}{4} & \frac{7}{8} \end{bmatrix}$. (b) $T(Z_1) = \begin{bmatrix} 24 \\ 2 \end{bmatrix}$, $T(Z_2) = \begin{bmatrix} 9 \\ -8 \end{bmatrix}$.

(c) $\begin{bmatrix} 3 & 2 & 0 \\ 0 & 1 & -2 \end{bmatrix}$. (d) Same as (b).

13. (a) $\begin{bmatrix} 1 & 3 & 0 \\ 1 & -2 & 1 \\ 1 & 4 & 0 \end{bmatrix}$. (b) $\begin{bmatrix} 22 \\ -7 \\ 28 \end{bmatrix}$.

Exercise 6.6

1.

(a) $T + S(X) = \begin{bmatrix} x + y + w \\ 4x + y + w \\ x + 2y + 2z + w \end{bmatrix}$. (b) $(6T)(X) = \begin{bmatrix} 12x + 6y \\ 18x + 6w \\ 6x + 6y + 6z + 6w \end{bmatrix}$.

(c) $A = \begin{bmatrix} 2 & 1 & 0 & 0 \\ 3 & 0 & 0 & 1 \\ 1 & 1 & 1 & 1 \end{bmatrix}$, $B = \begin{bmatrix} -1 & 0 & 0 & 1 \\ 1 & 1 & 0 & 0 \\ 0 & 1 & 1 & 0 \end{bmatrix}$.

(d) $(A + B)X = \begin{bmatrix} x + y + w \\ 4x + y + w \\ x + 2y + 2z + w \end{bmatrix}$, $(6A)X = \begin{bmatrix} 12x + 6y \\ 18x + 6w \\ 6x + 6y + 6z + 6w \end{bmatrix}$.

(e) No, no.

3. (a) $\begin{bmatrix} 1 & 0 & 0 \\ 0 & 1 & 0 \end{bmatrix}$. (b) $\begin{bmatrix} 0 & 0 & 1 \\ 0 & 1 & 0 \end{bmatrix}$. (c) $\begin{bmatrix} 1 & 0 & 1 \\ 0 & 2 & 0 \end{bmatrix}$. (d) $\begin{bmatrix} r & 0 & 0 \\ 0 & r & 0 \end{bmatrix}$.

5. (a) $T \circ S(X) = \begin{bmatrix} 3x + 5y - 4z + 3w \\ 8x + 9y - 3z + 5w \end{bmatrix}$. (b) $[S]_{CC'} = \begin{bmatrix} 2 & 2 & -3 & 0 \\ 1 & 1 & 1 & 1 \\ 1 & 2 & 0 & 2 \end{bmatrix}$.

$[T]_{C'C''} = \begin{bmatrix} 1 & -1 & 2 \\ 2 & 3 & 1 \end{bmatrix}$, $[T \circ S]_{CC''} = \begin{bmatrix} 3 & 5 & -4 & 3 \\ 8 & 9 & -3 & 5 \end{bmatrix}$.

(c) $\begin{bmatrix} 1 & -1 & 2 \\ 2 & 3 & 1 \end{bmatrix} \begin{bmatrix} 2 & 2 & -3 & 0 \\ 1 & 1 & 1 & 1 \\ 1 & 2 & 0 & 2 \end{bmatrix} = \begin{bmatrix} 3 & 5 & -4 & 3 \\ 8 & 9 & -3 & 5 \end{bmatrix}$.

7. (a) Same as 5(a). (b) $[S]_{C_1 C_2} = \begin{bmatrix} 1 & 2 & -2 & -3 \\ 0 & -1 & 0 & -1 \\ 1 & 3 & 3 & 5 \end{bmatrix}$,

$[T]_{C_2 C_3} = \begin{bmatrix} -1 & -5 & -4 \\ 2 & 5 & 6 \end{bmatrix}$, $[T \circ S]_{C_1 C_3} = \begin{bmatrix} -5 & -9 & -10 & -12 \\ 8 & 17 & 14 & 19 \end{bmatrix}$.

(c) $\begin{bmatrix} -1 & -5 & -4 \\ 2 & 5 & 6 \end{bmatrix} \begin{bmatrix} 1 & 2 & -2 & -3 \\ 0 & -1 & 0 & -1 \\ 1 & 3 & 3 & 5 \end{bmatrix} = \begin{bmatrix} -5 & -9 & -10 & -12 \\ 8 & 17 & 14 & 19 \end{bmatrix}$.

Exercise 6.7

3. $I' \circ I = I$ since I' is an identity. Also, $I' \circ I = I'$ because I is an identity. Thus $I = I'$.

7. No inverse.

9. No inverse.

11. T is not onto its codomain.

13. T is not onto its codomain.

Exercise 6.8

1. $\{(1, 0, 0), (0, 1, 0), (0, 0, 1)\}$.

3. $\{(1, 0), (0, 1)\}$.

5. 3.

7. (a) By Corollary 6.4, $\text{rank}(A) = \text{rank}(T) = \text{rank}(C)$.

   (b) $\left\{ \begin{bmatrix} 1 \\ 0 \end{bmatrix}, \begin{bmatrix} 0 \\ 1 \end{bmatrix} \right\}$ is a basis for $\mathbf{R}_T$. Therefore, $\text{rank}(T) = 2$.

   (c) $A = \begin{bmatrix} 1 & 2 & -1 \\ 1 & 1 & 4 \end{bmatrix} \longrightarrow \begin{bmatrix} 1 & 0 & 9 \\ 0 & 1 & -5 \end{bmatrix}$. $\text{Rank}(A) = 2$.

   (d) $C = \begin{bmatrix} 2 & 1 & 0 \\ 1 & 0 & 3 \end{bmatrix} \longrightarrow \begin{bmatrix} 1 & 0 & 3 \\ 0 & 1 & -6 \end{bmatrix}$. $\text{Rank}(C) = 2$.

9. (a) $\text{Rank}(A) = 2$.

   (b) $T \begin{bmatrix} x \\ y \\ z \end{bmatrix} = \begin{bmatrix} 6x + 2y + 2z \\ 5x + y + 3z \\ 2x + 2y - 2z \end{bmatrix}$. (c) $S \begin{bmatrix} x \\ y \\ z \end{bmatrix} = \begin{bmatrix} x + 2y + z \\ 3x + 2y + 7z \\ 2x + 4y + 2z \end{bmatrix}$.

   (d) By Corollary 6.4, $\text{rank}(T) = \text{rank}(A)$ and $\text{rank}(S) = \text{rank}(A)$. Therefore, $\text{rank}(T) = \text{rank}(S)$. $\text{Rank}(T) = 2$.

11. Row $\text{rank}(A) = $ dimension of the row space of $A \leq m$. Column $\text{rank}(A)$ = dimension of the column space of $A \leq n$. Since $\text{rank}(A) = $ row $\text{rank}(A) = $ column $\text{rank}(A)$, it follows that $\text{rank}(A)$ does not exceed the minimum of m and n.

13. $\text{Rank}(A) = 2$, $\text{rank}[A \,|\, B] = 2$. Consistent.

Exercise 6.9

1. (a) $P = \begin{bmatrix} 1 & 1 \\ 2 & 1 \end{bmatrix}$. (b) $P^{-1} = \begin{bmatrix} -1 & 1 \\ 2 & -1 \end{bmatrix}$.

   (c) $X = (-x + y)Y_1 + (2x - y)Y_2$.

3. (a) $P = \begin{bmatrix} 1 & 2 \\ -\frac{3}{2} & -4 \end{bmatrix}$. (b) $P^{-1} = Q = \begin{bmatrix} 4 & 2 \\ -\frac{3}{2} & -1 \end{bmatrix}$.

   (c) $Q \begin{bmatrix} x \\ (y - 4x)/2 \end{bmatrix} = \begin{bmatrix} y \\ (x - y)/2 \end{bmatrix}$; therefore, $X = yY_1 + ((x - y)/2)Y_2$.

5. (a) $P = \begin{bmatrix} 1 & 1 & 4 \\ 1 & 2 & 1 \\ 0 & 0 & 1 \end{bmatrix}$. (b) $P^{-1} = \begin{bmatrix} 2 & -1 & -7 \\ -1 & 1 & 3 \\ 0 & 0 & 1 \end{bmatrix}$.

   (c) $f = (2a - b - 7c)g_1 + (-a + b + 3c)g_2 + cg_3$.

7. (a) $A = \begin{bmatrix} 1 & -1 & 0 & 0 \\ 1 & 0 & 2 & 0 \\ 0 & 0 & 1 & 1 \\ 0 & 1 & 1 & 0 \end{bmatrix}.$ (b) $B = \begin{bmatrix} 0 & -\frac{1}{2} & -2 & -1 \\ 2 & 1 & 4 & 2 \\ 0 & -\frac{1}{2} & 0 & \frac{1}{2} \\ 0 & 1 & 2 & 1 \end{bmatrix}.$

(c) $P = \begin{bmatrix} 2 & 1 & 0 & 0 \\ 0 & 1 & 0 & 0 \\ 0 & 0 & 2 & 1 \\ 0 & 0 & 0 & 1 \end{bmatrix}.$ (d) $P^{-1} = \begin{bmatrix} \frac{1}{2} & -\frac{1}{2} & 0 & 0 \\ 0 & 1 & 0 & 0 \\ 0 & 0 & \frac{1}{2} & -\frac{1}{2} \\ 0 & 0 & 0 & 1 \end{bmatrix}.$

(f) $Y_1 + Y_2 + \frac{1}{2}Y_3 + Y_4.$ (g) $7E_1 + 3E_2 + (-3)E_3 + E_4.$

9. (a) $Y_1 = 2X_1 + X_3$ (b) $X_1 = -Y_1 + Y_2 - Y_3$
$Y_2 = 3X_1 + X_2 + 2X_3$ $X_2 = -3Y_1 + 2Y_2 - Y_3$
$Y_3 = X_2 + X_3$ $X_3 = 3Y_1 - 2Y_2 + 2Y_3$

11. $[T]_{C_2} = \begin{bmatrix} 4 & 6 \\ -\frac{3}{2} & -2 \end{bmatrix}.$

13. $C_2 = \left\{ \begin{bmatrix} 1 \\ 1 \\ 0 \end{bmatrix}, \begin{bmatrix} 1 \\ 2 \\ 0 \end{bmatrix}, \begin{bmatrix} 4 \\ 1 \\ 1 \end{bmatrix} \right\},$

$B = \begin{bmatrix} -19 & -26 & -75 \\ 10 & 15 & 33 \\ 3 & 4 & 12 \end{bmatrix},$ $T(Y_1) = -19Y_1 + 10Y_2 + 3Y_3$
$T(Y_2) = -26Y_1 + 15Y_2 + 4Y_3,$
$T(Y_3) = -75Y_1 + 33Y_2 + 12Y_3$

$T(Y_1) = \begin{bmatrix} 3 \\ 4 \\ 3 \end{bmatrix}, T(Y_2) = \begin{bmatrix} 5 \\ 8 \\ 4 \end{bmatrix}, T(Y_3) = \begin{bmatrix} 6 \\ 3 \\ 12 \end{bmatrix}.$

Exercise 6.10

3. (a) $P^{-1} = \begin{bmatrix} 2 & 0 & -1 \\ 1 & 1 & -1 \\ -1 & 0 & 1 \end{bmatrix}.$ (b) $B = \begin{bmatrix} 0 & 0 & -2 \\ -1 & \frac{1}{2} & -\frac{5}{2} \\ 1 & 0 & 3 \end{bmatrix}.$

(c) $B^5 = \begin{bmatrix} -30 & 0 & -62 \\ -31 & \frac{1}{32} & -63 + \frac{1}{32} \\ 31 & 0 & 63 \end{bmatrix}.$

7. $a_n = 0, b_n = (1/2^{n-1})a_0 + (1/2^n)b_0, c_n = (1 - 1/2^{n-1})a_0 + (1 - 1/2^n)b_0 + c_0.$

CHAPTER 7

Exercise 7.1

1. (a) 4 and 1. The nonzero vectors in $\left\langle \begin{bmatrix} 1 \\ 2 \end{bmatrix} \right\rangle$ belong to 4, and the nonzero

vectors in $\left\langle \begin{bmatrix} 1 \\ -1 \end{bmatrix} \right\rangle$ belong to 1.

(b) $C_2 = \left\{ \begin{bmatrix} 1 \\ 2 \end{bmatrix}, \begin{bmatrix} 1 \\ -1 \end{bmatrix} \right\}$. (c) $D = \begin{bmatrix} 4 & 0 \\ 0 & 1 \end{bmatrix} = [T]_{C_2}$.

3. (a) 2, -1. The nonzero vectors in $\langle f_1 : 1 + 2x \rangle$ belong to 2. The nonzero vectors in $\langle f_2 : 2 + x \rangle$ belong to -1.

(b) $C_2 = \left\{ \begin{bmatrix} 1 \\ 2 \end{bmatrix}, \begin{bmatrix} 2 \\ 1 \end{bmatrix} \right\}$. (c) $D = \begin{bmatrix} 2 & 0 \\ 0 & -1 \end{bmatrix} = [T]_{C_2}$.

5. (a) 0. The eigenvectors belonging to 0 are the nonzero vectors in the space $\langle f_1 : 1 \rangle$. (b) No basis. (c) None.

7. (a) 0. The eigenvectors belonging to 0 are the vectors in $\langle f_1 : 1 \rangle - \{0\}$. (b) No basis. (c) None.

9. (a) 0, $\frac{3}{2}$, 3. The eigenvectors belonging to 3, $\frac{3}{2}$, and 0 are the nonzero vectors in the spaces $\left\langle \begin{bmatrix} 1 \\ 0 \\ 2 \end{bmatrix} \right\rangle, \left\langle \begin{bmatrix} 2 \\ 0 \\ 1 \end{bmatrix} \right\rangle$, and $\left\langle \begin{bmatrix} -2 \\ -9 \\ 2 \end{bmatrix} \right\rangle$, respectively.

(b) $C_2 = \left\{ \begin{bmatrix} 1 \\ 0 \\ 2 \end{bmatrix}, \begin{bmatrix} 2 \\ 0 \\ 1 \end{bmatrix}, \begin{bmatrix} -2 \\ -9 \\ 2 \end{bmatrix} \right\}$. (c) $D = \begin{bmatrix} 3 & 0 & 0 \\ 0 & \frac{3}{2} & 0 \\ 0 & 0 & 0 \end{bmatrix}$.

11. (a) $\begin{bmatrix} 0 & 1 \\ 1 & 0 \end{bmatrix}$. (b) 1 and -1 are eigenvalues. The eigenvectors belonging to 1 and -1 are the nonzero vectors in the spaces $\left\langle \begin{bmatrix} 1 \\ 1 \end{bmatrix} \right\rangle$ and $\left\langle \begin{bmatrix} -1 \\ 1 \end{bmatrix} \right\rangle$, respectively.

(c) $C_2 = \left\{ Y_1 = \begin{bmatrix} 1 \\ 1 \end{bmatrix}, Y_2 = \begin{bmatrix} -1 \\ 1 \end{bmatrix} \right\}$.

(d) Vectors along the Y_1 axis are mapped into themselves; vectors along the Y_2 axis are mapped into their opposites. If $\begin{bmatrix} a \\ b \end{bmatrix}$ is the coordinate vector of X relative to C_2, then $T(X)$ has coordinates $\begin{bmatrix} a \\ -b \end{bmatrix}$ relative to this coordinate system.

Exercise 7.2

1. (a) 1 and $\frac{1}{2}$ are eigenvalues. The nonzero vectors in $\left\langle \begin{bmatrix} 1 \\ 0 \\ 0 \end{bmatrix}, \begin{bmatrix} 0 \\ 0 \\ 1 \end{bmatrix} \right\rangle$ and $\left\langle \begin{bmatrix} 1 \\ -2 \\ 1 \end{bmatrix} \right\rangle$ belong to 1 and $\frac{1}{2}$, respectively.

(b) $D = \begin{bmatrix} 1 & 0 & 0 \\ 0 & 1 & 0 \\ 0 & 0 & \frac{1}{2} \end{bmatrix} = P^{-1}BP$ if $P^{-1} = \begin{bmatrix} 1 & \frac{1}{2} & 0 \\ 0 & \frac{1}{2} & 1 \\ 0 & -\frac{1}{2} & 0 \end{bmatrix}$ and $P = \begin{bmatrix} 1 & 0 & 1 \\ 0 & 0 & -2 \\ 0 & 1 & 1 \end{bmatrix}$.

3. (a) 1, -2, and 0. (b) The nonzero vectors in $\left\langle \begin{bmatrix} 1 \\ 0 \\ 0 \end{bmatrix} \right\rangle, \left\langle \begin{bmatrix} 0 \\ 1 \\ 0 \end{bmatrix} \right\rangle$ and $\left\langle \begin{bmatrix} 0 \\ 0 \\ 1 \end{bmatrix} \right\rangle$ belong to 1, -2, and 0, respectively. (c) $(-1)^3(\lambda - 1)(\lambda + 2)\lambda$.

5. (a) (i). (b) (ii). (c) (iii). (d) (i).

7. 0 is the only eigenvalue. Every nonzero vector is an eigenvector belonging to 0. Any basis for $\mathbf{R}_n$ is a basis of eigenvectors for the zero matrix.

9. (a) The eigenvalues are 2, -1, and 4. The nonzero vectors in $\left\langle \begin{bmatrix} 1 \\ 0 \\ 0 \end{bmatrix} \right\rangle$, $\left\langle \begin{bmatrix} 1 \\ 1 \\ 0 \end{bmatrix} \right\rangle$,

and $\left\langle \begin{bmatrix} 1 \\ 1 \\ 1 \end{bmatrix} \right\rangle$ belong to 2, -1, and 4, respectively.

(b) $D = \begin{bmatrix} 2 & 0 & 0 \\ 0 & -1 & 0 \\ 0 & 0 & 4 \end{bmatrix} = P^{-1}AP$ if $P^{-1} = \begin{bmatrix} 1 & -1 & 0 \\ 0 & 1 & -1 \\ 0 & 0 & 1 \end{bmatrix}$ and $P = \begin{bmatrix} 1 & 1 & 1 \\ 0 & 1 & 1 \\ 0 & 0 & 1 \end{bmatrix}$.

13. (a) $|A - \lambda I| = (1 - \lambda)(\lambda^2 - \lambda + \frac{5}{36})$.

(c) $P = \begin{bmatrix} 1 & 1 & -7 \\ 0 & -4 & 4 \\ 0 & 3 & 3 \end{bmatrix}$, $P^{-1} = \begin{bmatrix} 1 & 1 & 1 \\ 0 & -\frac{1}{8} & \frac{1}{6} \\ 0 & \frac{1}{8} & \frac{1}{6} \end{bmatrix}$, $P^{-1}AP = \begin{bmatrix} 1 & 0 & 0 \\ 0 & \frac{1}{6} & 0 \\ 0 & 0 & \frac{5}{6} \end{bmatrix}$.

(e) The limiting value of A is $\begin{bmatrix} 1 & 1 & 1 \\ 0 & 0 & 0 \\ 0 & 0 & 0 \end{bmatrix}$. After a long time it is virtually certain

that a machine will be broken beyond repair regardless of the initial condition.

Exercise 7.3

1. (a) $Q(X) = 2x^2 + 2(3)xy + 5y^2 = \begin{bmatrix} x & y \end{bmatrix} \begin{bmatrix} 2 & 3 \\ 3 & 5 \end{bmatrix} \begin{bmatrix} x \\ y \end{bmatrix}$.

(b) Not a quadratic form.

(c) $Q(X) = 0x^2 + 2(1)xy + 2(\frac{3}{2})xz + 0y^2 + 2(2)yz + 0z^2$

$= \begin{bmatrix} x & y & z \end{bmatrix} \begin{bmatrix} 0 & 1 & \frac{3}{2} \\ 1 & 0 & 2 \\ \frac{3}{2} & 2 & 0 \end{bmatrix} \begin{bmatrix} x \\ y \\ z \end{bmatrix}$.

(d) $Q(X) = 2x^2 + 2(\frac{3}{2})xy + 2(2)xz + 0y^2 + 2(3)yz + 3z^2$

$= \begin{bmatrix} x & y & z \end{bmatrix} \begin{bmatrix} 2 & \frac{3}{2} & 2 \\ \frac{3}{2} & 0 & 3 \\ 2 & 3 & 3 \end{bmatrix} \begin{bmatrix} x \\ y \\ z \end{bmatrix}$.

(e) $Q(X) = 3x^2 + 2(4)xy + 2(0)xz + 0y^2 + 2(0)yz + z^2$

$= \begin{bmatrix} x & y & z \end{bmatrix} \begin{bmatrix} 3 & 4 & 0 \\ 4 & 0 & 0 \\ 0 & 0 & 1 \end{bmatrix} \begin{bmatrix} x \\ y \\ z \end{bmatrix}$.

(f) Not a quadratic form.

3. $X^T \begin{bmatrix} 3 & 2 & 0 & 0 \\ 2 & 7 & 0 & 1 \\ 0 & 0 & 0 & 0 \\ 0 & 1 & 0 & 3 \end{bmatrix} X$

5. If $X^T = [x \quad y \quad z \quad w]$, then $Q(X) = 2x^2 + 3y^2 + z^2 + 4w^2$.

7. $B = \begin{bmatrix} 2 & 4 & 4 \\ 4 & 5 & 5 \\ 4 & 5 & 6 \end{bmatrix}$, $Q \begin{bmatrix} 1 \\ 2 \\ 3 \end{bmatrix} = 24 = [-1 \quad 2 \quad 1]B \begin{bmatrix} -1 \\ 2 \\ 1 \end{bmatrix} = [1 \quad 2 \quad 3]A \begin{bmatrix} 1 \\ 2 \\ 3 \end{bmatrix}$.

9. $B = \begin{bmatrix} 7 & 2 & 11 \\ 2 & 4 & 0 \\ 11 & 0 & 16 \end{bmatrix}$, $Q \begin{bmatrix} 1 \\ 1 \\ 2 \end{bmatrix} = 15 = [1 \quad 1 \quad 0]B \begin{bmatrix} 1 \\ 1 \\ 0 \end{bmatrix}$.

11. $P = (1/\sqrt{6}) \begin{bmatrix} \sqrt{2} & -2 & 0 \\ \sqrt{2} & 1 & -\sqrt{3} \\ \sqrt{2} & 1 & \sqrt{3} \end{bmatrix}$, $P^T P = \frac{1}{6} \begin{bmatrix} 6 & 0 & 0 \\ 0 & 6 & 0 \\ 0 & 0 & 6 \end{bmatrix} = I_3$.

Exercise 7.4

1. (a) $P^T A P = \frac{1}{5} \begin{bmatrix} 2 & 1 \\ -1 & 2 \end{bmatrix} \begin{bmatrix} 5 & 2 \\ 2 & 2 \end{bmatrix} \begin{bmatrix} 2 & -1 \\ 1 & 2 \end{bmatrix} = \begin{bmatrix} 6 & 0 \\ 0 & 1 \end{bmatrix}$.

(b) The matrix of R_θ relative to the natural basis for $\mathbf{R}_2$ is

$$[R_\theta] = \begin{bmatrix} \cos\theta & -\sin\theta \\ \sin\theta & \cos\theta \end{bmatrix}.$$

Also, $\cos\theta = \dfrac{Y_1 \cdot E_1}{\|Y_1\|\|E_1\|} = Y_1 \cdot E_1 = 2/\sqrt{5}$. Since $\sin^2\theta = 1 - \cos^2\theta$

and $\sin\theta > 0$, we obtain $\sin\theta = 1/\sqrt{5}$. So $R_\theta = \begin{bmatrix} 2/\sqrt{5} & -1/\sqrt{5} \\ 1/\sqrt{5} & 2/\sqrt{5} \end{bmatrix}$

$= P$. Hence $R_\theta(X) = PX$.

5. If $P = \begin{bmatrix} 1/\sqrt{2} & -1/\sqrt{2} \\ 1/\sqrt{2} & 1/\sqrt{2} \end{bmatrix}$, then $P^T A P = \text{diag}(1, -1)$.

7. If $P = \begin{bmatrix} \frac{3}{5} & -\frac{4}{5} \\ \frac{4}{5} & \frac{3}{5} \end{bmatrix}$, then $P^T A P = \text{diag}(26, 1)$.

9. If $P = \begin{bmatrix} 1/\sqrt{5} & -2/\sqrt{5} \\ 2/\sqrt{5} & 1/\sqrt{5} \end{bmatrix}$, then $P^T A P = \text{diag}(9, 4)$.

11. $u^2 - v^2 = 1$. The principal axes are determined by $\dfrac{1}{\sqrt{2}} \begin{bmatrix} 1 \\ 1 \end{bmatrix}$ and $\dfrac{1}{\sqrt{2}} \begin{bmatrix} -1 \\ 1 \end{bmatrix}$.

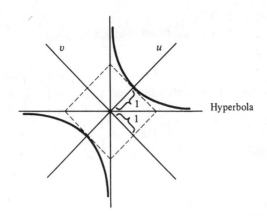

Hyperbola

13. $26u^2 + v^2 = 26$. The principal axes are determined by the orthonormal vectors $\begin{bmatrix} \frac{3}{5} \\ \frac{4}{5} \end{bmatrix}$ and $\begin{bmatrix} -\frac{4}{5} \\ \frac{3}{5} \end{bmatrix}$.

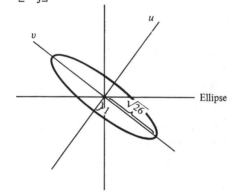

Ellipse

15. $9u^2 + 4v^2 = 36$. The principal axes are determined by the orthonormal vectors $\begin{bmatrix} 1/\sqrt{5} \\ 2/\sqrt{5} \end{bmatrix}$ and $\begin{bmatrix} -2/\sqrt{5} \\ 1/\sqrt{5} \end{bmatrix}$.

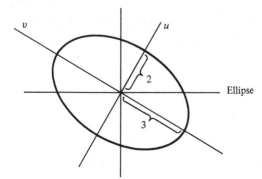

Ellipse

17. If $P = \frac{1}{3}\begin{bmatrix} -1 & -2 & 2 \\ -2 & 2 & 1 \\ 2 & 1 & 2 \end{bmatrix}$, then $P^T A P = \text{diag}(5, 8, 2)$.

19. $2u^2 + 5v^2 - w^2 = 10$, hyperboloid of one sheet. $\begin{bmatrix} u \\ v \\ w \end{bmatrix} = \frac{1}{3}\begin{bmatrix} -2x + y + 2z \\ x - 2y + 2z \\ 2x + 2y + z \end{bmatrix}$.

21. $u^2 + v^2 + 10w^2 = 10$, ellipsoid. $\begin{bmatrix} u \\ v \\ w \end{bmatrix} = \frac{1}{\sqrt{45}}\begin{bmatrix} 6x + 3z \\ -2x + 5y + 4z \\ -\sqrt{5}\,x - 2\sqrt{5}\,y + 2\sqrt{5}\,z \end{bmatrix}$

Exercise 7.5

1. (a) $\begin{bmatrix} 36 & -28 \\ -28 & 36 \end{bmatrix} - 8\begin{bmatrix} 10 & -6 \\ -6 & 10 \end{bmatrix} + 20\begin{bmatrix} 3 & -1 \\ -1 & 3 \end{bmatrix} - \begin{bmatrix} 16 & 0 \\ 0 & 16 \end{bmatrix} = \begin{bmatrix} 0 & 0 \\ 0 & 0 \end{bmatrix}$.

(b) $C(x) = x^2 - 6x + 8$. (c) $x^2 - 6x + 8$.

(d) $x^3 - 8x^2 + 20x - 16 = (x^2 - 6x + 8)(x - 2)$.

3. (a) $(x - 2)(x + 3)(x - 1)$ or $(x - 2)^2(x + 3)(x - 1)$
or $(x - 2)^2(x + 3)^2(x - 1)$ or $(x - 2)(x + 3)^2(x - 1)$.

(b) $(x - 2)(x + 3)(x - 1)$.

7. (a) $D_{C_2} = A = \begin{bmatrix} 0 & 0 & 0 \\ 1 & 0 & 0 \\ 0 & 1 & 0 \end{bmatrix}$. (b) $C(x) = -x^3$, $m(x) = x^3$, no.

9. $C(x) = -(x - 2)^3$, $m(x) = (x - 2)^2$. The eigenspace of A is the space $\left\langle \begin{bmatrix} -2 \\ 0 \\ 1 \end{bmatrix}, \begin{bmatrix} -2 \\ 1 \\ 0 \end{bmatrix} \right\rangle$, which has dimension 2. Therefore, A is not diagonalizable.

11. (a) $A = \begin{bmatrix} 0 & 0 & 0 & 0 \\ 1 & 0 & 0 & 0 \\ 0 & 1 & 0 & 0 \\ 0 & 0 & 1 & 0 \end{bmatrix}$. (b) $C(x) = x^4$, $m(x) = x^3$, A is not diagonalizable.

15. (a) $f_1: c_1e^{2x} + c_2e^{-x} + c_3e^{4x}$ (b) $f_1: 4e^{2x} + 2e^{-x} + 3e^{4x}$
$f_2: c_2e^{-x} + c_3e^{4x}$ $f_2: 2e^{-x} + 3e^{4x}$
$f_3: c_3e^{4x}$ $f_3: 3e^{4x}$

17. (a) $f_1: (1/\sqrt{5})(c_1e^{9x} - 2c_2e^{4x})$ (b) $f_1: 5e^{9x} - 10e^{4x}$
$f_2: (1/\sqrt{5})(2c_1e^{9x} + c_2e^{4x})$ $f_2: 10e^{9x} + 5e^{4x}$

19. (a) $f: c_1e^x + c_2e^{2x} + c_3e^{4x}$. (b) $f: 6e^x - 4e^{2x} + e^{4x}$.

CHAPTER 8

Exercise 8.1

1. (a) $\|A\| = \sqrt{(a-b)^2 + b^2}$, $\|C\| = \sqrt{5}$, $\|X_1\| = \|X_2\| = 1$.
   (b) $d(A, B) = \sqrt{(a - c - b + d)^2 + (b - d)^2}$, $d(C, D) = 1$,
   $d(X_1, X_2) = \sqrt{2}$.

3. (a) $\|A\| = \sqrt{b^2 + 4(a - b)^2 + c^2}$, $\|C\| = \sqrt{10}$,
   $\|X_1\| = 1 = \|X_2\| = \|X_3\|$.
   (b) $d(A, B) = \sqrt{(b - f)^2 + 4(a - e - b + f)^2 + (c - g)^2}$,
   $d(C, D) = 1$, $d(X_1, X_2) = \sqrt{2}$.

5. (a) $\|f\| = \sqrt{\dfrac{(b - a)^2}{4} + a^2}$, $\|h\| = \dfrac{\sqrt{37}}{2}$, $\|f_1\| = \|f_2\| = 1$.
   (b) $d(f, g) = \sqrt{\dfrac{(b - d - a + c)^2}{4} + (a - c)^2}$, $d(h, f_1) = \dfrac{3\sqrt{5}}{2}$,
   $d(f_1, f_2) = \sqrt{2}$.

7. (a) $\frac{3}{2}$. (b) $\sqrt{\frac{8}{15}}$. (c) $\sqrt{\frac{4}{3}}$.

9. (a) $\|a_1\| = |a_1|$. (b) $d(a_1, a_2) = |a_1 - a_2|$, $|a + b| \le |a| + |b|$,
   $\big| |a| - |b| \big| \le |a - b|$.

13. Yes.

Exercise 8.2

1. $\{(1/\sqrt{5})(1, 0, 2), (1/\sqrt{5})(2, 0, -1)\}$.

3. $\{(1/\sqrt{3})(1, 1, 1, 0), (1/\sqrt{15})(1, 1, -2, 3), (1/\sqrt{35})(1, -4, 3, 3),$
   $(1/\sqrt{7})(-2, 1, 1, 1)\}$.

5. $\{(1/\sqrt{5})(1, 2, 0), (1/\sqrt{6})(-2, 1, 1), (1/\sqrt{30})(2, -1, 5)\}$.

9. (b) $\sqrt{\displaystyle\int_{-\pi}^{\pi} (f - g)^2(x)\, dx}$.

13. $(6, 24, 12, 8) = 7Y_1 + \frac{13}{3}Y_2 + \frac{41}{6}Y_3 - \frac{7}{2}Y_4$.

Exercise 8.3

5. 32.65 years.

7. Using the fact that both $f(x) = x$ and $g_k(x) = x \cos kx$ are odd functions, we
   get $0 = a_0 = a_1 = a_2 = \ldots$, and the Fourier series of f is h, where $h(x)$
   $= 2(\sin x - \frac{1}{2} \sin 2x + \frac{1}{3} \sin 3x - \frac{1}{4} \sin 4x + - \ldots)$.

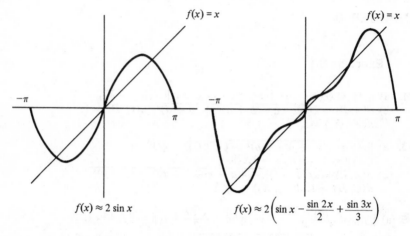

$$f(x) = x$$

$$f(x) \approx 2 \sin x$$

$$f(x) = x$$

$$f(x) \approx 2\left(\sin x - \frac{\sin 2x}{2} + \frac{\sin 3x}{3}\right)$$

11. $h(x) = \dfrac{1}{2} - \dfrac{2}{\pi}\left(\dfrac{\sin x}{1} + \dfrac{\sin 3x}{3} + \dfrac{\sin 5x}{5} + \dots\right)$

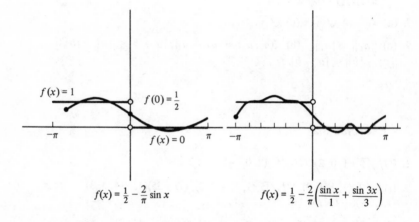

$$f(x) = 1 \qquad f(0) = \frac{1}{2}$$

$$f(x) = 0$$

$$f(x) = \frac{1}{2} - \frac{2}{\pi}\sin x$$

$$f(x) = \frac{1}{2} - \frac{2}{\pi}\left(\frac{\sin x}{1} + \frac{\sin 3x}{3}\right)$$

Index